Related Titles from AIP Conference Proceedings

666 The Emergence of Cosmic Structure: Thirteenth Astrophysics Conference
Edited by Stephen S. Holt and Christopher S. Reynolds, May 2003, 0-7354-0128-4

655 Particle Physics and Cosmology: Third Tropical Workshop on Particle Physics and Cosmology – Neutrinos, Branes, and Cosmology
Edited by José F. Nieves and Chung N. Leung, February 2003, 0-7354-0112-8

624 Cosmology and Elementary Particle Physics: Coral Gables Conference on Cosmology and Elementary Particle Physics
Edited by B. N. Kursunoglu, S. L. Mintz, and A. Perlmutter, July 2002, 0-7354-0073-3

575 Astrophysical Sources for Ground-Based Gravitational Wave Detectors
Edited by Joan M. Centrella, July 2001, 0-7354-0014-8

523 Gravitational Waves: Third Edoardo Amaldi Conference
Edited by Sydney Meshkov, June 2000, 1-56396-944-0

456 Laser Interferometer Space Antenna: Second International LISA Symposium on the Detection and Observation of Gravitational Waves in Space
Edited by William M. Folkner, December 1998, 1-56396-848-7

453 Particles, Fields, and Gravitation
Edited by Jakub Rembielinski, December 1998, 1-56396-837-1

To learn more about these titles, or the AIP Conference Proceedings Series, please visit the webpage **http://proceedings.aip.org/proceedings**

COSMOLOGY AND GRAVITATION

X[th] Brazilian School of Cosmology and Gravitation
25th Anniversary (1977-2002)

Mangaratiba, Rio de Janeiro, Brazil
29 July - 9 August 2002

EDITORS
Mário Novello
Santiago E. Perez Bergliaffa

CBPF, Rio de Janeiro, Brazil

SPONSORING ORGANIZATIONS
Fundação de Amparo à Pequisa do Rio de Janeiro - FAPERJ
Centro Brasileiro de Pesquisas Físicas - CBPF
Conselho Nacional de Desenvolvimento Cientifico e Tecnológico - CNPq
Coordenação de Aperfeiçoamento de Pessoal do Ensino Superior - CAPES
International Center for Relativistic Astrophysics - ICRA

Melville, New York, 2003
AIP CONFERENCE PROCEEDINGS ■ VOLUME 668

Editors:

Mário Novello
Santiago E. Perez Bergliaffa

Centro Brasileiro de Pesquisas Físicas
Rua Dr. Xavier Sigaud 150
URCA - 22290-180 Rio de Janeiro-RJ
BRAZIL

E-mail: novello@cbpf.br
sepb@cbpf.br

Authorization to photocopy items for internal or personal use, beyond the free copying permitted under the 1978 U.S. Copyright Law (see statement below), is granted by the American Institute of Physics for users registered with the Copyright Clearance Center (CCC) Transactional Reporting Service, provided that the base fee of $20.00 per copy is paid directly to CCC, 222 Rosewood Drive, Danvers, MA 01923. For those organizations that have been granted a photocopy license by CCC, a separate system of payment has been arranged. The fee code for users of the Transactional Reporting Service is: 0-7354-0131-4/03/$20.00.

© 2003 American Institute of Physics

Individual readers of this volume and nonprofit libraries, acting for them, are permitted to make fair use of the material in it, such as copying an article for use in teaching or research. Permission is granted to quote from this volume in scientific work with the customary acknowledgment of the source. To reprint a figure, table, or other excerpt requires the consent of one of the original authors and notification to AIP. Republication or systematic or multiple reproduction of any material in this volume is permitted only under license from AIP. Address inquiries to Office of Rights and Permissions, Suite 1NO1, 2 Huntington Quadrangle, Melville, NY 11747-4502; phone: 516-576-2268; fax: 516-576-2450; e-mail: rights@aip.org.

L.C. Catalog Card No. 2003105094
ISBN 0-7354-0131-4
ISSN 0094-243X

Printed in the United States of America

CONTENTS

Preface vii

Photographs ix

Methodology of Observational Cosmology 1
    J. G. Bartlett

New Perspectives in Physics and Astrophysics from the Theoretical Understanding of Gamma-Ray Bursts 16
    R. Ruffini, C. L. Bianco, P. Chardonnet, F. Fraschetti, L. Vitagliano, and S.-S. Xue

Chaotic Phenomena in Astrophysics and Cosmology 108
    V. G. Gurzadyan

String and M-Theory Cosmology 125
    E. J. Copeland

Canonical Quantization of General Relativity: The Last 18 Years in a Nutshell 141
    J. Pullin

The Strong-Coupling Expansion and the Singularities of the Perturbative Expansion 154
    N. F. Svaiter

Space and Spacetime 173
    M. Lachièze-Rey

P-Branes, Extra Dimensions, and Their Observational Windows 187
    V. N. Melnikov

Experimental Status of Corrections to Newtonian Gravity Inspired by the Extra Dimensional Physics 198
    V. M. Mostepanenko

Variable Cosmological Term 204
    I. Dymnikova

Cosmology from Topological Defects 226
    A. Gangui

On the Possible Role of Massive Neutrinos in Cosmological Structure Formation 263
    M. Lattanzi, R. Ruffini, and G. Vereshchagin

Effective Geometry 288
    M. Novello and S. E. Perez Bergliaffa

List of Participants 301

Author Index 305

Preface

This volume contains the set of lectures presented at the Xth Brazilian School of Cosmology and Gravitation (BSCG). The meeting took place at Mangaratiba, a small village by the sea, located 74 km to the south of Rio de Janeiro.

The BSCG has a tradition of 25 years among the practitioners and students in the fields of Gravitation, Cosmology, Astrophysics, and Field Theory. Most of the more relevant issues in those areas were covered in the various editions of the BSCG. An example of this richness is given by the topics presented in the tenth edition, which range from (almost purely) theoretical matters (like Quantum Gravity) to the latest observational developments (as for instance those in Observational Cosmology).

During the two weeks, a set of eight lectures was presented, plus five special seminars. People from 19 countries attended the lectures and seminars.

Mario Novello
Santiago E. Perez Bergliaffa

Acknowledgements

We would like to acknowledge financial support from FAPERJ (Fundação de Amparo à Pequisa do Rio de Janeiro), CBPF (Centro Brasileiro de Pesquisas Físicas), CNPq (Conselho Nacional de Desenvolvimento Cientifico e Tecnológico), CAPES (Coordenação de Aperfeiçoamento de Pessoal do Ensino Superior), and ICRA (International Center for Relativistic Astrophysics).

Our warmest thanks to the secretary of the Xth BSCG, Ms. Mônica Ramalho, for her work and patience. Many thanks also to Sonia Ribeiro da Silva Ferreira, and Deise Monte Rego.

International Organizing Committee

- M. Novello (CBPF, Brazil).
- E. Elbaz ((Lyon U., France).
- E. Kolb (Fermilab, USA).
- V. Melnikov (Institute of Metrology, Russia).
- R. Triay (CTP/ U. de Marseille, France)
- S. E. Perez Bergliaffa (CBPF, Brazil).

Methodology of Observational Cosmology

J.G. Bartlett

APC – Université Paris 7
mailing address: Collège de France
11 pl. Marcelin Berthelot
75231 Paris Cedex 05, France

These lectures focus on the methods of observational cosmology and their physical basis. After a brief reminder of the essential elements of the Big Bang model, we look at some of the basic methods for surveying the homogeneous universe. The second half of the lectures examines the perturbed universe – large–scale structure and the cosmic microwave background anisotropies. In the last few years we have seen the emergence of a "standard" cosmological model with coherent parameters. The goal of the course is to understand the rather vast array of observational approaches used to establish and test this model in detail [1].

I. INTRODUCTION

Although of clearly ancient origin, cosmology as the modern physical science currently practiced is relatively young. It begins with the introduction of general relativity (GR), and in fact represents one of the extremely few applications of GR amenable to an exact solution (FRW spacetime). It is worth emphasizing perhaps the term *application*, because in cosmology we are not involved in the construction of a fundamental physical theory, but rather concerned with the elaboration of an ever more detailed model of events in the history of the universe. Nevertheless, cosmology has had, and will continue to have, important consequences for fundamental physical theory, and this due to the impressively vast range of physics employed: from quantum gravity and the most esoteric particle theories, to the astrophysics of star formation.

The specific subject of these lectures is observational cosmology, whose driving fundamental question is, *How do we measure and survey the universe?* We will walk a line between theory and observation, focusing on the basis of the observational approaches used to answer this question. Theory will serve to interpret observations, develop observational methods and to identify key questions. Our working model, based on the Big Bang scenario, has come to be known in recent years as the "standard" cosmological model for its success in explaining what we observe. Our primary goal will be to understand the basis of and observational support for this model, as well as the general way we test it. We shall find that modern cosmology exhibits that satisfying characteristic of a mature science, namely, a fruitful interplay between theory and observations. In practice, we will be concerned with more recent events in the history of the universe, essentially those occurring after the production of the primordial perturbations (e.g., Inflation). Details of the production of perturbations and of the state of the early universe we leave to other lecturers, and concentrate instead on issues related to the nature of spacetime and of the matter distribution on large scales, e.g., the cosmological parameters, dark matter, galaxy and large-scale structure formation.

A natural subject division is one between the study of the homogeneous universe and that of perturbations from homogeneity. It is fortunate indeed that the smallness (on large scales) of cosmological fluctuations permits such a perturbative approach. We thus begin by surveying the homogeneous universe, its contents and descriptive parameters. During this first part, we develop the basic concepts of the Big Bang model and discuss how to test it and how to measure its fundamental parameters. We then turn in the second part to the perturbations, elaborating our model to explain large–scale structure (LSS) and its evolution. The way we describe LSS will guide both our theoretical constructions as well as our observational methods. The two parts are not independent, and we shall learn much about one part from study of the other.

A set of good general references for many of the topics covered in these lectures includes:

- T. Padmanabhan 1993, Structure Formation in the Universe, Cambridge University Press (Cambridge, UK)

- J.A. Peacock 1999, Cosmological Physics, Cambridge University Press, Cambridge UK

- P.J.E. Peebles 1993, Principles of Physical Cosmology, Princeton Series in Physics, Princeton University Press (Princeton, NJ)

- P.J.E. Peebles 1980, The Large–Scale Structure of the Universe, Princeton Series in Physics, Princeton University Press (Princeton, NJ)

- P.J.E. Peebles 1970, Physical Cosmology, Princeton Series in Physics, Princeton University Press (Princeton, NJ)

- S. Weinberg 1972, Gravitation and Cosmology, John Wiley & Sons (New York, NY)

Before beginning in earnest, we should take the opportunity on this anniversary to look at the progress made since the first Brazilian School of Cosmology and Gravitation held 25 years ago. At the time of the first School, for instance, the inflation scenario had yet to be proposed, as well as the cold dark matter (CDM) model. The cosmic microwave background (CMB) dipole had been recently discovered, but the observational proof of the thermal spectrum was yet to be had, and the first detection of temperature anisotropies was still many years off. Systematic redshift surveys were just beginning, and CCD and infrared cameras that were to revolutionize the field were coming. Save it to say that much has been accomplished in 25 years!

II. DESCRIBING THE HOMOGENEOUS UNIVERSE: THE MODEL AND ITS PHYSICAL BASIS

What exactly is the physical system that we are to model in cosmology? This is our first question, and it requires some observational input. Until 1920s, the universe was taken to be the observed system of stars. It was already appreciated at the time that this system is flattened and also limited in radial extent, as so wonderfully demonstrated today by the famous COBE–DIRBE composite IR image [25]. Although it is obvious how to see anisotropy in the distribution of a set of objects, such as stars, it is perhaps less so for its radial extent. A key method used in Astronomy for addressing this point is based on *number counts*. The counts refer to the surface density of objects on the sky as a function of their observed brightness. For a uniform and non–evolving population, the counts $dN/d\Omega \sim f^{-3/2} \sim 10^{0.6m}$, where f is the observed flux and m the corresponding magnitude[26]. The star counts do not follow this law (even when corrected for extinction effects), indicating that our stellar system is limited in radial extent.

It was in this context that Einstein first proposed what today is known as the Cosmological Principle: *the universe is spatially uniform and isotropic over large scales*, and applied it to construct (as we do below) a suitable metric for the universe. Satisfying for its prohibition of any preferred point or direction in space, its validity must of course reside solely on observational confirmation. Although, as we have seen, there was little observational support when first introduced by Einstein (who was apparently motivated rather by Machs principle and wished to avoid regions devoid of and far from all matter), today the observational coherence of the model built on the Cosmological Principle is certainly to be strongly counted in its favor. We shall have several opportunities to discuss this point in detail, but we may immediately cite the isotropy of the cosmic microwave background, galaxy counts and the distribution of galaxies in space, all as observational support for the Cosmological Principle. Consider, for example, the number counts test applied to the galaxy distribution. This was one of the earliest systematic extragalactic studies, first undertaken by Hubble (in the late 1920s and 1930s) shortly after it was realized that the nebulea were galactic systems like our Milky Way. Very recently, the Sloan Digital Sky Survey (SDSS) has published the bright star and galaxy counts. The latter beautifully show the "0.6 magnitude" law expected of a (at least) locally non-evolving and homogeneous population[20]. This is no perhaps no surprise, but it is worth mention because the SDSS data provide the first bright–end counts using electronic rather than photographic techniques, which are subject to particular photometric difficulties.

A. The FRW Metric and its Physical Interpretation

Notice that the Cosmological Principle is a statement about the homogeneity of space, and not spacetime. We are thus called on to construct a metric **g** that describes a spacetime with homogeneous and isotropic *spatial* hypersurfaces. To find explicit expressions for the components $g_{\mu\nu}$, we must choose a set of coordinates, and we shall choose them such that $g_{00} = 1$ and $g_{0i} = 0$ (four coordinate conditions). It is already clear from this construction that our coordinates will correspond to measurements made by a set of freely moving observers. The space part of the metric must conform to the assumed uniformity, so that any time dependence will be the same in all three spatial coordinates and may be factored out: $ds^2 = g_{\mu\nu}dx^\mu dx^\nu = dt^2 - a^2(t)\tilde{g}_{ij}(\vec{x})dx^i dx^j$. Furthermore, we are guaranteed that there exists a spatial coordinate system in which $\tilde{g}_{ij}dx^i dx^j = f(r)dr^2 + r^2(d\theta^2 + \sin^2\theta d\phi^2)$, explicitly manifesting isotropy. By direct calculation, the function $f(r)$ can be related to the curvature of the spatial hypersurfaces, κ ($= 1/\mathcal{R}^2_{\text{curv}}$, where $\mathcal{R}_{\text{curv}}$ is the curvature radius): $f(r) = 1/(1 - \kappa r^2)$. Homogeneity requires constant curvature, which leads finally to the well–known Friedmann–Walker–Robertson (FRW) metric

$$ds^2 = dt^2 - a^2(t)\left[\frac{dr^2}{1 - \kappa r^2} + r^2(d\theta^2 + \sin^2\theta d\phi^2)\right] \quad (1)$$

written in *comoving coordinates*. There are only two parameters, the *scale factor* $a(t)$ describing temporal evolution, and the *curvature of space* κ (or equivalently $\mathcal{R}_{\text{curv}}$), which may be positive (spherical space), negative (hyperbolic space), or zero (flat space). Notice that spherical space is finite ($r < \mathcal{R}_{\text{curv}}$), while hyperbolic and flat spaces are infinite in extent.

As usual in GR, once the components of the metric have been specified (coordinate conditions given), it befalls us to interpret the corresponding coordinates in physical terms (observations and observers). Our suspicion that comoving coordinates may be interpreted in terms of freely-moving observers can be checked using the geodesic equation of motion, the solution for which implies that an observer placed at an arbitrary point, labeled by comoving coordinates $\vec{x}$, at initial time t_i with zero coordinate 3-velocity will remain fixed at $\vec{x}$ for all later times $t > t_i$ (a good exercise in GR manipulation.). A simple physical argument is even more economical: what direction would the acquired velocity point in, given that the universe appears completely uniform to the observer in the comoving frame! Comoving coordinates are so named precisely because they correspond to a set of observers comoving with the expansion. The coordinates $(t, \vec{x})$ of an event in spacetime are the time t measured by the clock of the comoving observer called $\vec{x}$ (three numbers are needed to label each and every observer) whose worldline passes through the event[27]. Finally, note that the proper spatial distance between comoving observers evolves with time according to the scale factor as $dl \propto a(t)$ (which explains its name). This of course is the expansion (or contraction) of the universe. We now turn attention to this evolution, which we also need to understand the propagation of light signals and, hence, all potential observations.

B. Matter Description and the Field Equations

Evolution of the metric [the function $a(t)$] is given by the gravitational field equations, with a source term equal to the matter stress–energy tensor **T**. It seems reasonable to model the matter (both radiation and non-relativistic matter) as a perfect fluid with energy density $\rho(t)$, pressure $p(t)$ and 4- velocity **u** (only functions of time):

$$T_{\alpha\beta} = (\rho + p)u_\alpha u_\beta - p g_{\alpha\beta} \tag{2}$$

In the FRW coordinate frame, where the universe appears uniform, the 3-velocity of the matter must vanish (which direction would it point in?), and the 4-velocity takes the form $\mathbf{u} = (1, 0, 0, 0)$. A collisional fluid or gas can be described by an isotropic pressure p related to the energy density by an equation-of-state $p[\rho]$; this will apply to the baryons, for example. In the special case where its 3-velocity is zero, a cold collisionless fluid (e.g., cold dark matter) may also be modeled as a perfect fluid, with zero pressure[28].

Once the form of the source has been chosen, the field equations follow by straightforward calculation, a worthwhile exercise in GR that should be performed by everyone at least once. One finds only two equations, one corresponding to the time-time component and the other corresponding to three identical space- space equations (the time-space component yielding the identity $0 = 0$):

$$H^2 \equiv \left(\frac{\dot{a}}{a}\right)^2 = \frac{8\pi G \rho}{3} - \frac{\kappa}{a^2} + \frac{\Lambda}{3} \tag{3}$$

$$\frac{\ddot{a}}{a} = -\frac{4\pi G}{3}(\rho + 3p) + \frac{\Lambda}{3} \tag{4}$$

These are the Friedmann equations, and H is the *Hubble parameter*. We observe the universe in *expansion*, so we shall be particularly concerned with the root of the first equation with $\dot{a} > 0$; $a(t)$ *increases with time*.

There are two equations for the single function $a(t)$ and, furthermore, only one is second order in time as expected for a classical dynamic equation. This is because the field equations also impose energy conservation on the source: $T^{\alpha\beta}_{;\beta} = 0$, which when explicitly written out is

$$\dot{\rho} + 3\frac{\dot{a}}{a}(\rho + p) = 0 \tag{5}$$

Only two of this trio of equations are independent; usually one works with the first Friedmann equation (with the Hubble parameter) and the energy conservation equation, in part because both are first order in time. Note also that there are three functions to be found – $a(t)$, $\rho(t)$ and p; the equation–of–state is required to close the system of equations.

The nature of a particular kind of matter, specified in this case by its equation–of–state, determines its evolution with the scale factor: for non–relativistic matter, with $p << \rho$, Eq. (5) simply expresses mass conservation $\rho \propto 1/a^3(t)$, while for highly relativistic matter, with $p = \rho/3$, we find $\rho \propto 1/a^4$. Vacuum energy (or a cosmological constant) is a more exotic form of matter with $p = -\rho = const$. The different dependence on scale factor implies that the different forms of matter dominate at different times in cosmic history. Excluding very early and exotic times (before primordial nucleosynthesis), the universe passed from a radiation–dominated phase to a matter–dominated one, and may very well be in a period today of vacuum energy domination. Each of these epochs is dominated by a different gravitational source, and so follows a distinct temporal evolution:

Solutions:

- (Non–relativistic) matter domination: $a(t) \propto t^{2/3}$
- Radiation domination: $a(t) \propto t^{1/2}$
- Vacuum energy domination (cosmological constant): $a(t) \propto e^{Ht}$ ($H = const$)
- Curvature domination: $a(t) \propto t$.

C. The Cosmological Parameters

The cosmological parameters describing the homogeneous universe are all to be found in the (first) Friedmann equation, which at times well after the radiation–dominated era can be written

$$H^2 = \frac{8\pi G}{3}\rho_m - \kappa(1+z)^2 + \frac{\Lambda}{3}$$
$$\text{OR}$$
$$= H_0^2\left[\Omega_M(1+z)^3 + \Omega_\kappa(1+z)^2 + \Omega_\Lambda\right] \tag{6}$$

where we introduce the *Hubble constant* H_0 as the value of H today, at time t_0, and the density parameters $\Omega_M \equiv \rho_m/\rho_c$, where $\rho_c \equiv 3H_0^2/8\pi G$ is the known as the *critical density* (the matter density **today** that would correspond to a flat universe with zero cosmological constant [$\kappa = \Lambda = 0$], the oft–called critical universe); the *curvature parameter* $\Omega_\kappa \equiv -\kappa/H_0^2$ (note the minus sign); and the *vacuum density parameter* $\Omega_\Lambda \equiv \Lambda/3H_0^2$. The matter density parameter may be further separated into $\Omega_M = \Omega_B + \Omega_C$, with contributions from the baryons and (cold) dark matter, respectively. *It is important to note that all of the parameters are defined by their values today.* The energy density of photons (the cosmic microwave background) and of primordial neutrinos are not usually associated with a density parameter, because we know that the cosmic microwave background is an excellent blackbody of temperature $T = 2.725 \pm 0.001$ K[6] and we believe/suppose that, although unobserved to date, the three light neutrino species follow non–degenerate Fermi distributions at temperature $T \approx 1.9$ K. Finally, it should be remarked that Eq. (6) implies a constraint among the the density parameters: $1 = \Omega_M + \Omega_\kappa + \Omega_\Lambda$. This may be viewed as reducing the number of independent parameters by one, or, and more importantly, a way to test the basic underlying Friedmann equation in the real universe.

D. Observing Distant Objects

One of the primary aims of observational cosmology is the determination of the fundamental cosmological parameters just mentioned, and one of the direct ways to do so is by distance measurements. Particular attention must be given to the nature of observation when discussing distant objects in the universe. We observe the universe through light, and so null geodesics will play a central role: $ds^2 = dt^2 - a^2(t)dr^2/(1-\kappa r^2)$ (isotropy demands that light rays travel radially from source to destination – why would they turn!). The *comoving distance* r to a source a redshift z may be calculated from

$$\int_0^r \frac{dr}{\sqrt{1-\kappa r^2}} = \int_{t_e}^{t_o} \frac{dt}{a(t)} = \frac{1}{a_0}\int_0^{z_e} \frac{dz}{H(z)} \tag{7}$$

where $H^2(z) = H_0^2[\Omega_M(1+z)^3 + \Omega_\kappa(1+z)^2 + \Omega_\Lambda]$ is given by the Friedmann equation. The comoving distance to an object thus depends on the cosmological parameters, which implies that distance measurements can be used to constrain the model.

Let dl be the proper distance between the ends of a distant object (a meter stick) at redshift z and corresponding comoving distance $r(z)$. Light from the two ends travels along radial null geodesics to reach us today at approximately the same time [$(1+z)dl$ is taken to be small compared to r]. Thus the apparent angle sub-tended by dl is $d\theta = (1+z)dl/r(z)$, implying an *angular distance* of

$$D_{\text{ang}}(z) \equiv \frac{dl}{d\theta} = \frac{a_0 r(z)}{(1+z)} \tag{8}$$

where $r(z)$ is the comoving distance given in Eq. (7).

Now consider the *luminosity distance*, defined by $D_{\text{lum}}(z) \equiv \sqrt{L/4\pi f}$, where L is the intrinsic luminosity of a distant source at redshift z and f is the received flux today. Conservation of photons in a gravitational field implies $L dt_e/\hbar\omega_e = 4\pi f a_0^2 r^2(z) dt/\hbar\omega$, with dt_e and dt the emitted and received periods (both source and detector are taken to be comoving), $r(z)$ the comoving distance to the source at redshift z, and ω_e and ω the emitted and received frequencies ($\omega_e = \omega(1+z)$). Putting all this together provides an expression for the luminosity distance

$$D_{\text{lum}}(z) = a_0 r(z)(1+z) = D_{\text{ang}}(z)(1+z)^2 \qquad (9)$$

Note that the angular and luminosity distances differ, unlike the case of static Euclidean space, due to the redshifting of the photons.

III. SURVEYING THE HOMOGENEOUS UNIVERSE: CONTENTS AND MEASURES

How we make our measurements obviously depends on what we are given to observe, or in other words, on the contents of the universe. This is a particularly relevant point in cosmology, where we have no direct control over the physical system under study; it is a purely observational science. The observable structure of the universe is built on the galaxies and their distribution. They come in different forms, but they also display remarkable regularity that must reflect the physics of galaxy formation. They are distributed in a hierarchy of structures, ranging from dwarf galaxies, through large clusters containing thousands of galaxies, to the general large–scale distribution in space. Today, we have good reason to believe that the galaxies outline (perhaps in some complicated way) the distribution of a more important, but more difficult to directly observe, component, the dark matter. Cosmology has a two–fold interest in galaxies: we would like to explain their origin, evolution and properties within the context of a general cosmological scenario and, secondly, we may use them as tracers of this more elusive dark matter. In these lectures, we will be primarily concerned with the second aspect, but it should be noted that the first is taking a prominent role in cosmology, thanks to recent progress in both modeling and observational means (large telescopes and IR surveys in particular).

It is instructive to consider a more familiar example and system, namely, the stars and our Milk Way. Observationally, galactic structure is built on the stars. Despite their rich diversity, stars display remarkable regularity, such as the Main Sequence, and much may be learned about our galaxy from its stellar population, but not all – there are other important components of our Galaxy, such as the interstellar medium.

In this section on the homogeneous universe, we are interested in using galaxies and galaxy clusters as individual objects to measure the local universe. Their large–scale distribution in space will be examined in later sections, when we discuss density perturbations. After first summarizing some of their important properties, we then turn attention to using these objects, on an individual basis, to measure the local universe, to constrain the cosmological parameters and to test our model.

A. Galaxies

Galaxies come in a variety of forms, from spherical systems known as ellipticals to the cartwheel spirals. The traditional classification is the Hubble sequence, which progresses from gas–poor ellipticals (E) and S0s (also called *early type galaxies*) through to gas rich spirals (Sp) and irregulars (Irr) (also collectively called *late type galaxies*). The mix of galaxies changes significantly with local galaxy density, with late type spirals dominating the field population and early type E and S0s becoming dominant in cluster cores, in a trend known as the morphology–density relation. Ellipticals appear to contain rather old, and therefore cool and red, stellar populations, and they satisfy a very well–defined color–magnitude relation. Spirals and irregulars, on the other hand, are actively forming new stars from their gas reservoirs and therefore contain an important population of young, massive stars emitting at short (blue and UV) wavelengths.

Two fundamental galactic quantities are clearly mass and luminosity. The galaxy *luminosity function* ϕ characterizes the number of galaxies as a function of their intrinsic luminosity:

$$\text{no.gals/volume} = \phi\left(\frac{L}{L_*}\right) \frac{dL}{L_*}$$

where L_* is a characteristic luminosity. Based on observations of cluster galaxies, Schechter proposed in 1976 what has now become a universal form for the luminosity function, applicable to both cluster and field galaxies:

$$\phi\left(\frac{L}{L_*}\right) = \phi_* \left(\frac{L}{L_*}\right)^\alpha e^{-L/L_*} \qquad (10)$$

with ϕ_* a characteristic density. The field galaxy population appears rather well described in the blue by $\alpha \sim -1.2$, $M_{b_j} \sim -19.6 - 5\log h$ and $\phi_* \sim 0.015 h^{-3}$ Mpc^{-3}, as borne out by a number of studies, including the most recent 2dF and SDSS results[4, 12].

Determining galaxy masses is more difficult and it raises the important issue of *dark matter*. The mass of a galaxy may be estimated by studying its internal dynamics. For rotating spiral galaxies, the dynamics of the disk is characterized by the rotation curve – the rotation velocity as a function of radius from the center – observable through the systematic Doppler shift of spectral lines (in particular the 21cm hyperfine neutral Hydrogen transition that can be traced out to large galactic radii). For ellipticals, which generally have no systematic internal motion, one uses the stellar velocity dispersion measured by the width of observed stellar spectral lines. Supposing that mass traces light, the observed light profile of a spiral may be used to predict a specific rotation curve, within an overall normalization given by an unknown *mass–to–light* ratio. Observations can thus be used to determine this ratio, at least for those parts of a galaxy where there is light to measure velocities. The general result for spirals (ellipticals) falls around $\sim 1 M_\odot/L_\odot$ ($\sim 5 M_\odot/L_\odot$), which compares favorably with expectations based on stellar population models of Spirals (Peebles 1993)[8]. The big problem, however, is that this rotation curve fitting only works in the innermost regions of spiral galaxies. While the light profile falls–off exponentially in the outer regions, the rotation curve remains flat out to the farthest observable radii. This indicates that globally spirals are dominated by *dark matter*. A flat rotation curve implies that galaxy mass increases linearly with radius, and since no end to these flat rotation curves has yet been seen, we in fact have no idea of the mass of isolated spiral galaxies; we've never observed their edge! We will return shortly to the importance of the characteristic mass–to–light ration of galaxies.

B. Clusters

Galaxy clusters reside on the massive end of virialized objects in the clustering hierarchy. They range from poor groups containing ~ 10 galaxies to the enormous rich clusters with several hundreds. Of the three main cluster constituents, the stellar mass of member galaxies contributes the least to the overall mass (only $\sim 2\%$ in rich clusters). A hot intracluster gas, heated to ~ 5 keV on infall into the cluster during formation, contributes up to five times more in rich systems, for a total of $\sim 10-20\%$. Most of the cluster mass, however, is in an unknown form of dark matter. These results are found by studying the cluster galaxies and their dynamics, X–ray emission from the hot gas, and gravitational lensing of background sources by the cluster potential.

C. The Local Distance Scale: H_0

Galaxies are our beacons for surveying the universe. To our immediate interest, both ellipticals and spirals follow clearly defined relations that permit their use as distance indicators: the fundamental plane (FP) in the former case, and the Tully–Fisher relation (TF) in the later. The FP, for example, is a relation between three observables: a distance–dependent effective radius r_e, and the distance–independent stellar velocity dispersion σ (obtained from the width of stellar spectral lines) and surface brightness I_e: $\log r_e = \alpha \log \sigma + \beta \log I_e + \gamma$, where $\alpha \sim 1.24$, $\beta \sim -0.8$[10] and γ depends on the galaxy distance. The relation is called the FP because it defines a plane in the 3D space spanned by these observables. This is an empirical relationship established by looking at galaxies in a cluster, and hence all at the same distance.

Once established, the relation can be *calibrated* using the known distance to few nearby ellipticals, and subsequently used itself as a distance indicator (a so–called secondary distance indicator) over much larger scales. Calibration of this kind of empirical relation (there exist others, e.g., the TF relation, surface brightness fluctuations, planetary nebulae, SNIa, etc...) was precisely the goal of the HST Key Project on the Extragalactic Distance Scale, which measured the distance to a set of nearby galaxies and groups using Cepheid variables (which follow a known period–luminosity relation and represent one of the central, early rungs of the distance latter). The calibration provided by Key Project[7] produces a beautiful linear Hubble law, consistent among all the important secondary calibrators, and a value for the Hubble constant of $H_0 = 72\pm$ km/s/Mpc.

D. Mass–to–Light Ratios

One of the classic ways to estimate the matter density Ω_M is via the mass–to–light ratio M/L of galaxies. The method in fact has many inherent problems, but it is nonetheless useful, as well as representative of observational approaches used in cosmology. We may obtain an estimate of the matter density by multiplying the M/L ratio by the mean luminosity density j, which in turn is given as an integral over the observed luminosity function. The Schechter function yields

$$j = \int \frac{dL}{L_*} L \phi\left(\frac{L}{L_*}\right) = \phi_* L_* (1+\alpha)!$$

Putting in typical parameter values, multiplying by an M/L averaged over a representative mix of galactic populations and then dividing by the critical density gives an estimate of the *stellar mass* density, $\Omega_* \sim 0.002$[8]. This does not include the dark matter in galaxy halos and that we know begins to dominate in the outer regions of galaxies.

When discussing the rotation curves of spirals, we ran into the problem that the M/L ratio increases as we approach the observable edge of galaxies. The global mass–to–light ratio including the dark halo is unknown. We can try to circumvent this difficulty by using clusters. The entire mass of a galaxy, including its dark halo, is incorporated into a cluster, where if affects the observable internal cluster dynamics. A typical galaxy mass–to–light ratio can then be calculated by measuring the total cluster light (an integral over the luminosity function of member galaxies) and dividing by the total cluster mass. Two difficulties appear in practice: what is the best way to measure cluster mass and how to correct for the fact that the cluster galaxy population differs from that of the field, where we estimate the universal luminosity density.

One of the more extensive studies of this kind was carried out using the CNOC cluster survey. Internal dynamics were studied in a sample of clusters for which a large number of galaxy redshifts were obtained. Applying the virial theorem, a mass–to–light ratio could be obtained for each sample cluster, with the overall result that $M/L \sim 300\ h$ (in the r–band). This is much larger than the estimated stellar mass–to–light ratio, confirming the idea that galaxies are dark–matter dominated, and it raises the estimate for the mass density contributed by matter associated with galaxies to $\Omega_M \sim 0.2$[5].

Another method similar to the traditional mass–to–light ratio relies on the observed *baryon fraction*, or baryon mass–to–total mass ratio, of galaxy clusters. As pointed out by White et al. [19], we expect this ratio, $f_b \equiv M_b/M_{tot}$, to be the same as the universal value Ω_B/Ω_M, at least at the virial radius of clusters. This follows because we imagine that clusters collapse from a representative volume filled with the universal mixture of baryons and dark matter, and whatever else there may be. We obtain a lower limit to the baryon content of clusters by summing the luminous matter associated with the member galaxy stellar populations (stellar mass–to–light ratios from population models) and the hot X–ray emitting gas strewn throughout the cluster volume. In rich systems, as we saw previously, the intergalactic gas dominates. X–ray observations give us directly the gas content of a cluster; they also may be used to deduce the total cluster mass, if one assumes that the gas is in hydrostatic equilibrium in the gravitational potential well. This appears reasonable according to simulations of clusters in standard cosmological scenarios (caution is always advisable!). Once again, this give us a lower limit to the possible baryon content of clusters, for there could in principle be other forms of dark baryons. The method is nevertheless very powerful, because a lower limit on f_b yields an upper limit on Ω_M: using the baryon density established according to both the light element abundances and the CMB temperature anisotropies, we obtain $\Omega_M < \Omega_B/f_b$. A straightforward glance at the literature shows that the favored upper limits, often treated as a measurement, fall around $\Omega_M \sim 0.3$ (e.g., [11]); the method is not without problems, in particular related to the total mass estimate and the extrapolation of the gas density out beyond direct X–ray imaging to the virial radius[17].

E. Supernovae

One of the important secondary distance indicators touched upon while discussing the measurement of the Hubble constant was the luminosity of type Ia supernovae. This type of SN (identifiable by its spectral features) appears to have a nearly constant intrinsic peak luminosity, as comparison to other distance indicators and the tightness of their magnitude–redshift relation suggests. Their large intrinsic brightness means that they may be found at large distances, making them a key cosmological probe.

The fact that they can be found at large distances means that they can be used to constrain not only the Hubble constant (the small z, linear part of the D_{lum}), but potentially Ω_M and Ω_Λ through the detailed form of D_{lum} at higher redshift. Two teams of researchers, the Supernova Cosmology Project (SCP) [22] and the High–z Supernova Project [21], have undertaken searches for high redshift supernovae with this aim. The searches are carried out by observing a set of fields at several weeks interval. Comparison of the images reveals a set of SN candidates that are then followed–up with spectroscopic and detailed multi–color photometric observations. The spectra are used to confirm the SN origin of the source and to identify its type and find its redshift. The color data provide light curves that are used to estimate the peak luminosity[29] and correct it for eventual dust obscuration (*redding*). The SCP, for example, has found more than 40 SNIa out to redshift 0.8.

Both teams have published important constraints on Ω_M and Ω_Λ indicating that $\Omega_\Lambda > 0$, i.e., that there is a non–zero vacuum energy density driving an accelerated expansion [15, 16]. This is one of the most important cosmological results in the last few years. It's importance not only for cosmology, but for fundamental physics should be the cause for critical reflection. The result appears robust, but it would be extremely valuable to have independent confirmation. We will discuss the issue further below when discussing other cosmological constraints and the so–called *concordance model*.

IV. THE PERTURBED UNIVERSE: LARGE–SCALE STRUCTURE

This section begins the second part of the course, where we turn attention to deviations from the homogeneous model. These deviations are fortunately small and allow a perturbative approach. On most scales and at most times, for example, linear perturbation theory is sufficient. It is only at more recent epochs and on small scales that non–linear gravitational effects become important. But even this non–linear regime may be studied, using numerical simulations (and even, in some cases, good analytical approximations), as taking place on the more general, homogeneous background. Our two main areas of study will be large–scale structure and CMB anisotropies. We begin with the former because simple Newtonian gravity applies. The anisotropies, on the other hand, involve super-horizon scales and relativistic fluids, requiring a more elaborate treatment.

For many studies in cosmology, galaxies play the simple role of point masses, although they have the essential virtue of being luminous point masses. Their spatial distribution is an intricate hierarchical pattern embedding the virialized galaxy groups and clusters in its lacy large–scale distribution. Beautiful examples of this hierarchy are the Lick, APM, 2dF[24] and SDSS [23] surveys. The most natural interpretation of the galaxy distribution is that it reflects the overall matter distribution, or in other words, density perturbations. We are thus brought to study at length the *evolution of density perturbations* in the expanding universe. Linear theory is sufficient for large scales where the perturbations are small (a statement that will be short quantified), but *non–linear* evolution is important on smaller scales and for the formation of collapsed objects, such as galaxy clusters. While it seems obvious that the galaxy distribution should in some way reflect the underlying matter distribution, the exact relation between the two distributions is perhaps non–trivial. Does a given galaxy number density perturbation correspond to an equal matter density perturbation? The suspicion that this may not always be the case is summed up by the question, *does mass trace light?*. The question appears all the more relevant once one realizes that the matter directly observable by its luminous emission represents only a tiny fraction of the total mass density of the universe. The issue of mass v.s. galaxy distribution falls under the general heading of **biasing**; we say that galaxies could very well represent a *biased* tracer of the underlying matter distribution.

A central issue is how to interpret the cosmological principle of uniformity and isotropy in the presence of these perturbations. One way is to model galaxy clustering as a *homogeneous statistical process*, which is to say that some physical mechanism (e.g., Inflation) produced perturbations with the same statistical characteristics at all points in space, favoring neither location or direction. The mathematical object adapted to this kind of model is a **random field** $\delta(\vec{x})$, whose value at any point $\vec{x}$ is a random variable[30]. In this light, the goal of observations is to discover the statistics of the (random) density field, in order to understand the physics of the mechanism responsible for the perturbations. The cosmological principle imposes two conditions on the perturbation density field: 1/ its statistics must be homogeneous (a stationary random field) and 2/ regions separated by large distance must be uncorrelated (an ergotic field). The second condition, sometimes called the *fair sample hypothesis*, follows because there seems little practical value of discussing a homogeneous statistical process, as interpretation of the cosmological principle, if its only realization shows no indication of homogeneity on any scale! In any case, we are building a model whose final validity rests on its ability to explain what we observe, as we will discuss in the following.

In conclusion, at the center of our model is the concept that

the observed density field represents a homogeneous random field that approaches uniformity on large scales

and the overall picture that one should keep in mind is that at some early time, say by the end of Inflation, the initial density perturbation field is in place, and subsequent gravitational evolution serves to map the initial statistics of the field into the its statistics at any later time[31].

The construction and practical application to observations of appropriate statistical measures will occupy an important part of our discussion. Furthermore, three–dimensional galaxy surveys use redshift as the equivalent of radial distance. However, the observed redshift confounds the general expansion with galaxy peculiar velocity. Clustering measures in *redshift space* are therefore distorted relative to clustering in *real space*. These **redshift–space distortions** are in fact quite useful, for they tell us about the dynamics of galaxies and permit a measurement of the matter density parameter.

The main goals of this discussion of large-scale structure are to understand the origin of the form of the density power spectrum (to be defined) expected in standard models, its general success in explaining current observations, and the conclusions that one therein draws concerning the cosmological model.

A. Evolution of Density Perturbations

In this section we briefly touch on the wonderfully rich physics of perturbation evolution in the expanding universe. The discussion here is restricted to the density field at late times and on scales within the cosmic horizon; conditions where the Newtonian approximation applies. We thus model the cosmic plasma as a Newtonian fluid of mass density ρ and pressure p. The equations of evolution consist of the continuity equation (conservation of mass), the Euler equation (Newton's second law) and the Newton–Poisson equation for the gravitational potential Φ:

$$\dot{\rho} + \nabla \cdot (\rho \vec{u}) = 0$$
$$\rho \frac{d\vec{u}}{dt} = \rho \cdot \vec{u} + \rho(\vec{u} \cdot \nabla)\vec{u} = -\rho \nabla \Phi - \nabla p$$
$$\nabla^2 \Phi = 4\pi G \rho$$

They are give in terms of proper coordinates $(\vec{r}, t)$, and $\vec{u} = d\vec{r}/dt$. It is convenient rewrite these expressions in terms of expanding coordinates (i.e., the comoving coordinates) $(\vec{x}, t)$, with $\vec{r} = a(t)\vec{x}$, and using the peculiar velocity (relative to the Hubble flow) $\vec{v} = \vec{u} - \dot{a}\vec{x} = a(t)\dot{\vec{x}}$. This is done in standard texts. We give here the final linearized result. The following expressions assume that the *density contrast* $\delta \equiv (\rho - \bar{\rho})/\bar{\rho}$, the pressure perturbations $\delta_{\rm p} \equiv (p - \bar{p})/\bar{p}$ and the gravitational perturbation $\phi = \Phi - \bar{\Phi}$ are all small, where the barred quantities are universal mean values. Neglecting therefore quadratic terms in these quantities yields the linear theory equations:

$$\dot{\delta} + \frac{1}{a}\nabla \cdot \vec{v} = 0 \tag{11}$$

$$\dot{\vec{v}} + \frac{\dot{a}}{a}\vec{v} = -\frac{1}{a}\nabla\phi - \frac{1}{\bar{\rho}a}\nabla\delta_{\rm p} \tag{12}$$

$$\nabla^2 \phi = 4\pi G \bar{\rho} a^2 \delta \tag{13}$$

Solutions to these equations describe the linear evolution of matter perturbations in the Newtonian limit. Like any vector, the velocity may be decomposed into a rotational, $\nabla \cdot \vec{v}_{\rm rot} = 0$, and an irrotational, $\nabla \times \vec{v}_{\rm irr} = 0$ part. Taking the curl of the second equation immediately tells us that vorticity is damped by the expansion. We henceforth ignore the rotational part and assume that $\vec{v}$ is irrotational. The three linear equations may then be combined to yield a single, second order dynamic equation for the density contrast:

$$\ddot{\delta} + 2\frac{\dot{a}}{a}\dot{\delta} = 4\pi G \bar{\rho} \delta + \frac{1}{\bar{\rho}a^2}\nabla^2 \delta_{\rm p} \tag{14}$$

This equation will take us a far in our discussions of perturbation evolution in the standard model; it is accurate in the Newtonian limit, and its solutions will also provide us with qualitative insight into more complex (e.g., relativistic) cases. Finally, notice that the background model enters through the scale factor $a(t)$ and the background density.

We consider several instructive situations, the first being the evolution of pure "dust", i.e., non–relativistic matter with $p = 0$ in radiation–, matter–, curvature–, and cosmological constant–dominated backgrounds. There are two independent solutions to Eq. (14) for each situation. Only in the matter–dominated case do we find a *growing mode*, plus a decaying mode; in all other situations, we find a decaying mode and a constant (at best logarithmically–growing) mode. At late times when, for instance, the cosmological or curvature dominate the expansion, the density perturbations are frozen. The growing mode in the matter–dominated case is $\delta_{\rm g} \propto a(t)$.

Another situation of extreme interest is for $p \neq 0$. If we assume a simple equation–of–state and adiabatic density perturbations in which the speed of sound $c_{\rm s}^2 = \delta p/\delta \rho$, Eq. (14) becomes

$$\ddot{\delta} + 2\frac{\dot{a}}{a}\dot{\delta} = 4\pi G \bar{\rho} \delta + \frac{c_{\rm s}^2}{a^2}\nabla^2 \delta \tag{15}$$

This is more readily tackled in Fourier space, to which end we introduce the Fourier modes of the density contrast:

$$\delta(\vec{x}) = \frac{1}{(2\pi)^3} \int d^3k \, \tilde{\delta}(\vec{k}) e^{-i\vec{k}\cdot\vec{x}} \tag{16}$$

$$\tilde{\delta}(\vec{k}) = \int d^3x \, \delta(\vec{x}) e^{i\vec{k}\cdot\vec{x}} \tag{17}$$

In terms of Fourier modes, Eq. (15) becomes

$$\ddot{\tilde{\delta}} + 2\frac{\dot{a}}{a}\dot{\tilde{\delta}} = (4\pi G\bar{\rho} - \frac{c_s^2}{a^2}k^2)\tilde{\delta}$$

The right–hand–side clearly shows the opposing effects of gravity, driving collapse and growth of δ, and pressure. We define *Jeans mass* as the characteristic wavelength for which the right–hand–side vanishes: $a\lambda_J = \sqrt{\pi c_s^2/G\bar{\rho}}$. Scales larger than this ($\lambda > \lambda_J$) collapse, while smaller scale ($\lambda < \lambda_J$) modes oscillate as sound waves.

B. Random Fields

To describe the density perturbations as a stochastic process means that we will model the density contrast $\delta(\vec{x})$ as a random field. As mentioned above, this means that δ is a random variable at every point in space, and the goal of observation is then to deduce the statistics of the field δ. It must be imagined that our local volume[32] is but a single realization of the stochastic process producing the perturbations. In other words, what we observe is one realization of a (theoretical) ensemble of all possible local matter distributions consistent with the statistics of the producing mechanism. This permits us to define statistical measures as ensemble averages over possible realizations; these averages we will denote with the brackets $\langle \rangle$. We are furthermore supposing, following the cosmological principle, that the field is ergotic, so that spatial averages over a sufficiently large volume are representative of the ensemble averages used in the theory (fair sample hypothesis). This approach permits us to deduce the general statistical properties of the field with just our single realization.

One may apply a variety of statistical measures to the density field, but by far the most useful has proven to be the N–point correlation functions (or their equivalent in Fourier space; see below), defined as successive moments of the density field:

$$\langle \delta(\vec{x}) \rangle = 0 \tag{18}$$
$$\langle \delta(\vec{x_1})\delta(\vec{x_2}) \rangle = \xi(x_{12}) \tag{19}$$
$$\langle \delta(\vec{x_1})\delta(\vec{x_2})\delta(\vec{x_3}) \rangle = \zeta(x_{12}, x_{13}, x_{23}) \tag{20}$$

The first moment is simply the field average, which by definition is everywhere zero. The second moment is known as the 2–point, or auto– correlation function, usually simply called the correlation function. According to the cosmological principle, it must only be a function of the scalar distance between the two points x_1 and x_2; this guarantees that this statistical moment is independent of direction and absolute position. The third example is the 3–point correlation function, and to preserve uniformity, it can only be a function of the scalar separations between the three possible pairs of points. A complete description of the field requires the specification of the entire, infinite set of N–point functions.

A Gaussian random field is one for which the continuously infinite number of random variables making up the field follow an (enormous!) multivariate Gaussian distribution. Like any such normal process, it is uniquely specified by its average values and covariances. Since the former is everywhere zero, the Gaussian field is completely described by its correlation function. In the linear regime, we often assume that the density perturbations are Gaussian and work only with the correlation function. This assumption is motivated first of all by simplicity, and it remains consistent with current observational tests (on large scales in the linear regime); it is also consistent with simple inflation models. Gravitational evolution induces non–Gaussian statistics that become increasingly important on smaller scales. These effects are calculable, permitting a test of the Gaussian origin of the perturbations even on weakly non–linear scales.

Since Gaussian perturbations apparently provide an adequate description of the density field, the correlation function plays the central role in large–scale structure studies. Its counterpart in Fourier space is known as the power spectrum and is related to the correlation function as follows. We first define the Fourier modes of the density field

$$\tilde{\delta}(\vec{k}) \equiv \int d^3x \, e^{i\vec{k}\cdot\vec{x}} \delta(\vec{k}) \tag{21}$$

The modes $\delta(\vec{k})$ are clearly random variables for every wavevector $\vec{k}$. Their average $\langle \tilde{\delta} \rangle = 0$. Their second order moment $\langle \tilde{\delta}(\vec{k})\delta^*(\vec{k'}) \rangle = (2\pi)^3 \delta(\vec{k} - \vec{k'}) P(k)$ defines the power spectrum as

$$P(k) = \frac{1}{(2\pi)^3} \int d^3x \, e^{-i\vec{k}\cdot\vec{x}} \xi(x) \tag{22}$$

obtained by direct calculation. This relation, and in particular the fact that $P(k)$ depends only on the magnitude of the wavevector, is a direct consequence of the fact that ξ is only a function of separation, i.e., the cosmological principle once again. For Gaussian processes, either the correlation function or the power spectrum are sufficient to completely describe the field. They play a central role in all of large–scale structure studies.

C. The CDM–like Power Spectrum

An essential element of any physical cosmology model is a process of perturbation production. In the standard model, one usually invokes inflation, although in practice this simply means that the model assumes adiabatic Gaussian initial perturbations. The form of the final, observable power spectrum then depends on two important factors: 1/ the form of the initial power spectrum at the end of inflation and 2/ the nature of the dominant form of matter, which determines the gravitational evolution of the perturbations. We take up the second question in the next section and consider the first here.

What, then, should be the form of the initial power spectrum? In terms of conceptual economy, one prefers power–law spectra, $P(k) \propto k^n$, for their absence of any special scale; we want avoid introducing any explicit physical scale until required to do so by observations. Among all possible scale–free spectra, the case $n = 1$ is special in that it is *scale invariant* in the following sense. The Poisson equation relates the density power spectrum to the power spectrum of the Newtonian gravitational potential

$$\Delta_\phi^2 \equiv \frac{d\langle\phi^2\rangle}{d\ln k} = \frac{k^3 P_\phi(k)}{2\pi^2} \propto k^{n-1} \qquad (23)$$

which is constant for $n = 1$. In other words, the $n = 1$ spectrum has equal power in gravitational fluctuation on all scales. This spectrum is also known as the *Harrison–Zel'dovich(–Peebles)* power spectrum (HZ), after those who (independently) proposed it. One can easily show that for this spectrum the power of each density mode at horizon crossing is constant, i.e., $\Delta^2|_{\text{horizon crossing}} = const$. One may wonder why the perturbations are scale–invariant in the Newtonian potential, and not directly in the density perturbations. The Newtonian potential represents a perturbation to the homogeneous FRW metric corresponding to a curvature perturbation, so that what is actually scale invariant are the perturbations to the spatial curvature.

Note that we have motivated the form of the initial, or primordial, power spectrum without any mention of inflation. The above arguments are quite general, and the HZ spectrum was in fact proposed long before inflation. The fact that inflation predicts the same initial spectrum really only reflects the scale invariance inherent in the model. What we do gain with the inflation model is an example of a physical mechanism producing scale–invariant perturbations. On more general grounds, one can imagine (even with inflation) slightly more complicated initial power spectra. Power–law spectra are not even guaranteed in models where explicit scales appear. When analysing data, however, one usually restricts consideration to simple power–law spectra. As we shall see, current data do indeed strongly favor the scale–invariant value of $n = 1$. This impressive success should not (and will not!) go unremarked.

Now that we have a form for the initial power spectrum, we must consider its subsequent evolution. This evolution is for the most part approachable with our previous Newtonian results, at least qualitatively. The two situations in which we must call on GR are for perturbations outside the horizon and for perturbations in a relativistic fluid, such as the pressure–dominated baryon–photon fluid before recombination. We can work around the first difficulty by using the fact that Δ is constant at horizon crossing, and for the second situation, we know that effects due to pressure will be at least qualitatively the same as in the Newtonian approximation – they will prevent gravitational growth. These are all the ingredients needed to understand the overall form of the CDM–like power spectrum.

In the standard model, the universal plasma is a multi-component fluid consisting of baryons, electrons, photons, dark matter and (decoupled) neutrinos. The dark matter is decoupled from the baryon–photon fluid and evolves essentially on its own, interacting with the other components only through gravitation. It dominates the mass budget to the extent that the matter density perturbations are just the dark matter perturbations. Since it is decoupled and assumed to be cold (Cold Dark Matter), there are no thermal or pressure effects in the dark matter fluid; perturbations evolve by gravity alone. Our Newtonian solutions thus apply to the CDM perturbations, once inside the horizon. As we saw, the perturbations are (essentially) frozen during radiation domination and only grow after matter domination. From the fact that the amplitude at horizon crossing is constant, we deduce that there must be a break in the matter power spectrum corresponding to the physical scale of the horizon at equality. Below this scale, the perturbations are roughly constant, while above they reflect the initial power–law spectrum. This is shown in Figure 1. The shape of the matter power spectrum is fixed after equality, only the amplitude changing thereafter.

The effects of evolution are independent of the initial conditions, so we may encode the change from initial to final, evolved power spectrum by a *transfer function* $T(k)$ that depends only on the cosmological parameters:

$$P(k) = T^2(k) P_{\text{initial}}(k) \qquad (24)$$

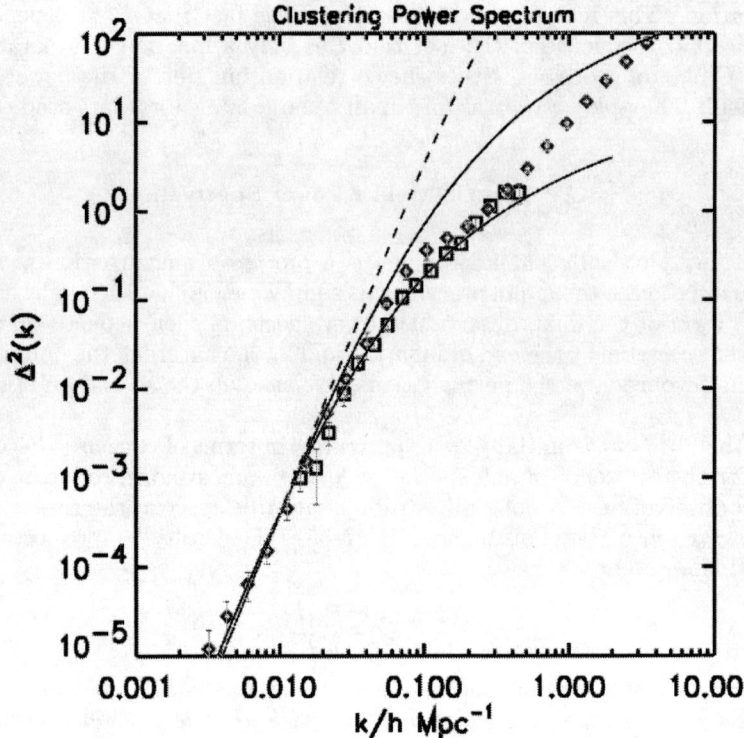

FIG. 1: CDM–like power spectra compared to clustering data. The solid lines are CDM model spectra showing the break induced by the radiation–matter transition away from the initial power law, indicated by the dashed line ($n = 1$). Two examples are given for different values of the shape parameter $\Gamma = \Omega h$, controlling the epoch of equality. The blue squares give the linear matter power spectrum reconstructed by Peacock and Dodds, while the red diamonds show the real–space power spectrum from the APM survey (kindly provided by C.M. Baugh). The agreement in the linear regime (covered by the blue squares) for the favored $\Gamma = 0.2$ model is good.

There is a transfer function for each component of the cosmic fluid. Several formula for the CDM transfer function have been proposed in the literature as fits to full–blown evolution codes. Although not the most accurate, the form proposed by Bardeen et al. [2] is representative:

$$T(k) = \frac{\ln(1+2.34q)}{2.34q} \left[1 + 3.89q + (16.9q)^2 + (5.46q)^3 (6.71q)^4\right]^- 1/2 \qquad (25)$$

where $q \equiv k/(\Omega_M h^2 \text{Mpc}^{-1})$ It involves the *shape parameter* $\Gamma \equiv \Omega_M h$ reflecting the change in the equality epoch with the matter density. The relevant quantity here is really the physical matter density $\propto \Omega_M h^2$, but since all scales are measured in units of the Hubble constant ($k \propto h^{-1}$), the observable parameter is in fact Γ.

It should be strongly emphasized that the predicted power spectrum has a physical scale imposed by gravitational evolution, despite the fact that the initial conditions where scale free. The first and foremost question, then, is whether or not this scale is actually observed; this is a crucial test of the model. The squares and diamonds in Figure 1 show two different data sets. Using a variety of different clustering data, Peacock and Dodds[13] reconstructed the linear matter power spectrum that is given by the blue squares. The red diamonds correspond to the APM galaxy real–space power spectrum[3], as kindly supplied by C.M. Baugh. It seems relatively clear that the data do indeed indicate the kind of break predicted by the CDM model. Furthermore, we can use the exact form of the power spectrum to deduce the shape parameter, and hence the density parameter. In essence, we are using the location of the break to identify the epoch of matter domination, which of course depends on the matter density. As is evident from the figure, lower values of Γ are preferred by this constraint. Careful analysis, e.g., Percival et al. confirms this in giving the constraint $\Gamma = 0.20 \pm 0.03$. Using the HST Key Project value for the Hubble constant of $h = 0.72 \pm 0.08$, this implies $\Omega_M \sim 0.28 \pm 0.05$, or, taking a a healthy view of the errors, we conclude that $\Omega_M \sim 0.3$.

D. Biasing and Non–linear Effects

In this discussion, we have conveniently forgotten the troublesome question of galaxy biasing. The data shown in the figure assume a linear bias, which is to say that light clustering traces matter clustering to within an overall multiplicative factor know as the *bias factor b*: $\xi_{gg} = b^2 \xi$, where ξ_{gg} is the galaxy correlation function (or cluster–cluster correlation function, etc...) and ξ is the correlation function of the matter perturbations. This seems reasonable (which means that if it is not, then our life is much more complicated than we want!) on large scales, in the linear regime. There are in fact ways to directly test this assumption, by measuring higher order correlations in the galaxy distribution. We only quote here the result that over the scales considered (e.g., $> 5h^{-1}$ Mpc), a linear bias appears consistent with the observations [18].

The issue of biasing is closely related to non–linear gravitational effects. Important on small scales, say below a few Mpc, non–linear evolution must be studied with N–body simulations. Fortunately, a simple analytic prescription has also been developed and favorably tested against N–body simulations. This prescription was first proposed by Hamilton et al. [9], and extensively used by Peacock and Dodds [13] (in particular, for the reconstruction shown in Figure 1). Our comparison of the CDM–like power spectrum with clustering data was restricted to the linear regime in order to avoid complications from non–linear evolution. However, once the comparison in the linear regime fixes the (linear) power spectrum, we are obliged to extend the comparison into the non–linear regime, to compare the predicted full non–linear spectrum to the data. It is here that we once again encounter the issue of biasing. What what finds is that no CDM–like spectrum can fit the observed clustering data on all scales with just a simple linear bias. It must be more complicated on small scales or the model is wrong; and the unsettling fact is that the galaxies should be anti–biased on small scales, flying face against traditional beliefs (which don't amount to much more than beliefs, so this is not serious).

How are we to know the nature of biasing in a particular model? The only way is with N–body simulations in which we can identify galaxies. Galaxy identification in simulations is in practice extremely difficult because it depends on very poorly known physics, such as star formation within galactic–sized halos. One approach is to completely ignore baryonic physics and just concentrate on the dark matter. In simulations with sufficiently high mass resolution, individual galactic halos can be identified, even within the cores of cluster halos. One obtains the model bias by comparing the halo clustering to the overall dark matter clustering. The happy result is that the halos are indeed anti–biased, and have a correlation function consistent with the observed galaxy function. One must nevertheless level the criticism that no account is made for the luminous evolution of the galaxy–sized halos. This issue is potentially insolvable from first principles (at least until more is known about how stars form), but some phenomenological prescriptions, the so–called semi–analytic modes (SAMs), brave the attempt and have proven quite useful. In brief, a simple linear bias appears reasonable and compatible with data on large scales, while on smaller scales the exact nature of the biasing and its evolution still needs to be flushed out a bit more. The bottom line is that the CDM–like power spectrum with its characteristic physical scale imposed by cosmological evolution is remarkable success in explaining the observed galaxy distribution.

E. Redshift–space Distortions

Before closing our discussion of galaxy clustering, I would like to examine the important subject of redshift–space distortions. These arise from the fact that one uses redshift as a surrogate for distance when mapping the galaxy distribution. In a pure Hubble flow, this would give an accurate representation of the 3D distribution, but in reality galaxy peculiar velocities contribute inseparably to the observed redshift. A map of (θ, ϕ, z) thus gives a distorted view of the real–space (θ, ϕ, r) distribution. This distortion may be measured, as we shall now see, and in fact contains important cosmological information.

The correlation function quantifies clustering in terms of galaxy pairs (hence the name 2–point function). Figure (2) shows a typical pair of galaxies (on the left–hand–side) separated by real–space distance $r = |\vec{r}_1 - \vec{r}_2|$, where the vectors $\vec{r}_1$ and $\vec{r}_2$ represent the individual galaxy distances from the observer, at the origin. Each galaxy has its own peculiar velocity, as indicated by the red arrows. We observe this pair on the sky separated by an angle θ, or perpendicular distance $r_p = \theta D$, where D is the angular–diameter distance. We also know the redshift separation Δz, which we translate to an effective real–space distance $\pi = H^{-1} z$. The correlation function is thus determined (by counting pairs) over the (r_p, π) plane, as shown on the right–hand–side of the figure[33].

Now comes the crucial (and beautiful) step where we employ the cosmological principal. If there were no redshift–space distortion, i.e., if all peculiar velocities where zero and we had a direct map in real–space, the contours of the correlation function would be perfect circles (right–hand–side) according to the cosmological principle that the

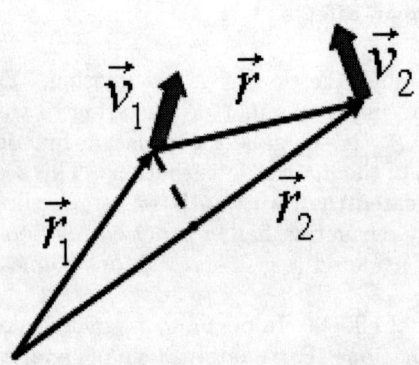 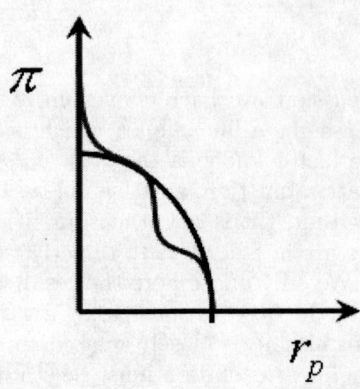

FIG. 2: Redshift–space distortions. The diagram on the left shows the situation in real space when observing a pair of galaxies separated by $\vec{r}$; the observer is taken to be at the origin. Each galaxy has its own peculiar velocity $\vec{v}_i$ relative to the Hubble flow. These peculiar velocities distort the correlation function contours in the r_p–π plane as shown on the right.

clustering depends only on the separation between pairs. Peculiar velocities distort our redshift–space view and pull the observed contours off the perfect circle in a manner similar to that shown by the red line in the figure. There are two main sources of peculiar velocities: the motion of galaxies within virialized structures, such as groups and clusters, and the general *infall* on large scales due to linear (coherent) growth of perturbations. The qualitative effect of each may be understood with simple arguments. On average, galaxies that are in reality physically close will be stretched out along the line–of–sight in redshift space by their virial motions. This will tend to pull out the contours along the π axis at small r_p, as shown in the figure. The general infall, on the other hand, tends to reduce the observed redshift difference and flatten the contours along the *pi* direction at larger values of r_p. These distortions can be measured and removed precisely because we know that the real–space contours must be circular. Measuring the distortion tells us about the mean relative peculiar velocity of pairs, and this without any need for a (problematic) distance indicator.

How to use this information on the peculiar velocities? We clearly expect the peculiar velocities to be related to the same density perturbations that we are measuring with the galaxy correlation function. The latter quantifies the galaxy overdensity relative to the mean, so any relation between the peculiar velocity and the measured correlation function will involve the mean cosmic matter density, Ω_M. Since we obtain both the real–space correlation function and the peculiar velocities by removing the redshift–space distortion, we have a dynamical method for constraining the density parameter. The traditional approach used a relation known as the cosmic virial theorem, based on the idea that the clustering hierarchy on small scales was *stable*. In order to avoid the aforementioned uncertainties related to galaxy biasing in the non–linear regime, one prefers today to instead focus on infall on larger scales, where a simple relation may be established using linear theory. The preference is also one of practicality – in the 1970s galaxy catalogs where not large enough to measure the distortion due to infall, while today we have some beautiful examples, such as the 2dF r_p–π diagram [14]. In slightly more detail, linear theory indicates that the relevant relation involves the factor $\beta = \Omega_M^{0.6}/b$, where b is the linear bias factor for galaxies. Since we are in the linear regime, we have more confidence that this simple linear bias is applicable[34]. The results from the 2dF redshift–space diagram give a value of $\beta = 0.4 - 0.5$, consistent with $\Omega_M \sim 0.3$ if $b \sim 1$, which in fact is indicated by the higher order clustering moments[18], as well as the comparison of the 2dF spectrum with the CMB anisotropies.

---

[1] These written notes only comprise part of the course, up through the discussion of large–scale structure; the CMB anistropies and final discussion are unfortunately missing from these proceedings. The author apologies for this fact and assumes full responsability for their absence in this volume.
[2] J.M. Bardeen, J.R. Bond, N. Kaiser and A.S. Szalay, Astrophys. J. **304**, 15 (1986)
[3] C.M. Baugh and G. Efstathiou, Mon. Not. R. Astron. Soc. **265**, 145 (1993)
[4] M.R. Blanton, J. Dalcanton, D. Eisenstein et al. Astron. J. **121**, 2358 (2001)
[5] R.G. Carlberg, H,K.C. Yee, E. Ellingson et al., Astrophys. J. **462**, 32 (1996)
[6] D.J. Fixsen and J.C. Mather, Astrophys. J. **581**, 817 (2002)
[7] W.L. Freedman, B.F. Madore, B.K. Gibson et al., Astrophys. J. **553**, 47 (2001)
[8] M. Fukugita, C.J. Hogan and P.J.E. Peebles, Astrophys. J. **503**, 518 (1998)
[9] A.J.S. Hamilton, P. Kumar, E. Lu and A. Matthews, Astrophys. J. **374** L1 (1991)
[10] I. Jorgensen, F. Marijn and P. Kjaergaard, Mon. Not. R. Astron. Soc. **280**, 167 (1996)

[11] J.J. Mohr, B. Mathiesen and A.E. Evrard, Astrophys. J. **517**, 627
[12] P. Norberg, S. Cole, C.M. Baugh et al., Mon. Not. R. Astron. Soc. **336**, 907 (2002)
[13] J.A. Peacock and S.J. Dodds, Mon. Not. R. Astron. Soc. **267**, 1020 (1994)
[14] J.A. Peacock, S. Cole, P. Norberg et al., Nature **410**, 169 (2001)
[15] S. Perlmutter, G. Aldering, G. Goldhaber et al., Astrophys. J. **517**, 565 (1999)
[16] A.G. Riess, A.V. Filippenko, P. Challis et al., Astron. J. **116**, 1009 (1998)
[17] R. Sadat and A. Blanchard, Astron. & Astrophys **371**, 19 (2001)
[18] L. Verde, A.F. Heavens, W.J. Percival et al., Mon. Not. R. Astron. Soc. **335**, 432 (2002)
[19] S.D.M. White, J.F. Navarro, A.E. Evrard and C.S. Frenk, Nature **366**, 429 (1993)
[20] N. Yasuda, M. Fukugita, V.K. Narayanan et al., Astron. J. **122**, 1104 (2001)
[21] http://cfa-www.harvard.edu/cfa/oir/Research/supernova/HighZ.html
[22] http://www-supernova.lbl.gov/
[23] http://www.sdss.org/
[24] ttp://www.mso.anu.edu.au/2dFGRS/
[25] http://space.gsfc.nasa.gov/astro/cobe/dirbe_image.html
[26] The magnitude is defined as $m = -2.5 \log(f/f_o)$, where f_o is some constant depending on the magnitude system.
[27] This construction works only so long as worldlines of freely moving observers don't cross. This is clearly the case in a pure FRW spacetime.
[28] The form of Eq. (2) is not applicable in general to a collisionless gas. In the absence of collisions, momentum flow is not isotropized and the stress-energy tensor becomes anisotropic and obtains off-diagonal terms. The medium must instead be modeled by the Boltzmann equation. In the limit of small thermal (cold matter) and bulk velocities, the general description reduces to that of a perfect fluid with zero pressure – must be the case, for nothing is left but the energy density of the fluid!
[29] An empirical relation between peak luminosity and lightcurve shape is often used to correct the magnitudes and reduce the scatter in the Hubble diagram.
[30] The position vector $\vec{x}$ can be viewed as a continuous index.
[31] This is the case in the standard cosmological model where only passive gravitational evolution takes place. Alternate theories, such as defect theories, exist where perturbations are continuously generated and subject also to non-gravitational influences; these are known as active perturbations.
[32] Which we approximate as a spatial hypersurface, ignoring evolutionary effects for nearby objects
[33] Since the separations that we are considering for clustering are small relative to the cosmic horizon, we make the approximation that the two galaxies are essentially at the same cosmic distance, or redshift. Thus we evaluate the angular-diameter distance at the mean z of the pair. In addition, we assume that the catalog is local and use the present value of the Hubble constant in the conversion to radial distance. Since $D(z) \propto H^{-1}$, the uncertainty in H affects both axes of the (r_p, π) diagram in the same manner, and hence is unimportant to the discussion.
[34] This relation assumes that the velocity field is irrotational, which is reasonable given that vorticity is damped by the expansion; this is a common assumption in large-scale structure work.

New perspectives in physics and astrophysics from the theoretical understanding of Gamma-Ray Bursts

Remo Ruffini,[1,2,*] Carlo Luciano Bianco,[1,2,†] Pascal Chardonnet,[1,3,‡] Federico Fraschetti,[1,4,§] Luca Vitagliano,[1,2,¶] and She-Sheng Xue[1,2,**]

[1]*ICRA — International Center for Relativistic Astrophysics.*
[2]*Dipartimento di Fisica, Università di Roma "La Sapienza", Piazzale Aldo Moro 5, I-00185 Roma, Italy.*
[3]*Université de Savoie, LAPTH - LAPP, BP 110, F74941 Annecy-le-Vieux Cedex, France.*
[4]*Università di Trento, Via Sommarive 14, I-38050 Povo (Trento), Italy.*

If due attention is given in formulating the basic equations for the Gamma-Ray Burst (GRB) phenomenon and in performing the corresponding quantitative analysis, GRBs open a main avenue of inquiring on totally new physical and astrophysical regimes. This program is very likely one of the greatest computational efforts in physics and astrophysics and cannot be actuated using shortcuts. A systematic approach is needed which has been highlighted in three basic new paradigms: the relative space-time transformation (RSTT) paradigm (Ruffini et al. [143]), the interpretation of the burst structure (IBS) paradigm (Ruffini et al. [144]), the GRB-supernova time sequence (GSTS) paradigm (Ruffini et al. [145]). From the point of view of fundamental physics new regimes are explored: (1) the process of energy extraction from black holes; (2) the quantum and general relativistic effects of matter-antimatter creation near the black hole horizon; (3) the physics of ultrarelativisitc shock waves with Lorentz gamma factor $\gamma > 100$. From the point of view of astronomy and astrophysics also new regimes are explored: (i) the occurrence of gravitational collapse to a black hole from a critical mass core of mass $M \gtrsim 10 M_\odot$, which clearly differs from the values of the critical mass encountered in the study of stars "catalyzed at the endpoint of thermonuclear evolution" (white dwarfs and neutron stars); (ii) the extremely high efficiency of the spherical collapse to a black hole, where almost 99.99% of the core mass collapses leaving negligible remnant; (iii) the necessity of developing a fine tuning in the final phases of thermonuclear evolution of the stars, both for the star collapsing to the black hole and the surrounding ones, in order to explain the possible occurrence of the "induced gravitational collapse". New regimes are as well encountered from the point of view of nature of GRBs: (I) the basic structure of GRBs is uniquely composed by a proper-GRB (P-GRB) and the afterglow; (II) the long bursts are then simply explained as the peak of the afterglow (the E-APE) and their observed time variability is explained in terms of inhomogeneities in the interstellar medium (ISM); (III) the short bursts are identified with the P-GRBs and the crucial information on general relativistic and vacuum polarization effects are encoded in their spectra and intensity time variability. A new class of space missions to acquire information on such extreme new regimes are urgently needed.

Contents

I. Introduction	18
II. Summary of the main results	20
A. The physical and astrophysical background	20
B. The Relative Space-Time Transformations: the RSTT paradigm and current scientific literature	23
C. The EMBH Theory	24
D. The GRB 991216 as a prototypical source	24
E. The interpretation of the burst structure: the IBS paradigm and the different eras of the EMBH theory	25
F. The Best fit of the EMBH theory to the GRB 991216: the global features of the solution	29
G. The explanation of the "long bursts" and the identification of the proper gamma ray burst(P-GRB)	32

---

[*]Electronic address: ruffini@icra.it
[†]Electronic address: bianco@icra.it
[‡]Electronic address: chardonnet@icra.it
[§]Electronic address: fraschetti@icra.it
[¶]Electronic address: vitagliano@icra.it
[**]Electronic address: xue@icra.it

H.	On the power-laws and beaming in the afterglow of GRB 991216.	34
I.	Substructures in the E-APE due to inhomogeneities in the Interstellar medium	34
J.	The definition of the equitemporal surfaces (EQTS) and the afterglow delayed intensity as a function of the viewing angle	35
K.	The E-APE temporal substructures taking into account the off-axis emission	38
L.	The observation of the iron lines in GRB 991216: on a possible GRB-supernova time sequence	39

III. The zeroth Era: the process of gravitational collapse and the formation of the dyadosphere 40

IV. The hydrodynamics and the rate equations for the plasma of e^+e^--pairs 43

V. The equations leading to the relative space-time transformations 45

VI. The numerical integration of the hydrodynamics and the rate equations 48
  A. The Livermore code 48
  B. The Rome code 49

VII. The Era I: the PEM pulse 50

VIII. The Era II: the interaction of the PEM pulse with the remnant of the progenitor star 52

IX. The Era III: the PEMB pulse 54

X. The identification of the free parameters of the EMBH theory 54

XI. The approach to transparency: the thermodynamical quantities 58

XII. The P-GRBs and the "short bursts". The end of the injector phase. 59

XIII. The Era IV: the ultrarelativistic and relativistic regimes in the afterglow 61

XIV. The Era V: the approach to the nonrelativistic regimes in the afterglow 62

XV. The best fit of the EMBH theory to the GRB 991216: the global features of the solution 63

XVI. The explanation of the "long bursts" and the identification of the proper gamma ray burst (P-GRB) 64

XVII. Considerations on the P-GRB spectrum and the hardness of the short bursts 66

XVIII. Approximations and power laws in the description of the afterglow 68
  A. The approximate expression of the hydrodynamic equations 68
  B. The approximate expression of the emitted flux 69
  Phase A 70
  Points P – the two maxima of the energy flux 71
  Phase B – the "golden value" $n = -1.6$ 71
  Phase C 72
  Phase D 73

XIX. The power-law index of the afterglow and inferences on beaming in GRBs 73

XX. Substructures in the E-APE due to inhomogeneities in the Interstellar medium 75

XXI. Considerations on the relativistic beaming angles and on the arrival time 77

XXII. The emission process taking off-axis contributions into account 81

XXIII. The E-APE temporal substructures taking into account the off-axis emission 83

XXIV. On the instantaneous spectrum of GRBs 86

XXV.	The observation of the iron lines in GRB 991216: on a possible GRB-Supernova time sequence	87
XXVI.	General considerations on the EMBH formation	90
XXVII.	Some propaedeutic analysis for the dynamical formation of the EMBH	91
XXVIII.	Contribution of the EMBH model to the black hole theory	98
XXIX.	Conclusions	101
	References	104

I. INTRODUCTION

In understanding new astrophysical phenomena, the solution has been found as soon as the energy source of the phenomena has been identified. This has been the case for pulsars (see Hewish et al. [78]) where the rotational energy of the neutron star was identified as the energy source (see e.g. Gold [67, 68]). Similarly, in binary X-ray sources the accretion process from a normal companion star in the deep potential well of a neutron star or a black hole has clearly pointed to the gravitational energy of the accreting matter as the basic energy source and all the main features of the light curves of the sources have been clearly understood (Giacconi & Ruffini [65]). In this spirit, our work in the field of Gamma-Ray Bursts (GRBs) has focused to identify the energy extraction process from the black hole (Christodoulou & Ruffini [29]) as the basic energy sources for the GRB phenomenon: a distinguishing feature of this process is a theoretically predicted energetics of the source all the way up to $1.8 \times 10^{54} \left(M_{BH}/M_\odot \right)$ ergs for $3.2 M_\odot \leq M_{BH} \leq 7.2 \times 10^6 M_\odot$ (Damour & Ruffini [32]). In particular, the very specific process of the formation of a "dyadosphere", during the process of gravitational collapse leading to a black hole endowed with electromagnetic structure (EMBH), has been indicated as originating and giving the initial boundary conditions of the onset of the GRB process (Preparata et al. [123], Ruffini [136]). Our model has been referred as "the EMBH model for GRBs", although the EMBH physics only determines the initial boundary conditions of the GRB process by specifying the physical parameters and spatial extension of the neutral electron positron plasma originating the phenomenon.

Traditionally, following the observations of the *Vela* (Strong [174]) and *CGRO*[1] satellites, GRBs have been characterized by few parameters such as the fluence, the characteristic duration (T_{90} or T_{50}) and the global time averaged spectral distribution (Band et al. [4]). With the observations of *BeppoSAX*[2] and the discovery of the afterglow, and the consequent optical identification, the distance of the GRB source has been determined and consequently the total energetics of the source has been added as a crucial parameter.

The observed energetics of GRBs, coinciding for spherically symmetric explosions with the ones theoretically predicted in (Damour & Ruffini [32]), has convinced us to develop in full details the EMBH model. For simplicity, we have considered the vacuum polarization process occurring in an already formed Riessner-Nordström black hole (Preparata et al. [123], Ruffini [136]), whose dyadosphere has an energy E_{dya}. It is clear, however, that this is only an approximation to the real dynamical description of the process of gravitational collapse to an EMBH. In order to prepare the background for attacking this extremely complex dynamical process, we have clarified some basic theoretical issues, necessary to be implemented prior to the description of the fully dynamical process of gravitational collapse to an EMBH (Cherubini et al. [27], Ruffini & Vitagliano [155, 156], see section XXVII). We have then described the following five eras in our model. *Era I*: the e^+e^- pairs plasma, initially at $\gamma = 1$, expands away from the dyadosphere as a sharp pulse (the PEM pulse), reaching Lorentz gamma factor of the order of 100 (Ruffini et al. [141]). *Era II*: the PEM pulse, still optically thick, engulfs the remnant left over in the process of gravitational collapse of the progenitor star with a drastic reduction of the gamma factor; the mass M_B of this engulfed baryonic material is expressed by the dimensionless parameter $B = M_B c^2 / E_{dya}$ (Ruffini et al. [142]). *Era III*: the newly formed pair-electromagnetic-baryonic (PEMB) pulse, composed of e^+e^- pair and of the electrons and baryons of the engulfed material, self-propels itself outward reaching in some sources Lorentz gamma factors of 10^3-10^4; this era stops when the transparency condition is reached and the emission of the proper-GRB (P-GRB) occurs (Bianco et al. [12]). *Era IV*: the resulting accelerated baryonic matter (ABM) pulse, ballistically expanding after the transparency condition has been reached, collides at ultrarelativistic velocities with the baryons and electrons of the interstellar matter (ISM)

---

[1] see http://cossc.gsfc.nasa.gov/batse/
[2] see http://www.asdc.asi.it/bepposax/

which is assumed to have a average constant number density, giving origin to the afterglow. *Era V*: this era represents the transition from the ultrarelativistic regime to the relativistic and then to the non relativistic ones (Ruffini et al. [148]).

Our approach differs in many respect from the ones in the current literature. The major difference consists in the appropriate theoretical description of all the above five eras, as well as in the evaluation of the process of vacuum polarization originating the dyadosphere. The dynamical equations as well as the description of the phenomenon in the laboratory time and the time sequence carried by light signals recorded at the detector have been explicitly integrated (see e.g. Tab. I and Ruffini et al. [148, 152]). In doing so we have also corrected a basic conceptual mistake, common to all the current works on GRBs, which led to the wrong spacetime parametrization of the GRB phenomenon, preempting all these theoretical works from their predictive power. The description of the inner engine originating the GRBs has never been addressed in the necessary details in the literature. In this sense neither the specific boundary conditions originating in the dyadosphere nor the needed solutions of the relativistic hydrodynamic and pair equations for the first three eras described above have been considered. Only the treatment of the afterglow has been widely considered in the literature by the so-called "fireball model" (see e.g. Mészáros & Rees [91, 93], Piran [115], Rees & Mészáros [127] and references therein).

However, also in the description of the afterglow, which is represented by the two conceptually and technically simplest eras in our model, there are major differences between the works in the literature and our approach:

a) Processes of synchrotron radiation and inverse Compton as well as an adiabatic expansion in the source generating the afterglow are usually adopted in the current literature. On the contrary, in our approach a "fully radiative" condition is systematically adopted in the description of the X-ray and γ-ray emission of the afterglow. The basic microphysical emission process is traced back to the physics of shock waves as considered by Zel'dovich & Rayzer [192]. A special attention is given to identify such processes in the comoving frame of the shock front generating the observed spectra of the afterglow (see Ruffini et al. [149]).

b) In the literature the variation of the gamma Lorentz factor during the afterglow is expressed by a unique power-law of the radial co-ordinate of the source and a similar power-law relation is assumed also between the radial coordinate of the source and the asymptotic observer frame time. Such simple approximations appear to be quite inadequate and do contrast with the almost hundred pages summarizing the needed computations which we recall in the rest of this article. In our approach the dynamical equations of the source are integrated self-consistently with the constitutive equations relating the observer frame time to the laboratory time and the boundary conditions are adopted and uniquely determined by each previous era of the GRB source (see e.g. Ruffini et al. [147, 148, 149, 152]).

c) At variance with the many power-laws for the observed afterglow flux found in the literature, our treatment naturally leads to a "golden value" for the power-law index $n = -1.6$. The fit of the EMBH model to the observed afterglow data fixes the only two free parameters of our theory: the E_{dya} and the B parameter, measuring the remnant mass left over by the gravitational collapse of the progenitor star (Ruffini et al. [147, 148, 149, 152]).

It is not surprising that such large differences in the theoretical treatment have led to a different interpretation of the GRB phenomenon as well as to the identification of new fundamental physical regimes. The introduction of new interpretative paradigms has been necessary and the theory has been confirmed by the observation to extremely high accuracy.

In particular from the definition of the complete space-time coordinates of the GRB phenomenon as a function of the radial coordinate, the comoving time, the laboratory time, the arrival time and the arrival time at the detector, expressed in Tab. I, it has been concluded that in no way a description of a given era is possible in the GRB phenomena without the knowledge of the previous ones. Therefore the afterglow as such cannot be interpreted unless all the previous eras have been correctly computed and estimated. It has also become clear that a great accuracy in the analysis of each era is necessary in order to identify the theoretically predicted features with the observed ones. If this is done, the GRB phenomena presents an extraordinary and extremely precise correspondence between the theoretically predicted features and the observations leading to the exploration of totally new physical and astrophysical process with unprecedented accuracy. This has been expressed in the relative space-time transformation (RSTT) paradigm: "the necessary condition in order to interpret the GRB data, given in terms of the arrival time at the detector, is the knowledge of the *entire* worldline of the source from the gravitational collapse. In order to meet this condition, given a proper theoretical description and the correct constitutive equations, it is sufficient to know the energy of the dyadosphere and the mass of the remnant of the progenitor star" (Ruffini et al. [143]).

Having determined the two independent parameters of the EMBH model, namely E_{dya} and B, by the fit of the afterglow we have introduced a new interpretative paradigm for the burst structure: the IBS paradigm (Ruffini et al. [144]). In it we reconsider the relative roles of the afterglow and the burst in the GRBs by defining in this complex phenomenon two new phases:

1) the *injector phase* starting with the process of gravitational collapse, encompassing the above Eras I, II, III and ending with the emission of the Proper-GRB (P-GRB);

2) the *beam-target phase* encompassing the above Eras IV and V giving rise to the afterglow. In particular in the afterglow three different regimes are present for the average bolometric intensity : one increasing with arrival time, a second one with an Extended Afterglow Peak Emission (E-APE) and finally one decreasing as a function of the arrival time. Only this last one appears to have been considered in the current literature (Ruffini et al. [144]).

The EMBH model allows, in the case of GRB 991216, to compute the intensity ratio of the afterglow to the P-GRB ($1.45 \cdot 10^{-2}$), and the arrival time of the P-GRB ($8.413 \cdot 10^{-2}$s) as well as the arrival time of the peak of the afterglow (19.87s) (see Figs. 12,6,11). The fact that the theoretically predicted intensities coincide within a few percent with the observed ones and that the arrival time of the P-GRB and the peak of the afterglow also do coincide within a tenth of millisecond with the observed one can be certainly considered a clear success of the predictive power of the EMBH model.

As a by-product of this successful analysis, we have reached the following conclusions:
a) The most general GRB is composed by a P-GRB, an E-APE and the rest of the afterglow. The ratio between the P-GRB and the E-APE intensities is a function of the B parameter.
b) In the limit B=0 all the energy is emitted in the P-GRB. These events represent the "short burst" class, for which no afterglows has been observed.
c) The "long bursts" do not exist, they are just part of the afterglow, the E-APEs.

We are currently verifying these theoretical predictions on the following GRBs: GRB 991216, GRB 980425, GRB 970228, GRB 980519. It is very remarkable that, although the energetics of GRB 980425 (see Fig. 12) differs from the one of GRB 991216 by roughly five orders of magnitude, the model applies also to this case with success. Furthermore from these analysis we can claim that both in the case of GRB 991216 and in the case of GRB 980425 there is not significant departure from spherical symmetry.

While this analysis of the average bolometric intensity of GRB was going on in the radial approximation, we have proceeded to the full non-radial approximation, taking into account all the relativistic corrections for the off-axis emission from the spherically symmetric expansion of the ABM pulse (Ruffini et al. [147, 152]). We have so defined the temporal evolution of the ABM pulse visible area (see Fig. 13), as well as the equitemporal surfaces (see Fig. 13) (Ruffini et al. [147, 152]).

We have then addressed the issue whether the fast temporal variations observed in the so-called long bursts, on time scales as short as fraction of a second (Ruffini et al. [147]), can indeed be explained as an effect of inhomogeneities in the interstellar medium.

We are making further progress in identifying the basic mechanisms of energy release in the afterglow by presenting a new theoretical formalism which as a function of only one parameter fits the entire spectral distribution of the X-ray and γ-ray radiation in GRB 991216 (Ruffini et al. [149]).

Finally the GRB-supernova time sequence (GSTS) paradigm introduces the concept of *induced supernova explosion* in the supernovae-GRB association (Ruffini et al. [145]) leading to the very novel possibility of a process of gravitational collapse induced on a companion star in a very special evolution phase by the GRB explosion.

Before concluding, we also present some theoretical developments which have been motivated by preparing the analysis of the general relativistic effects during the process of gravitational collapse itself and we also show how such results motivated by GRB studies have already generated new results in the fundamental understanding of black hole physics.

In the next section we briefly summarize the main results and we will then give the summary of the treatment in the following sections. For the complete details we refer to the quoted papers.

II. SUMMARY OF THE MAIN RESULTS

A. The physical and astrophysical background

Gamma-ray bursts (GRBs) are rapidly fueling one of the broadest scientific pursuit in the entire field of science, both in the observational and theoretical domains. Following the discovery of GRBs by the Vela satellites (Strong [174]), the observations from the Compton satellite and BATSE had shown the isotropic distribution of the GRBs strongly suggesting a cosmological nature for their origin. It was still through the data of BATSE that the existence of two families of bursts, the "short bursts" and the "long bursts" was presented, opening an intense scientific dialogue on their origin still active today, see e.g. Schmidt [169] and section XII.

An enormous momentum was gained in this field by the discovery of the afterglow phenomena by the BeppoSAX satellite and the optical identification of GRBs which have allowed the unequivocal identification of their sources at cosmological distances (see e.g. Costa [31]). It has become apparent that fluxes of 10^{54} erg/s are reached: during the peak emission the energy of a single GRB equals the energy emitted by all the stars of the Universe (see e.g. Ruffini [137]).

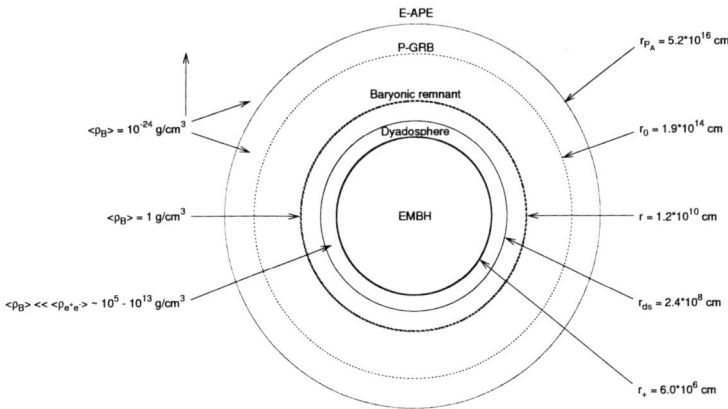

Figure 1: Selected events in the EMBH theory are represented. For each one the values of the energy density of the medium and the distances from the EMBH, in the laboratory frame and in logarithmic scale, are given.

From an observational point of view, an unprecedented campaign of observations is at work using the largest deployment of observational techniques from space with the satellites CGRO-BATSE, Beppo-SAX, Chandra[3], R-XTE[4], XMM-Newton[5], HETE-2[6], as well as the HST[7], and from the ground with optical (KECK[8], VLT[9]) and radio (VLA[10]) observatories. The further possibility of examining correlations with the detection of ultra high energy cosmic rays, UHECR for short, and in coincidence neutrinos should be reachable in the near future thanks to developments of AUGER[11] and AMANDA[12] (see also Halzen [73]).

From a theoretical point of view, GRBs offer comparable opportunities to develop entire new domains in yet untested directions of fundamental science. For the first time within the theory based on the vacuum polarization process occurring in an electromagnetic black hole, the EMBH theory, see Fig. 1, the opportunity exists to theoretically approach the following fundamental issues:

1. The extremely relativistic hydrodynamic phenomena of an electron-positron plasma expanding with sharply varying gamma factors in the range 10^2 to 10^4 and the analysis of the very high energy collision of such an expanding plasma with baryonic matter reaching intensities 10^{38} larger than the ones usually obtained in Earth-based accelerators.

2. The bulk process of vacuum polarization created by overcritical electromagnetic fields, in the sense of Heisenberg, Euler (Heisenberg & Euler [77]) and Schwinger (Schwinger [171]). This longly sought quantum ultrarelativistic effect has not been yet unequivocally observed in heavy ion collision on the Earth (see e.g. Ganz et al. [61], Heinz et al. [76], Leinberger et al. [87, 88]). The difficulty of the heavy ion collision experiments appears to be that the overcritical field is reached only for time scales of the order $\hbar/m_p c^2$, which is much shorter than the characteristic time for the e^+e^- pair creation process which is of the order of $\hbar/m_e c^2$, where m_p and m_e are respectively the proton and the electron mass. It is therefore very possible that the first appearance of such an effect occurs in the present general relativistic context: in the strong electromagnetic fields developed in astrophysical conditions during the process of gravitational collapse to an EMBH, where no problem of confinement exists.

3. A novel form of energy source: the extractable energy of a black hole. The enormous energies released almost instantly in the observed GRBs, points to the possibility that for the first time we are witnessing the release of the extractable energy of an EMBH, during the process of gravitational collapse itself. This problem presents

---

[3] see http://chandra.harvard.edu/
[4] see http://heasarc.gsfc.nasa.gov/docs/xte/
[5] see http://xmm.vilspa.esa.es/
[6] see http://space.mit.edu/HETE/
[7] see http://www.stsci.edu/
[8] see http://www2.keck.hawaii.edu:3636/
[9] see http://www.eso.org/projects/vlt/
[10] see http://www.aoc.nrao.edu/vla/html/VLAhome.shtml
[11] see http://www.auger.org/
[12] see http://amanda.berkeley.edu/amanda/amanda.html

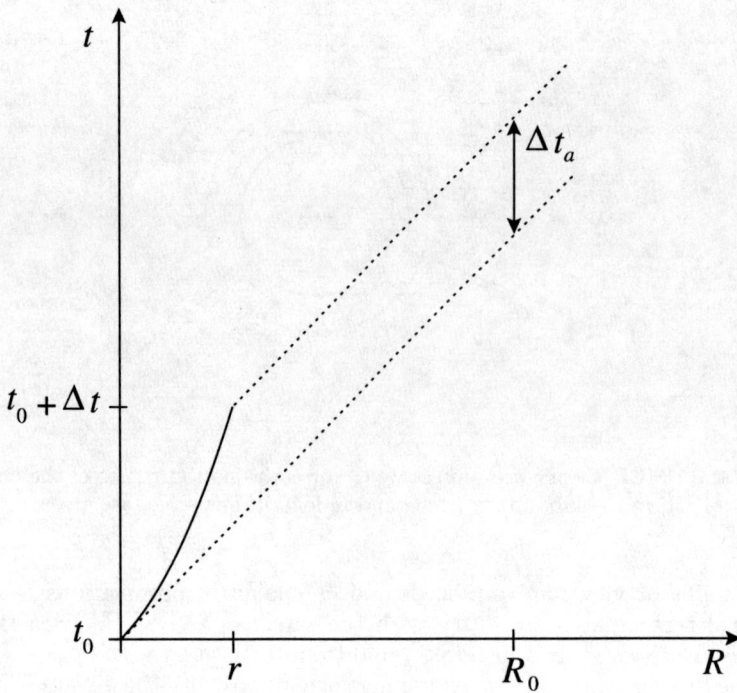

Figure 2: This qualitative diagram illustrates the relation between the laboratory time interval Δt and the arrival time interval Δt_a for a pulse moving with velocity v in the laboratory time (solid line). We have indicated here the case where the motion of the source has a nonzero acceleration. The arrival time is measured using light signals emitted by the pulse (dotted lines). R_0 is the distance of the observer from the EMBH, t_0 is the laboratory time corresponding to the onset of the gravitational collapse, and r is the radius of the expanding pulse at a time $t = t_0 + \Delta t$. See also Ruffini et al. [143].

still some outstanding theoretical issues in black hole physics. Having progressed in some of these issues (see Cherubini et al. [27], Ruffini & Vitagliano [155, 156], Ruffini et al. [158]) we can now compute and have the opportunity to study all general relativistic as well as the associated ultrahigh energy quantum phenomena as the horizon of the EMBH is approached and is being formed (see section XXVII).

It is clear that in approaching such a vast new field of research, implying previously unobserved relativistic and quantum regimes, it is not possible to proceed *as usual* with an uncritical comparison of observational data to theoretical models within the classical schemes of astronomy and astrophysics. Some insight to the new approach needed can be gained from past experience in the interpretation of relativistic effects in high energy particle physics as well as from the explanation of some observed relativistic effects in the astrophysical domain. Those relativistic regimes, both in physics and astrophysics, are however much less extreme than those encountered now in GRBs.

There are three major new features in relativistic systems which have to be properly taken into account:

1. Practically all data on astronomical and astrophysical systems is acquired by using photon arrival times. It was Einstein [46] at the very initial steps of special relativity who cautioned about the use of such an arrival time analysis and stated that when dealing with objects in motion proper care should be taken in defining the time synchronization procedure in order to construct the correct space-time coordinate grid (see Fig. 2). It is not surprising that as soon as the first relativistic bulk motion effects were observed their interpretations within the classical framework of astrophysics led to the concept of "superluminal" motion. These were observations of extragalactic radio sources, with gamma factors ~ 10 (Biretta et al. [15]) and of microquasars in our own galaxy with gamma factor ~ 5 (Mirabel & Rodriguez [101]). It has been recognized (Rees [126]) that no "superluminal" motion exists if the prescriptions indicated by Einstein are used in order to establish the correct space-time grid for the astrophysical systems. In the present context of GRBs, where the gamma factor can easily surpass 10^2 and is very highly varying, this approximation breaks down (Bianco et al. [12], Ruffini et al. [143, 152]). The direct application of classical concepts in this context would lead to enormous "superluminal" behaviors (see e.g. Tab. I). An approach based on classical arrival time considerations as sometimes done in the current literature completely subverts the causal relation in the observed astrophysical phenomenon.

2. One of the clear successes of relativistic field theories has been the understanding of the role of four-momentum

conservation laws in multiparticle collisions and decays such as in the reaction: $n \to p + e^- + \bar{\nu}_e$. From the works of Pauli and Fermi it became clear how in such a process, contrary to the case of classical mechanics, it is impossible to analyze a single term of the decay, the electron or the proton or the neutrino or the neutron, out of the context of the global point of view of the relativistic conservation of the total four momentum of the system. This in turn involves the knowledge of the system during the entire decay process. These rules are routinely used by workers in high energy particle physics and have become part of their cultural background. If we apply these same rules to the case of the relativistic system of a GRB it is clear that it is just impossible to consider a part of the system, e.g. the afterglow, without taking into account the general conservation laws and whole relativistic history of the entire system. Especially since in astrophysics the "somewhat pathological" arrival time coordinate is basically used (see Fig. 2). The description of the afterglow alone, as has been given at times in the literature, indeed possible within the framework of classical astronomy and astrophysics, is not viable in a relativistic astrophysics context where the space-time grid necessary for the description of the afterglow depends on the entire previous relativistic part of the worldline of the system (see also section XV).

3. The lifetime of a process has not an absolute meaning as special and general relativity have shown. It depends both on the inertial reference frame of the laboratory and of the observer and on their relative motion. Such a phenomenon, generally expressed in the "twin paradox", has been extensively checked and confirmed to extremely high accuracy as a byproduct of the elementary particle physics (g-2) experiment (see e.g. van Dick [177]). This situation is much more extreme in GRBs due to the very large (in the range 10^2–10^4) and time varying (on time scales ranging from fractions of seconds to months) gamma factors between the comoving frame and the far away observer (see Fig. 8). Moreover in the GRB context such an observer is also affected by the cosmological recession velocities of its local Lorentz frame.

B. The Relative Space-Time Transformations: the RSTT paradigm and current scientific literature

Here are some of the reasons why we have presented a basic relative space-time transformation (RSTT) paradigm (Ruffini et al. [143]) to be applied prior to the interpretation of GRB data.

The first step is the establishment of the governing equations relating:
a) The comoving time of the pulse (τ)
b) The laboratory time (t)
c) The arrival time at the detector (t_a)
d) The arrival time at the detector corrected for cosmological expansion (t_a^d)

The book-keeping of the four different times and corresponding space variables must be done carefully in order to keep the correct causal relation in the time sequence of the events involved.

As formulated the RSTT paradigm contains two parts: the first one is a necessary condition, the second one a sufficient condition. The first part reads: "the necessary condition in order to interpret the GRB data, given in terms of the arrival time at the detector, is the knowledge of the *entire* worldline of the source from the gravitational collapse".

Clearly such an approach is in contrast with articles in the current literature which emphasize either some too qualitative description of the sources and the quantitative description of the sole afterglow era. In this quantitative description they oversimplify the relations between the radial coordinate of the source and its gamma Lorentz factor as well as the relation between the radial coordinate and the arrival time using power-law relations which do not correctly take into account the complexity of the problem.

In the current literature several attempts have addressed the issue of the sources of GRBs. They include scenarios of binary neutron stars mergers (see e.g. Eichler et al. [45], Mészáros & Rees [91, 92], Narayan et al. [103]), black hole / white dwarf (Fryer et al. [56]) and black hole / neutron star binaries (Mészáros & Rees [95], Paczyński [107]), hypernovae (see Paczyński [109]), failed supernovae or collapsars (see MacFadyen & Woosley [89], Woosley [189]), supranovae (see Vietri & Stella [180, 181]). Only those based on binary neutron stars have reached the stage of a definite model and detailed quantitative estimates have been made. In this case, however, various problems have surfaced: in the general energetics which cannot be greater than $\sim 3 \times 10^{52}$ erg, in the explanation of "long bursts" (see Salmonson et al. [161], Wilson et al. [186]), and in the observed location of the GRB sources in star forming regions (see Bloom et al. [20]). In the remaining cases attention was directed to a qualitative analysis of the sources without addressing the overall problem from the source to the observations. Also generally missing are the necessary details to formulate the equations of the dynamical evolution of the system and to develop a complete theory to be compared with the observations.

Other models in the literature have addressed the problem of only fitting the data of the afterglow observations by simple power-laws. They are separated into two major classes:

The "internal shock model", introduced by Rees & Mészáros [127], by far the most popular one, has been developed in many different aspects, e.g. by Fenimore [48], Fenimore et al. [49], Paczyński & Xu [108], Sari & Piran [164]. The underlying assumption is that all the variabilities of GRBs in the range $\Delta t \sim 1$ ms up to the overall duration T of the order of 50 s are determined by a yet undetermined "inner engine". The difficulties of explaining the long time scale bursts by a single explosive model has evolved into a subclass of approaches assuming an "inner engine" with extended activity (see e.g. Piran [116] and references therein).

The "external shock model", see e.g. Cavallo & Rees [24], Mészáros & Rees [93], Shemi & Piran [172], is less popular today. Paradoxically, some of the authors who have qualitatively highlighted distinctive features of this model have later disclaimed its validity (see e.g. Mészáros & Rees [97], Piran [115], Rees & Mészáros [127] and references therein). Possibly they were carried to this extreme conclusion by an impressive sequence of mistakes they made in implementing the basic physical processes of the model. This model relates the GRB light curves and time variabilities to interactions of a single thin blast wave with clouds in the external medium. The interesting possibility has been also recognized within this model, that GRB light curves "are tomographic images of the density distribution of the medium surrounding the sources of GRBs" (Dermer & Mitman [41]), see also Dermer et al. [40], Dermer [42] and references therein. In this case, the structure of the burst is assumed not to depend directly on the "inner engine" (see e.g. Piran [116] and references therein).

All these works encounter the above mentioned difficulty: they present either a purely qualitative or phenomenological or a piecewise description of the GRB phenomenon. By neglecting the earlier phases, the relation of the space-time grid to the photon arrival time is not properly estimated. To tell more explicitly, their clocks are out of the proper synchronization and the theory is emptied of any predictive power!

We will explicitly show in the following how an unified description naturally leads to the identification of new characteristic features both in the burst and afterglow of GRBs. Our theory, in respect to the afterglow description, can be generally considered an "external shock model" and fits most satisfactorily all the observations.

C. The EMBH Theory

In a series of papers, we have developed the EMBH theory (Ruffini [136]) which has the advantage, despite its simplicity, that all eras following the process of gravitational collapse are described by precise field equations which can then be numerically integrated.

Starting from the vacuum polarization process *à la* Heisenberg-Euler-Schwinger (Heisenberg & Euler [77], Schwinger [171]) in the overcritical field of an EMBH first computed in Damour & Ruffini [32], we have developed the dyadosphere concept (Preparata et al. [123]).

The dynamics of the e^+e^--pairs and electromagnetic radiation of the plasma generated in the dyadosphere propagating away from the EMBH in a sharp pulse (PEM pulse) has been studied by the Rome group and validated by the numerical codes developed at Livermore Lab (Ruffini et al. [141]).

The collision of the still optically thick e^+e^--pairs and electromagnetic radiation plasma with the baryonic matter of the remnant of the progenitor star has been again studied by the Rome group and validated by the Livermore Lab codes (Ruffini et al. [142]). The further evolution of the sharp pulse of pairs, electromagnetic radiation and baryons (PEMB pulse) has been followed for increasing values of the gamma factor until the condition of transparency is reached (Bianco et al. [12]).

As this PEMB pulse reaches transparency the proper GRB (P-GRB) is emitted (Ruffini et al. [144]) and a pulse of accelerated baryonic matter (the ABM pulse) is injected into the interstellar medium (ISM) giving rise to the afterglow.

D. The GRB 991216 as a prototypical source

In the early phases of development of our model, the EMBH theory was developed from first principles by the EMBH uniqueness theorem (Ruffini & Wheeler [140]), the energetics of black hole (Christodoulou & Ruffini [29]) as well as the quantum description of the vacuum polarization process in overcritical electromagnetic fields (Damour & Ruffini [32]). Turning now to the afterglow, the variety of physical situations that can possibly be encountered are very large and far from unique: the description from first principles is just impossible. We have therefore proceeded to properly identify what we consider a prototypical GRB source and to develop a theoretical framework in close correspondence with the observational data.

The criteria which have guided us in the selection of the GRB source to be used as a prototype before proceeding to an uncritical comparison with the theory are expressed in the following. It is now clear, since the observations of GRB 980425, GRB 991216, GRB 970514 and GRB 980326 that the afterglow phenomena can present, especially in

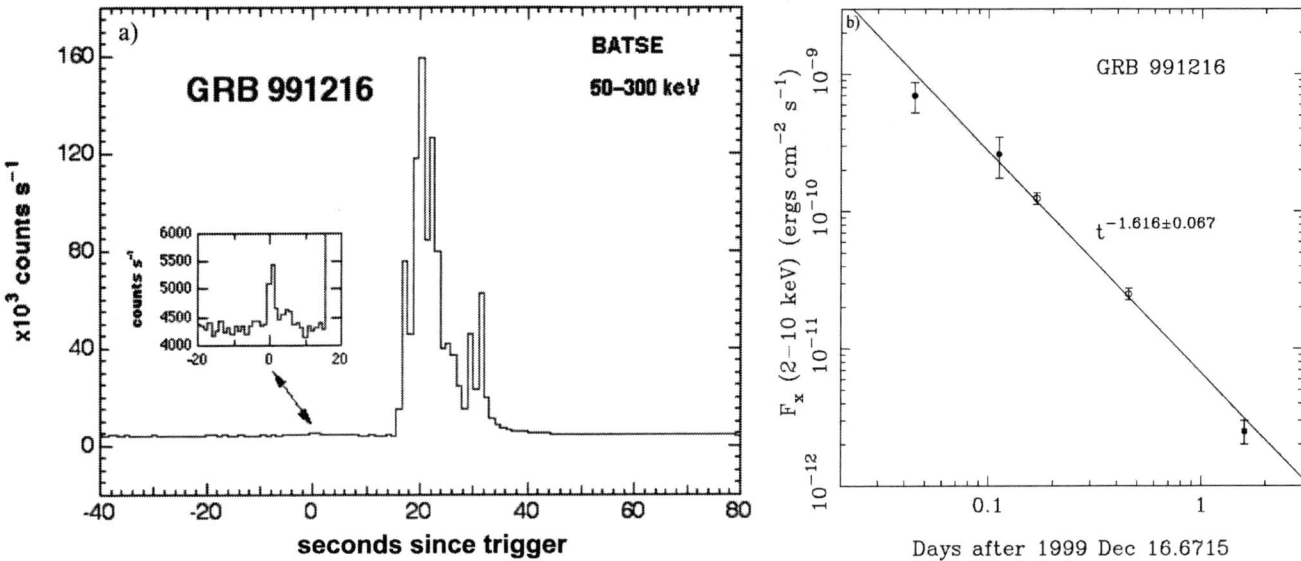

Figure 3: **a)** The peak emission of GRB 991216 as seen by BATSE (reproduced from BATSE Rapid Burst Response [6]); **b)** The afterglow emission of GRB 991216 as seen by XTE and Chandra (reproduced from Halpern et al. [72]).

the optical and radio wavelengths, features originating from phenomena spatially and causally distinct from the GRB phenomena. There is also the distinct possibility that phenomena related to a supernova can be erroneously attributed to a GRB. This problem has been clearly addressed by the GRB supernova time sequence (GSTS) paradigm in which the time sequence of the events in the GRB supernova phenomena has been outlined (Ruffini et al. [145]). This has led to the novel concept of an induced supernova (Ruffini et al. [145]). This problem will be addressed in a forthcoming paper (Ruffini et al. [151]).

In view of these considerations we have selected GRB 991216 as a prototypical case (see Fig. 3) for the following reasons:

1. GRB 991216 is one of the strongest GRBs in X-rays and is also quite general in the sense that it shows relevant cosmological effects. It radiates mainly in X-rays and in γ-rays and less than 3% is emitted in the optical and radio bands (see Halpern et al. [72]).

2. The excellent data obtained by BATSE on the burst (BATSE Rapid Burst Response [6]) is complemented by the data on the afterglow acquired by Chandra (Piro et al., [119]) and RXTE (Corbet & Smith [30]). Also superb data have been obtained from spectroscopy of the iron lines (Piro et al., [119]).

3. A value for the slope of the energy emission during the afterglow as a function of time has been obtained: $n = -1.64$ (Takeshima et al. [175]) and $n = -1.616 \pm 0.067$ (Halpern et al. [72]).

E. The interpretation of the burst structure: the IBS paradigm and the different eras of the EMBH theory

The comparison of the EMBH theory with the data of the GRB 991216 and its afterglow has naturally led to a new paradigm for the interpretation of the burst structures (IBS paradigm)) of GRBs (Ruffini et al. [144]). The IBS paradigm reads: *"In GRBs we can distinguish an injector phase and a beam-target phase. The injector phase includes the process of gravitational collapse, the formation of the dyadosphere, as well as Era I (the PEM pulse), Era II (the engulfment of the baryonic matter of the remnant) and Era III (the PEMB pulse). The injector phase terminates with the P-GRB emission. The beam-target phase addresses the interaction of the ABM pulse, namely the beam generated during the injection phase, with the ISM as the target. It gives rise to the E-APE and the decaying part of the afterglow"*. The detailed presentations of these results are a major topic in this article.

We recall that the **injector phase** starts from the moment of gravitational collapse and encompasses the following eras:

The zeroth Era: the formation of the dyadosphere. In section III we review the basic scientific results which lie at the basis of the EMBH theory: the black hole uniqueness theorem, the mass formula of an EMBH, the process of vacuum polarization in the field of an EMBH. We also point out how after the discovery of the GRB afterglow the

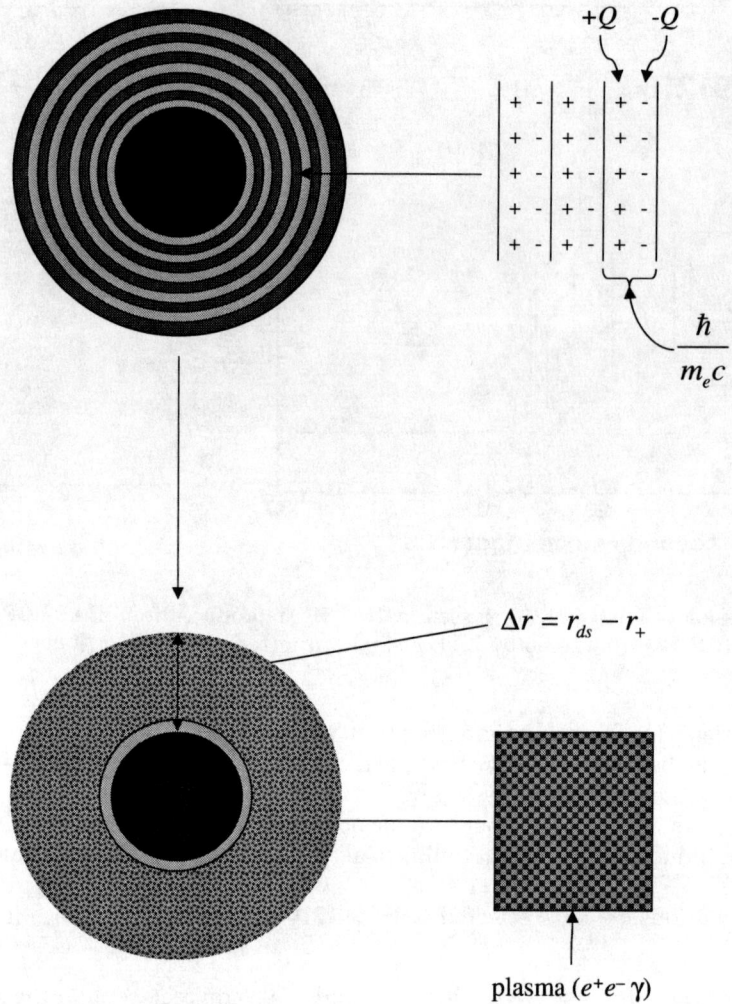

Figure 4: The dyadosphere of a Reissner-Nordström black hole can be represented as constituted by a concentric set of shells of capacitors, each one of thickness $\hbar/m_e c$ and producing a number of e^+e^- pairs of the order of $\sim Q/e$ on a time scale of 10^{-21} s, where Q is the EMBH charge. The shells extend in a region Δr, from the horizon r_+ to the dyadosphere outer radius $r_{\rm ds}$ (see text). The system evolves to a thermalised plasma configuration.

reexamination of these results has led to the novel concept of the dyadosphere of an EMBH. We have investigated this concept in the simplest possible case of an EMBH depending only on two parameters: the mass and charge, corresponding to the Reissner-Nordström spacetime. We recall the definition of the energy E_{dya} of the dyadosphere as well as the spatial distribution and energetics of the e^+e^- pairs. See Fig. 4. We return in section XXVII to the theoretical development of the time varying process lasting less than a second in the process of a realistic gravitational collapse. In reality the vacuum polarization process will lead to a final uncharged black hole, but the analysis based on a Reissner-Nordström black hole is an excellent approximation to the description of this phenomenon (Ruffini et al. [159]).

In order to analyse the time evolution of the dyadosphere we give in the three following sections the theoretical background for the needed equations.

In section IV we give the general relativistic equations governing the hydrodynamics and the rate equations for the plasma of e^+e^--pairs.

In section V we give the governing equations relating the comoving time τ to the laboratory time t corresponding to an inertial reference frame in which the EMBH is at rest and finally to the time measured at the detector t_a which, to finally get t_a^d, must be corrected to take into account the cosmological expansion.

In section VI we describe the numerical integration of the hydrodynamical equations and the rate equation developed by the Rome and Livermore groups. This entire research program could never have materialized without the fortunate interaction between the complementary computational techniques developed by these two groups. The validation of

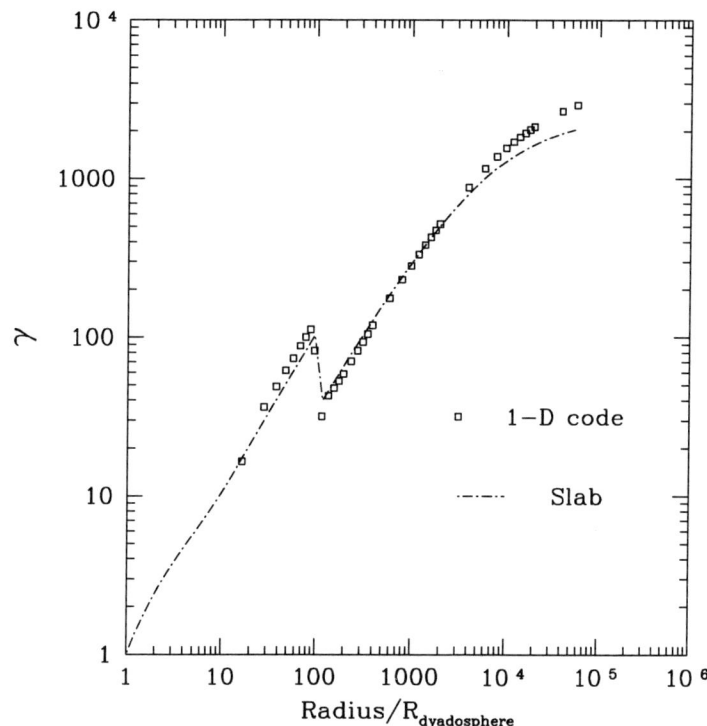

Figure 5: Comparison of gamma factor for the one-dimensional (1-D) hydrodynamic calculations (Livermore code) and slab calculations (Rome code) as a function of the radial coordinate (in units of dyadosphere radius) in the laboratory frame. The calculations show an excellent agreement.

the results of the Rome group by the fully general relativistic Livermore codes has been essential both from the point of view of the validity of the numerical results and the interpretation of the scientific content of the results.

The Era I: the PEM pulse. In section IV by the direct comparison of the integrations performed with the Rome and Livermore codes we show that among all possible geometries the e^+e^- plasma moves outward from the EMBH reaching a very unique relativistic configuration: the plasma self-organizes in a sharp pulse which expands in the comoving frame exactly by the amount which compensates for the Lorentz contraction in the laboratory frame. The sharp pulse remains of constant thickness in the laboratory frame and self-propels outwards reaching ultrarelativistic regimes, with gamma factors larger than 10^2, in a few dyadosphere crossing times. We recall that, in analogy with the electromagnetic (EM) pulse observed in a thermonuclear explosion on the Earth, we have defined this more energetic pulse formed of electron-positron pairs and electromagnetic radiation a pair-electromagnetic-pulse or PEM pulse.

The Era II: We describe the interaction of the PEM pulse with the baryonic remnant of mass M_B left over from the gravitational collapse of the progenitor star. We give the details of the decrease of the gamma factor and the corresponding increase in the internal energy during the collision. The dimensionless parameter $B = M_B c^2/E_{dya}$ which measures the baryonic mass of the remnant in units of the E_{dya} is introduced. This is the second fundamental free parameter of the EMBH theory.

The Era III: We describe in section IX the further expansion of the e^+e^- plasma, after the engulfment of the baryonic remnant of the progenitor star. By direct comparison of the results of integration obtained with the Rome and the Livermore codes it is shown how the pair-electromagnetic-baryon (PEMB) plasma further expands and self organizes in a sharp pulse of constant length in the laboratory frame (see Fig. 5). We have examined the formation of this PEMB pulse in a wide range of values $10^{-8} < B < 10^{-2}$ of the parameter B, the upper limit corresponding to the limit of validity of the theoretical framework developed.

In section X it is shown how the effect of baryonic matter of the remnant, expressed by the parameter B, is to smear out all the detailed information on the EMBH parameters. The evolution of the PEMB pulse is shown to depend only on E_{dya} and B: the PEMB pulse is degenerate in the mass and charge parameters of the EMBH and rather independent of the exact location of the baryonic matter of the remnant.

In section XI the relevant thermodynamical quantities of the PEMB pulse, the temperature in the different frames and the e^+e^- pair densities, are given and the approach to the transparency condition is examined. Particular attention is given to the gradual transfer of the energy of the dyadosphere E_{dya} to the kinetic energy of the baryons

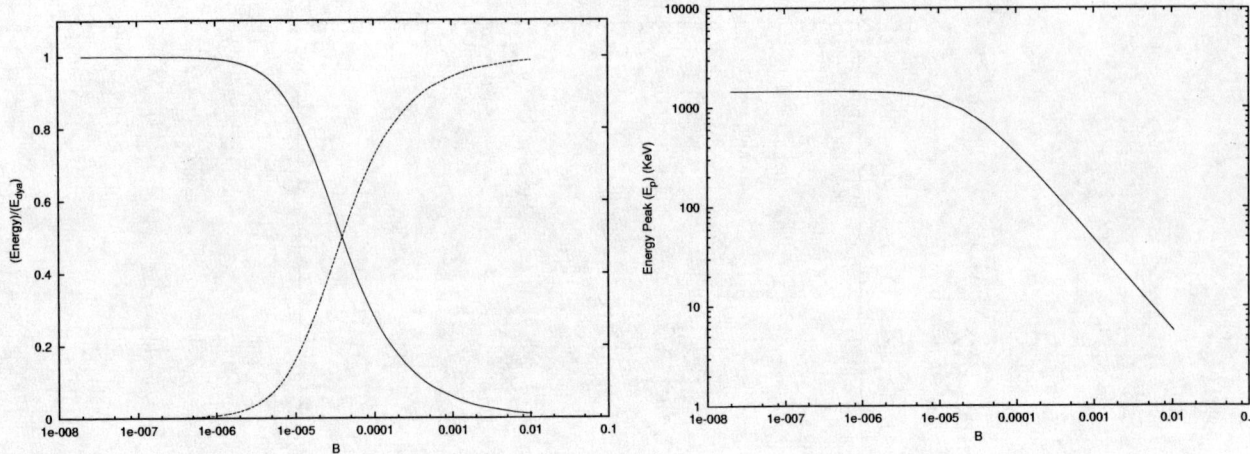

Figure 6: **Left)** At the transparent point, the energy radiated in the P-GRB (the solid line) and (the dashed line) the final kinetic energy of baryonic matter, $E_{Baryons}$, in units of the total energy of the dyadosphere (E_{dya}), are plotted as functions of the B parameter. **Right)** The energy corresponding to the peak of the photon number spectrum in the P-GRB as measured in the laboratory frame is plotted as function of the B parameter.

$E_{Baryons}$ during the optically thick part of the PEMB pulse.

In section XII, as the condition of transparency is reached, the injector phase is concluded with the emission of a sharp burst of electromagnetic radiation and an accelerated beam of highly relativistic baryons. We recall that we have respectively defined the radiation burst (the proper GRB or for short P-GRB) and the accelerated-baryonic-matter (ABM) pulse. By computing for a fixed value of the EMBH different PEMB pulses corresponding to selected values of B in the range $[10^{-8}-10^{-2}]$, it has been possible to obtain a crucial universal diagram which is reproduced in Fig.6. In the limit of $B \to 10^{-8}$ or smaller almost all E_{dya} is emitted in the P-GRB and a negligible fraction is emitted in the kinetic energy $E_{Baryons}$ of the baryonic matter and therefore in the afterglow. On the other hand in the limit $B \to 10^{-2}$ which is also the limit of validity of our theoretical framework, almost all E_{dya} is transferred to $E_{Baryons}$ and gives origin to the afterglow and the intensity of the P-GRB correspondingly decreases. We have identified the limiting case of negligible values of B with the process of emission of the so called "short bursts". A complementary result reinforcing such an identification comes from the thermodynamical properties of the P-GRB: the hardness of the spectrum decreases for increasing values of B, see Fig. 6.

The injector phase is concluded by the emission of the P-GRB and the ABM pulse, as the condition of transparency is reached.

The **beam-target phase**, in which the accelerated baryonic matter (ABM) generated in the injector phase collides with the ISM, gives origin to the afterglow. Again for simplicity we have adopted a minimum set of assumptions:

1. The ABM pulse is assumed to collide with a constant homogeneous interstellar medium of number density $n_{ism} \sim 1 cm^{-3}$. The energy emitted in the collision is assumed to be instantaneously radiated away (fully radiative condition). The description of the collision and emission process is done using spherical symmetry, taking only the radial approximation neglecting all the delayed emission due to off-axis scattered radiation.

2. Special attention is given to numerically compute the power of the afterglow as a function of the arrival time using the correct governing equations for the space-time transformations in line with the RSTT paradigm.

3. Finally some approximate solutions are adopted in order to obtain the determination of the power law exponents of the afterglow flux and compare and contrast them with the observational results as well as with the alternative results in the literature.

We first consider the above mentioned radial approximation and a spherically symmetric distribution in order to concentrate on the role of the correct space-time transformations in the RSTT paradigm and illustrate their impact on the determination of the power law index of the afterglow. This topic has been seriously neglected in the literature. We then turn to the fully relativistic analysis of the off-axis emission and of the temporal structure in the long bursts (see also Ruffini et al. [147] and sections XXI–XXII) and of their spectral distribution (see also Ruffini et al. [149] and section XXIV). Details of the role of beaming are going to be discussed elsewhere (Ruffini et al. [150]).

We can now turn to the two eras of the beam-target phase:

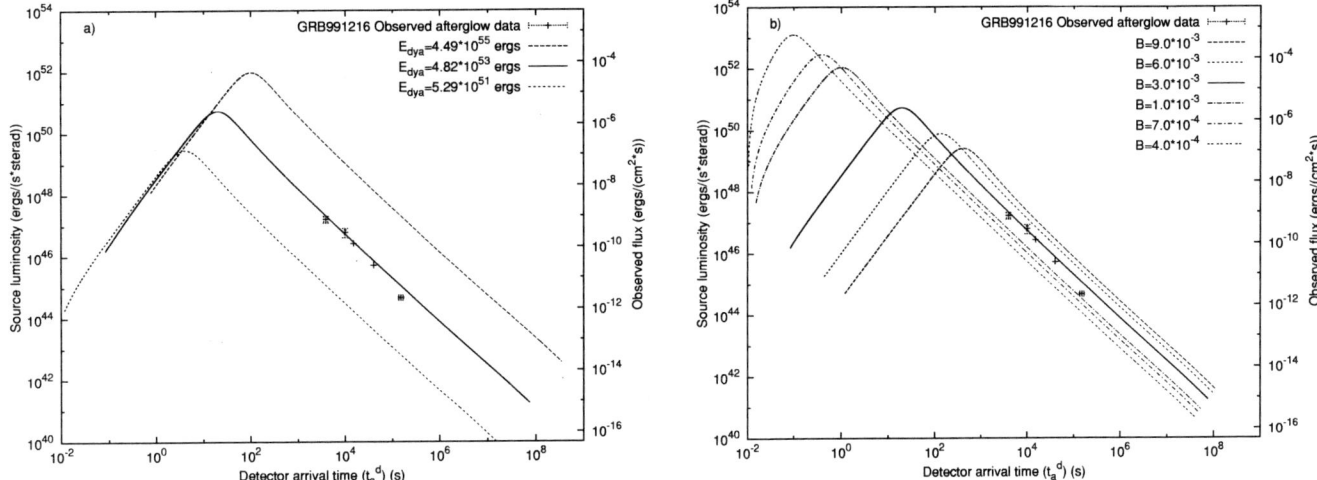

Figure 7: a) Afterglow luminosity computed for an EMBH of $E_{dya} = 5.29 \times 10^{51}$ erg, $E_{dya} = 4.83 \times 10^{53}$ erg, $E_{dya} = 4.49 \times 10^{55}$ erg and $B = 3 \times 10^{-3}$. b) for the $E_{dya} = 4.83 \times 10^{53}$, we give the afterglow luminosities corresponding respectively to $B = 9 \times 10^{-3}, 6 \times 10^{-3}, 3 \times 10^{-3}, 1 \times 10^{-3}, 7 \times 10^{-4}, 4 \times 10^{-4}$.

The Era IV: the ultrarelativistic and relativistic regimes in the afterglow. In section XIII the hydrodynamic relativistic equations governing the collision of the ABM pulse with the interstellar matter are given in the form of a set of finite difference equations to be numerically integrated. Expressions for the internal energy developed in the collision as well as for the gamma factor are given as a function of the mass of the swept up interstellar material and of the initial conditions. In section XVIII the infinitesimal limit of these equations is given as well as analytic power-law expansions in selected regimes.

The Era V: the approach to the nonrelativistic regimes in the afterglow. In section XIV it is stressed that this last era often discussed in the current literature can be described by the same equations used for era IV.

Having established all the governing equations for all the eras of the EMBH theory, we can proceed to compare and contrast the predictions of this theory with the observational data.

F. The Best fit of the EMBH theory to the GRB 991216: the global features of the solution

As expressed in section XV, we have proceeded to the identification of the only two free parameters of the EMBH theory, E_{dya} and B, by fitting the observational data from R-XTE and Chandra on the decaying part of the GRB 991216 afterglow. The afterglow appears to have three different parts: in the first part the luminosity increases as a function of the arrival time, it then reaches a maximum and finally monotonically decreases. In Fig. 7, we show how such a fit is actually made and how changing the two free parameters affects the intensity and the location in time of the peak of the afterglow. The best fit is obtained for $E_{dya} = 4.83 \times 10^{53}\, erg$ and $B = 3 \times 10^{-3}$.

Having determined the two free parameters of the theory, we have integrated the governing equations corresponding to these values and then obtained for the first time the complete history of the gamma factor from the moment of gravitational collapse to the latest phases of the afterglow observations (see Fig. 8). This diagram clearly shows the inadequacy of considering a simple power-law relation $\gamma \propto r^{-3/2}$ for the relation between the radius of the source and its Lorentz gamma factor as assumed in the large majority of current papers on GRBs (see e.g. Panaitescu & Mészáros [113], Piran [115], Sari [162, 163], Sari et al. [165], Waxman [185] and references therein). Actually, such a power-law behaviour is never found to exist.

We have also determined the different regimes encountered in the relation between the laboratory time and the detector arrival time within the RSTT paradigm compared and contrasted with the ones in the current literature (see Fig. 9). The solid curve is computed using the exact formula prescribed by the RSTT paradigm (see Eq.(37) in section V)

$$t_a^d = (1+z)\left(t - \int_0^t \frac{\sqrt{\gamma^2(t')-1}}{\gamma(t')} dt' - \frac{r_{ds}}{c}\right).$$

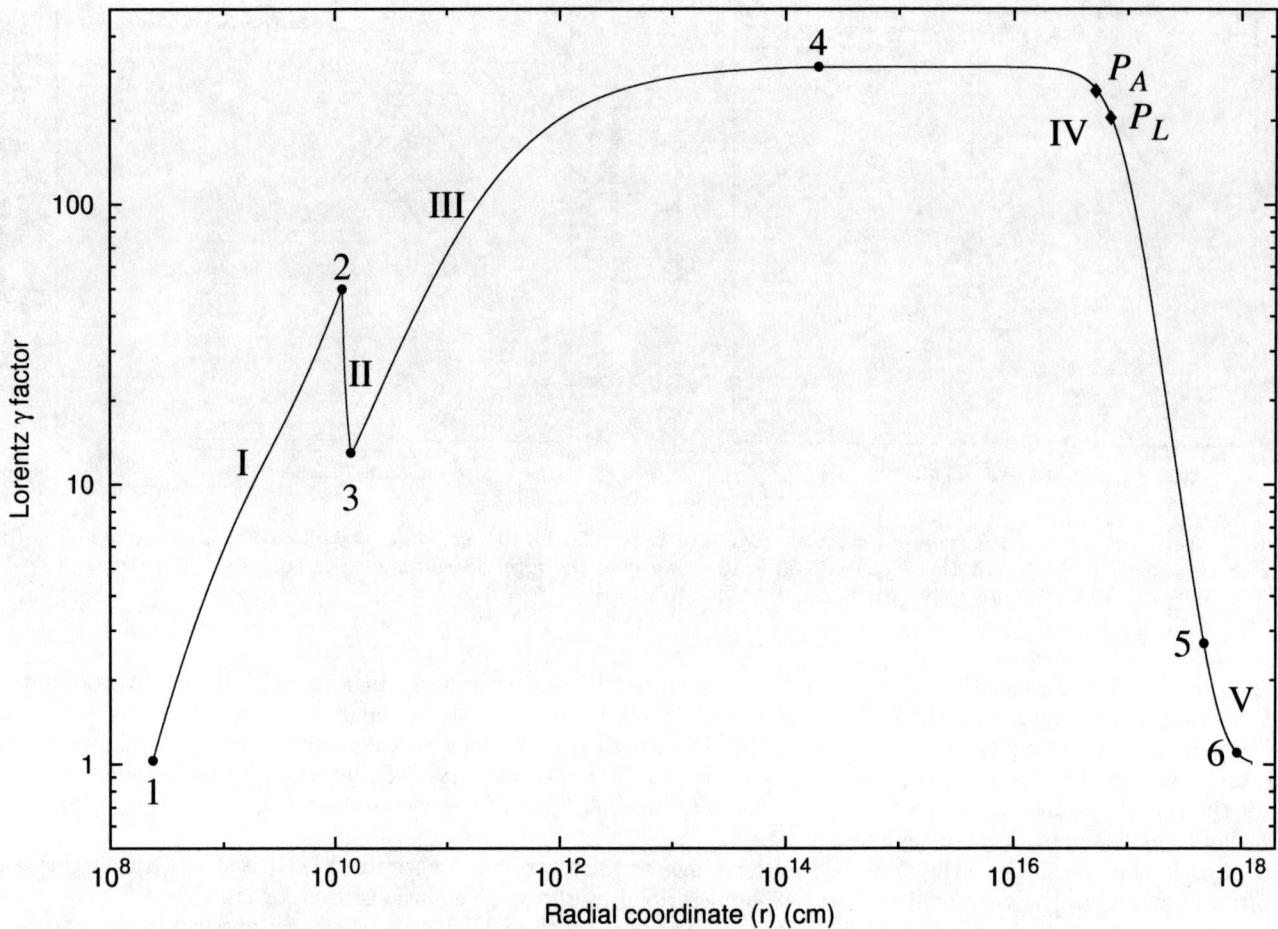

Figure 8: The theoretically computed gamma factor for the parameter values $E_{dya} = 4.83 \times 10^{53}$ erg, $B = 3 \times 10^{-3}$ is given as a function of the radial coordinate in the laboratory frame. The corresponding values in the comoving time, laboratory time and arrival time are given in Tab. I. The different eras indicated by roman numerals are illustrated in the text (see sections VII,VIII,IX,XIII,XIV), while the points 1,2,3,4,5 mark the beginning and end of each of these eras. The points P_L and P_A mark the maximum of the afterglow flux, respectively in emission time and in arrival time (see Ruffini et al. [144] and sections XIII,XVIII). The point 6 is the beginning of Phase D in Era V (see sections XIV,XVIII). At point 4 the transparency condition is reached and the P-GRB is emitted. This diagram clearly shows the inadequacy of considering a simple power-law relation $\gamma \propto r^{-3/2}$ for the relation between the radius of the source and its Lorentz gamma factor as assumed in the large majority of current papers on GRBs (see e.g. Panaitescu & Mészáros [113], Piran [115], Sari [162, 163], Sari et al. [165], Waxman [185] and references therein). Actually, such a power-law behaviour is never found to exist.

The dashed-dotted curve is computed using the approximate formula (see Eq.(41))

$$t_a^d = (1+z) \frac{t}{2\gamma^2(t)},$$

often used in the current literature (see e.g. Fenimore et al. [47], Piran [115], Sari [162, 163], Waxman [185] and references therein). The difference between the solid line and the dashed-dotted line clearly shows the inadequacy of using such an approximate relation. We like to stress that the difference between the above two curves is especially marked in the afterglow region. Note that this difference as been estimated assuming in both curves the correct relation between the Lorentz gamma factor and the radial coordinated of the source given in Fig. 8. In the case that the wrong relation $\gamma \propto r^{-3/2}$ is adopted as done in the literature (see e.g. Panaitescu & Mészáros [113], Piran [115], Sari [162, 163], Sari et al. [165], Waxman [185] and references therein) the discrepancy between the two curves will be much larger. It is anyway clear that, even knowing quantitatively the exact Lorentz gamma factor curve reported in Fig. 8, the use of the approximate relation given in Eq.(41) is enough to miss the correct clock synchronization and to obtain a wrong value for the power-law index n in the decaying phases of the afterglow (see sections XVIII–XIX and Tab. II).

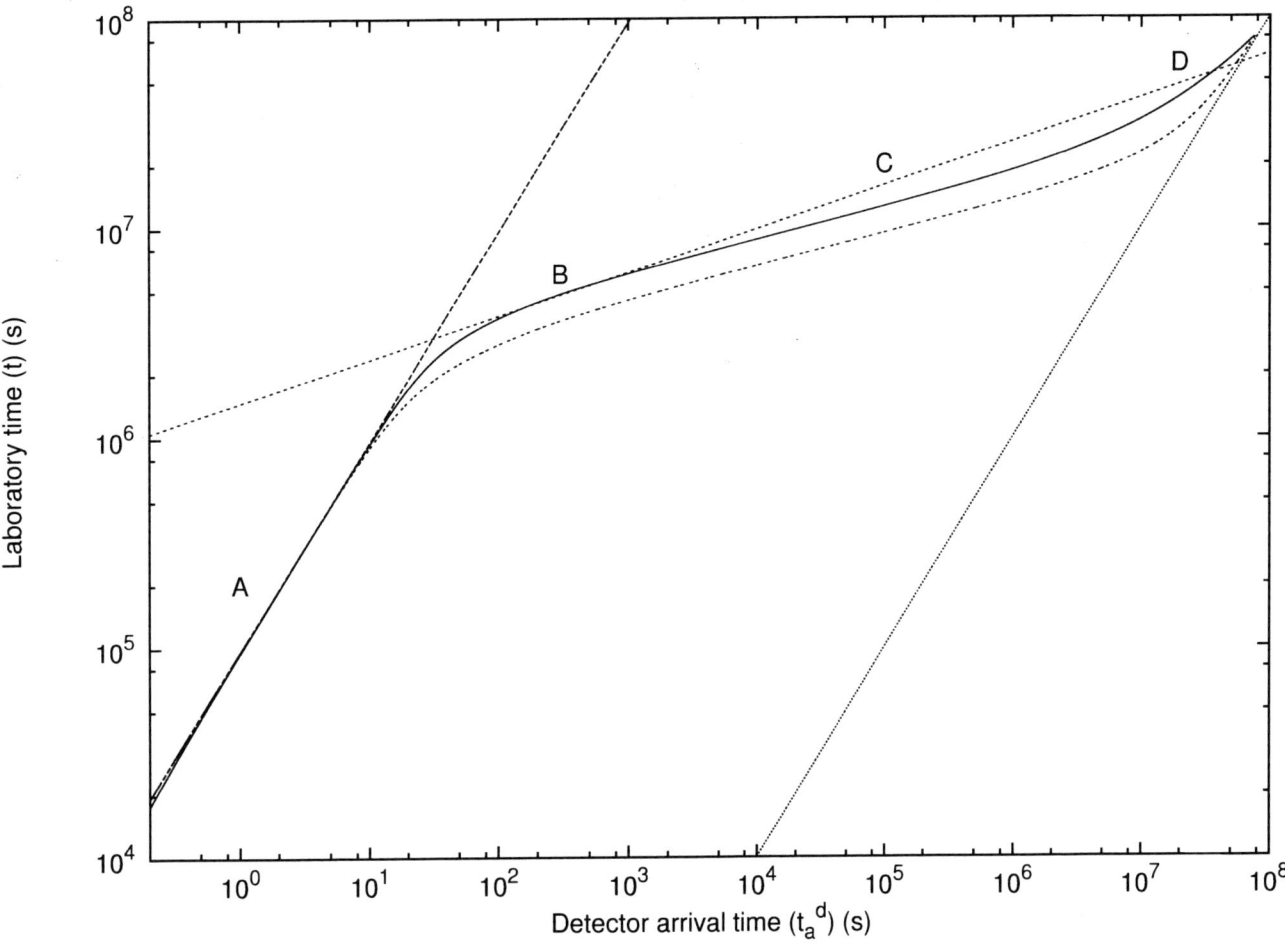

Figure 9: Relation between the arrival time (t_a^d) measured at the detector and the laboratory time (t) measured at the GRB source. The solid curve is computed using the exact formula prescribed by the RSTT paradigm $t_a^d = (1+z)\left(t - \int_0^t \frac{\sqrt{\gamma^2(t')-1}}{\gamma(t')}dt' - \frac{r_{ds}}{c}\right)$ (see Eq.(37) in section V). The dashed-dotted curve is computed using the approximate formula $t_a^d = (1+z)\left(t/2\gamma^2(t)\right)$ (see Eq.(41)) often used in the current literature (see e.g. Fenimore et al. [47], Piran [115], Sari [162, 163], Waxman [185] and references therein). The difference between the solid line and the dashed-dotted line clearly shows the inadequacy of using such an approximate relation. We like to stress that the difference between the above two curves is especially marked in the afterglow region. Note that this difference as been estimated assuming in both curves the correct relation between the Lorentz gamma factor and the radial coordinated of the source given in Fig. 8. In the case that the wrong relation $\gamma \propto r^{-3/2}$ is adopted as done in the literature (see e.g. Panaitescu & Mészáros [113], Piran [115], Sari [162, 163], Sari et al. [165], Waxman [185] and references therein) the discrepancy between the two curves will be much larger. It is anyway clear that, even knowing quantitatively the exact Lorentz gamma factor curve reported in Fig. 8, the use of the approximate relation given in Eq.(41) is enough to miss the correct clock synchronization and to obtain a wrong value for the power-law index n in the decaying phases of the afterglow (see sections XVIII–XIX and Tab. II). We distinguish four different phases. **Phase A**: There is a linear relation between t and t_a^d, given by Eq.(137) in the text (dashed line). **Phase B**: There is an "effective" power-law relation between t and t_a^d, given by Eq.(142) (dotted line). **Phase C**: No analytic formula holds and the relation between t and t_a^d has to be directly computed by the integration of the complete equations of energy and momentum conservation (Eqs.(107,108)). **Phase D**: As the gamma factor approaches $\gamma = 1$, the relation between t and t_a^d asymptotically goes to $t = t_a^d$ (light gray line). See also Ruffini et al. [143].

To be more explicit, from the result given in Figs. 8–9 follows that all existing GRB models, with the exception of ours, have the wrong spacetime coordinatization of the GRB phenomenon and they therefore lack the fundamental toola to compare the theoretical prediction in the laboratory time to the observations carried out in the asymptotic photon arrival time. This extreme situation affects all considerations on GRBs: as an example, all the considerations on the afterglow slopes, which drastically depend on the functional dependence between the laboratory time and the photon arrival time, are drastically affected (see subsection II H below and Tab. II). In turn, all the considerations

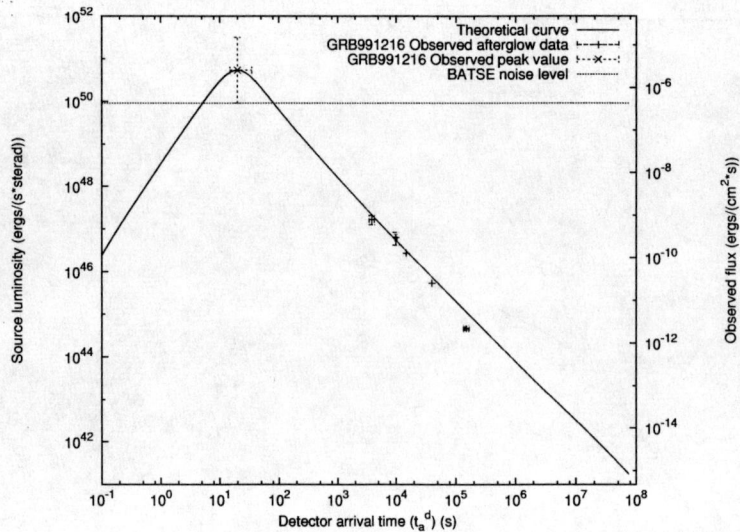

Figure 10: Best fit of the afterglow data of Chandra, RXTE as well as of the range of variability of the BATSE data on the major burst, by a unique afterglow curve leading to the parameter values $E_{dya} = 4.83 \times 10^{53}\,erg$, $B = 3 \times 10^{-3}$. The horizontal dotted line indicates the BATSE noise threshold. On the left axis the luminosity is given in units of the energy emitted at the source, while the right axis gives the flux as received by the detectors.

about the possible existence of beaming in GRBs inferred from the afterglow slopes are in this circumstance deprived of any meaning.

We have thus determined the entire space-time grid of the GRB 991216 by giving (see Tab. I) the radial coordinate of the GRB phenomenon as a function of the four coordinate time variables. A quick glance to Tab. I shows how the extreme relativistic regimes at work lead to enormous superluminal behaviour (up to $10^5 c$!) if the classical astrophysical concepts are adopted using the arrival time as the independent variable. In turn this implies that any causal relation based on classical astrophysics and the arrival time data, as at times found in the current GRB literature, is incorrect.

G. The explanation of the "long bursts" and the identification of the proper gamma ray burst(P-GRB)

In section XVI, having determined the two free parameters of the EMBH theory, we analyze the theoretical predictions of this theory for the general structure of GRBs. The first striking result, illustrated in Fig. 10, shows that the peak of the afterglow emission coincides both in intensity and in arrival time ($19.87\,s$) with the average emission of the long burst observed by BATSE. For this we have introduced the new concept of *extended afterglow peak emission (E-APE)*. Once the proper space-time grid is given (see Tab. I) it is immediately clear that the E-APE is generated at distances of 5×10^{16} cm from the EMBH. The long bursts are then identified with the E-APEs and are not bursts at all: they have been interpreted as bursts only because of the high threshold of the BATSE detectors (see Fig. 10). Thus the long standing unsolved problem of explaining the long GRBs (see e.g. Piran [116], Salmonson et al. [161], Wilson et al. [186]) is radically resolved.

Still in section XVI, the search for the identification of the P-GRB in the BATSE data is described. This identification is made using the two fundamental diagrams shown in Fig. 11. Having established the value of $E_{dya} = 4.83 \times 10^{53}\,erg$ and of $B = 3 \times 10^{-3}$, it is possible from the dashed line and the solid line in Fig. 11 to evaluate the ratio of the energy $E_{P\text{-}GRB}$ emitted in the P-GRB to the energy $E_{Baryons}$ emitted in the afterglow corresponding to the determined value of B, see the vertical line in Fig. 11. We obtain $E_{P\text{-}GRB}/E_{Baryons} = 1.58 \times 10^{-2}$, which gives $E_{P\text{-}GRB} = 7.54 \times 10^{51}\,erg$. Having so determined the theoretically expected intensity of the P-GRB, a second fundamental observable parameter, which is also a function of E_{dya} and B, is the arrival time delay between the P-GRB and the peak E-APE, determined in Fig. 11. From Tab. I, we have that the detector arrival time of the P-GRB occurs at $8.41 \times 10^{-2}\,s$, corresponding to a radial coordinate of $1.94 \times 10^{14}\,cm$, a comoving time of $21.57\,s$, a laboratory time of $6.48 \times 10^3\,s$ and an arrival time of $4.21 \times 10^{-2}\,s$. At this point, the gamma factor is 310.1. The peak of the E-APE occurs at a detector arrival time of $19.87\,s$, corresponding to a radial coordinate of $5.18 \times 10^{16}\,cm$, a comoving time of $5.85 \times 10^3\,s$, a laboratory time of $1.73 \times 10^6\,s$ and an arrival time of $9.93\,s$ (see Tab. I). The delay between the P-GRB and the peak of the E-APE is therefore $19.78\,s$, see Fig. 11. The theoretical prediction on the

Table I: Gamma factors for selected events and their space-time coordinates. The points marked $1,2,3,4,5,6,P_L,P_A$ are the same reported in Fig. 8, while the point F is the endpoint of the simulation. It is particularly important to read the last column, where the apparent motion in the radial coordinate, evaluated in the arrival time at the detector, leads to an enormous "superluminal" behaviour, up to $9.55 \times 10^4\, c$. This illustrates well the impossibility of using such a classical estimate in regimes with gamma factors up to 310.1.

Point	$r(cm)$	$\tau(s)$	$t(s)$	$t_a(s)$	$t_a^d(s)$	γ	"Superluminal" $v \equiv \frac{r}{t_a^d}$
	**The Injector Phase**						
1	2.354×10^8	0.0	0.0	0.0	0.0	1.000	0
	1.871×10^9	1.550×10^{-2}	5.886×10^{-2}	4.312×10^{-3}	8.625×10^{-3}	10.08	$7.23c$
	4.486×10^9	2.141×10^{-2}	1.463×10^{-1}	4.523×10^{-4}	9.046×10^{-3}	20.26	$16.5c$
	7.080×10^9	2.485×10^{-2}	2.329×10^{-1}	4.594×10^{-3}	9.187×10^{-3}	30.46	$25.7c$
	9.533×10^9	2.715×10^{-2}	3.148×10^{-1}	4.627×10^{-3}	9.253×10^{-3}	40.74	$34.4c$
	1.162×10^{10}	2.868×10^{-2}	3.845×10^{-1}	4.644×10^{-3}	9.288×10^{-3}	49.70	$41.7c$
2	1.162×10^{10}	2.868×10^{-2}	3.845×10^{-1}	4.644×10^{-3}	9.288×10^{-3}	49.70	$41.7c$
	1.186×10^{10}	2.889×10^{-2}	3.923×10^{-1}	4.646×10^{-3}	9.292×10^{-3}	38.06	$42.6c$
	1.234×10^{10}	2.949×10^{-2}	4.083×10^{-1}	4.655×10^{-3}	9.311×10^{-3}	24.21	$44.2c$
	1.335×10^{10}	3.144×10^{-2}	4.423×10^{-1}	4.706×10^{-3}	9.413×10^{-3}	15.14	$47.3c$
	1.389×10^{10}	3.279×10^{-2}	4.603×10^{-1}	4.753×10^{-3}	9.506×10^{-3}	12.94	$48.7c$
3	1.389×10^{10}	3.279×10^{-2}	4.603×10^{-1}	4.753×10^{-3}	9.506×10^{-3}	12.94	$48.7c$
	2.326×10^{10}	5.208×10^{-2}	7.733×10^{-1}	5.369×10^{-3}	1.074×10^{-2}	20.09	$72.2c$
	6.913×10^{10}	9.694×10^{-2}	2.304	6.086×10^{-3}	1.217×10^{-2}	50.66	$1.89 \times 10^2 c$
	1.861×10^{11}	1.486×10^{-1}	6.206	6.446×10^{-3}	1.289×10^{-2}	100.1	$4.82 \times 10^2 c$
	9.629×10^{11}	3.112×10^{-1}	32.12	6.978×10^{-3}	1.396×10^{-2}	200.3	$2.30 \times 10^3 c$
	3.205×10^{13}	3.958	1.069×10^3	1.343×10^{-2}	2.685×10^{-2}	300.1	$3.98 \times 10^4 c$
	1.943×10^{14}	21.57	6.481×10^3	4.206×10^{-2}	8.413×10^{-2}	310.1	$7.70 \times 10^4 c$
	**The Beam-Target Phase**						
4	1.943×10^{14}	21.57	6.481×10^3	4.206×10^{-2}	8.413×10^{-2}	310.1	$7.70 \times 10^4 c$
	6.663×10^{15}	7.982×10^2	6.481×10^3	1.164	2.328	310.0	$9.55 \times 10^4 c$
	2.863×10^{16}	3.114×10^3	9.549×10^5	5.057	10.11	300.0	$9.45 \times 10^4 c$
	4.692×10^{16}	5.241×10^3	1.565×10^6	8.775	17.55	270.0	$8.92 \times 10^4 c$
P_A	5.177×10^{16}	5.853×10^3	1.727×10^6	9.933	19.87	258.5	$8.69 \times 10^4 c$
	5.878×10^{16}	6.791×10^3	1.961×10^6	11.82	23.63	240.0	$8.30 \times 10^4 c$
	6.580×10^{16}	7.811×10^3	2.195×10^6	14.03	28.06	220.0	$7.82 \times 10^4 c$
P_L	7.025×10^{16}	8.506×10^3	2.343×10^6	15.66	31.32	207.0	$7.48 \times 10^4 c$
	7.262×10^{16}	8.895×10^3	2.422×10^6	16.61	33.23	200.0	$7.29 \times 10^4 c$
	9.058×10^{16}	1.236×10^4	3.021×10^6	26.66	53.32	150.0	$5.67 \times 10^4 c$
	1.136×10^{17}	1.866×10^4	3.788×10^6	52.84	1.057×10^2	100.0	$3.58 \times 10^4 c$
	1.539×10^{17}	3.819×10^4	5.134×10^6	2.000×10^2	4.000×10^2	50.02	$1.28 \times 10^4 c$
	2.801×10^{17}	2.622×10^5	9.351×10^6	7.278×10^3	1.455×10^4	10.00	$6.42 \times 10^2 c$
	3.624×10^{17}	6.702×10^5	1.213×10^7	3.860×10^4	7.719×10^4	5.001	$1.57 \times 10^2 c$
	4.454×10^{17}	1.433×10^6	1.500×10^7	1.439×10^5	2.877×10^5	2.998	$51.6c$
5	4.454×10^{17}	1.433×10^6	1.500×10^7	1.439×10^5	2.877×10^5	2.998	$51.6c$
	4.830×10^{17}	1.928×10^6	1.635×10^7	2.381×10^5	4.762×10^5	2.500	$33.8c$
	5.390×10^{17}	2.873×10^6	1.844×10^7	4.643×10^5	9.285×10^5	2.000	$19.4c$
	6.422×10^{17}	5.387×10^6	2.271×10^7	1.291×10^6	2.581×10^6	1.500	$8.30c$
	1.034×10^{18}	2.903×10^7	5.002×10^7	1.552×10^7	3.103×10^7	1.054	$1.11c$
6	1.034×10^{18}	2.903×10^7	5.002×10^7	1.552×10^7	3.103×10^7	1.054	$1.11c$
	1.202×10^{18}	4.979×10^7	7.150×10^7	3.140×10^7	6.280×10^7	1.025	$6.38 \times 10^{-1} c$
F	1.248×10^{18}	5.706×10^7	7.894×10^7	3.731×10^7	7.461×10^7	1.000	$5.58 \times 10^{-1} c$

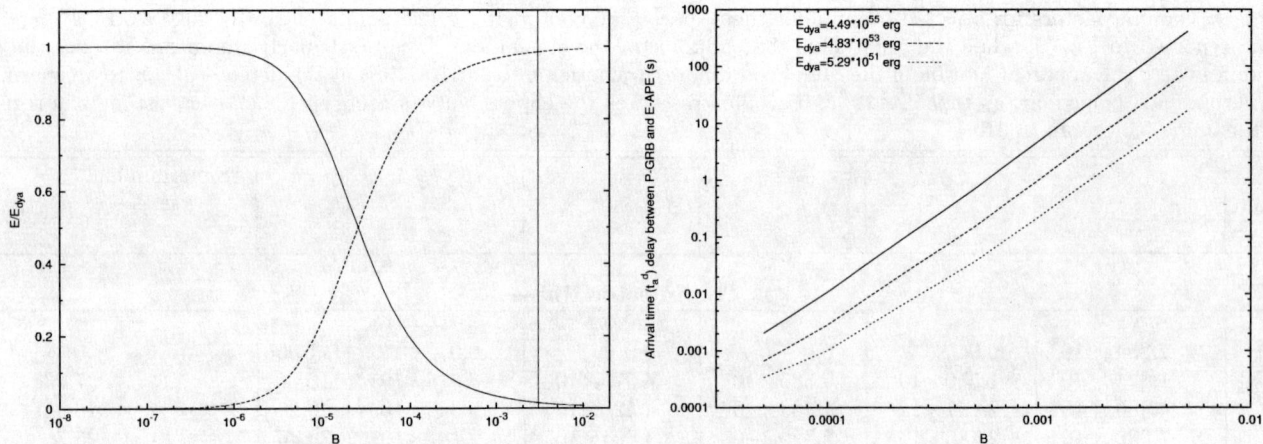

Figure 11: **Left)** Relative intensities of the E-APE (dashed line) and the P-GRB (solid line), as predicted by the EMBH theory corresponding to the values of the parameters determined in Fig. 10, as a function of B. Details are given in section XVI. The vertical line corresponds to the value $B = 3 \times 10^{-3}$. **Right)** The arrival time delay between the P-GRB and the peak of the E-APE is plotted as a function of the B parameter for three selected values of E_{dya}.

intensity and the arrival time uniquely identifies the P-GRB with the "precursor" in the GRB 991216 (see Fig. 3). Moreover, the hardness of the P-GRB spectra is also evaluated in this section. As pointed out in the conclusions, the fact that both the absolute and relative intensities of the P-GRB and E-APE have been predicted within a few percent accuracy as well as the fact that their arrival time has been computed with the precision of a few tenths of milliseconds, see Tab. I and Fig. 12, can be considered one of the major successes of the EMBH theory.

H. On the power-laws and beaming in the afterglow of GRB 991216.

In section XVIII a piecewise description of the afterglow by the expansion of the fundamental hydrodynamical equations given by Taub [176] and Landau & Lifshitz [86] have allowed the determination of a power-law index for the dependence of the afterglow luminosity on the photon arrival time at the detector. It is evident that the determination of the power-law index is very sensitive to the basic assumptions made for the description of the afterglow, as well as to the relations between the different temporal coordinates which have been clarified by the RSTT paradigm (see Ruffini et al. [143]). The different power-law indexes obtained are compared and contrasted with the ones in the current literature (see Tab. II and section. XIX). As a byproduct of this analysis, see also the conclusions, there is a perfect agreement between the observational data and the theoretical predictions, implying that the assumptions we have adopted for the description of the afterglow (see section XIII) must be necessarily all valid and therefore, in particular, there is no evidence for a beamed emission in GRB 991216.

We then summarize in Fig. 12 the results for the average bolometric luminosity of GRB 991216 with particular attention to the striking agreement, both in arrival time and in intensity, for the theoretically predicted structure of the P-GRB and the E-APE with the observational data. To show the generality of application of the EMBH theory, we have applied it also to GRB 980425 (see Ruffini [139]) and the excellent results are also shown, for comparsion, in Fig. 12.

I. Substructures in the E-APE due to inhomogeneities in the Interstellar medium

In section XX the role of the inhomogeneities in the interstellar matter has been analyzed in order to explain the observed temporal substructures in the BATSE data on GRB 991216. Having satisfactorily identified the average intensity distribution of the afterglow and the relative position of the P-GRB, in Ruffini et al. [146] we have addressed the issue whether the fast temporal variation observed in the so-called long bursts, on time scales as short as fraction of a second (see e.g. Fishman & Meegan [52]), can indeed be explained as an effect of inhomogeneities in the interstellar medium. Such a possibility was pioneered in the work by Dermer & Mitman [41], purporting that such a time variability corresponds to a tomographic analysis of the ISM. In order to probe the validity of such an explanation, we have first considered the simplified case of the radial approximation (Ruffini et al. [146]). The aim has been to explore the possibility of explaining the observed fluctuation in intensity on a fraction of a second as originated from

Table II: We compare and contrast the results on the power-law index n of the afterglow in the EMBH theory with other treatments in the current literature, in the limit of high energy and fully radiative conditions. The differences between the values of $-10/7 \sim -1.43$ (Dermer) and the results -1.6 in the EMBH theory can be retraced to the use of the two different approximation in the arrival time versus the laboratory time given in Fig. 9. See details in section XVIII.

	EMBH theory	Chiang & Dermer [28] Dermer et al. [40] Böttcher & Dermer [19]	Piran [115] Sari & Piran [167] Piran [116]	Vietri [179]	Halpern et al. [72]
Ultra-relativistic	$\gamma = \gamma_\circ$ $\gamma_\circ = 310.1$ $n = 2$	$\gamma = \gamma_\circ$ $n = 2$	$\gamma = \gamma_\circ$ $n \simeq 2$		
Relativistic	$\gamma \simeq r^{-3}$ $3.0 < \gamma < 258.5$ $n = -1.6$	$\gamma \sim r^{-3}$ $n = -\frac{10}{7} = -1.43$	$\gamma \sim r^{-3}$ $n = -\frac{5.5}{4} = -1.375$		$n > -1.47$
Non-relativistic	$n = -1.36$ $1.05 < \gamma < 3.0$			$n = -1.7$	
Newtonian	$n = -1.45$ $1 < \gamma < 1.05$				

inhomogeneities in ISM, typically of the order of 10^{16} due to apparent superluminal behaviour of roughly $10^5 c$. We have shown there that this approach is indeed viable: both the intensity variation and the time scale of the variability in the E-APE region can be explained by the interaction of the ABM pulse with inhomogeneities in the ISM, taking into due account the apparent superluminal effects. These effects, in turn, can be derived and computed self consistently from the dynamics of the source. We have then described the inhomogeneities of the ISM by an appropriate density profile (mask) of an ISM cloud. Of course at this stage, for simplicity, only the case of spherically symmetric "spikes" with over-density separated by low-energy regions, has been considered. Each spike has been assumed to have the spatial extension of 10^{15} cm. The cloud average density is $< n_{ism} > = 1$ particle/cm^3. In conclusion, from the data of Tab. I and the highly "superluminal" behaviour of the source in the region of the E-APE, it is concluded that the observed time variability in the intensity of the emission $(\Delta I / \bar{I}) \sim 5$ can be traced to inhomogeneities in the interstellar matter: $(\Delta n_{ism}/n_{ism}) \sim 5$. The typical size of the scattering region is estimated to be 5×10^{16} cm, and these are the typical sizes and density contrasts found in interstellar clouds. Since the emission of the E-APE occurs at typical dimensions of the order of 5×10^{16} cm, the observed inhomogeneities are probing the structure of the interstellar medium, and have nothing to do with the "inner engine" of the source.

The big issue was then open if all these results, obtained in the radial approximation, would still be valid in the more general case when off-axis emission in the description of the afterglow is taken into account. This is the reason why we have proceeded to the topic summarized in the next subsections (see Ruffini et al. [147]).

J. The definition of the equitemporal surfaces (EQTS) and the afterglow delayed intensity as a function of the viewing angle

While the analysis of the average bolometric intensity of GRB was going on in the radial approximation, we have proceeded to develop the full non-radial approximation, taking into account all the relativistic corrections for the off-axis emission from the spherically symmetric expansion of the ABM pulse (see Ruffini et al. [147, 152] and sections XXI–XXII). Photons emitted at the same time but at different angles of displacement from the line of sight

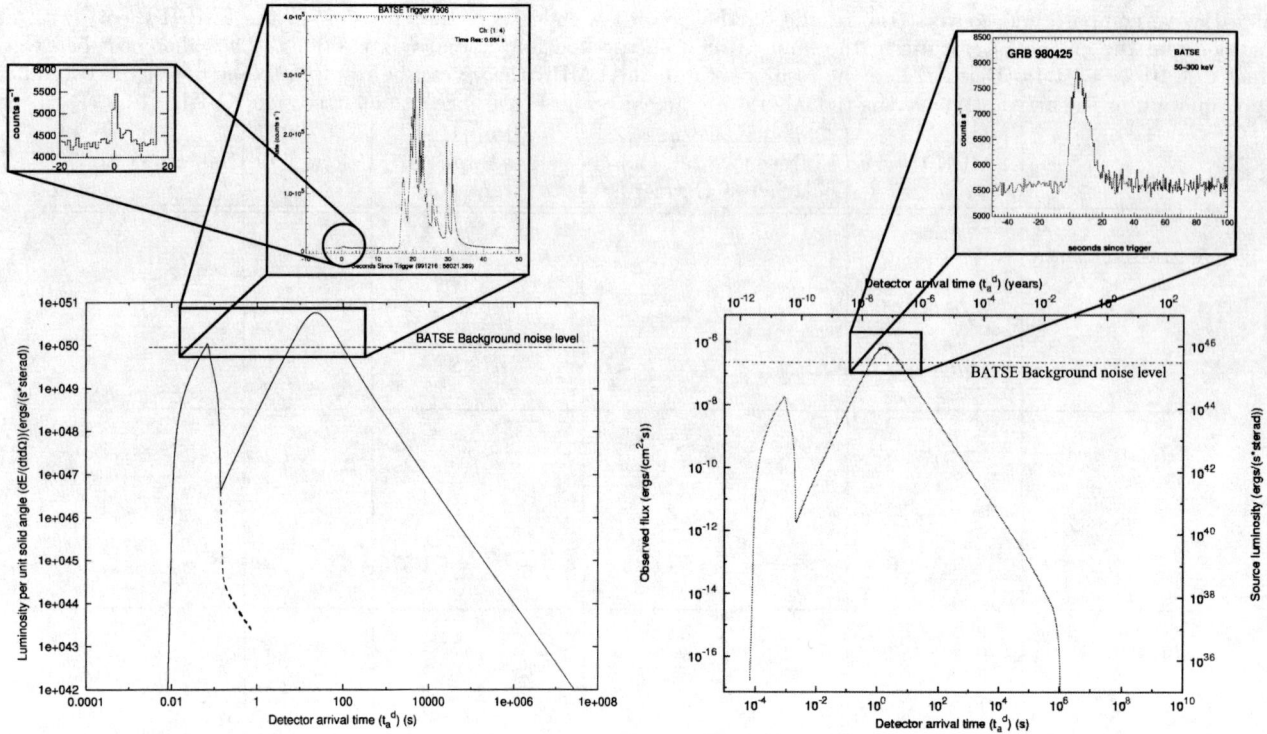

Figure 12: **Left)** The overall description of the EMBH theory applied to GRB 991216. The BATSE noise threshold is represented and the observations both of the P-GRB and of the E-APE are clearly shown in the subpanels. The continuos line in the picture represents the theoretical prediction of the EMBH model. **Right)** The same diagrams are represented for GRB 980425. Two aspects are especially important to be mentioned: a) in this source the theoretical prediction of the P-GRB intensity is lower than the BATSE noise treshold and is therefore unobservable and unobserved; b) the E-APE is especially smooth as a consequence of the low value of the gamma Lorentz factor (see also section XXIII and Ruffini [139]).

reach the detector at very different arrival times. Correspondingly, photons detected at the same arrival time are emitted at very different times and angles. We have so defined the temporal evolution of the ABM pulse visible area as well as the equitemporal surfaces (EQTS), i.e. the locus of points on the ABM pulse emitting surface corresponding to a constant value of the photon arrival time at the detector.

The very same difficulties found in the current literature, relating the laboratory time to the photon arrival time at the detector (see Figs. 8–9), still exists in the present context and are even magnified in the definition of the EQTS. In a classical article, Rees [126] expressed the relation between the laboratory time and the arrival time at the detector in order to explain observations in radio sources with a constant expansion velocity v and Lorentz gamma factor $\gamma \sim 5$. He pointed out the EQTS are ellipsoids of constant eccentricity v/c. In the current literature, the Rees approach has been adapted to the analysis of GRBs (see e.g. Fenimore et al. [47], Piran [115], Sari [162, 163], Waxman [185] and references therein). In addition to the very crucial relation between the laboratory time and the photon arrival time, which has not been properly treated, there have been a variety of other approximation and averaging processes on which we do not agree. Instead of specifically criticizing each assumption which we consider not correct, such comparison will be made in a forthcoming paper (Ruffini et al. [154]), we just report here in the following the results of the EQTS surfaces (see Fig. 13) obtained in conformity with the RSTT paradigm. In the present case of GRBs, the gamma factor is not only much larger than the one observed in radio sources, but is also strongly time varying (see Fig. 8). The Rees treatment has to be significantly improved to take into account the huge time variations in the Lorentz gamma factor: this is not just a technical point of modifying a formula by the introduction of a new integral. There is in the present context the crucial point expressed in the RSTT paradigm that the relation between the laboratory time and the arrival time at the detector is a function of all the the previous Lorentz gamma factors in the history of the source since $\gamma = 1$ (see Fig. 9). In the definition of each EQTS, therefore, the entire previous past history of the source does concur and the EQTS surfaces become therefore a very refined and sensitive test of the correct description of the entire spacetime evolution of the source. In this case, we no longer have ellipsoids of constant eccentricity $\frac{v}{c}$. Since the velocity is strongly varying from point to point, we have more complicated surfaces like the profiles reported in Fig. 13 where at every point there will be a tangent ellipsoid of a given eccentricity, but such an

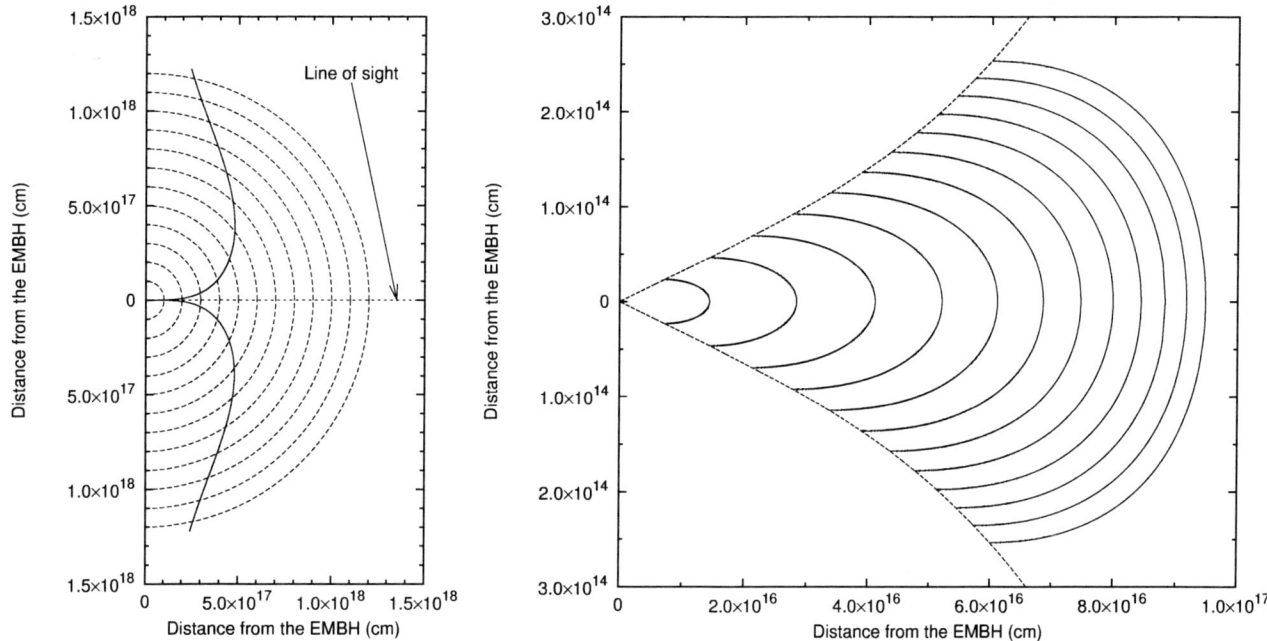

Figure 13: **Left)** This figure shows the temporal evolution of visible area of the ABM pulse. The dashed half-circles are the expanding ABM pulse at radii corresponding to different laboratory times. The black curve marks the boundary of the visible region. The EMBH is located at position (0,0) in this plot. Again, in the earliest GRB phases the visible region is squeezed along the line of sight, while in the final part of the afterglow phase almost all the emitted photons reach the observer. This time evolution of the visible area is crucial to the explanation of the GRB temporal structure. **Right)** Due to the extremely high and extremely varying Lorentz gamma factor, photons reaching the detector on the Earth at the same arrival time are actually emitted at very different times and positions. We represent here the surfaces of photon emission corresponding to selected values of the photon arrival time at the detector: the *equitemporal surfaces* (EQTS). Such surfaces differ from the ellipsoids described by Rees in the context of the expanding radio sources with typical Lorentz factor $\gamma \sim 4$ and constant. In fact, in GRB 991216 the Lorentz gamma factor ranges from 310 to 1. The EQTSes represented here (solid lines) correspond respectively to values of the arrival time ranging from $5\,s$ (the smallest surface on the left of the plot) to $60\,s$ (the largest one on the right). Each surface differs from the previous one by $5\,s$. To each EQTS contributes emission processes occurring at different values of the Lorentz gamma factor. The dashed lines are the boundaries of the visible area of the ABM pulse and the EMBH is located at position $(0,0)$ in this plot. Note the different scale on the two axes, indicating the very high EQTS "effective eccentricity". The time interval from $5\,s$ to $60\,s$ has been chosen to encompass the E-APE emission, ranging from $\gamma = 308.8$ to $\gamma = 56.84$.

ellipsoid varies in eccentricity from point to point (see Fig. 13 and section XXI). Any departure from the correct equation of motion strongly alters the EQTS surfaces and accordingly modifies all the results of the integrations based on the EQTS surfaces, e.g. the spectral distribution or the afterglow (Ruffini et al. [153]).

Having determined the EQTS surfaces we have computed the observed GRB flux at selected values of the photon arrival time at the detector, taking into due account the delayed contributions at different angles and we have presented the results in section XXII and Fig. 14.

We have then recomputed the afterglow emission of GRB 991216 taking into account all the effects due to this temporal spreading in the arrival time as well as the ones due to the dependency of the photon Doppler shift on the angle of displacement from the line of sight of the emission location (see section XXII). The result is reported in Fig. 14.

From now on all the afterglow intensities are estimated using this very complex and extensive numerical program which is rooted in all previous history of the source: the general considerations on simple analytic expansion expressed in section XVIII are kept only as an heuristic procedure as a guideline to comprehend these more complex results.

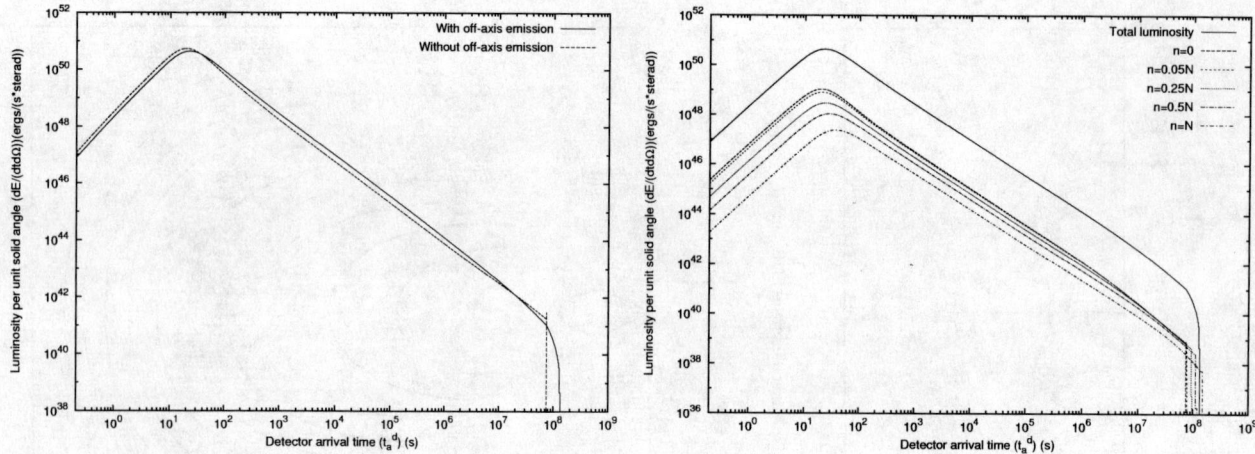

Figure 14: **Left)** The predicted afterglow curve for GRB 991216 assuming a constant ISM density equal to 1 particle/cm^3 and taking into account all the effects due to off-axis emission (solid line). For comparison we plot also the corresponding curve obtained in the simple radial approximation (dashed line). We see that this last curve falls sharply to zero when the ABM pulse reaches $\gamma = 1$, while the first one has a much smoother behavior due to the time delay in the arrival of the photons emitted at large ϑ. Recall that when γ tends to 1, the maximum allowed values of ϑ tend to 90°. **Right)** This figure shows how the radiation emitted from different angles contributes to the afterglow luminosity. The solid line on the top of the picture is the total luminosity as in the previous plots. The other dashed and dotted curves represent the radiation components corresponding to selected values of n in Eq.(181). From the upper to the lower one they corresponds respectively to $n = 0$, $n = 0.05N$, $n = 0.25N$, $n = 0.5N$, $n = N$, where in this plot $N = 200$. We can easily see that the radiation emitted at large angles ($n = N$) is time shifted with respect to that emitted near the line of sight ($n = 0$).

K. The E-APE temporal substructures taking into account the off-axis emission

Having determined the EQTS surfaces, we have reconsidered the E-APE temporal substructure taking into due account the off-axis emission contribution (see Fig. 15 and section XXIII).

We can distinguish two different regimes corresponding respectively to $\gamma > 150$ and to $\gamma < 150$. In the E-APE region ($\gamma > 150$) the GRB substructure intensities indeed correlate with the ISM inhomogeneities. In this limited region (see peaks A, B, C) the Lorentz gamma factor of the ABM pulse ranges from $\gamma \sim 304$ to $\gamma \sim 200$. The boundary of the visible region is smaller than the thickness ΔR of the inhomogeneities (see Figs. 15,13, Tab. IV and Ruffini et al. [147, 152]). Under these conditions the adopted spherical symmetry for the density spikes is not only mathematically simpler but also fully justified. The angular spreading is not strong enough to wipe out the signal from the inhomogeneity spike.

As we descend in the afterglow ($\gamma < 150$), a border-line case occurs at peak D where $\gamma \sim 140$. There the visible region is comparable to the thickness ΔR: to fit the observed data a three dimensional description would be necessary, breaking the spherical symmetry and making the computation more difficult, but we do not foresee any conceptual difficulty. For the peaks E and F we have $\gamma \sim 50$: under these circumstances the boundary of the visible region becomes much larger than the thickness ΔR. The spherically symmetric description of the inhomogeneities is already enough to prove the overwhelming effect of the angular spreading and no three dimensional description is needed (Ruffini et al. [147, 152]).

From our analysis we can conclude that Dermer's expectations do indeed hold for $\gamma > 150$. However, as the gamma factor drops from $\gamma \sim 150$ to $\gamma \sim 1$ the intensity due to the inhomogeneities markedly decreases due to the angular spreading (events E and F). The initial Lorentz factor of the ABM pulse $\gamma \sim 310$ decreases very rapidly to $\gamma \sim 150$ as soon as a fraction of a typical ISM cloud is engulfed (see Figs. 15,8, Tab. IV and Ruffini et al. [147, 152]). We conclude that the "tomography" is indeed effective, but uniquely in the first ISM region close to the source and for GRBs with $\gamma > 150$.

It is then clear that no information on the nature of the GRB source can be inferred by the analysis of the T_{90}, nor by the intensity variability structure of the so-called "long bursts": the only indirect information can be obtained from the value of Lorentz gamma factor, which has to be $\gamma > 150$ in presence of significant observed substructure. In this sense compare and contrast the two cases of GRB 991216 and GRB 980425 where the γ value in the E-APE is found to be $\gamma \sim 120$ (see Ruffini [139]). The intensity substructures in the E-APE only carry information on the structure of the ISM clouds.

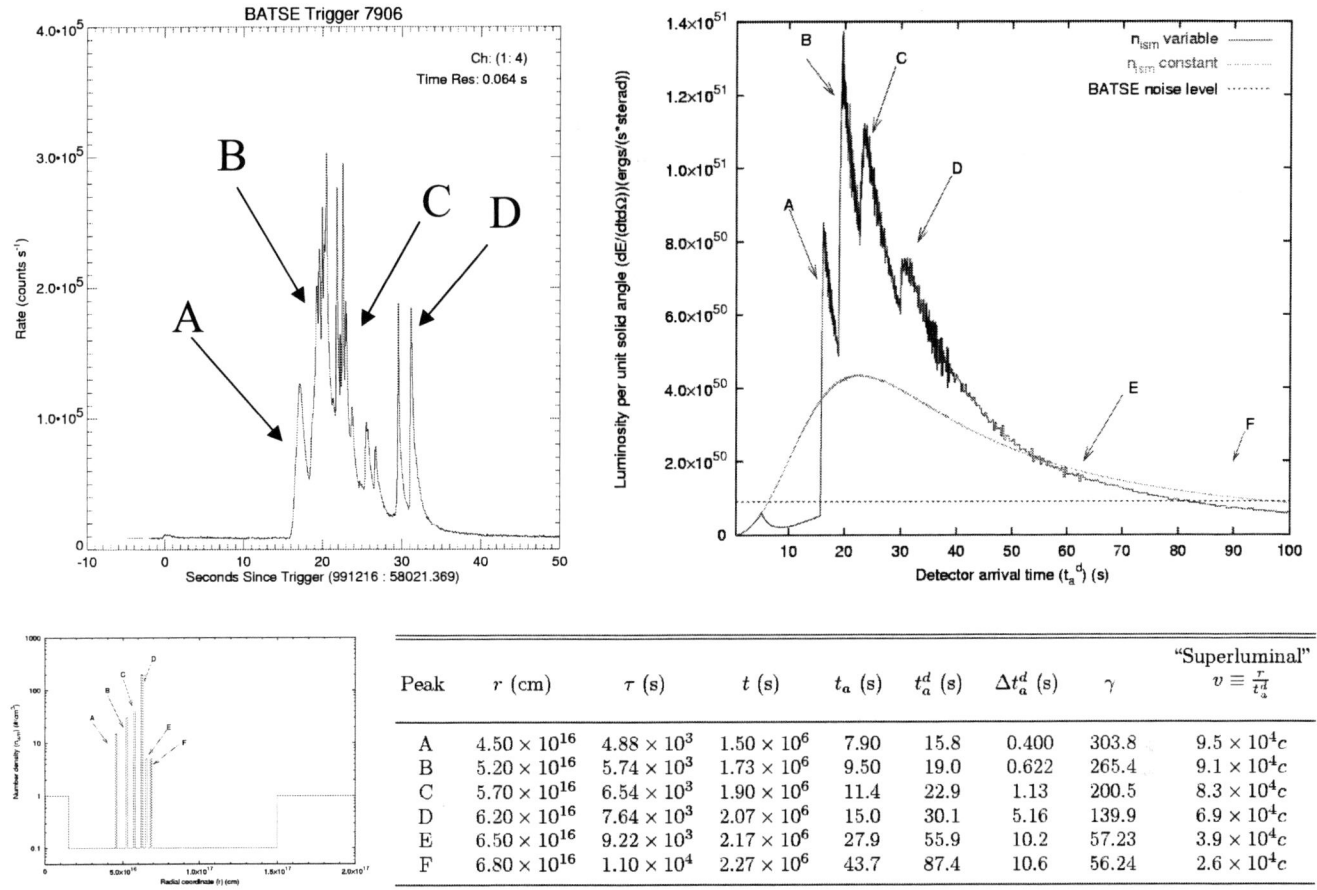

Figure 15: In this figure we summarize the main results of the fit obtained by the EMBH model for the E-APE intensity in the case of GRB 991216 taking into account all off-axis contributions. The upper two diagrams represent respectively the observational data and the corresponding theoretically computed results. On the lower left the "mask" of the spherically symmetric density inhomogeneities with average $<n_{ism}> = 1\,\text{particle/cm}^3$ is represented. The table summarizes all the parameters corresponding to the inhomogeneities including the vary large apparent superluminal effect up to $\sim 10^5 c$. Details in section 23.

L. The observation of the iron lines in GRB 991216: on a possible GRB-supernova time sequence

In section XXV the program of using GRBs to further explore the region surrounding the newly formed EMBH is carried one step further by using the observations of the emitted iron lines (Piro et al., [119]). This gives us the opportunity to introduce the GRB-supernova time sequence (GSTS) paradigm and to introduce as well the novel concept of an *induced supernova explosion*. The GSTS paradigm reads: *A massive GRB-progenitor star P_1 of mass M_1 undergoes gravitational collapse to an EMBH. During this process a dyadosphere is formed and subsequently the P-GRB and the E-APE are generated in sequence. They propagate and impact, with their photon and neutrino components, on a second supernova-progenitor star P_2 of mass M_2. Assuming that both stars were generated approximately at the same time, we expect to have $M_2 < M_1$. Under some special conditions of the thermonuclear evolution of the supernova-progenitor star P_2, the collision of the P-GRB and the E-APE with the star P_2 can induce its supernova explosion.*

Using the result presented in Tab. I and in all preceding sections, the GSTS paradigm is illustrated in the case of GRB 991216. Some general considerations on the nature of the supernova progenitor star are also advanced.

Some general considerations on the EMBH formation are presented in section XXVI. The general conclusions are presented in section XXIX.

We now proceed to a more detailed presentation of the results and we refer to the already published material for the complete details.

III. THE ZEROTH ERA: THE PROCESS OF GRAVITATIONAL COLLAPSE AND THE FORMATION OF THE DYADOSPHERE

We first recall the three theoretical results which lie at the basis of the EMBH theory.

In 1971 in the article *"Introducing the Black Hole"* (Ruffini & Wheeler [140]), the theorem was advanced that the most general black hole is characterized uniquely by three independent parameters: the mass-energy M, the angular momentum L and the charge Q making it an EMBH. Such an ansatz, which came to be known as the "uniqueness theorem" has turned out to be one of the most difficult theorems to be proven in all of physics and mathematics. The progress in the proof has been authoritatively summarized by Carter [23]. The situation can be considered satisfactory from the point of view of the physical and astrophysical considerations. Nevertheless some fundamental mathematical and physical issues concerning the most general perturbation analysis of an EMBH are still the topic of active scientific discussion (Bini et al. [14]).

In 1971 it was shown that the energy extractable from an EMBH is governed by the mass-energy formula (Christodoulou & Ruffini [29]),

$$E_{BH}^2 = M^2 c^4 = \left(M_{\mathrm{ir}} c^2 + \frac{Q^2}{2\rho_+} \right)^2 + \frac{L^2 c^2}{\rho_+^2}, \qquad (1)$$

with

$$\frac{1}{\rho_+^4} \left(\frac{G^2}{c^8} \right) (Q^4 + 4L^2 c^2) \leq 1, \qquad (2)$$

where

$$S = 4\pi \rho_+^2 = 4\pi (r_+^2 + \frac{L^2}{c^2 M^2}) = 16\pi \left(\frac{G^2}{c^4} \right) M_{\mathrm{ir}}^2, \qquad (3)$$

is the horizon surface area, M_{ir} is the irreducible mass, r_+ is the horizon radius and ρ_+ is the quasi-spheroidal cylindrical coordinate of the horizon evaluated at the equatorial plane. Extreme EMBHs satisfy the equality in Eq.(2). Up to 50% of the mass-energy of an extreme EMBH can in principle be extracted by a special set of transformations: the reversible transformations (Christodoulou & Ruffini [29]).

In 1975, generalizing some previous results of Zaumen [190], and Gibbons [66], Damour & Ruffini [32] showed that the vacuum polarization process *à la* Heisenberg-Euler-Schwinger (Heisenberg & Euler [77], Schwinger [171]) created by an electric field of strength larger than

$$\mathcal{E}_c = \frac{m_e^2 c^3}{\hbar e} \qquad (4)$$

can indeed occur in the field of a Kerr-Newmann EMBH. Here m_e and e are respectively the mass and charge of the electron. There Damour and Ruffini considered an axially symmetric EMBH, due to the presence of rotation, and limited themselves to EMBH masses larger then the upper limit of a neutron star for astrophysical applications. They purposely avoided all complications of black holes with mass smaller then the dual electron mass of the electron $\left(m_e^\star = \frac{c\hbar}{G m_e} = \frac{m_{Planck}^2}{m_e} \right)$ which may lead to quantum evaporation processes (Hawking [75]). They pointed out that:

1. The vacuum polarization process can occur for an EMBH mass larger than the maximum critical mass for neutron stars all the way up to $7.2 \times 10^6 M_\odot$.

2. The process of pair creation occurs on very short time scales, typically $\frac{\hbar}{m_e c^2}$, and is an almost perfect reversible process, in the sense defined by Christodoulou-Ruffini, leading to a very efficient mechanism of extracting energy from an EMBH.

3. The energy generated by the energy extraction process of an EMBH was found to be of the order of 10^{54} erg, released almost instantaneously. They concluded at the time *"this work naturally leads to a most simple model for the explanation of the recently discovered γ-ray bursts"*.

After the discovery of the afterglow of GRBs and the determination of the cosmological distance of their sources we noticed the coincidence between the theoretically predicted energetics and the observed ones in Damour & Ruffini [32]: we returned to our theoretical results developing some new basic theoretical concepts (Preparata et al. [122, 123], Ruffini [136], Ruffini et al. [141, 142]), which have led to the EMBH theory.

As a first simplifying assumption we have developed our considerations in the absence of rotation with spherically symmetric distributions. The space-time is then described by the Reissner-Nordström geometry, whose spherically symmetric metric is given by

$$d^2s = g_{tt}(r)d^2t + g_{rr}(r)d^2r + r^2 d^2\theta + r^2 \sin^2\theta d^2\phi ,\qquad(5)$$

where $g_{tt}(r) = -\left[1 - \frac{2GM}{c^2 r} + \frac{Q^2 G}{c^4 r^2}\right] \equiv -\alpha^2(r)$ and $g_{rr}(r) = \alpha^{-2}(r)$.

The first new result we obtained is that the pair creation process does not occur at the horizon of the EMBH: it extends over the entire region outside the horizon in which the electric field exceeds the critical value given by Eq. 4. Since the electric field in the Reissner-Nordström geometry has only a radial component given by (see Ruffini [135])

$$\mathcal{E}(r) = \frac{Q}{r^2},\qquad(6)$$

this region extends from the horizon radius

$$r_+ = 1.47 \cdot 10^5 \mu (1 + \sqrt{1-\xi^2})\text{ cm}\qquad(7)$$

out to an outer radius (Ruffini [136])

$$r^\star = \left(\frac{\hbar}{mc}\right)^{\frac{1}{2}} \left(\frac{GM}{c^2}\right)^{\frac{1}{2}} \left(\frac{m_\text{p}}{m}\right)^{\frac{1}{2}} \left(\frac{e}{q_\text{p}}\right)^{\frac{1}{2}} \left(\frac{Q}{\sqrt{G}M}\right)^{\frac{1}{2}} = 1.12 \cdot 10^8 \sqrt{\mu\xi}\text{ cm},\qquad(8)$$

where we have introduced the dimensionless mass and charge parameters $\mu = \frac{M}{M_\odot}$, $\xi = \frac{Q}{(M\sqrt{G})} \leq 1$, see Fig. 4.

The second new result has been to realize that the local number density of electron and positron pairs created in this region as a function of radius is given by

$$n_{e^+e^-}(r) = \frac{Q}{4\pi r^2 \left(\frac{\hbar}{mc}\right) e}\left[1 - \left(\frac{r}{r^\star}\right)^2\right],\qquad(9)$$

and consequently the total number of electron and positron pairs in this region is

$$N^\circ_{e^+e^-} \simeq \frac{Q - Q_c}{e}\left[1 + \frac{(r^\star - r_+)}{\frac{\hbar}{mc}}\right],\qquad(10)$$

where $Q_c = \mathcal{E}_c r_+^2$.

The total number of pairs is larger by an enormous factor $r^\star/(\hbar/mc) > 10^{18}$ than the value Q/e which a naive estimate of the discharge of the EMBH would have predicted. Due to this enormous amplification factor in the number of pairs created, the region between the horizon and $r^\star$ is dominated by an essentially high density neutral plasma of electron-positron pairs. We have defined this region as the dyadosphere of the EMBH from the Greek duas, duadsos for pairs. Consequently we have called $r^\star$ the dyadosphere radius $r^\star \equiv r_\text{ds}$ (Preparata et al. [122, 123], Ruffini [136]). The vacuum polarization process occurs as if the entire dyadosphere are subdivided into a concentric set of shells of capacitors each of thickness $\hbar/m_e c$ and each producing a number of e^+e^- pairs on the order of $\sim Q/e$ (see Fig. 4). The energy density of the electron-positron pairs is given by

$$\epsilon(r) = \frac{Q^2}{8\pi r^4}\left(1 - \left(\frac{r}{r_\text{ds}}\right)^4\right),\qquad(11)$$

(see Figs. 2–3 of Preparata et al. [122]). The total energy of pairs converted from the static electric energy and deposited within the dyadosphere is then

$$E_\text{dya} = \frac{1}{2}\frac{Q^2}{r_+}\left(1 - \frac{r_+}{r_\text{ds}}\right)\left[1 - \left(\frac{r_+}{r_\text{ds}}\right)^4\right].\qquad(12)$$

As we will see in the following this is one of the two fundamental parameters of the EMBH theory (see Fig. 17). In the limit $\frac{r_+}{r_\text{ds}} \to 0$, Eq.(12) leads to $E_\text{dya} \to \frac{1}{2}\frac{Q^2}{r_+}$, which coincides with the energy extractable from EMBHs by reversible processes ($M_\text{ir} = $ const.), namely $E_{BH} - M_\text{ir} = \frac{1}{2}\frac{Q^2}{r_+}$ (Christodoulou & Ruffini [29]), see Fig. 16. Due to the

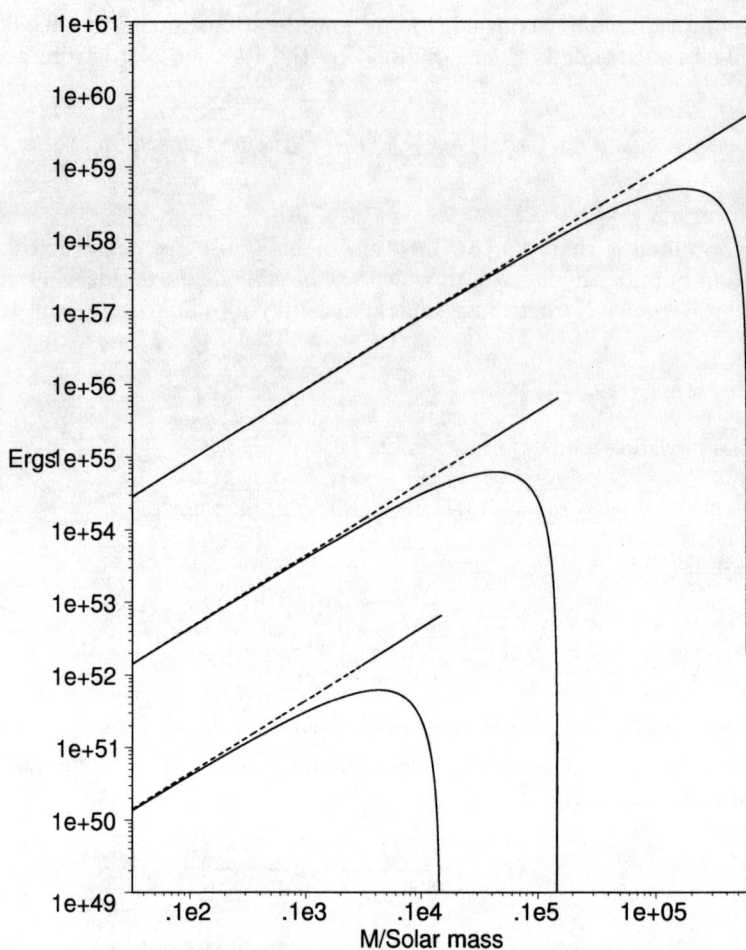

Figure 16: The energy extracted by the process of vacuum polarization is plotted (solid lines) as a function of the mass M in solar mass units for selected values of the charge parameter $\xi = 1, 0.1, 0.01$ (from top to bottom) for an EMBH, the case $\xi = 1$ reachable only as a limiting process. For comparison we have also plotted the maximum energy extractable from an EMBH (dotted lines) given by eq. (1). Details in Preparata et al. [124].

very large pair density given by Eq.(9) and to the sizes of the cross-sections for the process $e^+e^- \leftrightarrow \gamma + \gamma$, the system is expected to thermalize to a plasma configuration for which

$$n_{e^+} = n_{e^-} \sim n_\gamma \sim n^\circ_{e^+e^-}, \tag{13}$$

where $n^\circ_{e^+e^-}$ is the total number density of e^+e^--pairs created in the dyadosphere (see Preparata et al. [122, 123]).

The third new result which we have introduced for simplicity is that for a given E_{dya} we have assumed either a constant average energy density over the entire dyadosphere volume, or a more compact configuration with energy density equal to the peak value. These are the two possible initial conditions for the evolution of the dyadosphere (see Fig. 17).

These three old and three new theoretical results permit a good estimate of the general energetics processes originating in the dyadosphere, assuming an already formed EMBH. In reality, if the data become accurate enough, the full dynamical description of the dyadosphere formation mentioned above will be needed in order to follow all the general relativistic effects and characteristic time scales of the approach to the EMBH horizon (Cherubini et al. [27], Ruffini & Vitagliano [155, 156], Ruffini et al. [158] see also section XXVI).

Below we shall concentrate on the dynamical evolution of the electron-positron plasma created in the dyadosphere. We shall first examine in the next three sections the governing equations necessary to approach such a dynamical description.

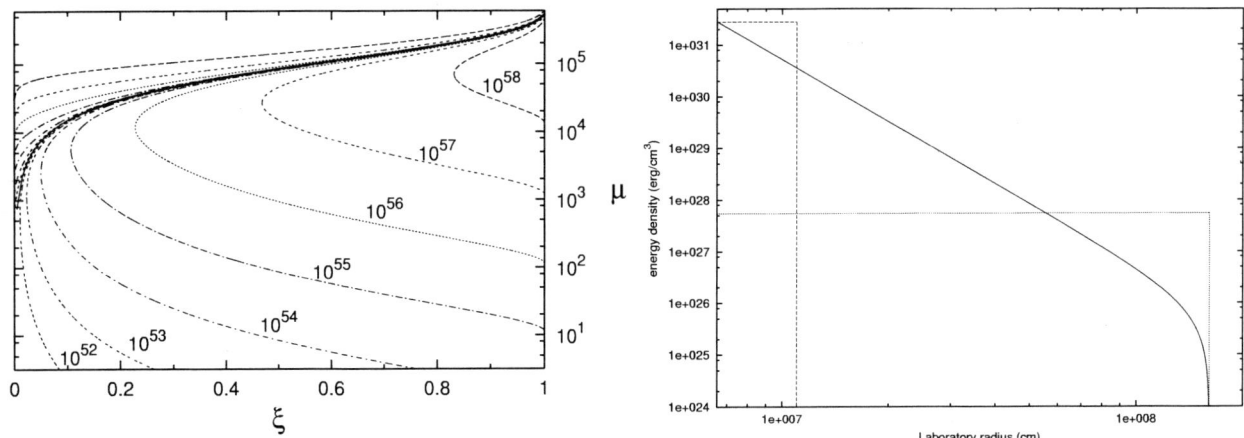

Figure 17: **Left)** Selected lines corresponding to fixed values of the E_{dya} are given as a function of the two parameters μ ξ, only the solutions below the continuous heavy line are physically relevant. The configurations above the continuous heavy lines correspond to unphysical solutions with $r_{ds} < r_+$. **Right)** Two different approximations for the energy density profile inside the dyadosphere. The first one (dashed line) fixes the energy density equal to its peak value, and computes an "effective" dyadosphere radius accordingly. The second one (dotted line) fixes the dyadosphere radius to its correct value, and assumes an uniform energy density over the dyadosphere volume. The total energy in the dyadosphere is of course the same in both cases. The solid curve represents the real energy density profile.

IV. THE HYDRODYNAMICS AND THE RATE EQUATIONS FOR THE PLASMA OF e^+e^--PAIRS

The evolution of the e^+e^--pair plasma generated in the dyadosphere has been treated in two papers (Ruffini et al. [141, 142]). We recall here the basic governing equations in the most general case in which the plasma fluid is composed of e^+e^--pairs, photons and baryonic matter. The plasma is described by the stress-energy tensor

$$T^{\mu\nu} = pg^{\mu\nu} + (p+\rho)U^\mu U^\nu, \qquad (14)$$

where ρ and p are respectively the total proper energy density and pressure in the comoving frame of the plasma fluid and U^μ is its four-velocity, satisfying

$$g_{tt}(U^t)^2 + g_{rr}(U^r)^2 = -1, \qquad (15)$$

where U^r and U^t are the radial and temporal contravariant components of the 4-velocity.

The conservation law for baryon number can be expressed in terms of the proper baryon number density n_B

$$\begin{aligned}(n_B U^\mu)_{;\mu} &= g^{-\frac{1}{2}}(g^{\frac{1}{2}}n_B U^\nu)_{,\nu} \\ &= (n_B U^t)_{,t} + \frac{1}{r^2}(r^2 n_B U^r)_{,r} = 0 .\end{aligned} \qquad (16)$$

The radial component of the energy-momentum conservation law of the plasma fluid reduces to

$$\frac{\partial p}{\partial r} + \frac{\partial}{\partial t}\left((p+\rho)U^t U_r\right) + \frac{1}{r^2}\frac{\partial}{\partial r}\left(r^2(p+\rho)U^r U_r\right) - \frac{1}{2}(p+\rho)\left[\frac{\partial g_{tt}}{\partial r}(U^t)^2 + \frac{\partial g_{rr}}{\partial r}(U^r)^2\right] = 0 . \qquad (17)$$

The component of the energy-momentum conservation law of the plasma fluid equation along a flow line is

$$\begin{aligned}U_\mu(T^{\mu\nu})_{;\nu} &= -(\rho U^\nu)_{;\nu} - p(U^\nu)_{;\nu}, \\ &= -g^{-\frac{1}{2}}(g^{\frac{1}{2}}\rho U^\nu)_{,\nu} - pg^{-\frac{1}{2}}(g^{\frac{1}{2}}U^\nu)_{,\nu} \\ &= (\rho U^t)_{,t} + \frac{1}{r^2}(r^2\rho U^r)_{,r} \\ &\quad + p\left[(U^t)_{,t} + \frac{1}{r^2}(r^2 U^r)_{,r}\right] = 0 .\end{aligned} \qquad (18)$$

Defining the total proper internal energy density ϵ and the baryonic mass density ρ_B in the comoving frame of the plasma fluid,

$$\epsilon \equiv \rho - \rho_B, \qquad \rho_B \equiv n_B mc^2 , \qquad (19)$$

and using the law (16) of baryon-number conservation, from Eq. (18) we have

$$(\epsilon U^\nu)_{;\nu} + p(U^\nu)_{;\nu} = 0 .\qquad (20)$$

Recalling that $\frac{dV}{d\tau} = V(U^\mu)_{;\mu}$, where V is the comoving volume and τ is the proper time for the plasma fluid, we have along each flow line

$$\frac{d(V\epsilon)}{d\tau} + p\frac{dV}{d\tau} = \frac{dE}{d\tau} + p\frac{dV}{d\tau} = 0 ,\qquad (21)$$

where $E = V\epsilon$ is the total proper internal energy of the plasma fluid. We express the equation of state by introducing a thermal index $\Gamma(\rho, T)$

$$\Gamma = 1 + \frac{p}{\epsilon} .\qquad (22)$$

We now turn to the second set of governing equations describing the evolution of the e^+e^- pairs. Letting n_{e^-} and n_{e^+} be the proper number densities of electrons and positrons associated with pairs and $n^b_{e^-}$ the proper number densities of ionized electrons, we clearly have

$$n_{e^-} = n_{e^+} = n_{\text{pair}}, \qquad n^b_{e^-} = \bar{Z} n_B,\qquad (23)$$

where n_{pair} is the number of e^+e^- pairs and $\bar{Z}$ the average atomic number $\frac{1}{2} < \bar{Z} < 1$ ($\bar{Z} = 1$ for hydrogen atom and $\bar{Z} = \frac{1}{2}$ for general baryonic matter). The rate equation for electrons and positrons gives,

$$\begin{aligned}(n_{e^+} U^\mu)_{;\mu} &= (n_{e^+} U^t)_{,t} + \frac{1}{r^2}(r^2 n_{e^+} U^r)_{,r} \\ &= \overline{\sigma v}\big[(n_{e^-}(T) + n^b_{e^-}(T)) n_{e^+}(T) \\ &\quad - (n_{e^-} + n^b_{e^-}) n_{e^+}\big],\end{aligned}\qquad (24)$$

$$\begin{aligned}(n_{e^-} U^\mu)_{;\mu} &= (n_{e^-} U^t)_{,t} + \frac{1}{r^2}(r^2 n_{e^-} U^r)_{,r} \\ &= \overline{\sigma v}\left[n_{e^-}(T) n_{e^+}(T) - n_{e^-} n_{e^+}\right],\end{aligned}\qquad (25)$$

$$\begin{aligned}(n^b_{e^-} U^\mu)_{;\mu} &= (n^b_{e^-} U^t)_{,t} + \frac{1}{r^2}(r^2 n^b_{e^-} U^r)_{,r} \\ &= \overline{\sigma v}\left[n^b_{e^-}(T) n_{e^+}(T) - n^b_{e^-} n_{e^+}\right],\end{aligned}\qquad (26)$$

where $\overline{\sigma v}$ is the mean of the product of the annihilation cross-section and the thermal velocity of the electrons and positrons, $n_{e^\pm}(T)$ are the proper number densities of electrons and positrons associated with the pairs, given by appropriate Fermi integrals with zero chemical potential, and $n^b_{e^-}(T)$ is the proper number density of ionized electrons, given by appropriate Fermi integrals with non-zero chemical potential μ_e at an appropriate equilibrium temperature T. These rate equations can be reduced to

$$\begin{aligned}(n_{e^\pm} U^\mu)_{;\mu} &= (n_{e^\pm} U^t)_{,t} + \frac{1}{r^2}(r^2 n_{e^\pm} U^r)_{,r} \\ &= \overline{\sigma v}\left[n_{e^-}(T) n_{e^+}(T) - n_{e^-} n_{e^+}\right],\end{aligned}\qquad (27)$$

$$(n^b_{e^-} U^\mu)_{;\mu} = (n^b_{e^-} U^t)_{,t} + \frac{1}{r^2}(r^2 n^b_{e^-} U^r)_{,r} = 0,\qquad (28)$$

$$Frac \equiv \frac{n_{e^\pm}}{n_{e^\pm}(T)} = \frac{n^b_{e^-}(T)}{n^b_{e^-}}.\qquad (29)$$

Equation (28) is just the baryon-number conservation law (16) and (29) is a relationship satisfied by $n_{e^\pm}, n_{e^\pm}(T)$ and $n^b_{e^-}, n^b_{e^-}(T)$.

The equilibrium temperature T is determined by the thermalization processes occurring in the expanding plasma fluid with a total proper energy density ρ governed by the hydrodynamical equations (16,17,18). We have

$$\rho = \rho_\gamma + \rho_{e^+} + \rho_{e^-} + \rho^b_{e^-} + \rho_B,\qquad (30)$$

where ρ_γ is the photon energy density, $\rho_B \simeq m_B c^2 n_B$ is the baryonic mass density which is considered to be non-relativistic in the range of temperature T under consideration, and $\rho_{e^\pm}$ is the proper energy density of electrons and positrons pairs given by

$$\rho_{e^\pm} = \frac{n_{e^\pm}}{n_{e^\pm}(T)} \rho_{e^\pm}(T),\qquad (31)$$

where $n_{e\pm}$ is obtained by integration of Eq.(27) and $\rho_{e\pm}(T)$ is the proper energy density of electrons(positrons) obtained from zero chemical potential Fermi integrals at the equilibrium temperature T. On the other hand ρ_{e-}^b is the energy density of the ionized electrons coming from the ionization of baryonic matter

$$\rho_{e-}^b = \frac{n_{e-}^b}{n_{e-}^b(T)} \rho_{e-}^b(T), \qquad (32)$$

where n_{e-}^b is obtained by integration of Eq.(28) and $\rho_{e-}(T)$ is the proper energy density of ionized electrons obtained from an appropriate Fermi integral of non-zero chemical potential μ_e at the equilibrium temperature T.

Having intrinsically defined the equilibrium temperature T in Eq.(30), we can also analogously evaluate the total pressure

$$p = p_\gamma + p_{e+} + p_{e-} + p_{e-}^b + p_B, \qquad (33)$$

where p_γ is the photon pressure, $p_{e\pm}$ and p_{e-}^b are given by

$$p_{e\pm} = \frac{n_{e\pm}}{n_{e\pm}(T)} p_{e\pm}(T), \qquad (34)$$

$$p_{e-}^b = \frac{n_{e-}^b}{n_{e-}^b(T)} p_{e-}^b(T), \qquad (35)$$

the pressures $p_{e\pm}(T)$ are determined by zero chemical potential Fermi integrals, and $p_{e-}^b(T)$ is the pressure of the ionized electrons, evaluated by an appropriate Fermi integral of non-zero chemical potential μ_e at the equilibrium temperature T. In Eq.(33), the ion pressure p_B is negligible by comparison with the pressures $p_{\gamma,e\pm,e-}(T)$, since baryons and ions are expected to be nonrelativistic in the range of temperature T under consideration. Finally using Eqs.(30,33) we compute the thermal factor Γ of the equation of state (22).

It is clear that the entire set of equations considered above, namely Eqs.(16,17,18) with equation of state given by Eq.(22) and the rate equation (27), have to be integrated satisfying the total energy conservation for the system. The boundary conditions adopted here are simply purely ingoing conditions at the horizon and purely outgoing conditions at radial infinity. The calculation is initiated by depositing a proper energy density (11) between the Reissner-Nordström horizon radius r_+ and the dyadosphere radius r_{ds}, following the approximation presented in Fig.16 The total energy deposited is given by Eq.(12).

V. THE EQUATIONS LEADING TO THE RELATIVE SPACE-TIME TRANSFORMATIONS

In order to relate the above hydrodynamic and pair equations with the observations we need the governing equations relating the comoving time to the laboratory time corresponding to an inertial reference frame in which the EMBH is at rest and finally to the time measured at the detector, which must also include the effect of the cosmological expansion. These transformations have been the object of the Relative space-time Transformations (RSTT) Paradigm, (Ruffini et al. [143]).

For signals emitted by a pulse moving with velocity v in the laboratory frame (see also Ruffini et al. [143]), we have the following relation between the interval of arrival time Δt_a and the corresponding interval of laboratory time Δt (see Fig. 18):

$$\Delta t_a = \left(t_0 + \Delta t + \frac{R_0 - r}{c}\right) - \left(t_0 + \frac{R_0}{c}\right) = \Delta t - \frac{r}{c}. \qquad (36)$$

For simplicity in what follows we indicate by t_a the interval of arrival time measured from the reception of a light signal emitted at the onset of the gravitational collapse. Analogously, t indicates the laboratory time interval measured from the time of the gravitational collapse. In this case, Eq.(36) can be written simply as:

$$t_a = t - \frac{r}{c} = t - \frac{\int_0^t v(t')\,dt' + r_{ds}}{c} = t - \int_0^t \frac{\sqrt{\gamma^2(t') - 1}}{\gamma(t')}dt' - \frac{r_{ds}}{c}, \qquad (37)$$

where, as usual, $\gamma(t') = 1/\sqrt{1 - v^2(t')/c^2}$ and the dyadosphere radius r_{ds} is the value of r at $t = 0$. It is important to stress that, although there is the presence of the Lorentz gamma factor, Eq.(37) is not a Lorentz transformation,

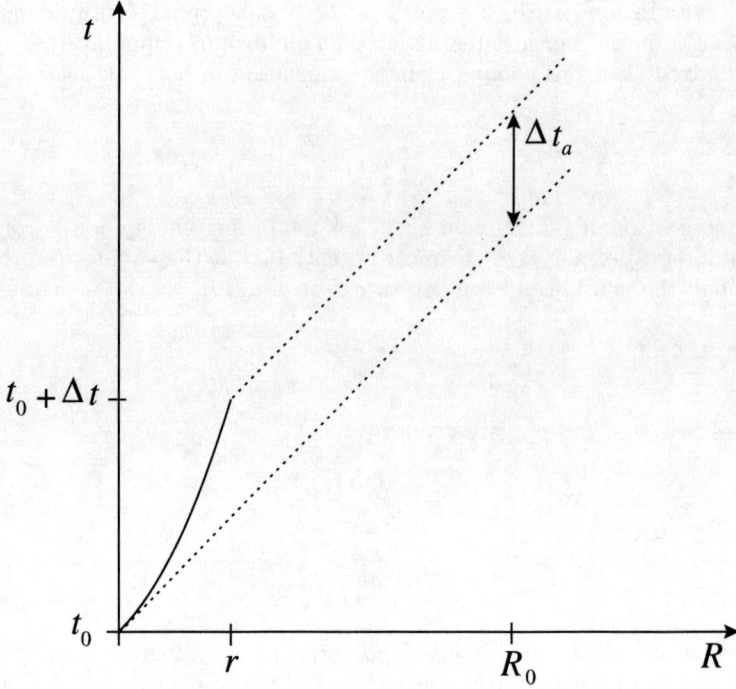

Figure 18: This qualitative diagram illustrates the relation between the laboratory time interval Δt and the arrival time interval Δt_a for a pulse moving with velocity v in the laboratory time (solid line). We have indicated here the case where the motion of the source has a nonzero acceleration. The arrival time is measured using light signals emitted by the pulse (dotted lines). R_0 is the distance of the observer from the EMBH, t_0 is the laboratory time corresponding to the onset of the gravitational collapse, and r is the radius of the expanding pulse at a time $t = t_0 + \Delta t$. See also Ruffini et al. [143].

which by its own nature is linear and refers to a specific value of the Lorentz gamma factor at a given laboratory time. The transformation in Eq.(37) is nonlinear in the Lorentz gamma factor and do depend on all the values of the gamma factor of the source from the time $t = 0$ to the laboratory time t. This transformation is the price to pay to relate the laboratory time t, relativistically correct, to the "highly pathological" time usually considered by the astronomers, even in the case of object moving close to the speed of light, against the correct synchronization procedures established by Einstein in his classical paper of 1905 (Einstein [46]). We consider here only the photons emitted along the line of sight from the external surface of the pulse. The arrival time spreading due to the angular dependence and that due to the thickness has also been given (see section XXI and Ruffini et al. [147, 150]). The solution of Eq.(37) has the expansion:

$$t_a = t - \frac{r_{ds}}{c} - \frac{v(0)}{c}t - \frac{1}{2}\frac{v'(0)}{c}t^2 - \frac{1}{3}\frac{v''(0)}{c}t^3 - \ldots, \tag{38}$$

so the relation between t_a and t in the specific case of GRBs is very highly nonlinear: it is sufficient to recall that in the early GRB phases we are witnessing the strongest acceleration ever recorded in the universe, since the PEM pulse goes from Lorentz factor $\gamma = 1$ to Lorentz factor $\gamma = 1000$ in 10^2 seconds in the laboratory time (see section VII). The series in Eq.(38) will definitely converge, but the number of terms needed to reach a good approximation will strongly depend on the variability of the functions around the initial values $\gamma = 1$. It is clear that the precise knowledge of t_a as a function of the laboratory time, which is indeed essential for any physical interpretation of GRB data, depends on the definite integral given in Eq.(37) whose limits in the laboratory time extend from the onset of the gravitational collapse to the time t relevant for the observations. Such an integral depends on all previous values of the Lorentz gamma factor in the history of the source and is not generally expressible by a simple linear relation or even by any explicit analytic relation since we are dealing with processes with variable gamma factor unprecedented in the entire realm of physics (see Figs. 8 and Fig. 9). This is the crucial point of the RSTT paradigm (Ruffini et al. [143]) and this is the reason why we have spent a very large amount of work to develop the exact equations of motion of all different eras of the GRB phenomenon, starting from the onset of gravitational collapse and the creation of dyadosphere (see the following sections). It is clear then that, in order to express the arrival time t_a and the radial coordinate of the source at the start of the afterglow phase, we need the explicit knowledge of all the previous eras of the GRB phenomenon, starting from $\gamma = 1$ (Ruffini et al. [143]).

What has been currently done in the literature, is an extremely different approach. First they have assumed γ constant. Therefore Eq.(37) has been modified in:

$$t_a = t - \frac{\sqrt{\gamma^2 - 1}}{\gamma} \int_0^t dt' - \frac{r_{ds}}{c} \simeq t - \frac{\sqrt{\gamma^2 - 1}}{\gamma} t, \qquad (39)$$

where in the last approximation the contribution of the initial size of the source has been neglected. Even the validity of this last approximation has to be actually carefully verified since it is only valid in the late phases of the GRB expansion. They have further assumed $\gamma \gg 1$ and obtained:

$$t_a \simeq t - \left(1 - \frac{1}{2\gamma^2}\right) t = \frac{t}{2\gamma^2}. \qquad (40)$$

At this stage, they emphasize the existence of a linear relation between the arrival time t_a and the laboratory time t. After this they proceed in two different directions. One to assume (see e.g. Fenimore et al. [47], Fenimore [48], Fenimore et al. [49], Sari & Piran [164], Waxman [185])

$$t_a = t / \left(2\gamma^2(t)\right), \qquad (41)$$

concurrently advancing the belief that the relation between the arrival time and the laboratory time does not depend from an integral on all the previous values of the gamma Lorentz factor of the source but from the instantaneous value of the gamma Lorentz factor at the time t, much like in a Lorentz transformation. This claim is clearly absurd from a physical point of view.

They further assume (see e.g. Panaitescu & Mészáros [112], Piran [115], Sari [162, 163] and references therein)

$$\delta t_a = \delta t / \left(2\gamma^2(t)\right) \text{ or, alternatively}, dt_a = dt / \left(2\gamma^2(t)\right), \qquad (42)$$

and they proceed to develop all the observable quantities of the GRB phenomenon by integrating using the "differential" given in Eq.(42), reaching clearly meaningless results. As we show later, this also leads to the unfortunate attempt to obtain the gamma Lorentz factor and its time variability from the astrophysical data of the afterglow, neglecting all previous GRB source history what is clearly physically and astrophysically impossible.

Having established the correct relations between the laboratory time t and the arrival time t_a in Eq.(37), we now proceed to relate the time in the laboratory frame t to the time in the detector frame t_a^d. We have to do one additional step: the two frames are related by a transformation which is a function of the cosmological expansion. We recall that the geometry of the space-time of the universe is described by the Robertson-Walker metric:

$$ds^2 = dt^2 - \mathcal{R}^2(t) \left(\frac{dr^2}{1 - kr^2} + r^2 d\vartheta^2 + r^2 \sin\vartheta^2 d\varphi^2\right), \qquad (43)$$

where $\mathcal{R}(t)$ is the cosmic scale factor and k is a constant related to the curvature of the three-dimensional space ($k = 0, +1, -1$ corresponds to flat, close and open space respectively). The wavelength of an electromagnetic wave traveling from the point $P_1(t_1, r_1, \vartheta_1, \varphi_1)$ to the point $P_o(t_o, r_o, \vartheta_o, \varphi_o)$ where the observer is located is related to the red-shift parameter z by

$$z = \frac{\lambda_o - \lambda_1}{\lambda_1}, \qquad (44)$$

where λ_o is the wavelength of the radiation for the observer and λ_1 for the emitter. We have the following general relation:

$$1 + z = (1 + z_u)(1 + z_o)(1 + z_s), \qquad (45)$$

where z is the total redshift due to the motion of the source z_s, the motion of the observer z_o and the cosmological redshift z_u. In the following we will assume $z_o \ll 1$ and $z_s \ll 1$ so $z = z_u$. In terms of the scale factor $\mathcal{R}(t)$ the relation (44) gives

$$\frac{\lambda_o}{\lambda_1} = \frac{\mathcal{R}(t_o)}{\mathcal{R}(t_1)} = 1 + z = \frac{\omega_1}{\omega_0} \qquad (46)$$

where ω_1 and ω_0 are the frequencies associated to λ_1 and λ_0 respectively. This frequency ratio then relates the time elapsing at the source with the time elapsing at the detector due to the cosmological expansion.

We can now define the corrected arrival time t_a^d measured at the detector, which is related to t_a, clearly defined by Eq.(37), by

$$t_a^d = t_a\,(1+z),\qquad(47)$$

where z is the cosmological redshift of the GRB source. In the case of GRB 991216 we have $z \simeq 1.00$.

The observed flux is the flux which crosses the surface $4\pi(\mathcal{R}(t_o)r)^2$ but this flux is lower by a factor $1+z$ due to the redshift energy of the photons and by another factor $1+z$ due to the fact that the number of photons at reception is less than the number at emission. Thus we can define a luminosity distance by:

$$d_L^2 = \mathcal{R}_o^2 r^2 (1+z)^2.\qquad(48)$$

Then the observed flux is related to the absolute luminosity of the GRB by the following relation:

$$l = \frac{L}{4\pi d_L^2},\qquad(49)$$

where the luminosity distance d_L is simply related to the proper distance $d_p = \mathcal{R}_o r$ by $d_L = d_p(1+z)$. The observed total fluence f is related to the total energy E of the GRB by the following relation:

$$f = \frac{E(1+z)}{4\pi d_L^2}\qquad(50)$$

Then the cosmological effect is taken into account by the definition of the proper distance $\mathcal{R}_o r$ which depends on the cosmological parameters: the Hubble constant $H_\circ = \dot{\mathcal{R}}(t_\circ)/\mathcal{R}(t_\circ)$ at time $t_\circ$ and the matter density $\rho_\circ$ or $\Omega_M = \rho_\circ/\rho_{crit}$, where $\rho_{crit} = \frac{3H_\circ^2}{8\pi G}$.

The computation of the proper distance is then simply given by the relation :

$$d_p = \frac{c}{H_o}\int_0^z \frac{dz}{F(z)},\qquad(51)$$

where $F(z) = \sqrt{\Omega_M(1+z)^3}$.

In the case of the Friedman flat universe, $\Omega_M = 1$ and we have:

$$d_p(z) = \frac{2c}{H_o}\left[1 - \frac{1}{\sqrt{1+z}}\right].\qquad(52)$$

So the measurement of the redshift gives us the luminosity distance via a cosmological scenario. With the measurement of the flux we can deduce the proper luminosity of the burst and from the measurement of the total fluence the total energy so we are then able to find the E_{dya}.

VI. THE NUMERICAL INTEGRATION OF THE HYDRODYNAMICS AND THE RATE EQUATIONS

A. The Livermore code

A computer code (Wilson et al. [187, 188]) has been used to evolve the spherically symmetric general relativistic hydrodynamic equations starting from the dyadosphere (Ruffini et al. [141]).

We define the generalized gamma factor γ and the radial 3-velocity in the laboratory frame V^r

$$\gamma \equiv \sqrt{1+U^r U_r},\qquad V^r \equiv \frac{U^r}{U^t}.\qquad(53)$$

From Eqs.(5, 15), we then have

$$(U^t)^2 = -\frac{1}{g_{tt}}(1+g_{rr}(U^r)^2) = \frac{1}{\alpha^2}\gamma^2.\qquad(54)$$

Following Eq.(19), we also define

$$E \equiv \epsilon\gamma,\qquad D \equiv \rho_B\gamma,\quad \text{and}\quad \tilde{\rho} \equiv \rho\gamma\qquad(55)$$

so that the conservation law of baryon number (16) can then be written as

$$\frac{\partial D}{\partial t} = -\frac{\alpha}{r^2}\frac{\partial}{\partial r}(\frac{r^2}{\alpha}DV^r). \tag{56}$$

Eq.(18) then takes the form,

$$\frac{\partial E}{\partial t} = -\frac{\alpha}{r^2}\frac{\partial}{\partial r}(\frac{r^2}{\alpha}EV^r) - p\left[\frac{\partial \gamma}{\partial t} + \frac{\alpha}{r^2}\frac{\partial}{\partial r}(\frac{r^2}{\alpha}\gamma V^r)\right]. \tag{57}$$

Defining the radial momentum density in the laboratory frame

$$S_r \equiv \alpha(p+\rho)U^t U_r = (D+\Gamma E)U_r, \tag{58}$$

we can express the radial component of the energy-momentum conservation law given in Eq.(17) by

$$\begin{aligned}
\frac{\partial S_r}{\partial t} &= -\frac{\alpha}{r^2}\frac{\partial}{\partial r}(\frac{r^2}{\alpha}S_r V^r) - \alpha\frac{\partial p}{\partial r} \\
&\quad - \frac{\alpha}{2}(p+\rho)\left[\frac{\partial g_{tt}}{\partial r}(U^t)^2 + \frac{\partial g_{rr}}{\partial r}(U^r)^2\right] \\
&= -\frac{\alpha}{r^2}\frac{\partial}{\partial r}(\frac{r^2}{\alpha}S_r V^r) - \alpha\frac{\partial p}{\partial r} \\
&\quad - \alpha\left(\frac{M}{r^2} - \frac{Q^2}{r^3}\right)\left(\frac{D+\Gamma E}{\gamma}\right)\left[\left(\frac{\gamma}{\alpha}\right)^2 + \frac{(U^r)^2}{\alpha^4}\right].
\end{aligned} \tag{59}$$

In order to determine the number-density of e^+e^- pairs, we turn to Eq.(27). Defining the e^+e^--pair density in the laboratory frame $N_{e^\pm} \equiv \gamma n_{e^\pm}$ and $N_{e^\pm}(T) \equiv \gamma n_{e^\pm}(T)$, where the equilibrium temperature T has been obtained from Eqs.(30) and (31), and using Eq.(54), we rewrite the rate equation given by Eq.(27) in the form

$$\frac{\partial N_{e^\pm}}{\partial t} = -\frac{\alpha}{r^2}\frac{\partial}{\partial r}(\frac{r^2}{\alpha}N_{e^\pm}V^r) + \overline{\sigma v}(N_{e^\pm}^2(T) - N_{e^\pm}^2)/\gamma^2, \tag{60}$$

These equations are integrated starting from the dyadosphere distributions given in Fig. 17 and assuming as usual ingoing boundary conditions on the horizon of the EMBH.

B. The Rome code

In the following we recall a zeroth order approximation of the fully relativistic equations of the previous section (Ruffini et al. [141]):
(i) Since we are mainly interested in the expansion of the e^+e^- plasma away from the EMBH, we neglect the gravitational interaction.
(ii) We describe the expanding plasma by a special relativistic set of equations.
(iii) In contrast with the previous treatment where the evolution of the density profiles given in Fig. 17 are followed in their temporal evolution leading to a pulse-like structure, selected geometries of the pulse are a priori adopted and the correct one validated by the complete integration of the equations given by the Livermore codes.

Analogously to Eq.(21), from Eq.(16) we have along each flow line in the general case in which baryonic matter is present

$$\frac{d(n_B \mathcal{V})}{d\tau} = 0. \tag{61}$$

For the expansion of a shell from its initial volume $\Delta \mathcal{V}_\circ$ to the volume $\Delta \mathcal{V}$, we obtain

$$\frac{n_B^\circ}{n_B} = \frac{\Delta \mathcal{V}}{\Delta \mathcal{V}_\circ} = \frac{\Delta V \gamma(r)}{\Delta V_\circ \gamma_\circ(r)}, \tag{62}$$

where ΔV is the volume of the shell in the laboratory frame, related to the proper volume $\Delta \mathcal{V}$ in the comoving frame by $\Delta V = \gamma(r)\Delta \mathcal{V}$, where $\gamma(r)$ defined in Eq.(53) is the gamma factor of the shell at the radius r.

Similarly from Eq.(21), using the equation of state (22), along the flow lines we obtain

$$d\ln\epsilon + \Gamma d\ln V = 0. \tag{63}$$

Correspondingly we obtain for the internal energy density ϵ along the flow lines

$$\frac{\epsilon_\circ}{\epsilon} = \left(\frac{\Delta V}{\Delta V_\circ}\right)^\Gamma = \left(\frac{\Delta \mathcal{V}}{\Delta \mathcal{V}_\circ}\right)^\Gamma \left(\frac{\gamma(r)}{\gamma_\circ(r)}\right)^\Gamma, \tag{64}$$

where the thermal index Γ given by (22) is a slowly-varying function with values around 4/3. It can be computed for each value of ϵ, p as a function of ΔV.

The overall energy conservation requires that the change of the internal proper energy of a shell is compensated by a change in its bulk kinetic energy. We then have (Ruffini et al. [141])

$$dK = [\gamma(r) - 1](dE + \rho_B dV). \tag{65}$$

In order to model the relativistic expansion of the plasma fluid, we assume that E and D as defined by Eq.(55) are constant in space over the volume ΔV. As a consequence the total energy conservation for the shell implies (Ruffini et al. [141])

$$(\epsilon_\circ + \rho_B^\circ)\gamma_\circ^2(r)\Delta \mathcal{V}_\circ = (\epsilon + \rho_B)\gamma^2(r)\Delta \mathcal{V}, \tag{66}$$

which leads the solution

$$\gamma(r) = \gamma_\circ(r)\sqrt{\frac{(\epsilon_\circ + \rho_B^\circ)\Delta \mathcal{V}_\circ}{(\epsilon + \rho_B)\Delta \mathcal{V}}}. \tag{67}$$

Corresponding to Eq.(60) we obtain the equation for the evolution of the $e^\pm$ number-density as seen by an observer in the laboratory frame

$$\frac{\partial}{\partial t}(N_{e^\pm}) = -N_{e^\pm}\frac{1}{\Delta \mathcal{V}}\frac{\partial \Delta \mathcal{V}}{\partial t} + \overline{\sigma v}\frac{1}{\gamma^2(r)}(N_{e^\pm}^2(T) - N_{e^\pm}^2). \tag{68}$$

Eqs.(62), (64), (67) and (68) are a complete set of equations describing the relativistic expansion of the shell. If we now turn from a single shell to a finite distribution of shells, we can introduce the average values of the proper internal-energy, baryon-mass, baryon-number and pair-number densities ($\bar\epsilon, \bar\rho_B, \bar n_B, \bar n_{e^\pm}$) and $\bar E \equiv \bar\gamma\bar\epsilon$, $\bar D \equiv \bar\gamma\bar\rho_B$, $\bar N_{e^\pm} \equiv \bar\gamma(r)\bar n_{e^\pm}$ for the PEM-pulse, where the average $\bar\gamma$-factor is defined by

$$\bar\gamma = \frac{1}{\mathcal{V}}\int_\mathcal{V}\gamma(r)d\mathcal{V}, \tag{69}$$

and $\mathcal{V}$ is the total volume of the shell in the laboratory frame. The corresponding equations are given in Ruffini et al. [141]. Having defined all its governing equations we can now return to the description of the different eras of the GRB phenomena.

VII. THE ERA I: THE PEM PULSE

We have assumed that, following the gravitational collapse process, a region of very low baryonic contamination exists in the dyadosphere all the way to the remnant of the progenitor star.

Recalling Eq.(9) the limit on such baryonic contamination, where ρ_{B_c} is the mass-energy density of baryons, is given by

$$\rho_{B_c} \ll m_p n_{e^+e^-}(r) = 3.2 \cdot 10^8 \left(\frac{r_{ds}}{r}\right)^2 \left[1 - \left(\frac{r}{r_{ds}}\right)^2\right] (g/cm^3). \tag{70}$$

Near the horizon $r \simeq r_+$, this gives

$$\rho_{B_c} \ll m_p n_{e^+e^-}(r) = 1.86 \cdot 10^{14}\left(\frac{\xi}{\mu}\right)(g/cm^3), \tag{71}$$

and near the radius of the dyadosphere r_{ds}:

$$\rho_{B_c} \ll m_p n_{e^+e^-}(r) = 3.2 \cdot 10^8 \left[1 - \left(\frac{r}{r_{ds}}\right)^2\right]_{r \to r_{ds}} (g/cm^3). \tag{72}$$

Such conditions can be easily satisfied in the collapse to an EMBH, but not necessarily in a collapse to a neutron star.

Consequently we have solved the equations governing a plasma composed solely of e^+e^--pairs and electromagnetic radiation, starting at time zero from the dyadosphere configurations corresponding to constant density in Fig. 17. The Livermore code (Ruffini et al. [141]) has shown very clearly the self organization of the expanding plasma in a very sharp pulse which we have defined as the pair-electromagnetic pulse (PEM pulse), in analogy with the EM pulse observed in nuclear explosions. In order to further examine the structure of the PEM pulse with the simpler procedures of the Rome codes we have assumed (Ruffini et al. [141]) three alternative patterns of expansion of the PEM pulse on which to try the simplified special relativistic treatment and then compared the results with the fully general relativistic hydrodynamical results:

- Spherical model: we assume the radial component of the four-velocity $U_r(r) = U\frac{r}{\mathcal{R}}$, where U is the radial component of the four-velocity at the moving outer surface $r = \mathcal{R}(t)$ of the PEM pulse and the $\bar{\gamma}$-factor and the velocity V_r are

$$\begin{aligned}\bar{\gamma} &= \frac{3}{8U^3}\left[2U(1+U^2)^{\frac{3}{2}} - U(1+U^2)^{\frac{1}{2}}\right.\\ &\quad\left. - \ln(U+\sqrt{1+U^2})\right], \quad V_r = \frac{U_r}{\bar{\gamma}};\end{aligned} \tag{73}$$

this distribution expands keeping an uniform density profile which decreases with time similar to a portion of a Friedmann Universe.

- Slab 1: we assume $U(r) = U_r = $ const., the constant width of the expanding slab $\mathcal{D} = R_\circ$ in the laboratory frame of the PEM pulse, while $\bar{\gamma}$ and V_r are

$$\bar{\gamma} = \sqrt{1+U_r^2}, \quad V_r = \frac{U_r}{\bar{\gamma}}; \tag{74}$$

this distribution does not need any averaging process.

- Slab 2: we assume a constant width $R_2 - R_1 = R_\circ$ of the expanding slab in the comoving frame of the PEM pulse, while $\bar{\gamma}$ and V_r are

$$\bar{\gamma} = \sqrt{1+U_r^2(\tilde{r})}, \quad V_r = \frac{U_r}{\bar{\gamma}}, \tag{75}$$

This distribution needs an averaging procedure and $R_1 < \tilde{r} < R_2$, i.e. $\tilde{r}$ is an intermediate radius in the slab.

These different assumptions lead to three different distinct slopes for the monotonically increasing $\bar{\gamma}$-factor as a function of the radius (or time) in the laboratory frame, having assumed for the energy of dyadosphere $E_{dya} = 3.1 \times 10^{54}$ erg (see Fig. 19). In principle, we could have an infinite number of models by defining arbitrarily the geometry of the expanding fluid in the special relativistic treatment given above. To find out which expanding pattern of PEM pulses is the physically realistic one, we need to compare and contrast the results of our simplified models (performed in Rome) with the numerical results based on the hydrodynamic Eqs.(56,57,59) (obtained at Livermore) (Ruffini et al. [141]). Details of the iterative method used to solve the special relativistic equation can be found in Ruffini et al. [141].

It is manifest from the results (see Fig. 19) that the slab 1 approximation (constant thickness in the laboratory frame) is in excellent agreement with the Livermore results (open squares).

The remarkable validation of the special relativistic treatment of the PEM pulse (Ruffini et al. [141]), allows us to easily estimate the related quantities of physical and astrophysical interest in the model, like the e^+e^--pair densities as a function of the laboratory time, the temperature of the plasma in the comoving and laboratory frames, the reheating ratio as a function of the e^+e^--pair annihilation for a variety of initial conditions (Ruffini et al. [141]).

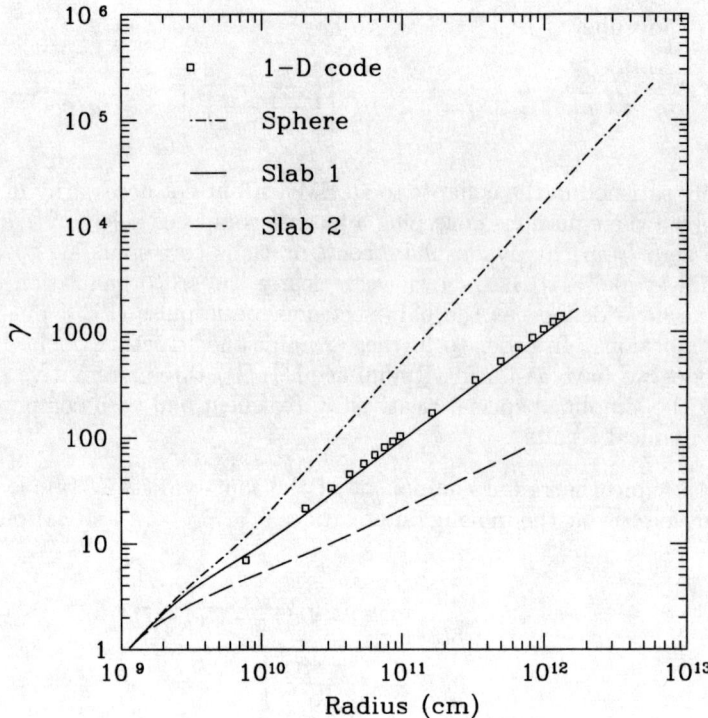

Figure 19: gamma factor as a function of radius. Three models for the expansion pattern of the PEM-pulse are compared with the results of the one dimensional hydrodynamic code for an energy of dyadosphere $E_{dya} = 3.1 \times 10^{54}$ erg. The 1-D code has an expansion pattern that strongly resembles that of a shell with constant thickness in the laboratory frame.

VIII. THE ERA II: THE INTERACTION OF THE PEM PULSE WITH THE REMNANT OF THE PROGENITOR STAR

The PEM pulse expands initially in a region of very low baryonic contamination created by the process of gravitational collapse. As it moves further out the baryonic remnant (see Fig. 1) of the progenitor star is encountered. As discussed in section XXVI below, the existence of such a remnant is necessary in order to guarantee the overall charge neutrality of the system: the collapsing core has the opposite charge of the remnant and the system as a whole is clearly neutral. The number of extra charges in the baryonic remnant negligibly affects the overall charge neutrality of the PEM pulse (Ruffini [138], Ruffini et al. [158]).

The baryonic matter remnant is assumed to be distributed well outside the dyadosphere in a shell of thickness Δ between an inner radius $r_{\rm in}$ and an outer radius $r_{\rm out} = r_{\rm in} + \Delta$ at a distance from the EMBH at which the original PEM pulse expanding in vacuum has not yet reached transparency. For the sake of an example we choose

$$r_{\rm in} = 100 r_{\rm ds}, \quad \Delta = 10 r_{\rm ds}. \tag{76}$$

The total baryonic mass $M_B = N_B m_p$ is assumed to be a fraction of the dyadosphere initial total energy ($E_{\rm dya}$). The total baryon-number N_B is then expressed as a function of the dimensionless parameter B given by

$$B = \frac{N_B m_p c^2}{E_{\rm dya}}, \tag{77}$$

where B is a parameter in the range $10^{-8} - 10^{-2}$ and m_p is the proton mass. We shall see below the paramount importance of B in the determination of the features of the GRBs. We will see in section X the sense in which B and E_{dya} can be considered to be the only two free parameters of the EMBH theory for the entire GRB family, the so called "long bursts". We shall see in section XII that for the so called "short bursts" the EMBH theory depends on the two other parameters μ, ξ, since in that case $B = 0$. The baryon number density n_B° is assumed to be a constant

$$\bar{n}_B^\circ = \frac{N_B}{V_B}, \quad \bar{\rho}_B^\circ = m_p \bar{n}_B^\circ c^2. \tag{78}$$

As the PEM pulse reaches the region $r_{in} < r < r_{out}$, it interacts with the baryonic matter which is assumed to be at rest. In our simplified quasi-analytic model we make the following assumptions to describe this interaction:

- the PEM pulse does not change its geometry during the interaction;
- the collision between the PEM pulse and the baryonic matter is assumed to be inelastic,
- the baryonic matter reaches thermal equilibrium with the photons and pairs of the PEM pulse.

These assumptions are valid if: (i) the total energy of the PEM pulse is much larger than the total mass-energy of baryonic matter M_B, $10^{-8} < B < 10^{-2}$, (ii) the ratio of the comoving number density of pairs and baryons at the moment of collision $n_{e^+e^-}/n_B^\circ$ is very high (e.g., $10^6 < n_{e^+e^-}/n_B^\circ < 10^{12}$) and (iii) the PEM pulse has a large value of the gamma factor ($100 < \bar\gamma$).

In the collision between the PEM pulse and the baryonic matter at $r_{out} > r > r_{in}$, we impose total conservation of energy and momentum. We consider the collision process between two radii r_2, r_1 satisfying $r_{out} > r_2 > r_1 > r_{in}$ and $r_2 - r_1 \ll \Delta$. The amount of baryonic mass acquired by the PEM pulse is

$$\Delta M = \frac{M_B}{V_B}\frac{4\pi}{3}(r_2^3 - r_1^3), \tag{79}$$

where M_B/V_B is the mean-density of baryonic matter at rest. The conservation of total energy leads to the estimate of the corresponding quantities before (with "$\circ$") and after such a collision

$$(\Gamma\bar\epsilon_\circ + \bar\rho_B^\circ)\bar\gamma_\circ^2 \mathcal{V}_\circ + \Delta M = (\Gamma\bar\epsilon + \bar\rho_B + \frac{\Delta M}{\mathcal{V}} + \Gamma\Delta\bar\epsilon)\bar\gamma^2 \mathcal{V}, \tag{80}$$

where $\Delta\bar\epsilon$ is the corresponding increase of internal energy due to the collision. Similarly the momentum-conservation gives

$$(\Gamma\bar\epsilon_\circ + \bar\rho_B^\circ)\bar\gamma_\circ U_r^\circ \mathcal{V}_\circ = (\Gamma\bar\epsilon + \bar\rho_B + \frac{\Delta M}{\mathcal{V}} + \Gamma\Delta\bar\epsilon)\bar\gamma U_r \mathcal{V}, \tag{81}$$

where the radial component of the four-velocity of the PEM pulse is $U_r^\circ = \sqrt{\bar\gamma_\circ^2 - 1}$ and Γ is the thermal index. We then find

$$\Delta\bar\epsilon = \frac{1}{\Gamma}\left[(\Gamma\bar\epsilon_\circ + \bar\rho_B^\circ)\frac{\bar\gamma_\circ U_r^\circ \mathcal{V}_\circ}{\bar\gamma U_r \mathcal{V}} - (\Gamma\bar\epsilon + \bar\rho_B + \frac{\Delta M}{\mathcal{V}})\right], \tag{82}$$

$$\bar\gamma = \frac{a}{\sqrt{a^2 - 1}}, \qquad a \equiv \frac{\bar\gamma_\circ}{U_r^\circ} + \frac{\Delta M}{(\Gamma\bar\epsilon_\circ + \bar\rho_B^\circ)\bar\gamma_\circ U_r^\circ \mathcal{V}_\circ}. \tag{83}$$

These equations determine the gamma factor $\bar\gamma$ and the internal energy density $\bar\epsilon = \bar\epsilon_\circ + \Delta\bar\epsilon$ in the capture process of baryonic matter by the PEM pulse.

The effect of the collision of the PEM pulse with the remnant leads to the following results (Ruffini et al. [142]) as a function of the B parameter defined in Eq.(77):
1) an abrupt decrease of the gamma factor given by

$$\gamma_{coll} = \gamma_\circ \frac{1 + B}{\sqrt{\gamma_\circ^2(2B + B^2) + 1}}, \tag{84}$$

where $\gamma_\circ$ is the gamma factor of the PEM pulse prior to the collision and B is given by Eq.(77),
2) an increase of the internal energy in the comoving frame E_{coll} developed in the collision given by

$$\frac{E_{coll}}{E_{dya}} = \frac{\sqrt{\gamma_\circ^2(2B + B^2) + 1}}{\gamma_\circ} - \left(\frac{1}{\gamma_\circ} + B\right), \tag{85}$$

3) a corresponding reheating of the plasma in the comoving frame but not in the laboratory frame, an increase of the number of e^+e^- pairs and correspondingly an overall increase of the opacity of the pulse. See details in section XI.

IX. THE ERA III: THE PEMB PULSE

After the engulfment of the baryonic matter of the remnant the plasma formed of e^+e^--pairs, electromagnetic radiation and baryonic matter expands again as a sharp pulse, namely the PEMB pulse. The calculation is continued as the plasma fluid expands, cools and the e^+e^- pairs recombine until it becomes optically thin:

$$\int_R dr(n_{e^\pm} + \bar{Z}n_B)\sigma_T \simeq O(1), \tag{86}$$

where $\sigma_T = 0.665 \cdot 10^{-24} \text{cm}^2$ is the Thomson cross-section and the integration is over the radial interval of the PEMB pulse in the comoving frame. We have first explored the general problem of the PEMB pulse evolution by integrating the general relativistic hydrodynamical equations with the Livermore codes, for a total energy in the dyadosphere of 3.1×10^{54} erg and a baryonic shell of thickness $\Delta = 10r_{\text{ds}}$ at rest at a radius of $100r_{\text{ds}}$ and $B \simeq 1.3 \cdot 10^{-4}$.

In total analogy with the special relativistic treatment for the PEM pulse, presented in section VII (see also Ruffini et al. [141]), we obtain for the adiabatic expansion of the PEMB pulse in the constant-slab approximation described by the Rome codes the following hydrodynamical equations with $\rho_B \neq 0$

$$\frac{\bar{n}_B^\circ}{\bar{n}_B} = \frac{V}{V_\circ} = \frac{\mathcal{V}\bar{\gamma}}{\mathcal{V}_\circ \bar{\gamma}_\circ}, \tag{87}$$

$$\frac{\bar{\epsilon}_\circ}{\bar{\epsilon}} = \left(\frac{V}{V_\circ}\right)^\Gamma = \left(\frac{\mathcal{V}}{\mathcal{V}_\circ}\right)^\Gamma \left(\frac{\bar{\gamma}}{\bar{\gamma}_\circ}\right)^\Gamma, \tag{88}$$

$$\bar{\gamma} = \bar{\gamma}_\circ \sqrt{\frac{(\Gamma\bar{\epsilon}_\circ + \bar{\rho}_B^\circ)\mathcal{V}_\circ}{(\Gamma\bar{\epsilon} + \bar{\rho}_B)\mathcal{V}}}, \tag{89}$$

$$\frac{\partial}{\partial t}(N_{e^\pm}) = -N_{e^\pm}\frac{1}{\mathcal{V}}\frac{\partial \mathcal{V}}{\partial t} + \overline{\sigma v}\frac{1}{\bar{\gamma}^2}(N_{e^\pm}^2(T) - N_{e^\pm}^2). \tag{90}$$

In these equations ($r > r_{\text{out}}$) the comoving baryonic mass- and number densities are $\bar{\rho}_B = M_B/V$ and $\bar{n}_B = N_B/V$, where V is the comoving volume of the PEMB pulse.

We compare and contrast (see Fig. 5) the bulk gamma factor as computed from the Rome and Livermore codes, where excellent agreement has been found. This validates the constant-thickness approximation in the case of the PEMB pulse as well. On this basis we easily estimate a variety of physical quantities for an entire range of values of B.

For the same EMBH we have considered five different cases: a shell of baryonic mass with (1) $B \simeq 1.3 \cdot 10^{-4}$; (2) $B \simeq 3.8 \cdot 10^{-4}$; (3) $B \simeq 1.3 \cdot 10^{-3}$; (4) $B \simeq 3.8 \cdot 10^{-3}$; (5) $B \simeq 1.3 \cdot 10^{-3}$). The results of the integration given in detail in Ruffini et al. [142] show that for the first parameter range the PEMB pulse propagates as a sharp pulse of constant thickness in the laboratory frame, but already for $B \simeq 1.3 \cdot 10^{-2}$ the expansion of the PEMB pulse becomes much more complex and the constant-thickness approximation ceases to be valid; see Ruffini et al. [142] for details.

It is particularly interesting to evaluate the final value of the gamma factor of the PEMB pulse when the transparency condition given by Eq.(86) is reached as a function of B, see Fig. 20. For a given EMBH, there is a *maximum* value of the gamma factor at transparency. By further increasing the value of B the entire E_{dya} is transferred into the kinetic energy of the baryons; see also section XII. Details are given in Ruffini et al. [142].

In Fig. 20 we plot the gamma factor of the PEMB pulse versus the radius for different amounts of baryonic matter. The diagram extends to values of the radial coordinate at which the transparency condition given by Eq.(86) is reached. The "asymptotic" gamma factor

$$\bar{\gamma}_{\text{asym}} \equiv \frac{E_{\text{dya}}}{M_B c^2} \tag{91}$$

is also shown for each curve. The closer the gamma value approaches the "asymptotic" value (91) at transparency, the smaller the intensity of the radiation emitted in the burst and the larger the amount of kinetic energy left in the baryonic matter.

X. THE IDENTIFICATION OF THE FREE PARAMETERS OF THE EMBH THEORY

Within the approximation presented in section III the EMBH is characterized by two parameters: μ and ξ. The energy of the dyadosphere is expressed in terms of these two parameters by Eq.(12).

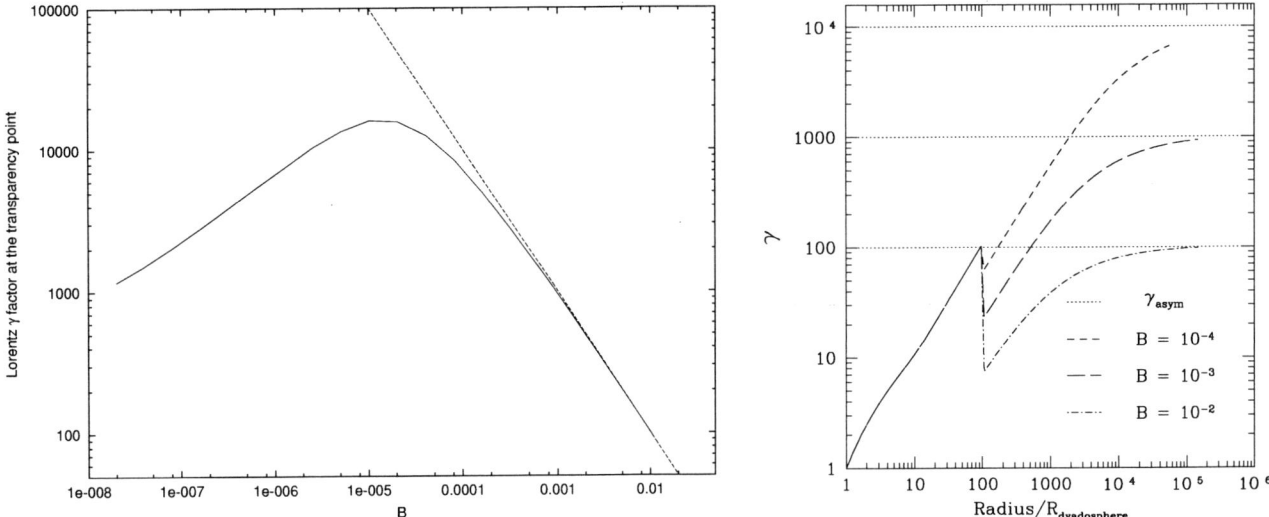

Figure 20: **Left)** The gamma factor (the solid line) at the transparent point is plotted as a function of the B parameter. The asymptotic value (the dashed line) $E_{\text{dya}}/(M_B c^2)$ is also plotted. **Right)** The gamma factors are given as functions of the radius in units of the dyadosphere radius for selected values of B for the typical case $E_{dya} = 3.1 \times 10^{54}$ erg. The asymptotic values $\gamma_{\text{asym}} = E_{\text{dya}}/(M_B c^2) = 10^4, 10^3, 10^2$ are also plotted. The collision of the PEM pulse with the baryonic remnant occurs at $r/r_{ds} = 100$ where the jump occurs and the PEMB pulse starts.

There is an entire family of EMBH solutions with different values of μ and ξ corresponding to the same value of E_{dya} (see Fig. 17). These solutions are physically different with respect to the density of electron-positron pair distributions given by Eq.(9), as well as to their energy density given by Eq.(11). A clear example of such a degeneracy is given in Fig. 21 where the two limiting energy density profiles approximating the dyadosphere as introduced in Fig. 17 are given for three different EMBH configurations corresponding to the same value of $E_{dya} = 3.1 \times 10^{54}$ erg. The three configurations correspond respectively to the three different pairs (μ, ξ): $(10, 0.76)$, $(10^2, 0.27)$, $(10^3, 0.10)$.

The corresponding dynamical evolution of the PEM pulse introduced in section VII and Ruffini et al. [141] is clearly different in the three cases. It is remarkable that when the collision with the remnant of the progenitor star is considered all these differences disappear. As usual (see section VIII) we describe the baryonic content of the remnant by the parameter B. The PEMB pulse generated after the collision with the baryonic matter depends uniquely on the two parameters E_{dya} and B. In Fig. 22 the temperature in the laboratory frame is given for the PEM pulse and the PEMB pulse corresponding to the three configurations of Fig.21 and $B = 4 \times 10^{-3}$. It is clear that while for the PEM pulse era the three configurations are markedly different, they do converge to a common behaviour in the PEMB pulse era.

If we turn now to the effect of the distance between the EMBH and the baryonic remnant, we see that this degeneracy is further extended: while the three PEM pulse eras are quite different, the common PEMB pulse era is largely insensitive to the location of the baryonic remnant, see Fig. 23. We have plotted the three gamma factors in the PEM pulse era corresponding to the different configurations of Fig. 21 and $B = 10^{-2}$, in the two cases the baryonic remnant is positioned at different distances from the EMBH.

If the PEM pulse has reached extreme relativistic regimes, the common value γ_{coll} to which the three gamma factors drop in the collision with the baryonic matter of the remnant can be simply expressed by the large gamma limit of Eq.(84)

$$\gamma_{coll} = \frac{B+1}{\sqrt{B^2 + 2B}}, \qquad (92)$$

while the internal energy E_{coll} developed in that collision is simply given by the corresponding limit of Eq.(85)

$$\frac{E_{coll}}{E_{dya}} = -B + \sqrt{B^2 + 2B}. \qquad (93)$$

This approximation applies when the final gamma factor at the end of the PEM pulse era is larger than γ_{coll}, upper panel in Fig. 23.

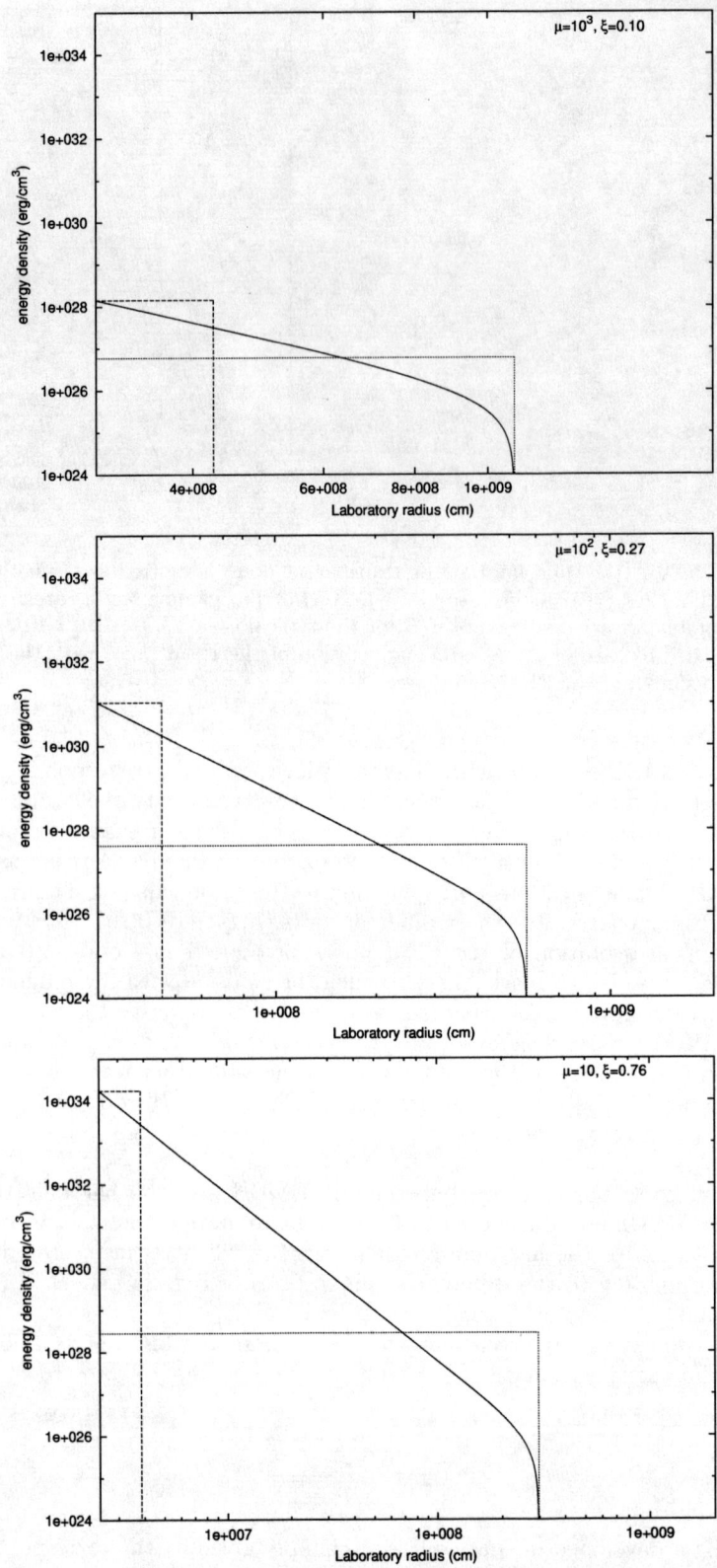

Figure 21: Three different dyadospheres corresponding to the same value of $E_{dya} = 3.1 \times 10^{54}$ erg and with different values of the two parameters μ and ξ are given. The three different configurations are markedly different in their spatial extent as well as in their energy-density distribution.

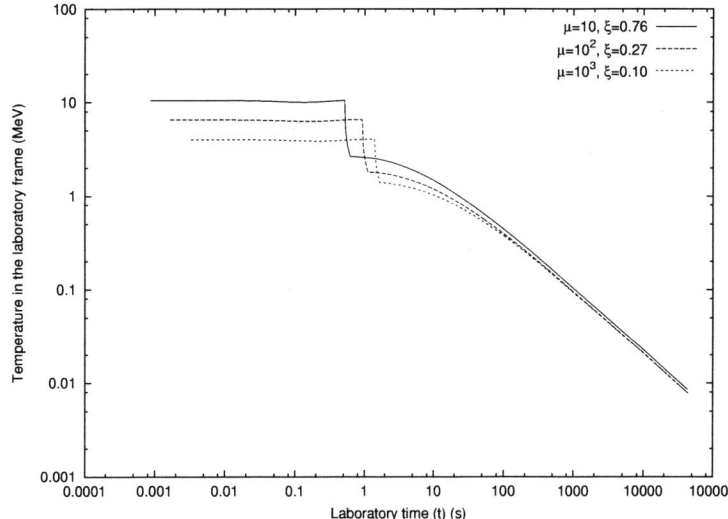

Figure 22: The temperature of the plasma during the PEM pulse and PEMB pulse eras, measured in the laboratory frame, corresponding to the three configurations presented in Fig. 21 is given as a function of the laboratory time. The three different curves converge to a common one in the PEMB pulse era, which is therefore only a function of the E_{dya} and B. The difference among the three curves in the early part of the PEMB pulse follows from having located the baryonic matter at a distance of $50(r_{ds} - r_+)$, which is different in the three cases. Such difference become negligible at large distances in the later phases of the evolution.

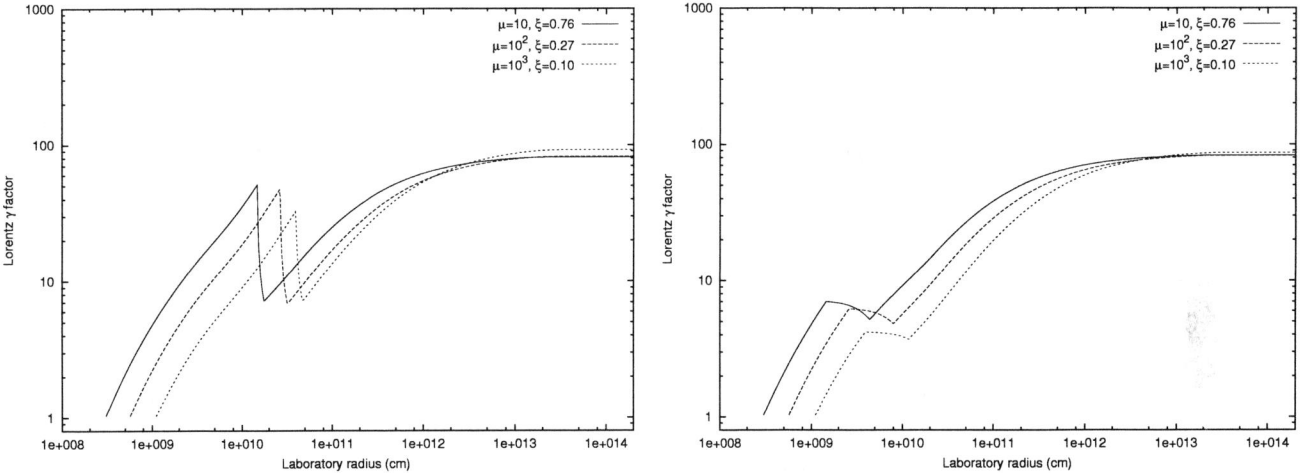

Figure 23: The gamma factors for the three configurations considered in Fig. 21 are given as a function of the radial coordinate in the laboratory frame. The two figures correspond to a baryonic remnant positioned respectively at $r_{in} = 50(r_{ds} - r_+)$ (left) and at $r_{in} = 5(r_{ds} - r_+)$ (right). Again the convergence to a common behaviour, uniquely a function of E_{dya} and B for the late stages of the PEMB pulse, is manifest.

Turning from these general considerations to the GRB data, this degeneracy in the PEMB pulse eras and their dependence on only two parameters E_{dya} and B has far reaching astrophysical implications for the identification of the source of GRBs. As we will see in the conclusions all the information obtainable from GRBs with a large value of the parameter B will lead to the determination of the above two parameters. An entire family of degenerate astrophysical solutions in the range of charges and masses given in Fig. 17 are possible. The direct knowledge of the mass and charge of the EMBH can only be gained from the PEM pulse or from GRBs with very small values of B — the so called "short bursts", see section XII and the conclusions.

XI. THE APPROACH TO TRANSPARENCY: THE THERMODYNAMICAL QUANTITIES

As the condition of transparency expressed by Eq.(86) is reached the *injector phase* terminates. The electromagnetic energy of the PEMB pulse is released in the form of free-streaming photons — the proper GRB. The remaining energy of the PEMB pulse is released as an accelerated-baryonic-matter (ABM) pulse.

We now proceed to the analysis of the approach to the transparency condition. It is then necessary to turn from the pure dynamical description of the PEMB pulse described in the previous sections to the relevant thermodynamic parameters. Also such a description at the time of transparency needs the knowledge of the thermodynamical parameters in all previous eras of the GRB.

As above we shall consider as a typical case an EMBH of $E_{dya} = 3.1 \times 10^{54}$ erg and $B = 10^{-2}$. The considerations will refer to a dyadosphere configuration described by the two limiting approximations shown in Fig. 17.

One of the key thermodynamical parameters is represented by the temperature of the PEM and PEMB pulses. It is given as a function of the radius both in the comoving and in the laboratory frames in Fig. 24. Before the collision the PEM pulse expands keeping its temperature in the laboratory frame constant while its temperature in the comoving frame falls (see Ruffini et al. [141]). In fact Eqs.(66,67) are equivalent to

$$\frac{d(\epsilon\gamma^2\mathcal{V})}{dt} = 0, \tag{94}$$

where the baryon mass-density is $\rho_B = 0$ and the thermal energy-density of photons and e^+e^--pairs is $\epsilon = \sigma_B T^4(1 + f_{e^+e^-})$, σ_B is the Boltzmann constant and $f_{e^+e^-}$ is the Fermi-integral for e^+ and e^-. This leads to

$$\epsilon\gamma^2\mathcal{V} = E_{\text{dya}}, \quad T^4\gamma^2\mathcal{V} = \text{const.} \tag{95}$$

Since e^+ and e^- in the PEM pulse are extremely relativistic, we have the equation of state $p \simeq \epsilon/3$ and the thermal index (22) $\Gamma \simeq 4/3$ in the evolution of PEM pulse. Eq.(95) is thus equivalent to

$$T^3\bar{\gamma}\mathcal{V} \simeq \text{const.} \tag{96}$$

These two equations (94) and (96) result in the constancy of the laboratory temperature $T\bar{\gamma}$ in the evolution of the PEM pulse.

It is interesting to note that Eqs.(95) and (96) hold as well in the cross-over region where $T \sim m_e c^2$ and e^+e^- annihilation takes place. In fact from the conservation of entropy it follows that asymptotically we have

$$\frac{(VT^3)_{T<m_e c^2}}{(VT^3)_{T>m_e c^2}} = \frac{11}{4}, \tag{97}$$

exactly for the same reasons and physics scenario discussed in the cosmological framework by Weinberg, see e.g. Eq. (15.6.37) of Weinberg (1972). The same considerations when repeated for the conservation of the total energy $\epsilon\gamma V = \epsilon\gamma^2\mathcal{V}$ following from Eq. (94) then lead to

$$\frac{(VT^4\gamma)_{T<m_e c^2}}{(VT^4\gamma)_{T>m_e c^2}} = \frac{11}{4}. \tag{98}$$

The ratio of these last two quantities gives asymptotically

$$T_\circ = (T\gamma)_{T>m_e c^2} = (T\gamma)_{T<m_e c^2}, \tag{99}$$

where $T_\circ$ is the initial average temperature of the dyadosphere at rest.

During the collision of the PEM pulse with the remnant we have an increase in the number density of e^+e^- pairs (see Fig. 24). This transition corresponds to an *increase* of the temperature in the comoving frame and a *decrease* of the temperature in the laboratory frame as a direct effect of the dropping of the gamma factor (see Fig. 20).

After the collision we have the further acceleration of the PEMB pulse (see Fig. 20). The temperature now decreases both in the laboratory and in the comoving frame (see Fig. 24). Before the collision the total energy of the e^+e^- pairs and the photons is constant and equal to E_{dya}. After the collision

$$E_{\text{dya}} = E_{\text{Baryons}} + E_{e^+e^-} + E_{\text{photons}}, \tag{100}$$

which includes both the total energy $E_{e^+e^-} + E_{\text{photons}}$ of the nonbaryonic components and the kinetic energy E_{Baryons} of the baryonic matter

$$E_{\text{Baryons}} = \bar{\rho}_B V(\bar{\gamma} - 1). \tag{101}$$

In Fig. 25 we plot both the total energy $E_{e^+e^-} + E_{\text{photons}}$ of the nonbaryonic components and the kinetic energy E_{Baryons} of the baryonic matter as functions of the radius for the typical case $E_{\text{dya}} = 3.1 \times 10^{54}$ erg and $B = 10^{-2}$. Further details are given in Ruffini et al. [142].

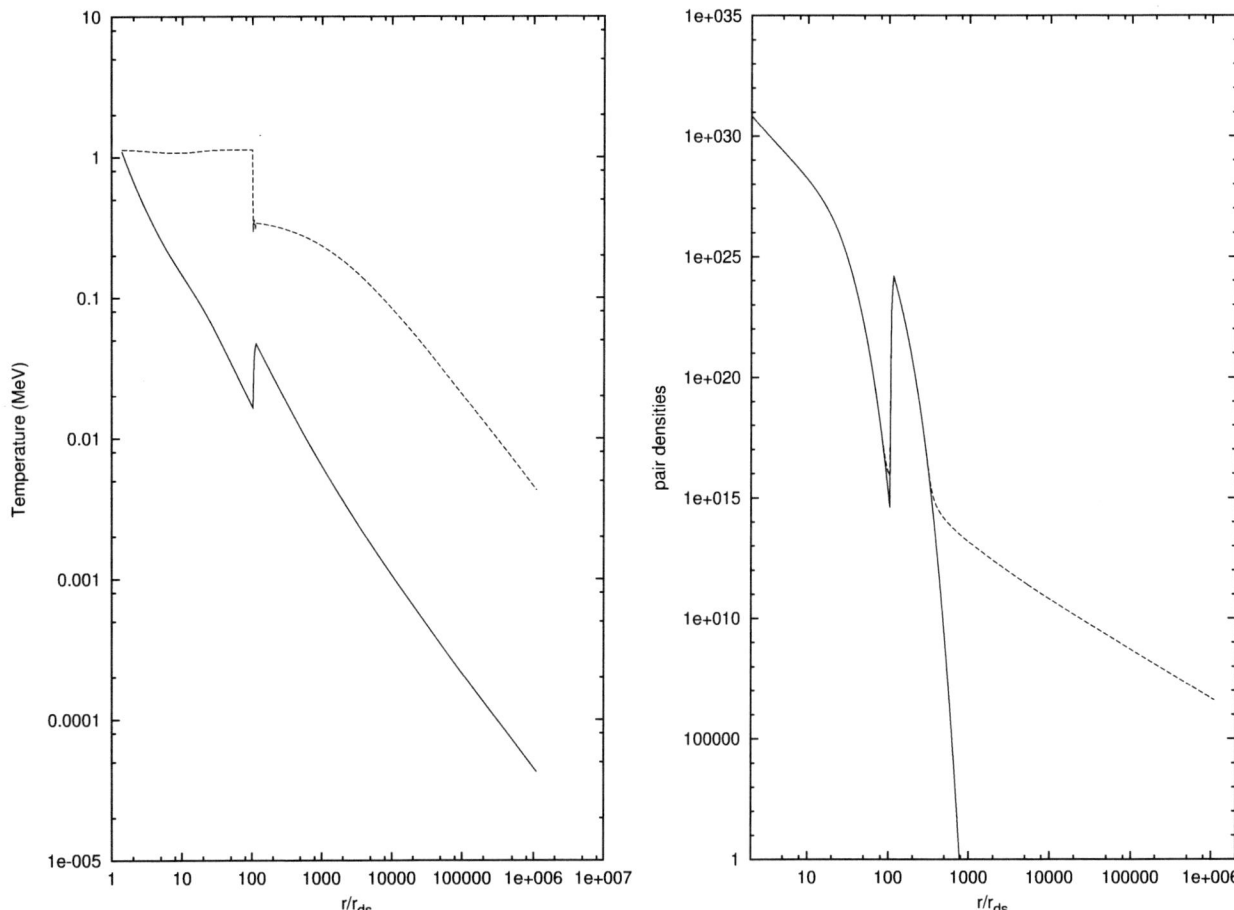

Figure 24: **Left)** The temperature of the plasma in the comoving frame T'(MeV) (the solid line) and in the laboratory frame $\bar{\gamma}T'$ (the dashed line) are plotted as functions of the radius in the unit of the dyadosphere radius r_{ds}. **Right)** The number densities $n_{e^+e^-}(T)$ (the solid line) computed by the Fermi integral and $n_{e^+e^-}$ (the dashed line) computed by the rate equation (see section IV) are plotted as functions of the radius. $T' \ll m_e c^2$, two curves strongly divergent due to e^+e^--pairs frozen out of the thermal equilibrium. The peak at $r \simeq 100 r_{\mathrm{ds}}$ is due to the internal energy developed in the collision.

XII. THE P-GRBS AND THE "SHORT BURSTS". THE END OF THE INJECTOR PHASE.

We now analyze the approach to the transparency condition given by Eq.(86). For selected values of B we give the energy $E_{P\text{-}GRB}$ of the P-GRB, and E_{Baryons} of the ABM pulse. We clearly have

$$E_{dya} = E_{P\text{-}GRB} + E_{\mathrm{Baryons}}. \tag{102}$$

Taking into account the results shown in Figs. 24–25, we can repeat all the considerations for selected values of B. We shall examine values of B ranging from $B = 10^{-8}$ only up to $B = 10^{-2}$: for larger values of B our constant slab approximation breaks down. We will see in the following that this range does indeed cover the most relevant observational features of the GRBs.

As clearly shown in Fig. 20 both the final value of the gamma factor and the radial coordinate at which the transparency condition is reached depend very strongly on B. Therefore a strong dependence on B is also found in the relative values of $E_{P\text{-}GRB}$ and E_{Baryons}.

We are now finally ready to give in Fig. 6 the crucial diagram representing the values of $E_{P\text{-}GRB}$ and E_{Baryons} in units of the E_{dya} as functions of B. This diagram, a universal one, is very important and is essential for the understanding of the GRB structure.

We find that for small values of B (around 10^{-8}) almost all the E_{dya} is emitted in the P-GRB (see also our previous paper Ruffini et al. [141]) and very little energy is left in the baryons. While for $B \simeq 10^{-2}$ roughly only 10^{-2} of the total initial energy of the dyadosphere is radiated away in the P-GRB and almost all energy is transferred to the baryons.

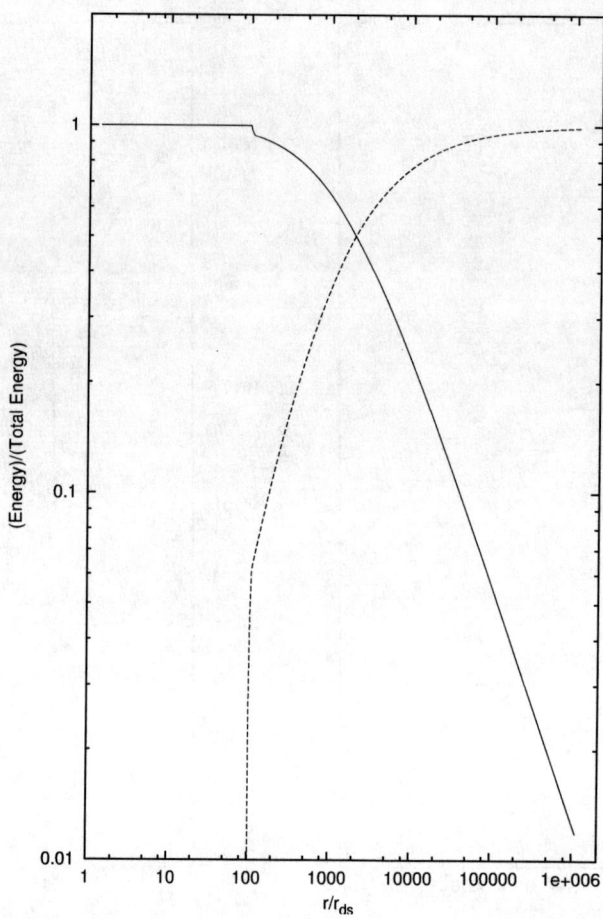

Figure 25: The energy of the non baryonic components of the PEMB pulse (the solid line) and the kinetic energy of the baryonic matter (the dashed line) in unit of the total energy are plotted as functions of the radius in the unit of the dyadosphere radius $r_{\rm ds}$.

This behaviour is at the heart of the fundamental difference between the so called *short bursts* and *long bursts*. We have proposed in Ruffini et al. [144] that the *short bursts* must be identified with the P-GRBs in the case of very small B. There are a variety of reasons supporting this identification:

1. For small values of B, $E_{\rm Baryons}$ is negligible, see Fig. 6, and consequently the intensity of the afterglow is also negligible and the entire energy E_{dya} is released into the P-GRB. This is clearly consistent with the absence of observed afterglows in the short bursts.

2. The temperature of the P-GRB in the laboratory frame $\bar\gamma T$ at the transparency point is a strongly decreasing function of B, see Fig. 6. $\bar\gamma T$ is related to the energy corresponding to the peak of the photon-number spectrum, as described in Ruffini et al. [141]. This is also in very good agreement with the observed decrease of the hardness ratio between the *short bursts* and the *long bursts* (Kouveliotou et al. [83]).

3. The time T_{90}, the duration of 90% of the energy emission as used in the current literature and discussed in Ruffini et al. [142] is plotted in Fig. 26 for selected values of E_{dya} and for different values of B.

Before concluding a word of caution is needed about how to use the above results: all these considerations are based on the drastic approximations in the description of the dyadosphere presented in section III, see also Fig. 21. This treatment is very appropriate in estimating the general dependence of the energy of the P-GRB, the kinetic energy of the ABM pulse and consequently the intensity of the afterglow. Especially powerful is the establishment of the dependence of $E_{P\text{-}GRB}$ and $E_{\rm Baryons}$ on B (see Fig. 6). As we will see in the next sections, this approximation is similarly powerful in determining the overall time structure of the GRB and especially the time of the release of the P-GRB with respect to the moment of gravitational collapse and the afterglow.

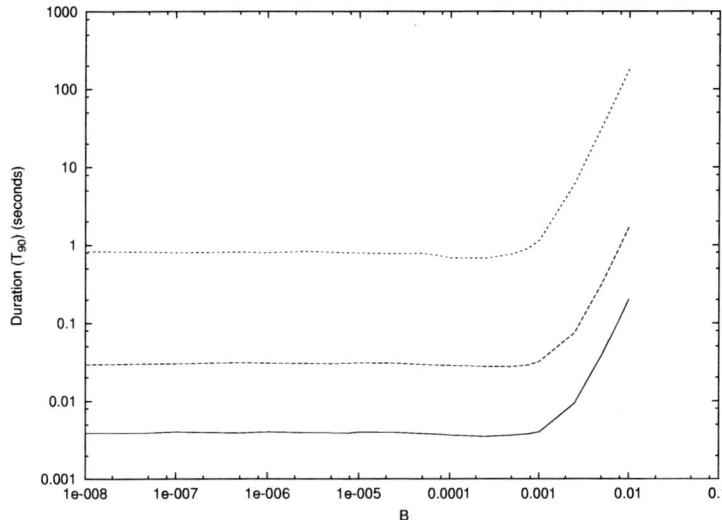

Figure 26: The duration computed with the T_{90} criterion is represented as a function of the B parameter for three selected EMBH respectively with $E_{dya} = 4.4 \times 10^{52}$ erg, $E_{dya} = 3.1 \times 10^{54}$ erg, $E_{dya} = 4.1 \times 10^{58}$ erg going from the lower curve to the upper one.

If, however, we turn to the detailed temporal structure of the P-GRB and its detailed spectral distribution, it is clear that the approximations given in section III is no longer valid. The detailed description of the formation of the dyadosphere as qualitatively expressed in Fig. 48 is now needed in all mathematical rigour with the full development of all its governing equations. Progress in this direction is being made at this moment (Cherubini et al. [27], Ruffini & Vitagliano [155, 156], Ruffini et al. [158]). This situation, however, provides a unique opportunity to follow in real time the general relativistic effects of the approach to the EMBH horizon as it occurs. In other words all direct general relativistic effects of the GRBs are encoded in the fine structure of the P-GRB. For the reasons given in section X the information on the EMBH mass and charge can only come from the short bursts.

This terminates the *injector phase*. We now turn to the *Beam-Target phase* in which the ABM pulse collides with the interstellar medium target and the afterglow is generated. We shall in the following sections review the basic theoretical treatment necessary for the description of these remaining eras and proceed then to the confrontation of the EMBH theory with the data.

XIII. THE ERA IV: THE ULTRARELATIVISTIC AND RELATIVISTIC REGIMES IN THE AFTERGLOW

In the introduction we have already expressed the basic assumptions which we have adopted for the description of the collision of the ABM pulse with the ISM. In analogy and by extension of the results obtained for the PEM and PEMB pulse cases, we also assume that the expansion of the ABM pulse through the ISM occurs keeping its width constant in the laboratory frame, although the results are quite insensitive to this assumption. We assume then that this interaction can be represented by a sequence of inelastic collisions of the expanding ABM pulse with a large number of thin and cold ISM spherical shells at rest with respect to the central EMBH. Each of these swept up shells of thickness Δr has a mass $\Delta M_{\rm ism}$ and is assumed to be located between two radial distances r_1 and r_2 (where $r_2 - r_1 = \Delta r \ll r_1$) in the laboratory frame. These collisions create an internal energy $\Delta E_{\rm int}$.

We indicate by $\Delta \epsilon$ the increase in the proper internal energy density due to the collision with a single shell and by ρ_B the proper energy density of the swept up baryonic matter. This includes the baryonic matter composing the remnant around the central EMBH, already swept up in the PEMB pulse formation, and the baryonic matter from the ISM swept up by the ABM pulse:

$$\rho_B = \frac{(M_B + M_{\rm ism})c^2}{V}. \tag{103}$$

Here V is the ABM pulse volume in the comoving frame, M_B is the mass of the baryonic remnant and $M_{\rm ism}$ is the

ISM mass swept up from the transparency point through the r in the laboratory frame:

$$M_{\text{ism}} = m_p n_{\text{ism}} \frac{4\pi}{3}\left(r^3 - r_\circ{}^3\right), \tag{104}$$

where m_p the proton mass and n_{ism} the number density of the ISM in the laboratory frame.

The energy conservation law in the laboratory frame at a generic step of the collision process is given by

$$\rho_{B_1}\gamma_1{}^2 \mathcal{V}_1 + \Delta M_{\text{ism}} c^2 = \left(\rho_{B_1}\frac{V_1}{V_2} + \frac{\Delta M_{\text{ism}} c^2}{V_2} + \Delta\epsilon\right)\gamma_2{}^2 \mathcal{V}_2, \tag{105}$$

where the quantities with the index "1" are calculated before the collision of the ABM pulse with an elementary shell of thickness Δr and the quantities with "2" after the collision, γ is the gamma factor and $\mathcal{V}$ the volume of the ABM pulse in the laboratory frame so that $V = \gamma \mathcal{V}$.

The momentum conservation law in the laboratory frame is given by

$$\rho_{B_1}\gamma_1 U_{r_1} \mathcal{V}_1 = \left(\rho_{B_1}\frac{V_1}{V_2} + \frac{\Delta M_{\text{ism}} c^2}{V_2} + \Delta\epsilon\right)\gamma_2 U_{r_2}\mathcal{V}_2, \tag{106}$$

where $U_r = \sqrt{\gamma^2 - 1}$ is the radial covariant component of the four-velocity vector (see Ruffini et al. [141, 142] and Eq.53).

We thus obtain

$$\Delta\epsilon = \rho_{B_1}\frac{\gamma_1 U_{r_1} \mathcal{V}_1}{\gamma_2 U_{r_2} \mathcal{V}_2} - \left(\rho_{B_1}\frac{V_1}{V_2} + \frac{\Delta M_{\text{ism}} c^2}{V_2}\right), \tag{107}$$

$$\gamma_2 = \frac{a}{\sqrt{a^2 - 1}}, \qquad a \equiv \frac{\gamma_1}{U_{r_1}} + \frac{\Delta M_{\text{ism}} c^2}{\rho_{B_1}\gamma_1 U_{r_1}\mathcal{V}_1}. \tag{108}$$

We can use for $\Delta\varepsilon$ the following expression

$$\Delta\varepsilon = \frac{E_{\text{int}_2}}{V_2} - \frac{E_{\text{int}_1}}{V_1} = \frac{E_{\text{int}_1} + \Delta E_{\text{int}}}{V_2} - \frac{E_{\text{int}_1}}{V_1} = \frac{\Delta E_{\text{int}}}{V_2} \tag{109}$$

because we have assumed a "fully radiative regime" and so $E_{\text{int}_1} = 0$. Substituting Eq.(108) in Eq.(107) and applying Eq.(109), we obtain:

$$\Delta E_{\text{int}} = \rho_{B_1} V_1 \sqrt{1 + 2\gamma_1 \frac{\Delta M_{\text{ism}} c^2}{\rho_{B_1} V_1} + \left(\frac{\Delta M_{\text{ism}} c^2}{\rho_{B_1} V_1}\right)^2} - \rho_{B_1} V_1 \left(1 + \frac{\Delta M_{\text{ism}} c^2}{\rho_{B_1} V_1}\right), \tag{110}$$

$$\gamma_2 = \frac{\gamma_1 + \frac{\Delta M_{\text{ism}} c^2}{\rho_{B_1} V_1}}{\sqrt{1 + 2\gamma_1 \frac{\Delta M_{\text{ism}} c^2}{\rho_{B_1} V_1} + \left(\frac{\Delta M_{\text{ism}} c^2}{\rho_{B_1} V_1}\right)^2}}. \tag{111}$$

These relativistic hydrodynamic (RH) equations have to be numerically integrated.

These are the actual set of equations we have integrated in the EMBH theory. In order to compare and contrast our results with the ones in the current literature, in section XVIII we have introduced the continuous limit of our equations and proceeded to have piecewise approximate power law solutions. We examine as well in section XX still under the above assumptions, the effects of a possible departure from homogeneity in the interstellar medium, still keeping the average density $n_{ism} = const$. Although these inhomogeneities are not relevant for the overall behaviour of the afterglow which we address here, they are indeed important for the actual observed flux and its temporal structures (see Ruffini et al. [146]). Also these considerations are affected by the angular spreading (Ruffini et al. [147]).

XIV. THE ERA V: THE APPROACH TO THE NONRELATIVISTIC REGIMES IN THE AFTERGLOW

The only reason for addressing this last era is that the issue of the approach to nonrelativistic behaviour has been extensively discussed in the literature. In our treatment these results do not show any particular problems and the relativistic equations of the previous section continue to hold. In the specific example of GRB 991216 we will present in section XVIII some analytic asymptotic expansions of these equations.

This concludes the exposition of the different eras of the EMBH theory. It goes without saying that for the description of each era, all the preceding eras must necessarily be known in order to determine the space-time grid in the laboratory frame and its relation to the arrival times as seen by a distant observer. This is the basic message expressed in the RSTT paradigm.

We can now turn to the comparison of the EMBH theory with the observational data.

XV. THE BEST FIT OF THE EMBH THEORY TO THE GRB 991216: THE GLOBAL FEATURES OF THE SOLUTION

For reasons already explained in the introduction, we use the GRB 991216 as a prototype. We will then later apply the EMBH theory to other GRBs. The relevant data of GRB 991216 are reproduced in Fig. 3: the data on the burst as recorded by BATSE Rapid Burst Response [6] and the data on the afterglow from the RXTE satellite (Corbet & Smith [30]) and the Chandra satellite (Piro et al., [119]), see also Halpern et al. [72].

The data fitting procedure relies on three basic assumption:

1. In the E-APE region, the source luminosity is mainly in the energy band 50–300 KeV, so we consider the flux observed by BATSE a good approximation of the total flux.

2. In the decaying part of the afterglow, we assume that during the R-XTE and Chandra observations the source luminosity is mainly in the energy band 2–10 KeV, so we can again assume that the flux observed by these satellites is a good approximation of the total one.

3. We have neglected in this paper the optical and radio emissions, since they are always negligible with respect to the X and γ ray fluxes. In fact, even in the latest afterglow phases up to where the X-ray data are available, they are one order of magnitude smaller then the X-ray flux.

These assumptions were initially adopted for the sake of simplicity, but have now also been justified on the basis of the spectral description of the afterglow (Ruffini et al. [149]).

As already emphasized in the previous sections, in the EMBH theory there are only two free parameters characterising the afterglow: the energy of the dyadosphere, E_{dya}, and the baryonic matter in the remnant of the progenitor star, parametrized by the dimensionless parameter B. The location of the remnant has been assumed $\sim 10^{10}$ cm. As discussed in Ruffini et al. [143] and section X, the results are rather insensitive to the actual density and location of the baryonic component but they are very sensitive to the value of B (Ruffini et al. [142]).

In Fig. 7 we present the actual first results of fitting our EMBH theory to the data from the R-XTE and Chandra satellites, corresponding to selected values of E_{dya} and B. There are three distinct features which are clearly evident as a function of the arrival time at the detector: an initial rising part in the afterglow luminosity which reaches a peak followed by a monotonically decreasing part.

We have then proceeded to fine tune the two parameters in Fig. 27. The main conclusions from our model are the following:

1) The slope of the afterglow in the region where the experimental data are present is $n = -1.6$ and is in perfect agreement with the observational data. The index n in this region is rather insensitive to the values of the parameters E_{dya} and B. The physical reason for this universality of the slope is rather remarkable since it depends on a variety of factors including the ultrarelativistic energy of the baryons in the ABM pulse, the assumption of constant average density in the ISM, the "fully radiative" conditions leading predominantly to X-ray emission, as well as all the different relativistic effects described in the RSTT paradigm (see also section XVIII).

2) The afterglow fit does not depend directly on the parameters μ, ξ but only through their combination E_{dya}. Thus there is a 1-parameter family of values of the pair (μ, ξ) allowed by a given viable value of E_{dya} (see Fig 17 and section X).

3) By fine tuning the parameters of the best fit of the luminosity profile and time evolution of the afterglow the following parameters have been found:

$$E_{dya} = 4.83 \times 10^{53} erg, \quad B = 3 \times 10^{-3} \ . \tag{112}$$

After fixing in Eq.(112) the two free parameters of the EMBH theory, modulo the mass-charge relationship which fixes E_{dya}, we can derive all the space-time parameters of the GRB 991216 (see Tab. I) as well as the explicit dependence of the gamma factor as a function of the radial coordinate (see Fig. 8).

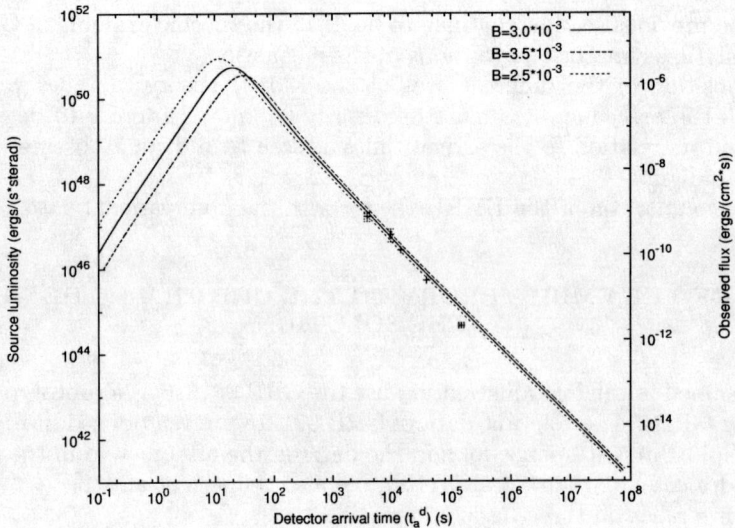

Figure 27: Fine tuning of the best fit of the afterglow data of Chandra, RXTE as well as of the range of variability of the BATSE data on the major burst by a unique afterglow curve leading to the parameter values $E_{dya} = 4.83 \times 10^{53} erg, B = 3 \times 10^{-3}$.

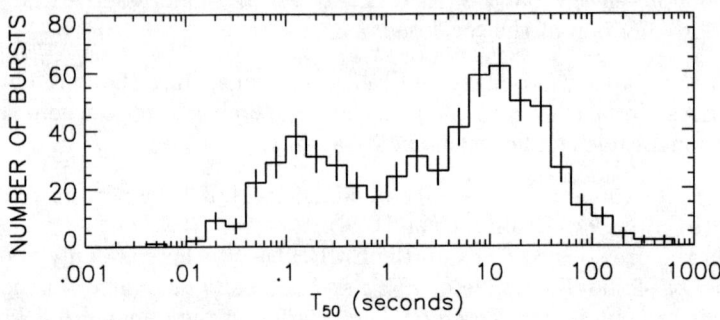

Figure 28: The distribution of the burst durations clearly shows two different classes of events: the "short bursts" and the "long bursts" (reproduced from Paciesas et al. [106]).

Of special interest is the fundamental diagram of Fig. 9. Its role is essential in interpreting all quantities measured in arrival time (the time of an observer in an inertial frame at the detector) and their relations to the ones measured in the laboratory time by an observer in an inertial frame at the GRB source. The two times are clearly related by light signals (see Fig. 18) and expressed by the integral Eq.(37) and are also affected by the cosmological expansion (see section V).

XVI. THE EXPLANATION OF THE "LONG BURSTS" AND THE IDENTIFICATION OF THE PROPER GAMMA RAY BURST (P-GRB)

Having determined the two free parameters of the EMBH theory, any other feature is a new prediction. An unexpected result soon became apparent, namely that the average luminosity of the main burst observed by BATSE can be fit by the afterglow curve (see Fig. 10). This led us to the identification of the long bursts observed by BATSE with the extended afterglow peak emission (E-APE). The peak of this E-APE occurs at $\sim 19.87\,s$ and its intensity and time scale are in excellent agreement with the BATSE observations (see also Ruffini et al. [146]). It is clear that this E-APE is *not* a burst, but is seen as such by BATSE due to its high noise threshold (see also Ruffini et al. [146]). Thus the outstanding unsolved problem of explaining the long GRBs (see e.g. Piran [116], Salmonson et al. [161], Wilson et al. [186]) is radically resolved: the so called "long bursts" do not exist, they are just E-APEs (see Fig. 28).

We now turn to the most cogent question to be asked: where does one find the burst which is emitted when the

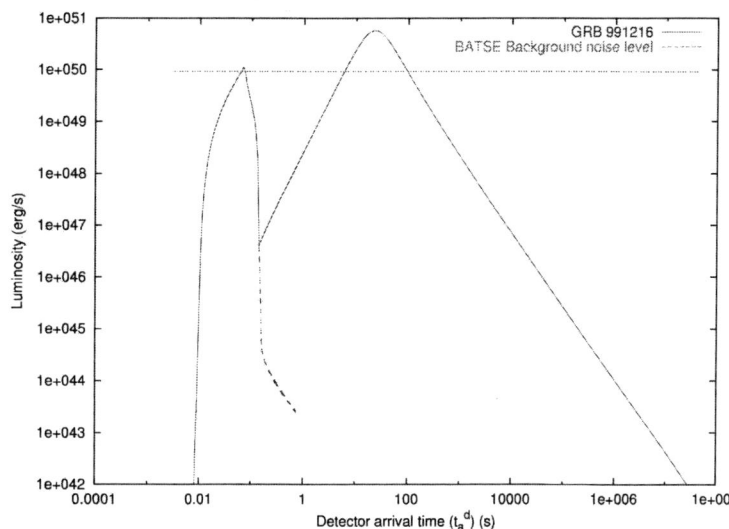

Figure 29: A qualitative diagram showing the full picture of the model, with both P-GRB and E-APE.

condition of transparency against Thomson scattering is reached? We have referred to this as the proper gamma ray burst (P-GRB) in order to distinguish it from the global GRB phenomena (see Bianco et al. [12], Ruffini et al. [143]). We are guided in this search by two fundamental diagrams (see Fig. 11 and Fig. 11):

1. In Ruffini et al. [142] it is shown that for a fixed value of E_{dya} the value of B uniquely determines the energy $E_{P\text{-}GRB}$ of the P-GRB and the kinetic energy $E_{Baryons}$ of the ABM pulse which gives origin to the afterglow (see Fig. 11). For the particular values of the parameters given in Eq. (112), we find

$$E_{P\text{-}GRB} = 7.54 \times 10^{51} erg, \quad E_{Baryons} = 9.43 \times 10^{52} erg \tag{113}$$

and then:

$$\frac{E_{P\text{-}GRB}}{E_{Baryons}} = 1.58 \times 10^{-2}. \tag{114}$$

2. One important additional piece of information comes from the differences in arrival time between the P-GRB and the peak of the E-APE, see Fig. 11. Using the results of this figure and the numerical values given in Tab. I, we can retrace the P-GRB by reading off the time parameters of point 4 in Fig. 8. Transparency is reached at $21.57\,s$ in comoving time at a radial coordinate $r = 1.94 \times 10^{14}$ cm in the laboratory frame and at $8.41 \times 10^{-2}\,s$ in arrival time at the detector.

All this, namely the energy predicted in Eq.(113) for the intensity of the burst and its time of arrival, leads to the unequivocal identification of the P-GRB with the apparently inconspicuous initial burst in the BATSE data. We have estimated from the BATSE data the ratio of the P-GRB to the E-APE over the noise threshold to be $\sim 10^{-2}$, in excellent agreement with the result in Eq. (114), see Fig. 29.

It is important to emphasize that the diagrams in Fig. 6 and Fig. 11 are not universal, but depend on the dyadosphere energy. The corresponding diagrams for three selected E_{dya} values ($E_{dya} = 5.29 \times 10^{51}$ erg, $E_{dya} = 4.83 \times 10^{53}$ erg and $E_{dya} = 4.49 \times 10^{55}$ erg) are given in Fig. 30a where we have plotted the energy of the P-GRB and of the E-APE as a function of B. The crossing of the intensity of P-GRB and E-APE occurs respectively at $B_1 = 6.0 \times 10^{-5}$, $B_2 = 2.5 \times 10^{-5}$ and $B_3 = 1.2 \times 10^{-5}$ where $B_1 > B_2 > B_3$. In Fig. 30b the same quantities are plotted as a function of the baryon mass M_B in units of solar masses and the opposite dependence occurs: $M_1 < M_2 < M_3$.

The physical reasons beyond these results is the following. We recall that the kinetic energy $E_{Baryons}$ and mass M_B of PEMB pulse are

$$E_{Baryons} = (\gamma - 1)M_B \quad M_B \equiv B E_{dya} \tag{115}$$

at the crossing point defined by

$$E_{Baryons} = E_{P\text{-}GRB} = \frac{1}{2}E_{dya}. \tag{116}$$

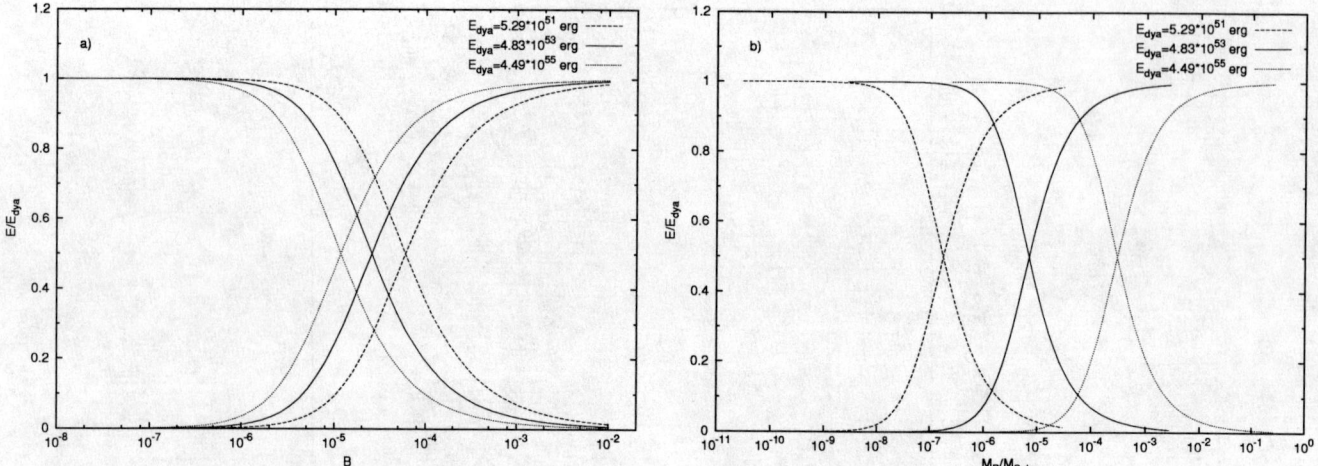

Figure 30: **a)** The same diagram of Fig. 6 is plotted for three different E_{dya} values: $E_{dya} = 5.29 \times 10^{51}$ erg (dashed lines), $E_{dya} = 4.83 \times 10^{53}$ erg (solid lines) and $E_{dya} = 4.49 \times 10^{55}$ erg (dotted lines). **b)** Same as in a) but plotted as a function of the baryonic mass M_B in units of solar masses instead of B.

From these two equations, we obtain

$$B = \frac{1}{2(\gamma_\circ - 1)} \simeq \frac{1}{2\gamma_\circ}, \qquad (117)$$

$\gamma_\circ$ is the Lorentz gamma factor of the PEMB pulse at the transparency point, where (see section XI)

$$(n_{pair} + n_B)\sigma_T \simeq n_B \sigma_T = 1, \qquad n_B = \frac{M_B}{4\pi r_\circ^2 \Delta \gamma_\circ}, \qquad (118)$$

Δ_t is the PEMB pulse thickness and $r_\circ$ the radial position at the transparency point. In addition, from the total energy conservation, we have

$$(\epsilon + n_B)\gamma_\circ^2 4\pi r_\circ^2 \Delta = const., \qquad (119)$$

where ϵ is the thermal energy of the PEMB pulse. In the regime $n_B \gg \epsilon$, we have

$$\gamma_\circ \simeq \frac{E_{dya}}{M_B}, \qquad (120)$$

and in the regime $n_B \ll \epsilon$, we have

$$\gamma_\circ \sim r_\circ. \qquad (121)$$

Considering the crossing point to occur in the second regime, we obtain at the crossing point

$$B \sim (E_{dya})^{-\frac{1}{4}}, \qquad M_B \sim (E_{dya})^{\frac{3}{4}}. \qquad (122)$$

These results are plotted in Figs. 31a–b. The agreement with the computed results is quite satisfactory. The differences can be attributed to the approximation adopted in Eq.(121) which is modified for high B values.

The conclusion is that for increasing E_{dya} also the baryonic mass corresponding to the cross increases, but in percentage it increases less than E_{dya}.

XVII. CONSIDERATIONS ON THE P-GRB SPECTRUM AND THE HARDNESS OF THE SHORT BURSTS

Regarding the P-GRB spectrum, the initial energy of the electron-positron pairs and photons in the dyadosphere for given values of the parameters can be easily computed following the work of Preparata et al. [123]. We obtain

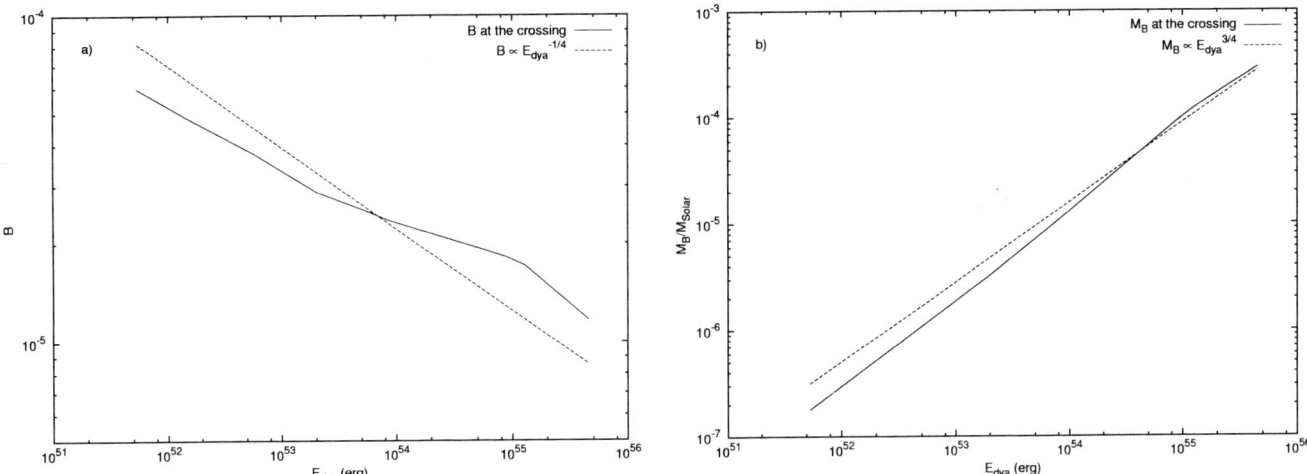

Figure 31: **a)** The B values corresponding to the crossings in Fig. 30a are plotted versus E_{dya} (solid line). The function $B \propto E_{dya}^{-1/4}$ obtained from a qualitative theoretical estimate (see Eq.(122)) is also plotted (dashed line). **b)** The M_B values corresponding to the crossings in Fig. 30b are plotted versus E_{dya} (solid line). The function $M_B \propto E_{dya}^{3/4}$ obtained from a qualitative theoretical estimate (see Eq.(122)) is also plotted (dashed line).

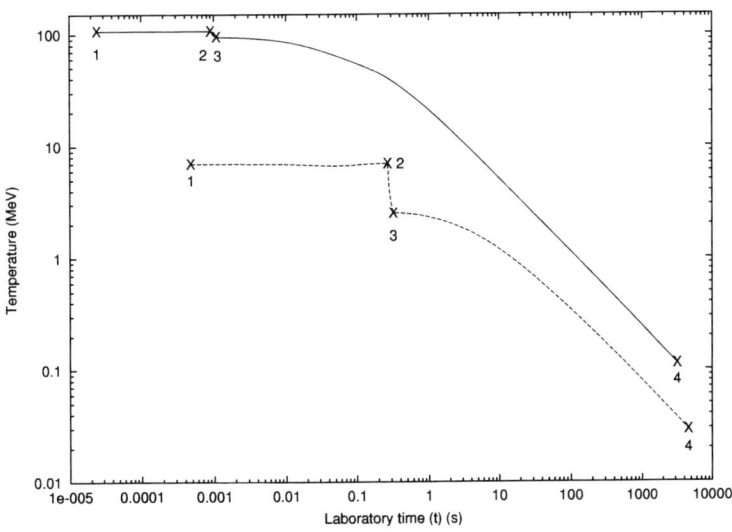

Figure 32: The temperature of the pulse in the laboratory frame for the first three eras of Fig. 1 of Ruffini et al. [143] is given as a function of the laboratory time. The numbers 1, 2, 3, 4 represent the beginning and end of each era. The two curves refer to two extreme approximations adopted in the description of the dyadosphere. Details are given in Ruffini et al. [142] and in section X.

respectively $T = 1.95$ MeV and $T = 29.4$ MeV in the two approximations we have used for the average energy density of the dyadosphere (see section X). It is then possible to follow in the laboratory frame the time evolution of the temperature of the electron-positron pairs and photons through the different eras, see Fig. 32. The condition of transparency is reached at temperatures in the range of $\sim 15 - 55$ KeV at the detector, in agreement with the BATSE results. We emphasize that in the limit of B going to 10^{-8} in which the P-GRB coincides with the "short bursts" the spectrum of the P-GRB becomes harder in agreement with the observational data (see Fig. 6 and Band et al. [4], Dermer et al. [39], Frontera et al. [55], Norris et al. [104]).

All the above are average values derived from the two approximations used in Fig. 17. If one wishes to compare the EMBH theoretical results with the fine temporal details of the observational data on the P-GRB, a departure from this average approach will be needed and the fully time varying relativistic analysis outlined in Fig. 48 applies as will be further discussed in section XXVI.

XVIII. APPROXIMATIONS AND POWER LAWS IN THE DESCRIPTION OF THE AFTERGLOW

In addition to the BATSE data, there is also clearly perfect agreement with the decaying part of the afterglow data from the RXTE and Chandra satellites.

We can also establish at this point a first set of conclusions on the luminosity power law index "n" which is a function depending strongly on the transformation $t \to t_a \to t_a^d$ (see Fig. 9). In the current literature such transformations and the corresponding n values are incorrect. Our theoretical value $n_{theo} = -1.6$ obtained for spherical symmetry for fully radiative conditions and constant density of the ISM is in agreement with observed $n_{obs} = -1.616 \pm 0.067$. No evidence of beaming is found in GRB 991216. We shall return to this point in the conclusions.

An extremely large number of papers in the literature deal with the power law index in the afterglow era. This issue has been particularly debated in connection with the aim of decreasing the energy requirements of GRBs by the effect of beaming (see e.g. Davies et al. [33], Mao & Yi [90]). It is currently very popular to infer the existence of beaming from the direct observations of breakings in the power-law index of the afterglow (see e.g. Dermer & Chiang [38], Gou et al. [70], Halpern et al. [72], Mészáros & Rees [94], Mészáros et al. [96], Panaitescu et al. [111], Panaitescu & Mészáros [114], Rhoads [128, 130], Sari, at al. [166]). Our aim here is to underline an often neglected point that the power law index of the afterglow is the result of a variety of factors including the very different regimes in the relation between the laboratory time t and the detector arrival time t_a^d presented in Fig. 9. No meaningful statements on the values of the power-law index of the afterglow can be made neglecting these necessary considerations expressed in the RSTT paradigm. This becomes particularly transparent from the power law expansion in the semianalytic treatments we present below. It is therefore not so surprising, as we will show in the next session, that the results obtained in the EMBH theory differ from the ones in the current literature.

A. The approximate expression of the hydrodynamic equations

We proceed to a first approximation and expand Eqs.(110, 111) to second order in the quantity

$$\frac{\Delta M_{\text{ism}} c^2}{\rho_{B_1} V_1} \ll 1. \tag{123}$$

We obtain the following expressions:

$$\Delta E_{\text{int}} = (\gamma_1 - 1)\Delta M_{\text{ism}} c^2 - \frac{1}{2}\frac{\gamma_1^2 - 1}{M_B + M_{\text{ism}}}(\Delta M_{\text{ism}})^2 c^2, \tag{124}$$

$$\Delta\gamma = -\frac{\gamma_1^2 - 1}{M_B + M_{\text{ism}}}\Delta M_{\text{ism}} + \frac{3}{2}\gamma_1 \frac{\gamma_1^2 - 1}{(M_B + M_{\text{ism}})^2}(\Delta M_{\text{ism}})^2, \tag{125}$$

where we set $\Delta\gamma \equiv \gamma_2 - \gamma_1$ and have used the fact that $\rho_{B_1} V_1 \equiv (M_B + M_{\text{ism}})c^2$. In the limit $\Delta E_{\text{int}} \to dE_{\text{int}}$, $\Delta\gamma \to d\gamma$, and $\Delta M_{\text{ism}} \to dM_{\text{ism}}$, neglecting also second order terms, where

$$dM_{\text{ism}} = 4\pi r^2 m_p n_{\text{ism}} dr = 4\pi r^2 m_p n_{\text{ism}} v dt, \quad v = \frac{dr}{dt}, \tag{126}$$

and where the ISM number density n_{ism} is assumed for simplicity to be $n_{\text{ism}} = 1\,\text{cm}^{-3}$, we obtain:

$$dE_{\text{int}} = (\gamma - 1)dM_{\text{ism}} c^2, \tag{127}$$

$$d\gamma = -\frac{\gamma^2 - 1}{M_B + M_{\text{ism}}}dM_{\text{ism}}. \tag{128}$$

Eqs.(127, 128) are limiting cases of Taub's hydrodynamical equations (Boccaletti et al. [21], Landau & Lifshitz [86], Taub [176]). They have been at times referred into the GRB literature as the Blandford-McKee equations (see Blandford & McKee [16]). It is clear that the application of these equations holds if Eq.(123) applies. The behaviour of $\frac{\Delta M_{\text{ism}} c^2}{\rho_{B_1} V_1}$ as a function of the radius when $M_{\text{ism}} \ll M_B$ is:

$$\frac{\Delta M_{\text{ism}} c^2}{\rho_{B_1} V_1} \sim \frac{r^2 \Delta r}{M_B}. \tag{129}$$

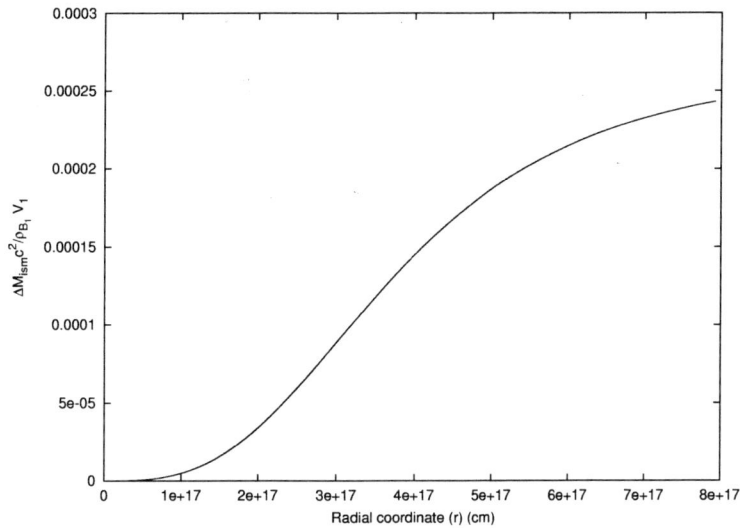

Figure 33: The factor $\frac{\Delta M_{\text{ism}} c^2}{\rho_{B_1} V_1}$ is represented as a function of the radial coordinate. It is manifestly an increasing function.

The condition $M_{\text{ism}} \ll M_B$ holds for GRB 991216 during the entire evolution of the system and so Eq.(123) is valid (see Fig. 33).

Eqs.(127,128) can be simply solved analytically (see e.g. Blandford & McKee [16]). We then have:

$$\gamma = \frac{(M_B + M_{\text{ism}})^2 + C}{(M_B + M_{\text{ism}})^2 - C}, \tag{130}$$

where

$$C = M_B{}^2 \frac{\gamma_\circ - 1}{\gamma_\circ + 1}, \tag{131}$$

where we recall that $r_\circ$ and $\gamma_\circ$ are the radial coordinate and the gamma factor at the transparency point and M_B is the initial baryonic mass of the ABM pulse.

Eq.(130) is a differential equation for $r(t)$, namely

$$1 - \left(\frac{dr}{cdt}\right)^2 = \left[\frac{(M_B + M_{\text{ism}})^2 + C}{(M_B + M_{\text{ism}})^2 - C}\right]^{-2}, \tag{132}$$

which can be integrated analytically with solution (see e.g. Abramowitz & Stegun [1])

$$2c\sqrt{C}\,(t - t_\circ) = (M_B - m_i^\circ)(r - r_\circ) + \frac{1}{4} m_i^\circ r_\circ \left[\left(\frac{r}{r_\circ}\right)^4 - 1\right] + \frac{Cr_\circ}{6m_i^\circ B^2} \ln\left[\frac{\left(B + \frac{r}{r_\circ}\right)^3}{B^3 + \left(\frac{r}{r_\circ}\right)^3} \frac{B^3 + 1}{(B+1)^3}\right] \\ + \frac{Cr_\circ}{3m_i^\circ B^2}\left[\sqrt{3}\arctan\frac{2\frac{r}{r_\circ} - B}{B\sqrt{3}} - \sqrt{3}\arctan\frac{2 - B}{B\sqrt{3}}\right], \tag{133}$$

where $m_i^\circ = \frac{4}{3}\pi m_p n_{\text{ism}} r_\circ^3$, $B = \left(\frac{M_B - m_i^\circ}{m_i^\circ}\right)^{1/3}$ and we recall that $t_\circ$ is the laboratory time at the transparency point. Clearly the fulfilment of Eq.(123) has to be checked to ensure the validity of this solution.

B. The approximate expression of the emitted flux

From Eqs.(127,128), it follows that the emitted flux in the laboratory frame is given by (see Fig. 34a)

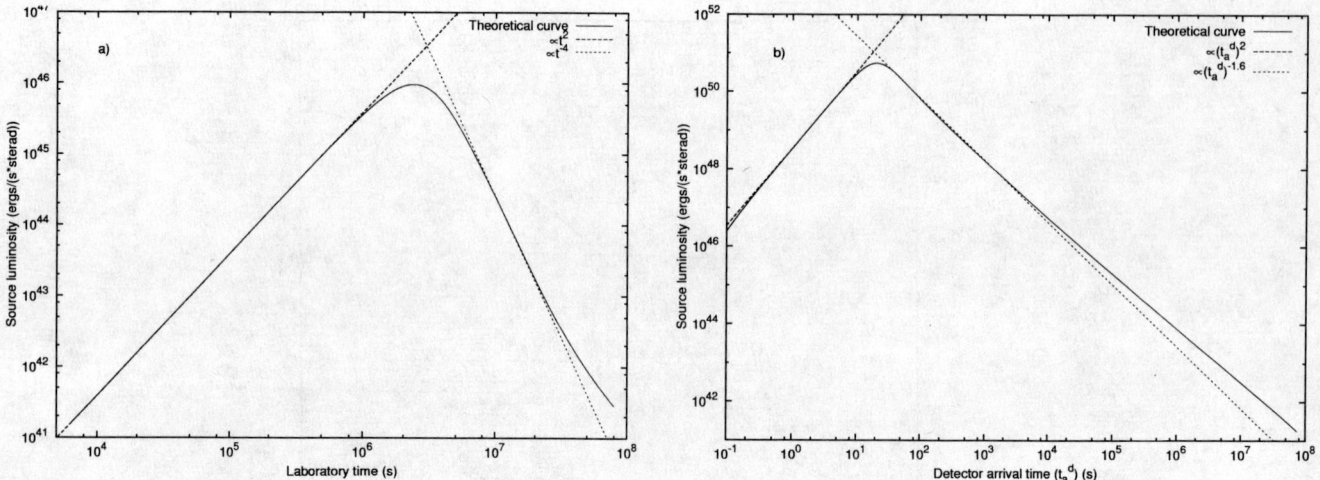

Figure 34: **a)** The GRB flux emitted in laboratory time. **b)** the flux emitted in the arrival time, measured by an observer at rest with respect to the detector (see section XVIII).

$$\frac{dE}{dt} = 4\pi r^2 n_{\rm ism} m_p v \gamma \left(\gamma - 1\right) c^2, \tag{134}$$

and the corresponding flux in detector arrival time (see Fig. 34b) by

$$\frac{dE}{dt_a^d} = \left[\frac{dt}{dt_a^d}\frac{dE}{dt}\right]_{t=t(t_a^d)} = 4\pi n_{\rm ism} m_p c^2 \left[v r^2 \gamma \left(\gamma - 1\right) \frac{dt}{dt_a^d}\right]_{t=t(t_a^d)}. \tag{135}$$

For the solution of these equations we distinguish four different phases (A–D). The first two correspond to era V.

Phase A

Just after the transparency condition is reached, the ISM matter involved is so small that we can approximately neglect the $M_{\rm ism}$ term in Eq.(130) and we have:

$$\gamma \simeq \gamma_\circ. \tag{136}$$

In the specific case of GRB 991216 we have $\gamma_\circ = 310.1$, $r_\circ = 1.94 \times 10^{14}\,cm$, $t_\circ = 6.48 \times 10^3\,s$, $t_{a_\circ} \simeq 4.21 \times 10^{-2}\,s$ and $t_{a_\circ}^d \simeq 8.41 \times 10^{-2}\,s$, where the index "$\circ$" refers to the quantities at the transparency point. We can then establish the following equation describing the ABM pulse motion in this phase: $r(t) = vt$ with $v \simeq c$. We can than use the following relation between laboratory time and arrival time:

$$t = 2\gamma_\circ^2 t_a = \frac{2\gamma_\circ^2}{1+z} t_a^d, \tag{137}$$

which is in perfect agreement with the full numerical computation (see Fig. 9).

We can substitute these equations into Eqs.(134,135), obtaining:

$$\frac{dE}{dt} \propto \gamma_\circ^2 n_{\rm ism} t^2 \tag{138}$$

in laboratory time and

$$\frac{dE}{dt_a^d} \propto \frac{\gamma_\circ^8 n_{\text{ism}}}{(1+z)^3} \left(t_a^d\right)^2 \tag{139}$$

in arrival time, assuming $\gamma(\gamma-1) \simeq \gamma^2$. The results of the numerical integration of Eqs.(107,108) are in perfect agreement with these approximations (see Fig. 34).

Points P – the two maxima of the energy flux

Since the contribution of the ISM mass in Eqs.(130–131) can no longer be neglected, the value of γ starts to significantly decrease (see Fig. 8) and the flux reaches a maximum value. We integrate Eq.(134) and Eq.(135) using Eq.(130) for γ, assuming $r(t) = vt$ with $v \simeq c$ and Eq.(137) for the relation between the laboratory time and the arrival time (see Figs. 35–9). We can now obtain the point where the emitted flux reaches its maximum. In general, the location of the maximum of the flux, point P in Ruffini et al. [143], will occur at different events, if considered in the arrival time (P_A) or in the laboratory time (P_L). In this second case, the point P_L is determined by equating to zero the first derivative of Eq.(134), and we have:

$$\gamma_{P_L} \simeq \frac{2}{3}\gamma_\circ, \quad \left.\frac{M_B}{M_{\text{ism}}}\right|_{P_L} \simeq 2\gamma_\circ, \tag{140}$$

which in the case of GRB 991216 gives $\gamma_{P_L} = 206.7$ and $\left.\frac{M_B}{M_{\text{ism}}}\right|_{P_L} \simeq 620.2$. The maximum of the observed flux is determined by equating to zero the first derivative of Eq.(135). We obtain:

$$\gamma_{P_A} \simeq \frac{5}{6}\gamma_\circ, \quad \left.\frac{M_B}{M_{\text{ism}}}\right|_{P_A} \simeq 5\gamma_\circ, \tag{141}$$

which in the case of GRB 991216 gives $\gamma_{P_A} \simeq 258.4$ and $\left.\frac{M_B}{M_{\text{ism}}}\right|_{P_A} \simeq 1550.5$.

The results of the numerical integration of Eqs.(107,108) are in perfect agreement with these approximations (see Fig. 34).

Phase B – the "golden value" $n = -1.6$

In this phase γ can no longer be considered constant and strongly decreases (see Fig. 8). M_{ism} is increasing, but v is still almost constant, equal to c. As a consequence, we can still say that $r(t) = vt$ with $v = c$, but the relation between laboratory time and arrival time given in Eq.(137) is no longer valid, and also Eq.(41) is no longer applicable in this phase (see Fig. 9). We can instead write the following "effective" relation:

$$t \propto \left(t_a^d\right)^{0.20}, \tag{142}$$

which is a result of a best fit of the numerical data in this region. Expanding the squares in Eq.(130), neglecting M_{ism}^2 with respect to M_B^2 but retaining the terms in M_{ism} and assuming $\gamma_\circ \gg 1$ we obtain:

$$\gamma \sim \frac{M_B}{M_{\text{ism}}} \sim \gamma_{P_L} \frac{r_{P_L}^3}{r^3} = \gamma_{P_L} \frac{t_{P_L}^3}{t^3}, \tag{143}$$

where r_{P_L} and t_{P_L} are the values of r and t at point P_L. Substituting this result into Eqs.(134), we obtain the emitted flux in the laboratory frame, given by

$$\frac{dE}{dt} \propto \gamma_P^2 t_P^6 n_{\text{ism}} t^{-4}, \tag{144}$$

and this is in good agreement with the full numerical computation (see Fig. 34).

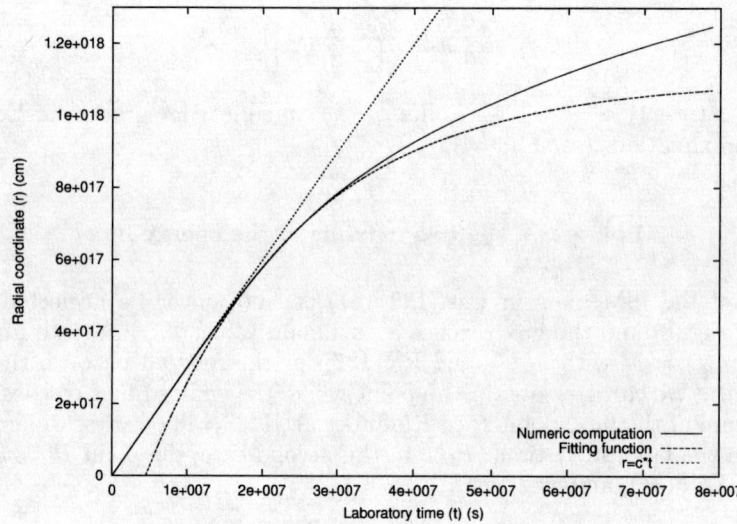

Figure 35: The exact numerical solution for $r(t)$ (solid line), together with the line $r = ct$ (dotted line) and the fitting function given in Eq.(149) (dashed line).

To obtain an analytic formula for the observed flux on the detector, we can still try to use the approximate relation between t and t_a^d given by Eq.(41):

$$t = 2\gamma(t)^2 t_a = \frac{2\gamma(t)^2}{1+z} t_a^d, \qquad (145)$$

where $\gamma(t)$ is given by Eq.(143). We obtain:

$$t = \left(\frac{2\gamma_{P_L}^2 t_{P_L}^6}{1+z} t_a^d\right)^{1/7}. \qquad (146)$$

Using this formula in Eq.(135), we finally obtain:

$$\frac{dE}{dt_a^d} \propto \frac{\gamma_P^{\frac{8}{7}} t_P^{\frac{24}{7}} n_{\text{ism}}}{(1+z)^{-\frac{17}{7}}} \left(t_a^d\right)^{-\frac{10}{7}} \qquad (147)$$

where we again assumed $\gamma(\gamma-1) \simeq \gamma^2$. This results are not in agreement with the observational data, because the power-law index for the observed flux is $-10/7 \simeq -1.43$, instead of the observed value -1.6.

This is a confirmation that Eq.(145) cannot be applied in this phase, as instead has been done by many authors in the current literature. We instead have to use Eq.(142). In fact, doing so we obtain the correct value:

$$\frac{dE}{dt_a^d} \propto n_{\text{ism}} \left(t_a^d\right)^{-1.6}, \qquad (148)$$

The results of the numerical integration of Eqs.(107,108) are in perfect agreement with these approximations (see Fig. 34), which implies that the approximate Eq.(127,128) can still be used in this regime, but not Eq.(41), which has to be replaced by an "effective" local power-law behaviour (see Eq.(142)).

Phase C

This new phase begins when γ has decreased so much that the approximation $r = ct$ is no longer valid (see Fig. 35). In the case of GRB 991216 this happens when $\gamma \simeq 3.0$, $t \simeq 1.5 \times 10^7$ s, $t_a^d \simeq 2.9 \times 10^5$ s and $r \simeq 4.4 \times 10^{17}$ cm. In this entire phase, $r(t)$ manifests the following behaviour typical of damped motion:

$$r(t) = \hat{r}\left(1 - e^{-\frac{t-t^*}{\tau}}\right), \qquad (149)$$

where $\hat{r}$, $t^\star$ and τ are constants that can be determined by the best fit of the numerical solution. In the present case of GRB 991216 we obtain:

$$\hat{r} \simeq 1.101 \times 10^{18} cm, \quad \tau \simeq 2.072 \times 10^7 s, \quad t^\star \simeq 4.52 \times 10^6 s. \tag{150}$$

It is important to note that this interesting behaviour, typical of a damped motion, does not lead to any power-law relationship for the emitted flux as a function of the laboratory time (see Fig. 34). However, if we look at the observed flux as a function of the detector arrival time, we see that a power-law relationship still can be established, fitting the numerical solution. The result is:

$$\frac{dE}{dt_a^d} \propto \left(t_a^d\right)^{-1.36}. \tag{151}$$

This quite unexpected result can be explained because the relation between t and t_a^d depends on $r(t)$ in a nonpower-law behaviour. This fact balances the complex behaviour of the emitted flux as a function of the laboratory time, leading finally again to a power-law behaviour arrival time.

In this last phase, however, the flux decreases markedly, and from the point of view of the GRB observations, the most relevant regions are phases A and B described above, as well as the peak separating them.

Phase D

This last phase starts when the system approaches a Newtonian regime. In the case of GRB 991216 this occurs when $\gamma \simeq 1.05$, $t \simeq 5.0 \times 10^7$ s, $t_a^d \simeq 3.1 \times 10^7$ s and $r \simeq 1.0 \times 10^{18}$ cm. In this phase $r(t)$ is again approaching a linear behaviour, due to the velocity decreasing less steeply than in Phase C. The emitted flux as a function of the laboratory time still does not show a power-law behaviour, while the observed flux as a function of detector arrival time does, with an index $n = -1.45$ (see Fig. 34).

XIX. THE POWER-LAW INDEX OF THE AFTERGLOW AND INFERENCES ON BEAMING IN GRBS

The results obtained in the previous sections have emphasized the relevance of the proper application of the RSTT paradigm to the determination of the power-law index of the afterglow. Particularly interesting is the subtle interplay between the different regimes in the relation between the laboratory time and the arrival time at the detector clearly expressed by Fig. 9 and the corresponding different regimes encountered in the first order expansion of the relativistic hydrodynamic equations of Taub [176] (see section XVIII). It is interesting to compare and contrast our treatment with selected results of the current literature, in order to illustrate some relevant points (see Tab. III). We will consider the results in the literature only with reference to the limiting case which we address in our work: the condition of fully radiative emission.

The first line of Tab. III describes the ultrarelativistic regime, corresponding to an increasing energy flux of the afterglow as a function of the arrival time (phase A in previous section). Our treatment and the results in the literature by Dermer et al. (see e.g. Böttcher & Dermer [19], Chiang & Dermer [28], Dermer et al. [40]) coincide. They agree as well with the results by Piran et al. (see e.g. Piran [115, 116], Sari & Piran [167]).

The second line corresponds to the relativistic regime, in which the energy flux of the afterglow, after having reached the maximum (point P in previous section), monotonically decreases (phase B in previous section). The dependence we have found of the gamma factor on the radial coordinate of the expanding ABM pulse does coincide with the one given by Dermer et al. and Piran et al. Our power law index n in this regime, which perfectly fits the data, however, is markedly different from the others. Particularly interesting is the difference between our results and those of Dermer et al: the two treatments coincide up to the last relation between the laboratory time and the arrival time at the detector. As explained in Eqs.(147-148), the two treatments differ in the approximation adopted in relating the laboratory time to the arrival time at the detector, illustrated in Fig. 9. Dermer et al. incorrectly adopted the approximation represented by the lower curve in Fig. 9 and consequently they do not find agreement with the observational data. We have not been able to retrace in the treatment by Piran et al. the steps which have led to their different results. Special mention must be made of a result stated by Halpern et al. [72], the last entry in line 2, that an absolute lower limit for the power-law index $n - 1.47$ can be established on theoretical grounds. Such a result, clearly not correct also on the basis of our analysis, has been erroneously used ti support the existence of beaming in GRBs, as we will see below.

Table III: We compare and contrast the results on the power-law index n of the afterglow in the EMBH theory with other treatments in the current literature, in the limit of high energy and fully radiative conditions. The differences between the values of $-10/7 \sim -1.43$ (Dermer) and the results -1.6 in the EMBH theory can be retraced to the use of the two different approximation in the arrival time versus the laboratory time given in Fig. 9. See details in section XVIII.

	EMBH theory	Chiang & Dermer [28] Dermer et al. [40] Böttcher & Dermer [19]	Piran [115] Sari & Piran [167] Piran [116]	Vietri [179]	Halpern et al. [72]
Ultra-relativistic	$\gamma = \gamma_\circ$ $\gamma_\circ = 310.1$ $n = 2$	$\gamma = \gamma_\circ$ $n = 2$	$\gamma = \gamma_\circ$ $n \simeq 2$		
Relativistic	$\gamma \simeq r^{-3}$ $3.0 < \gamma < 258.5$ $n = -1.6$	$\gamma \sim r^{-3}$ $n = -\frac{10}{7} = -1.43$	$\gamma \sim r^{-3}$ $n = -\frac{5.5}{4} = -1.375$		$n > -1.47$
Non-relativistic	$n = -1.36$ $1.05 < \gamma < 3.0$			$n = -1.7$	
Newtonian	$n = -1.45$ $1 < \gamma < 1.05$				

The third line in Tab. III is also interesting, treating the nonrelativistic limit (Phase C in previous section). This regime has been analysed by Vietri [179], avoiding the exact integration of the equations and relying on simple qualitative arguments. These results are not confirmed by the integration of the equations we have performed. This is an interesting case to be examined for its pedagogical consequences. Having totally neglected the relation between the laboratory time and the time of arrival at the detector, which we have illustrated in Fig. 9, and identifying $t_a^d \equiv t$, Vietri reaches a very different power law from our. Moreover, his solution brings to an underestimation of the radial coordinate: he estimated a radial coordinate of $1.1 \times 10^{15}\,cm$ at $t_a^d = 3.5 \times 10^4\,s$, while the exact computation shows a result greater than $3.0 \times 10^{17}\,cm$ (see Tab. I). On the other hand if one assumes, from the above mentioned identity $t_a^d \equiv t$, $t = 3.5 \times 10^4\,s$, one obtains a gamma factor of ~ 300 (see Tab. I) in total disagreement with the nonrelativistic approximation adopted by Vietri. Quite apart from this pedagogical value, this nonrelativistic phase is of little interest from the observational point of view, due to the smallness of the flux emitted.

For completeness, we have also shown our estimates of the index n as the Newtonian phase approaches in the last line of Tab. III.

The perfect agreement between our theoretically predicted value for the power-law index, n_{theo}, and the observed one, n_{obs},

$$n_{theo} = -1.6, \quad n_{obs} = -1.616 \pm 0.067, \tag{152}$$

confirms the validity of our major assumptions:

1. The fully radiative regime.

2. The constant average density of the ISM ($n_{ism} = 1\,proton/cm^3$).

3. The spherical symmetry of the emission and the absence of beaming in GRB 991216.

After the work of Mao & Yi [90] pointing to the possibility of introducing beaming to reduce the energetics of GRBs and after the discovery of the afterglow, many articles have appeared trying to obtain theoretical and observational evidence for beamed emission in GRBs. The observations have ranged from radio (see e.g. Frail et al. [53], Rol et al. [133]) to optical (see e.g. Garnavich, et al [62], Halpern et al. [72], Sagar et al. [160], Schaefer [168]) all the way to X-rays. Particular attention has been devoted to relating the existence of beaming to possible breaks in the light curve slope, generally expected at a value of the gamma factor

$$\gamma = \frac{1}{\vartheta_0},\qquad(153)$$

where ϑ_0 is the beam opening angle. There are many articles on this subject; to mention only the most popular ones, we recall Mészáros et al. [96], Panaitescu & Mészáros [114], Rhoads [128, 129, 130], Sari, at al. [166]. Far from having reached a standard formulation, these approaches differ from each other in the expected time at which the break should take place up to a factor of 20 (see e.g. Sari, at al. [166]). They differ as well for the opening angle of the beam, up to a factor of 3 (see e.g. Sari, at al. [166]). Disagreement still exists on the number of breaking points: two in the case of Panaitescu & Mészáros [114], one in the case of Sari, at al. [166], one again in the case of Rhoads [128, 129, 130] but differing in position from the one of Sari, at al. [166]. It has also been noticed that other authors have shown through numerical simulations that such a transition, if visible at all, is not very sharp (see e.g. Halpern et al. [72]).

Ample observational data have been obtained for the GRB 991216, in addition to the X-ray band, also in the optical and radio. For the reason mentioned at the beginning of section XV, we only address in this article the problem of the γ- and the X-ray emission. In that respect, the main article addressing the issue of beaming in the X-rays for GRB 991216 is the one of Halpern et al. [72]. The key argument is based on the theoretical inequality claimed to exist for the power-law index $n > -1.47$ (see above). The fact that the observed X-ray decay rate is found to be $n_{obs} = 1.616 \pm 0.067$ is interpreted by the authors as evidence for beaming. Moreover, the fact that the decay rate $n = -1.6$ has been observed before a steepening in the optical decay occurred at approximately 1 day of arrival time authorized an even more extreme proposal of a narrower beam in the X-rays within the optical beam.

It is clear from the entire treatment which we have presented and the results of the EMBH theory given by $n_{theo} = -1.6$ that there is no evidence for such a beaming, as already stated above. The motivation by Halpern et al. [72] stems from the incorrect theoretical assumption of the existence of a lower limit in the afterglow power-law index $n > -1.47$. From our theoretical analysis the existence of $n = -1.6$ is clear proof of isotropic emission in the GRB 991216 and a clear test of the complete relativistic treatment of the source. The fact that the break in the index should be "achromatic" and the absence of beaming in the X-rays imply an absence of beaming also in the optical and radio bands. The observed steepening in the optical decay has to find an alternative explanation. Although this is not the subject of our present work for the above mentioned reasons, we have found interesting the considerations by Panaitescu & Kumar [110], which find that "there are some major difficulties to apply a jet model to GRB 991216". They also state, still for GRB 991216, that "the steepening of the optical decay of a few days is not due to a jet effect, as suggested by Halpern et al. [72], but to the passage of a spectral break".

Concerning our own position on the possibility of beaming in GRBs, we would like just to remark that, from a preliminary analysis of beamed emission within the EMBH model, we have found some new features which are not encompassed by the results in the current literature, and they could become a distinctive signature for the discrimination of the existence or nonexistence of beaming (Ruffini et al. [150]). The study of the steepening in the optical and radio decay is addressed within the EMBH theory in Ruffini et al. [149].

XX. SUBSTRUCTURES IN THE E-APE DUE TO INHOMOGENEITIES IN THE INTERSTELLAR MEDIUM

The afterglow is emitted as the ABM pulse plows through the interstellar matter engulfing new baryonic material. In our previous articles we were interested in explaining the overall energetics of the GRB phenomena and in this sense, we have adopted the very simplified assumption that the interstellar medium is a constant density medium with $n_{ism} = 1/cm^3$. Consequently, the afterglow emission obtained is very smooth in time. We are now interested in seeing if in this framework we can also explain most of the time variability observed by BATSE (see e.g. Fishman & Meegan [52]), all of which except for the P-GRB should correspond to the beam-target phase in the IBS paradigm.

We pursue this treatment neglecting the angular spreading due to off-axis scattering in the radiation of the afterglow, which will be presented in sections XXI–XXIII.

Our goal is to focus in this simplified model on the basic energetic parameters as well as on the drastic consequences of the space-time variables expressed in the RSTT paradigm.

Having obtained the two results presented in Fig. 8 and Fig. 34, we can proceed to attack the specific problem of the time variability observed by BATSE.

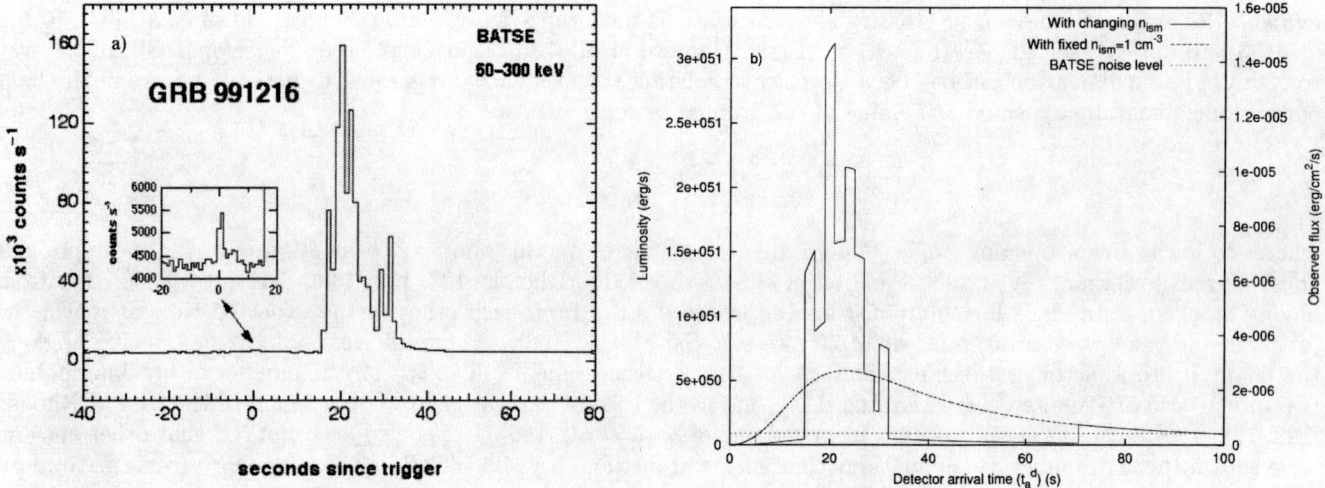

Figure 36: **a)** Flux of GRB 991216 observed by BATSE. The enlargement clearly shows the P-GRB (see Ruffini et al. [144]). **b)** Flux computed in the collision of the ABM pulse with an ISM cloud with the density profile given in Fig. 37. The dashed line indicates the emission from an uniform ISM with $n = 1 cm^{-3}$. The dotted line indicates the BATSE noise level.

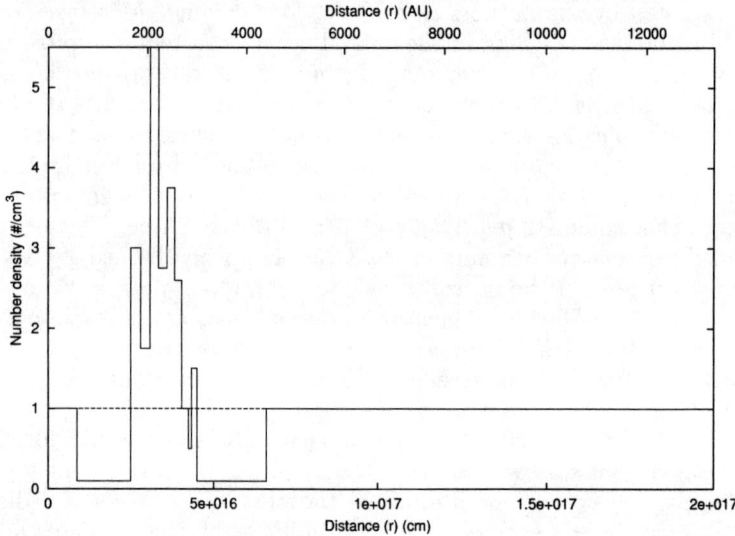

Figure 37: The density contrast of the ISM cloud profile introduced in order to fit the observation of the burst of GRB991216. The dashed line indicates the average uniform density $n = 1 cm^{-3}$.

The fundamental point is that in both regimes *the flux observed in the arrival time is proportional to the interstellar matter density*: any inhomogeneity in the interstellar medium $\Delta n_{ism}/\overline{n}_{ism}$ will lead correspondingly to a proportional variation in the intensity $\Delta I/\overline{I}$ of the afterglow. This result has been erroneously interpreted in the current literature as a burst originating in an unspecified "inner engine".

In particular, for the main burst observed by BATSE (see Fig. 36a) we have

$$\left(\Delta I/\overline{I}\right) = \left(\Delta n_{ism}/n_{ism}\right) \sim 5. \tag{154}$$

There are still a variety of physical circumstances which may lead to such density inhomogeneities.

The additional crucial parameter in understanding the physical nature of such inhomogeneities is the time scale of the burst observed by BATSE. Such a burst lasts $\Delta t_a \simeq 20s$ and shows substructures on a time scale of $\sim 1s$ (see Fig. 36a). In order to infer the nature of the structure emitting such a burst we must express these times scales in the laboratory time (see Ruffini et al. [143]). Since we are at the peak of the GRB we have $\gamma_{P_A} \sim 258.5$ (see Eq.(141))

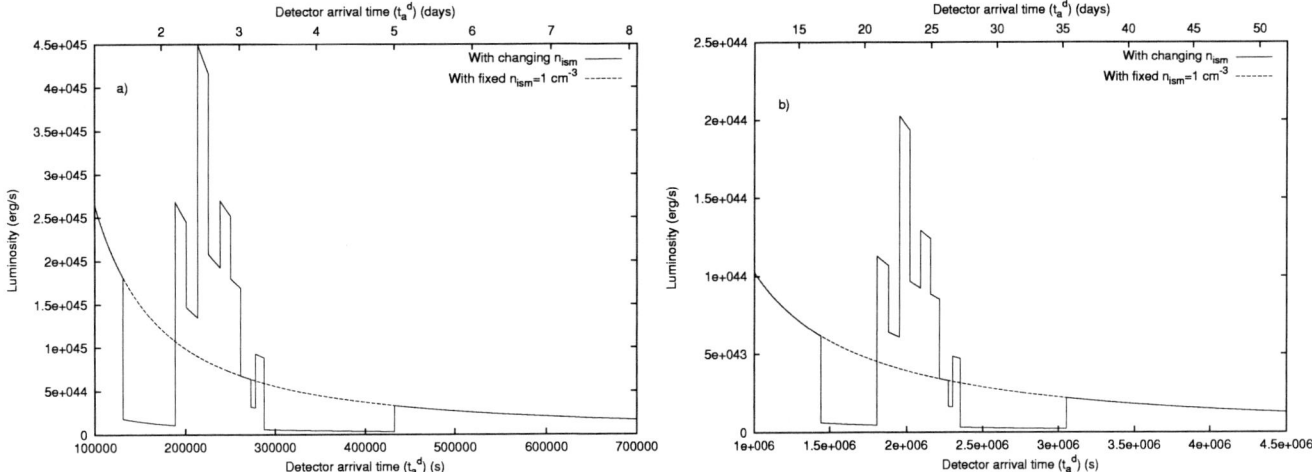

Figure 38: **a)** Same as Fig. 36b with the ISM cloud located at a distance of $3.17 \times 10^{17} cm$ from the EMBH, the time scale of the burst now extends to $\sim 1.58 \times 10^5 s$. **b)** Same as a) with the ISM cloud at a distance of $4.71 \times 10^{17} cm$ from the EMBH, the time scale of the burst now extends to $\sim 1.79 \times 10^6 s$.

and Δt_a corresponds in the laboratory time to an interval

$$\Delta t \sim 1.0 \times 10^6 s, \qquad (155)$$

which determines the characteristic size of the inhomogeneity creating the burst $\Delta L \sim 5.0 \times 10^{16} cm$ (see Tab. I and Fig. 9).

It is immediately clear from Eq.(154) and Eq.(155) that these are the typical dimensions and density contrasts corresponding to a small interstellar cloud. As an explicit example we have shown in Fig. 37 the density contrasts and dimensions of an interstellar cloud with an *average density* $<n> = 1/cm^3$. Such a cloud is located at a distance of $\sim 8.7 \times 10^{15} cm$ from the EMBH, gives rise to a signal similar to the one observed by BATSE (see Fig. 36b).

It is now interesting to see the burst that would be emitted, if our present approximation would still apply, by the interaction of the ABM pulse with the same ISM cloud encountered at later times during the evolution of the afterglow. Fig. 38a shows the expected structure of the burst at a distance $4.1 \times 10^{17} cm$, corresponding to an arrival time delay of ~ 2 days, where the gamma factor is now $\gamma_\star \sim 3.6$. It is interesting that the overall intensity would be smaller, the intensity ratio of the burst relative to the average emission would remains consistent with Eq.(154), but the time scales of the burst would be longer by a factor $\left(\frac{\gamma_{P_A}}{\gamma_\star}\right)^2 \simeq 5 \times 10^3$. Fig. 38b shows the corresponding quantities for the same ISM cloud located at a distance $6.4 \times 10^{17} cm$ from the EMBH, corresponding to an arrival time delay of ~ 1 month, where the gamma factor is ~ 1.5.

We are going to analyze in the coming sections the modifications of this basic theory by the effect of the angular spreading: it will increase the accuracy of the fit obtained in Fig. 36 and will wash away all the features at late arrival time in the afterglow (see Fig. 38).

XXI. CONSIDERATIONS ON THE RELATIVISTIC BEAMING ANGLES AND ON THE ARRIVAL TIME

We now generalize the results obtained in section V to consider also the effects due to the size of the emitting surface and of its curvature. The frequency ω and wave-vector $\mathbf{k}$ of photons emitted from the ABM pulse (see Fig. 39) expressed in the laboratory frame are:

$$\mathbf{k} = \frac{\omega}{c}\left(-\sin\vartheta\mathbf{u} + \cos\vartheta\mathbf{v}\right), \quad |\mathbf{k}| = \frac{\omega}{c}, \qquad (156)$$

where ϑ is the angle (in the laboratory frame) between the radial expansion velocity and the line of sight, $\mathbf{v}$ is a unit vector along the radial expansion velocity of the ABM pulse, and $\mathbf{u}$ is a unit vector orthogonal to $\mathbf{v}$ oriented toward rising ϑ. We are assuming here that $\mathbf{k}$ and R_T are parallel, also for photons emitted with $\vartheta \neq 0$, so that $\lambda \equiv \vartheta$.

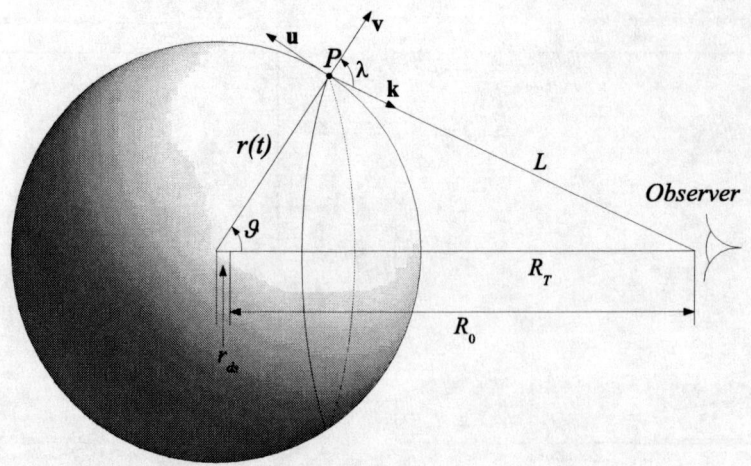

Figure 39: Qualitative description of the kinematics of the system. The big sphere is the expanding ABM pulse interacting with the ISM (not shown in the picture). The radius of the ABM pulse at time t is $r(t)$. The generic point P on the ABM pulse, from which the photon is emitted, corresponds to a displacement angle ϑ from the line of sight. L is the distance of P from the observer. R_T is the distance of the EMBH from the observer. r_{ds} is the dyadosphere radius. R_0 is defined by $R_0 \equiv R_T - r_{ds}$. $\mathbf{v}$ is a unit vector along the radial expansion velocity. $\mathbf{u}$ is a unit vector orthogonal to $\mathbf{v}$ oriented toward rising ϑ. $\mathbf{k}$ is the momentum of the photons emitted toward the observer. Note that we have assumed $\vartheta \equiv \lambda$, i.e. $\mathbf{k} \parallel R_T$ (see text).

This is clearly a good approximation, because the distance R_T corresponds to a redshift $z \sim 1$, while the radius of the emitting region is less than a light year in order of magnitude. Then the Lorentz boost along $\mathbf{v}$ to the comoving frame of the ABM pulse yields the corresponding comoving quantities:

$$\omega_\circ = \gamma \omega \left(1 - \frac{v}{c} \cos \vartheta \right), \quad \omega_\circ = |\mathbf{k}_\circ|\, c, \tag{157}$$

$$\mathbf{k}_\circ = -|\mathbf{k}| \sin \vartheta\, \mathbf{u} + \gamma\, |\mathbf{k}| \left(\cos \vartheta - \frac{v}{c} \right) \mathbf{v}, \tag{158}$$

In the comoving frame photons radiating out of the ABM pulse must have (see Eq.(158)):

$$\cos \vartheta \geq \frac{v}{c}, \tag{159}$$

because the component of the photon momentum in the comoving frame along the radial expansion velocity direction must be positive in order to escape. There will then be a maximum allowed ϑ value ϑ_{max} defined by $\cos \vartheta_{max} = (v/c)$ (see Figs. 40–41).

Due to the high value of the Lorentz gamma factor (~ 300) for the bulk motion of the expanding ABM pulse, the spherical waves emitted from its external surface appear extremely distorted to a distant observer. Let us indicate by t_a the arrival time at a detector of a photon emitted at a laboratory time t by the spherical surface of the relativistically expanding shell (see also section V). Photons arriving at the same time t_a will be emitted at different t as a function of the angle ϑ (see Fig. 39). The relation between t and t_a in the case of a constant $\gamma \sim 5$ for expanding radio sources was found by Rees (see Rees [126]):

$$t_a = t \left(1 - \frac{v}{c} \cos \vartheta \right). \tag{160}$$

For a constant expansion speed, the radius $r(t)$ of the source is given by:

$$r(t) = vt. \tag{161}$$

From Eqs.(160–161) we find the equation describing the "surface" emitting the photons detected at arrival time t_a:

$$r = \frac{v\, t_a}{1 - \frac{v}{c} \cos \vartheta}, \tag{162}$$

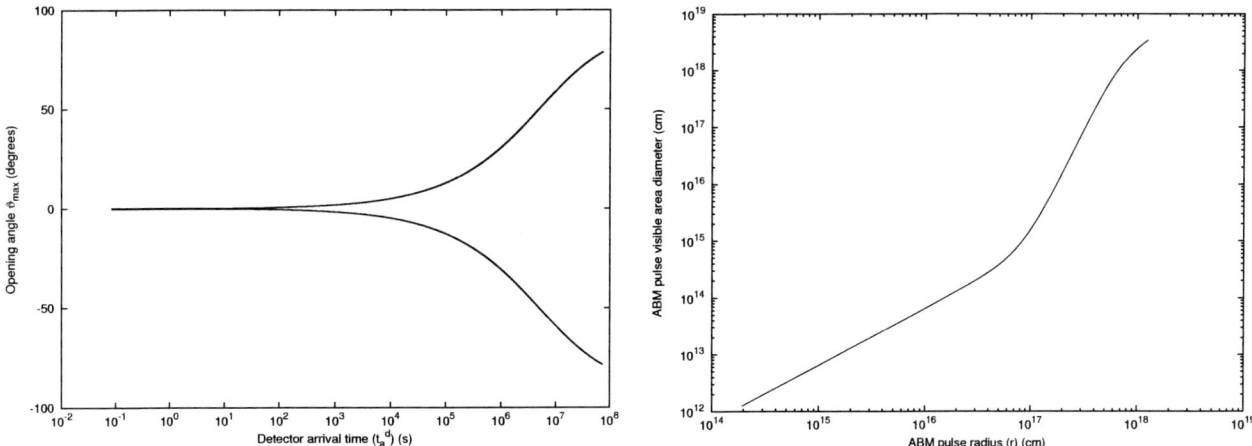

Figure 40: **Left)** Not all values of ϑ are allowed. Only photons emitted at an angle such that $\cos\vartheta \geq (v/c)$ can be viewed by the observer. Thus the maximum allowed ϑ value ϑ_{max} corresponds to $\cos\vartheta_{max} = (v/c)$. In this figure we represent ϑ_{max} (i.e. the angular amplitude of the visible area of the ABM pulse) in degrees as a function of the arrival time at the detector for the photons emitted along the line of sight (see text). In the earliest GRB phases $v \sim c$ and so $\vartheta_{max} \sim 0$. On the contrary, in the latest phases of the afterglow the ABM pulse velocity decreases and ϑ_{max} tends to the maximum possible value, i.e. 90°. **Right)** The diameter of the visible area is represented as a function of the ABM pulse radius. In the earliest expansion phases ($\gamma \sim 310$) ϑ_{max} is very small (see left pane and Fig. 41), so the visible area is just a small fraction of the total ABM pulse surface. On the other hand, in the final expansion phases $\vartheta_{max} \to 90°$ and almost all the ABM pulse surface becomes visible.

which describes an ellipsoid of eccentricity $\frac{v}{c}$ (see Rees [126]).

In our case the ABM pulse Lorentz gamma factor is not constant (see Fig. 8), and so we must generalize Eqs.(160,162) to nonconstant expansion velocity. This can be done using the geometry of Fig. 39. We set $t = 0$ when the plasma starts to expand, so that $r(0) = r_{\rm ds}$, i.e. the dyadosphere radius. Let a photon be emitted at time t from the point P. Its distance from the observer is L. The time it takes to arrive at the detector is of course $\frac{L}{c}$. Thus its arrival time, measured from the arrival of the first photon a time $\frac{R_0}{c}$ after its emission at $t = 0$, is:

$$t_{\rm a} = t + \frac{L}{c} - \frac{R_0}{c}, \tag{163}$$

where we have defined $t_{\rm a} = 0$ when a photon emitted at $t = 0$ and $\vartheta = 0$ reaches the observer. L is clearly given by:

$$L = \sqrt{R_{\rm T}^2 + r(t)^2 - 2 R_{\rm T} r(t) \cos\vartheta}, \tag{164}$$

where at any given value of emission time t, $\cos\vartheta$ can assume any value between $\left(\frac{v(t)}{c}\right)$ and 1 as noted above, where $v(t)$ is the expansion speed of the ABM pulse at time t (see Eq.(159)). Now $r(t)$ is less than one light year in order of magnitude while $R_{\rm T}$ corresponds to a redshift $z \sim 1$. Thus we can expand the right hand side of equation (164) in powers of $\frac{r(t)}{R_{\rm T}}$ to first order:

$$L \simeq R_{\rm T}\left(1 - \frac{r(t)}{R_{\rm T}}\cos\vartheta\right), \tag{165}$$

which corresponds to assuming L to be equal to its projection on the line of sight (see Fig. 39). Substituting (165) into (163) yields:

$$t_{\rm a} = t + \frac{r_{\rm ds}}{c} - \frac{r(t)}{c}\cos\vartheta, \tag{166}$$

where we have used the fact that $R_{\rm T} = R_0 + r_{\rm ds}$ (see Fig. 39). For $r(t)$ we can use the following expression:

$$r(t) = \int_0^t v(t')\, dt' + r_{\rm ds}, \tag{167}$$

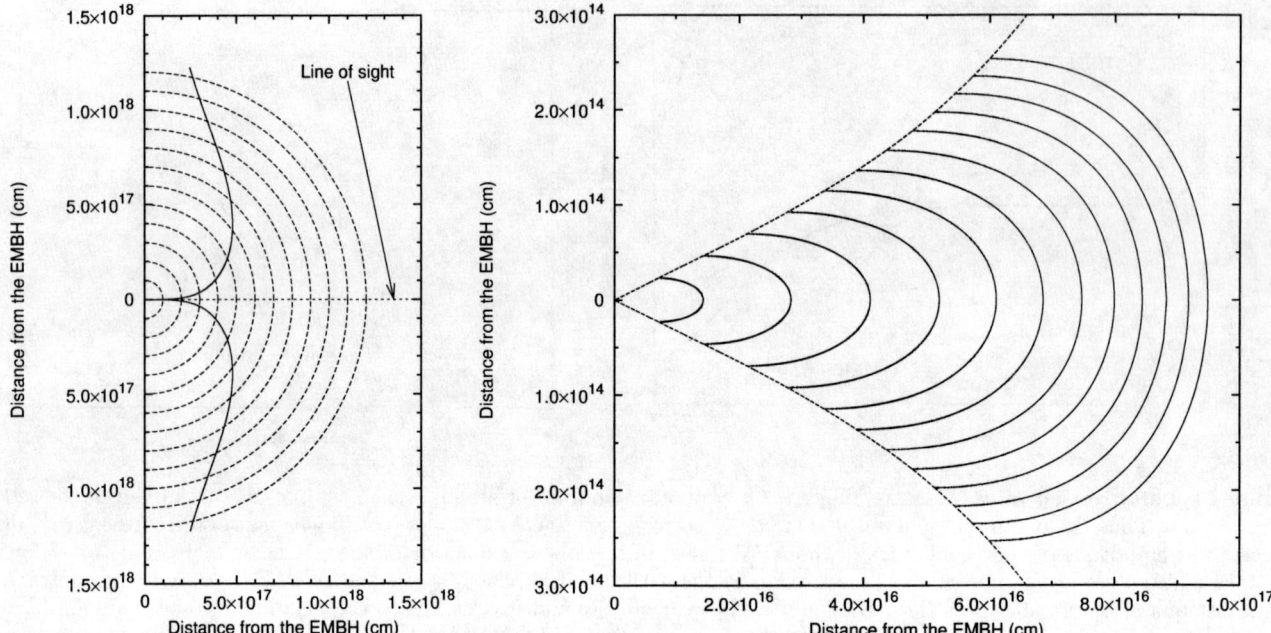

Figure 41: **Left)** This figure shows the temporal evolution of visible area of the ABM pulse. The dashed half-circles are the expanding ABM pulse at radii corresponding to different laboratory times. The black curve marks the boundary of the visible region. The EMBH is located at position (0,0) in this plot. Again, in the earliest GRB phases the visible region is squeezed along the line of sight, while in the final part of the afterglow phase almost all the emitted photons reach the observer. This time evolution of the visible area is crucial to the explanation of the GRB temporal structure. **Right)** Due to the extremely high and extremely varying Lorentz gamma factor, photons reaching the detector on the Earth at the same arrival time are actually emitted at very different times and positions. We represent here the surfaces of photon emission corresponding to selected values of the photon arrival time at the detector: the *equitemporal surfaces* (EQTS). Such surfaces differ from the ellipsoids described by Rees in the context of the expanding radio sources with typical Lorentz factor $\gamma \sim 4$ and constant. In fact, in GRB 991216 the Lorentz gamma factor ranges from 310 to 1. The EQTSes represented here (solid lines) correspond respectively to values of the arrival time ranging from $5\,s$ (the smallest surface on the left of the plot) to $60\,s$ (the largest one on the right). Each surface differs from the previous one by $5\,s$. To each EQTS contributes emission processes occurring at different values of the Lorentz gamma factor. The dashed lines are the boundaries of the visible area of the ABM pulse and the EMBH is located at position $(0,0)$ in this plot. Note the different scale on the two axes, indicating the very high EQTS "effective eccentricity". The time interval from $5\,s$ to $60\,s$ has been chosen to encompass the E-APE emission, ranging from $\gamma = 308.8$ to $\gamma = 56.84$.

so that equation (166) can be written in the form:

$$t_\mathrm{a} = t - \frac{\int_0^t v\left(t'\right) dt' + r_\mathrm{ds}}{c} \cos\vartheta + \frac{r_\mathrm{ds}}{c}, \qquad (168)$$

which reduces to Eq.(160) only if v is constant and r_ds is negligible with respect to $r(t)$.

Also from Eq.(168) we can obtain the equation describing the surface that emits the photons detected at an arrival time t_a. In this case, we no longer have ellipsoids of constant eccentricity $\frac{v}{c}$. Since the velocity is strongly varying from point to point, we have more complicated surfaces like the profiles reported in Fig. 41 where at every point there will be a tangent ellipsoid of a given eccentricity, but such an ellipsoid varies in eccentricity from point to point.

For a fixed time t of emission in Eq.(168), the allowed angular interval $\frac{v}{c} \leq \cos\vartheta \leq 1$ leads to a corresponding smearing of the arrival time t_a over the interval

$$\Delta t_a = \frac{r}{\gamma^2 c \left(1 + \frac{v}{c}\right)}. \qquad (169)$$

We need now to correct Eq.(168) for the cosmological expansion effects to get the wanted relation between t and t_a^d. We recall that (see section V)

$$t_a^d = (1+z)\, t_a\,, \qquad (170)$$

where z is the cosmological redshift. Our final relation is therefore:

$$t_a^d = (1+z)\left(t - \frac{\int_0^t v(t')\,dt' + r_{ds}}{c}\cos\vartheta + \frac{r_{ds}}{c}\right). \tag{171}$$

XXII. THE EMISSION PROCESS TAKING OFF-AXIS CONTRIBUTIONS INTO ACCOUNT

We now take into consideration the contributions of the off-axis emission to the afterglow to see if the previous positive results still hold and if some of the problems just stated can be overcome by a more detailed and relativistic treatment. The corresponding computation for the P-GRB structure will be presented elsewhere, where the time evolutions of the dyadosphere formation and its consequences on the P-GRB structures are presented following the work of Cherubini et al. [27], Ruffini & Vitagliano [155, 156], Ruffini et al. [158]. The effects on the P-GRB structure of the dyadosphere formation dominate those due to the angular spreading.

Following Eqs.(110–111), we recall that in the comoving frame of the expanding ABM pulse we suppose that the internal energy due to kinetic collision is instantly radiated away and that the corresponding emission is isotropic. As in section II, let $\Delta\varepsilon$ be the internal energy density developed in the collision. In the comoving frame the energy per unit of volume and per solid angle is simply

$$\left(\frac{dE}{dV\,d\Omega}\right)_\circ = \frac{\Delta\varepsilon}{4\pi} \tag{172}$$

due to the fact that the emission is isotropic in this frame. The total number of photons emitted is an invariant quantity independent of the frame used. Thus we can compute this quantity as seen by an observer in the comoving frame (which we denote with the subscript "o") and by an observer in the laboratory frame (which we denote with no subscripts). Doing this we find

$$\frac{dN_\gamma}{dt\,d\Omega\,d\Sigma} = \int_{shell}\left(\frac{dN_\gamma}{dt\,d\Omega\,d\Sigma}\right)_\circ \Lambda^{-3}\cos\vartheta, \tag{173}$$

where $\cos\vartheta$ comes from the projection of the elementary surface of the shell on the direction of propagation and $\Lambda = \gamma(1-\beta\cos\vartheta)$ is the Doppler factor introduced in the two following differential transformation

$$d\Omega_\circ = d\Omega \times \Lambda^{-2} \tag{174}$$

for the solid angle transformation and

$$dt_\circ = dt \times \Lambda^{-1} \tag{175}$$

for the time transformation. The integration in $d\Sigma$ is performed over the visible area of the ABM pulse at laboratory time t, namely with $0 \leq \vartheta \leq \vartheta_{max}$ and ϑ_{max} defined in section XXI (see Eq.(159) and Figs. 40–41). An extra Λ factor comes from the energy transformation:

$$E_\circ = E \times \Lambda. \tag{176}$$

See also Chiang & Dermer [28]. Thus finally we obtain:

$$\frac{dE}{dt\,d\Omega\,d\Sigma} = \int_{shell}\left(\frac{dE}{dt\,d\Omega\,d\Sigma}\right)_\circ \Lambda^{-4}\cos\vartheta. \tag{177}$$

Doing this we clearly identify $\left(\frac{dE}{dt\,d\Omega\,d\Sigma}\right)_\circ$ as the energy density in comoving frame up to a factor $\frac{v}{4\pi}$ (see Eq.(172)). Then we have:

$$\frac{dE}{dt\,d\Omega} = \int_{shell}\frac{\Delta\varepsilon}{4\pi}\,v\,\cos\vartheta\,\Lambda^{-4}\,d\Sigma, \tag{178}$$

where the integration in $d\Sigma$ is performed over the ABM pulse visible area at laboratory time t, namely with $0 \leq \vartheta \leq \vartheta_{max}$ and ϑ_{max} defined in section XXI.

Eq.(178) gives us the energy emitted toward the observer per unit solid angle and per unit laboratory time t in the laboratory frame. But what we really need is the energy emitted per unit solid angle and per unit detector arrival time t_a^d, so we must use the complete relation between t_a^d and t given in Eq.(171). First we have to multiply the integrand in Eq.(178) by the factor (dt/dt_a^d) to transform the energy density generated per unit of laboratory time t into the energy density generated per unit arrival time t_a^d. Then we have to integrate with respect to $d\Sigma$ over the *equitemporal surface* (EQTS, see section XXI) of constant arrival time t_a^d instead of the ABM pulse visible area at laboratory time t. The analog of Eq.(178) for the source luminosity in detector arrival time is then:

$$\frac{dE_\gamma}{dt_a^d d\Omega} = \int_{EQTS} \frac{\Delta\varepsilon}{4\pi} \, v \, \cos\vartheta \, \Lambda^{-4} \, \frac{dt}{dt_a^d} d\Sigma. \tag{179}$$

It is important to note that, in the present case of GRB 991216, the Doppler factor Λ^{-4} in Eq.(179) enhances the apparent luminosity of the burst, as compared to the intrinsic luminosity, by a factor which at the E-APE is in the range between 10^{10} and 10^{12}!

To perform the numerical integration of Eq.(179) we have implemented the following procedure for each fixed value of the laboratory time t:

1. We fix the laboratory time t.

2. We divide the interval of the allowed values $(v(t)/c) \leq \cos\vartheta \leq 1$ into N small steps, each one of amplitude

$$\Delta_N(\cos\vartheta) = \frac{1 - (v(t)/c)}{N}. \tag{180}$$

3. We select n directions defined by:

$$\cos\vartheta_n = 1 - n\Delta(\cos\vartheta), \tag{181}$$

where n is an integer, $0 \leq n \leq N$ and so $\vartheta_0 = 0$ and $\vartheta_n = \vartheta_{max}$.

4. For each ϑ_n we compute with Eq.(179) the contribution to the afterglow luminosity arising from an angular aperture corresponding to $\Delta_N(\cos\vartheta)$ around such a direction.

5. We compute for each value of n the corresponding values of the arrival time t_a^d using Eq.(171).

To obtain the total luminosity at arrival time t_a^d we sum together all the above contributions corresponding to the same t_a^d.

We first apply this treatment to the analysis of the afterglow using assumptions 1 and 2 of section. II, namely that the ISM density is constant $n_{ism} = <n_{ism}> = 1\,\text{particle}/\text{cm}^3$ and that the ABM is spherically symmetric.

Fig. 42 compares the new result for the afterglow luminosity as a function of the detector arrival time with the previous one obtained in section XV by neglecting off-axis emission. The main conclusions are:

1. The total energy emitted both in the radial approximation and in the full computation with the off-axis emission is conserved. This is a necessary condition for checking the consistency of the model.

2. The slope of the decreasing part of the afterglow is unchanged. We emphasize once more the great advantage of the radial approximation which has allowed to obtain an analytic expression for this slope.

3. The final phase of the afterglow ($\gamma < 2$) is largely affected by the late arrival of the radiation emitted at large angles. In fact in the radial approximation the luminosity goes abruptly to zero when γ reaches 1 while in the new complete treatment the behavior is much smoother due to the delayed arrival of the radiation emitted at large angles. Consequently, enforcing the energy conservation, in the rising part of the afterglow the luminosity in the new treatment is shown to be slightly smaller than in the radial case.

In order to acquire a better understanding of the effects of angular spreading, we have found it helpful to analyze the radiation emitted from selected angles ϑ between 0 and ϑ_{max}. This is in addition to the integration results presented in Fig. 42. In Fig. 42 we show the results of such an analysis plotting the contributions to the total luminosity corresponding to selected values of n in Eq.(181). We easily see that radiation emitted at large angles is time shifted with respect to that emitted near the line of sight. In fact the afterglow peak occurs later going to higher n values (see Fig. 42).

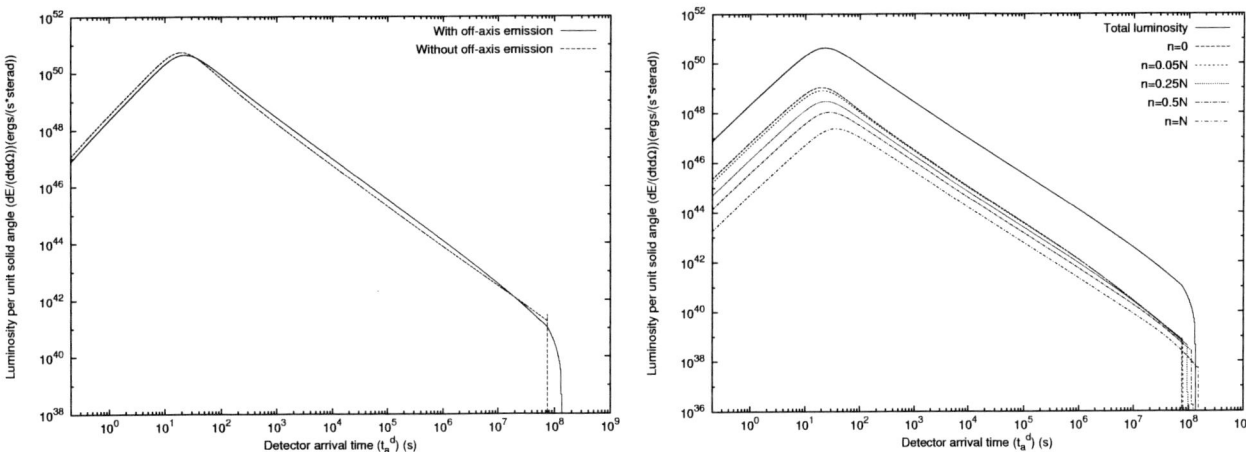

Figure 42: **Left)** The predicted afterglow curve for GRB 991216 assuming a constant ISM density equal to 1 particle/cm^3 and taking into account all the effects due to off-axis emission (solid line). For comparison we plot also the corresponding curve obtained in the simple radial approximation (dashed line). We see that this last curve falls sharply to zero when the ABM pulse reaches $\gamma = 1$, while the first one has a much smoother behavior due to the time delay in the arrival of the photons emitted at large ϑ. Recall that when γ tends to 1, the maximum allowed values of ϑ tend to 90°. **Right)** This figure shows how the radiation emitted from different angles contributes to the afterglow luminosity. The solid line on the top of the picture is the total luminosity as in the previous plots. The other dashed and dotted curves represent the radiation components corresponding to selected values of n in Eq.(181). From the upper to the lower one they corresponds respectively to $n = 0$, $n = 0.05N$, $n = 0.25N$, $n = 0.5N$, $n = N$, where in this plot $N = 200$. We can easily see that the radiation emitted at large angles ($n = N$) is time shifted with respect to that emitted near the line of sight ($n = 0$).

Table IV: For each ISM density peak represented in Fig. 43 we give the initial radius r, the corresponding comoving time τ, laboratory time t, arrival time at the detector t_a^d, diameter of the ABM pulse visible area d_v, Lorentz factor γ and observed duration Δt_a^d of the afterglow luminosity peaks generated by each density peak. In the last column, the apparent motion in the radial coordinate, evaluated in the arrival time at the detector, leads to an enormous "superluminal" behavior, up to $9.5 \times 10^4\,c$.

Peak	$r(cm)$	$\tau(s)$	$t(s)$	$t_a^d(s)$	$d_v(cm)$	$\Delta t_a^d(s)$	γ	"Superluminal" $v \equiv \frac{r}{t_a^d}$
A	4.50×10^{16}	4.88×10^3	1.50×10^6	15.8	2.95×10^{14}	0.400	303.8	$9.5 \times 10^4 c$
B	5.20×10^{16}	5.74×10^3	1.73×10^6	19.0	3.89×10^{14}	0.622	265.4	$9.1 \times 10^4 c$
C	5.70×10^{16}	6.54×10^3	1.90×10^6	22.9	5.83×10^{14}	1.13	200.5	$8.3 \times 10^4 c$
D	6.20×10^{16}	7.64×10^3	2.07×10^6	30.1	9.03×10^{14}	5.16	139.9	$6.9 \times 10^4 c$
E	6.50×10^{16}	9.22×10^3	2.17×10^6	55.9	2.27×10^{15}	10.2	57.23	$3.9 \times 10^4 c$
F	6.80×10^{16}	1.10×10^4	2.27×10^6	87.4	2.42×10^{15}	10.6	56.24	$2.6 \times 10^4 c$

XXIII. THE E-APE TEMPORAL SUBSTRUCTURES TAKING INTO ACCOUNT THE OFF-AXIS EMISSION

We are now ready to reconsider the problem of the ISM inhomogeneity generating the temporal substructures in the E-APE by integrating on the EQTS surfaces and improving on the considerations based on the purely radial approximation. We have created (see details in Ruffini et al. [152]) an ISM inhomogeneity "mask" (see Fig. 43 and Tab. IV) with the main criteria that the density inhomogeneities and their spatial distribution still fulfill $<n_{ism}>=$ 1 particle/cm^3.

The results are given in Fig. 44. We obtain, in perfect agreement with the observations:

1. the theoretically computed intensity of the A, B, C peaks as a function of the ISM inhomogneities;

2. the fast rise and exponential decay shape for each peak;

3. a continuous and smooth emission between the peaks.

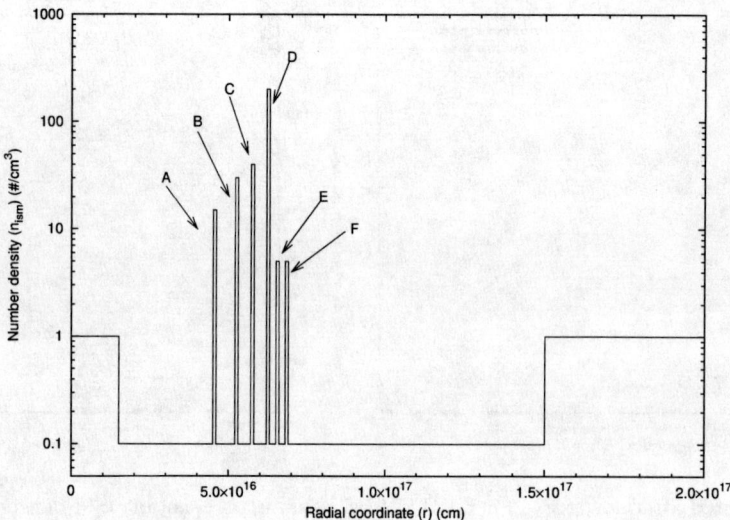

Figure 43: The density profile ("mask") of an ISM cloud used to reproduce the GRB 991216 temporal structure. As before, the radial coordinate is measured from the black hole. In this cloud we have six "spikes" with overdensity separated by low density regions. Each spike has the same spatial extension of 10^{15} cm. The cloud average density is $<n_{ism}>=1\,\mathrm{particle/cm^3}$.

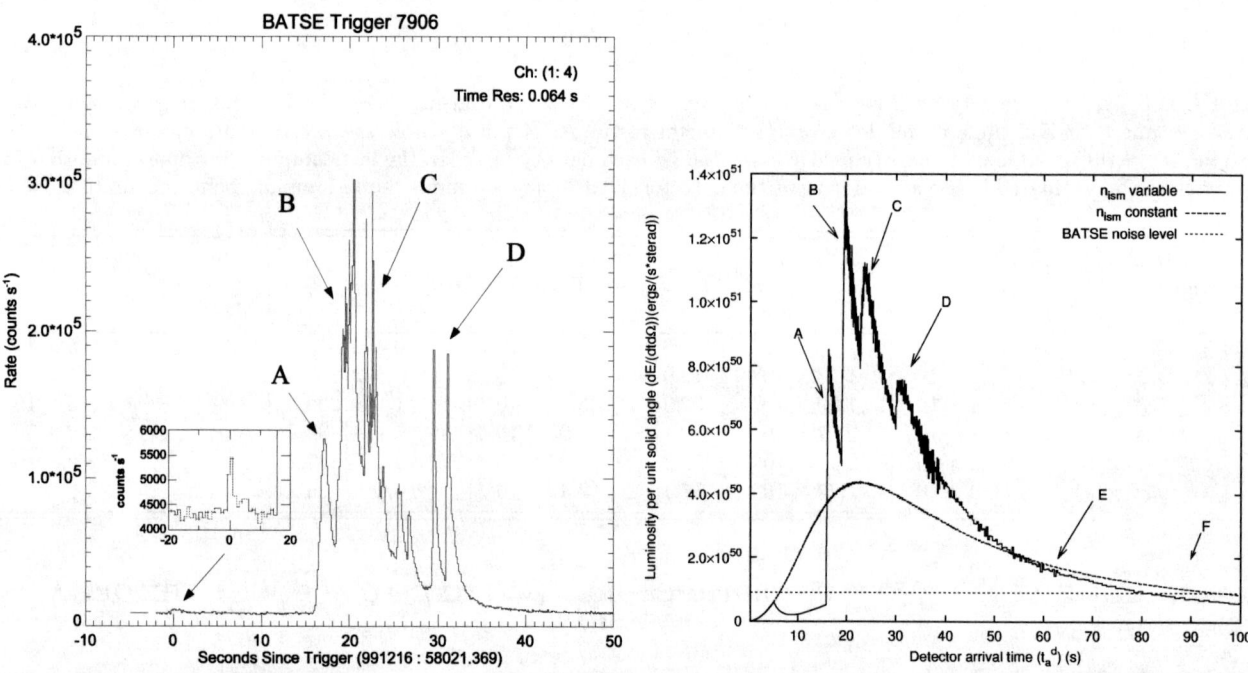

Figure 44: **Left)** The BATSE data on the E-APE of GRB 991216 (source: BATSE GRB light curves [5]) together with an enlargement of the P-GRB data (source: BATSE Rapid Burst Response [6]). For convenience each E-APE peak has been labeled by a different uppercase Latin letter. **Right)** The source luminosity connected to the mask in Fig. 43 is given as a function of the detector arrival time (solid "spiky" line) with the corresponding curve for the case of constant $n_{ism} = 1\,\mathrm{particle/cm^3}$ (dashed smooth line) and the BATSE noise level (dotted horizontal line). The "noise" observed in the theoretical curves is due to the discretization process adopted, described in Ruffini et al. [152], for the description of the angular spreading of the scattered radiation. For each fixed value of the laboratory time we have summed 500 different contributions from different angles. The integration of the equation of motion of this system is performed in 22, 314, 500 contributions to be considered. An increase in the number of steps and in the precision of the numerical computation would lead to a smoother curve.

Interestingly, the signals from shells E and F, which have a density inhomogeneity comparable to A, are undetectable. The reason is due to a variety of relativistic effects and partly to the spreading in the arrival time, which for A, corresponding to $\gamma = 303.8$ is $0.4\,s$ while for E (F) corresponding to $\gamma = 57.23$ (56.24) is of $10.2\,s$ ($10.6\,s$) (see Tab. IV and Ruffini et al. [147, 152]).

In the case of D, the agreement with the arrival time is reached, but we do not obtain the double peaked structure. The ABM pulse visible area diameter at the moment of interaction with the D shell is $\sim 1.0 \times 10^{15}$ cm, equal to the extension of the ISM shell (see Tab. IV and Ruffini et al. [147, 152]). Under these conditions, the concentric shell approximation does not hold anymore: the disagreement with the observations simply makes manifest the need for a more detailed description of the three dimensional nature of the ISM cloud.

The physical reasons for these results can be simply summarized: we can distinguish two different regimes corresponding in the afterglow of GRB 991216 respectively to $\gamma > 150$ and to $\gamma < 150$. For different sources this value may be slightly different. In the E-APE region ($\gamma > 150$) the GRB substructure intensities indeed correlate with the ISM inhomogeneities. In this limited region (see peaks A, B, C) the Lorentz gamma factor of the ABM pulse ranges from $\gamma \sim 304$ to $\gamma \sim 200$. The boundary of the visible region is smaller than the thickness ΔR of the inhomogeneities (see Fig. 41 and Tab. IV). Under this condition the adopted spherical approximation is not only mathematically simpler but also fully justified. The angular spreading is not strong enough to wipe out the signal from the inhomogeneity spike.

As we descend in the afterglow ($\gamma < 150$), the Lorentz gamma factor decreases markedly and in the border line case of peak D $\gamma \sim 140$. For the peaks E and F we have $\gamma \sim 50$ and, under these circumstances, the boundary of the visible region becomes much larger than the thickness ΔR of the inhomogeneities (see Fig. 41 and Tab. IV). A three dimensional description would be necessary, breaking the spherical symmetry and making the computation more difficult. However we do not need to perform this more complex analysis for peaks E and F: any three dimensional description would *a fortiori* augment the smoothing of the observed flux. The spherically symmetric description of the inhomogeneities is already enough to prove the overwhelming effect of the angular spreading (Ruffini et al. [152]).

On this general issue of the possible explanation of the observed substructures with the ISM inhomogeneities, there exists in the literature two extreme points of view: the one by Fenimore and collaborators (see e.g. Fenimore et al. [47], Fenimore [48], Fenimore et al. [49]) and Piran and collaborators (see e.g. Piran [115, 116], Piro et al., [119], Sari & Piran [164]) on one side and the one by Dermer and collaborators (Dermer [37], Dermer et al. [39], Dermer & Mitman [41]) on the other.

Fenimore and collaborators have emphasized the relevance of a specific signature to be expected in the collision of a relativistic expanding shell with the ISM, what they call a fast rise and exponential decay (FRED) shape. This feature is confirmed by our analysis (see peaks A, B, C in Fig. 44). However they also conclude, sharing the opinion by Piran and collaborators, that the variability observed in GRBs is inconsistent with causally connected variations in a single, symmetric, relativistic shell interacting with the ambient material ("external shocks") (Fenimore et al. [49]). In their opinion the solution of the short time variability has to be envisioned within the protracted activity of an unspecified "inner engine" (Sari & Piran [164]); see as well Mészáros & Rees [97, 98], Mészáros [99], Panaitescu & Mészáros [112], Rees & Mészáros [127].

On the other hand, Dermer and collaborators, by considering an idealized process occurring at a fixed $\gamma = 300$, have reached the opposite conclusions and they purport that GRB light curves are tomographic images of the density distributions of the medium surrounding the sources of GRBs (Dermer & Mitman [41]).

From our analysis we can conclude that Dermer's conclusions are correct for $\gamma \sim 300$ and do indeed hold for $\gamma > 150$. However, as the gamma factor drops from $\gamma \sim 150$ to $\gamma \sim 1$ (see Fig 8), the intensity due to the inhomogeneities markedly decreases also due to the angular spreading (events E and F). The initial Lorentz factor of the ABM pulse $\gamma \sim 310$ decreases very rapidly to $\gamma \sim 150$ as soon as a fraction of a typical ISM cloud is engulfed (see Tab. IV). We conclude that the "tomography" is indeed effective, but uniquely in the first ISM region close to the source and for GRBs with $\gamma > 150$.

One of the most striking feature in our analysis is clearly represented by the fact that the inhomogeneities of a mask of radial dimension of the order of 10^{17} cm give rise to arrival time signals of the order of $20\,s$. This outstanding result implies an apparent "superluminal velocity" of $\sim 10^5 c$ (see Tab. IV). The "superluminal velocity" here considered, first introduced in Ruffini et al. [143], refers to the motion along the line of sight. This effect is proportional to γ^2. It is much larger than the one usually considered in the literature, within the context of radio sources and microquasars (see e.g. Mirabel & Rodriguez [100]), referring to the component of the velocity at right angles to the line of sight (see details in Ruffini et al. [152]). This second effect is in fact proportional to γ (see Rees [126]). We recall that this "superluminal velocty" was the starting point for the enunciation of the RSTT paradigm (Ruffini et al. [143]), emphasizing the need of the knowledge of the *entire* past worldlines of the source. This need has been further clarified here in the determination of the EQTS surfaces (see Fig. 41 which indeed depend on an integral of the Lorentz gamma factor extended over the *entire* past worldlines of the source. In turn, therefore, the agreement between the observed structures and the theoretical predicted ones (see Figs. 3–44) is also an extremely stringent additional test on the values of the Lorentz gamma factor determined as a function of the radial coordinate within the EMBH theory (see Fig. 8).

XXIV. ON THE INSTANTANEOUS SPECTRUM OF GRBS

Variability on the shortest time scale ever observed in nature is the main message we have acquired from the theoretical understanding of GRB astrophysical phenomena (see sections V,VII–XI). This situation is made even more extreme by the fact that astronomical and astrophysical observations are carried out in the "pathological" time coordinate of the photon arrival time at the detector (see section V), whereby the first 10^4 seconds of the GRB phenomena are further compressed in ~ 0.1 seconds (see Tab. I) and further enhanced. The understanding that in these first 10^4 seconds four different physical eras of the GRB phenomena occur has led us to a sentiment of natural skepticism toward any global or average description of the GRB phenomenon. We start to realize that such average descriptions mediate on totally different physical processes and lead to very questionable results. Such skepticism was even strengthened as soon as we realized that the characteristic quantities usually adopted for the description of the bursts, e.g. T_{50} and T_{90}, which so many tried for years to explain within the context of the internal shock model (see e.g. Fenimore [48], Fenimore et al. [49], Paczyński & Xu [108], Piran [115], Rees & Mészáros [127], Sari & Piran [164] and references therein) were actually referring not at all to the bursts but to the extended emission from the peak of the afterglow: the E-APE! In this sense they were quite irrelevant for understanding the nature of the GRB source and were at most of interest for inquiring the structure of the ISM a few light months away from the source! It has been then with this sentiment of marked skepticism toward a global approach that we have started to consider the problem of the spectrum of GRBs and the validity of the band relation (Band et al. [4]). To attempt an integral description of the spectra of the GRBs extending over 10^6 seconds in arrival time is clearly meaningless. It mediates on two conceptually physically different phases of GRBs: the injector phase and the beam-target phase (Ruffini et al. [144]). In addition, in each of these phases many specific eras are present and each one of these eras needs due attention and can lead in principle to a different instantaneous spectrum. The fact that the spectral distribution observed by Band was a non-thermal one has been a very strong objection to consider any thermal spectrum. The situation became so extreme in the recent years that the sole appearance of a thermal spectrum in any part of a theoretical paper was considered a good reason for rejecting the paper by a refereed journal and to discard the validity of that work.

Having developed the very powerful theoretical tool of the EQTS surfaces (see section XXI and Ruffini et al. [147, 152]) and having been successful in having established the substructure of the E-APE, in addition to the features of the afterglow, we have decided to approach the instantaneous spectra of the GRBs in Ruffini et al. [149]. In the abstract of that paper, we summarize as follows the results: "*A theoretical attempt to identify the physical process at the basis of the afterglow emission of GRBs is presented, assuming GRB 991216 as a prototype. Such a physical process is identified in a mechanism leading to a thermal emission occurring in the comoving frame of the shock wave originating the GRBs. For the determination of the actually observed GRB luminosities and spectra at a given arrival time, the concept of equitemporal surfaces (EQTS, see Ruffini et al. [147]) has to be implemented: the final results comprehend an integration over an infinite number of planckian spectra, weighted by appropriate relativistic transformations, each one corresponding to a different viewing angle in the past light cone of the observer. The relativistic transformations have been computed on the ground of the knowledge of the already determined equations of motion of GRBs within the EMBH theory (Ruffini et al. [143, 144, 147]). The only free parameter of the present theory is then the dimension of the "effective cavity" where the thermalization process occurs. A precise fit ($\chi^2 \simeq 1.08$) of the observed luminosity in the 2–10 keV band of GRB 991216 is presented as well as a detailed estimate of the observed luminosity in the 50–300 keV band and of the expected one in the 10–50 keV band. The long awaited explanation of the observed hard-to-soft transition in GRBs is also presented*" (Ruffini et al. [149]). It is interesting that this theoretical result, which up to few years ago were hardly testable due to the paucity of photons collected by the detectors, have now become a necessity in order to interpret the splendid observational results of the new families of space observatory like Chandra and XMM (see e.g. Borozdin & Trudolyubov [22], Watson et al. [183, 184]).

Prior to our work, the possibility that the non-thermal looking spectrum of GRBs can be found as a superposition of a set of thermal blackbody spectrum was forcefully expressed in a simple paper by Blinnikov, Kozyreva & Panchenko [17]. These three authors have expressed in an analytic treatment that indeed the time integration of the black body planckian spectrum with a temperature varying with time following a simple power-law and expanding with another power-law can lead to a non-thermal spectrum in agreement with the observed Band relations. To obtain this result, they use two indexes for their qualitative analysis to be fitted by the observational data. Toward the end of their paper they finally quoted "*In reality, not only time, but also space integration takes place. As shown by Rees [126], (see also Drozdova & Panchenko [43], Sari [163]) in the case of an expanding emitting shell an observer simultaneously detects radiation produced in different moments of time (thus, with different temperatures) on the ellipsoidal or egg-like surface. The integration over this surface can give the same effect as the integration over time done in this paper, but we do not perform this here because the result strongly depends on the unknown geometry of the emitting surface"* (Blinnikov, Kozyreva & Panchenko [17]). This treatment which they outline but they discard due to the difficulty of defining the geometry of the EQTS is exactly what we have done. Our treatment has only one free parameter and can fit the data of GRBs in a range between a few seconds all the way up to 10^6 seconds. There is a basic observational feature between our treatment and the one by Blinnikov, Kozyreva & Panchenko [17]: their instantaneous spectral distribution has necessarily to be a blackbody one, while in our case is represented by an integration over an infinite number of planckian spectra, weighted by appropriate relativistic transformations, each one corresponding to a different viewing angle in the past light cone of the observer. The difference between such unique spectra should be simply discernible using the observations of XMM, Chandra, and of future space observatories.

XXV. THE OBSERVATION OF THE IRON LINES IN GRB 991216: ON A POSSIBLE GRB-SUPERNOVA TIME SEQUENCE

We have seen in the previous sections how the time structure of the E-APE gives information on the composition of the interstellar matter at distances of the order of 5×10^{16} cm from the source. We would like now to point out that the data on the iron lines from the Chandra satellite on the GRB 991216 (Piro et al., [119]) and similar observations from other sources (Amati et al. [3], Piro et al. [118], Piro et al., [119]) make it possible to extend this analysis to a larger distance scale, possibly all the way out to a few light years, and consequently probe the distribution of stars in the surroundings of the newly formed EMBH.

Most importantly, these considerations lead to a new paradigm for the interpretation of the supernova-GRB correlation (see Ruffini et al. [145]). Indeed a correlation between the occurrence of GRBs and supernova events exists and has been established by the works of Bloom et al. [18], Galama et al. [57, 58, 59], Kulkarni et al. [84], Piran [115], Piro et al. [117], Rhoads [130], van Paradijs et al. [178].

Such an association has been assumed to indicate that GRBs are generated by supernova explosions (see e.g. Kulkarni et al. [84]). In turn, such a point of view has implied further consequences: the optical and radio data of the supernova have been attributed to the GRB afterglow, and many theorists have tried to encompass these data and explain them as a genuine component of the GRB scenario.

We propose instead an alternative point of view implying a very clear distinction between the GRB phenomenon and the supernova: if relativistic effects presented in the RSTT paradigm are properly taken into account, then a kinematically viable explanation can be given of the supernova-GRB association. We still use GRB 991216 as a prototypical case.

The GRB-Supernova Time Sequence paradigm, which we have indicated for short as GSTS paradigm (see Ruffini et al. [145]), states that: *A massive GRB-progenitor star P_1 of mass M_1 undergoes gravitational collapse to an EMBH. During this process a dyadosphere is formed and subsequently the P-GRB and the E-APE are generated in sequence. They propagate and impact, with their photon and neutrino components, on a second supernova-progenitor star P_2 of mass M_2. Assuming that both stars were generated approximately at the same time, we expect to have $M_2 < M_1$. Under some special conditions of the thermonuclear evolution of the supernova-progenitor star P_2, the collision of the P-GRB and the E-APE with the star P_2 can induce its supernova explosion.*

Especially relevant to our paradigm are the following data from the Chandra satellite (see Piro et al., [119]):

1. At the arrival time of 37 hr after the initial burst there is evidence of iron emission lines for GRB 991216.

2. The emission lines are present during the entire observation period of 10^4 s. The iron lines could also have been produced earlier, before Chandra was observing. Thus the times used in these calculations are not unique: they do serve to provide an example of the scenario.

3. The emission lines appear to have a peak at an energy of 3.49 ± 0.06 keV which, at a redshift $z = 1.00 \pm 0.02$ corresponds to an hydrogen-like iron line at 6.97 keV at rest. This source does not appear to have any significant motion departing from the cosmological flow. The iron lines have a width of 0.23 keV consistent with a radial velocity field of $0.1c$. The iron lines are only a small fraction of the observed flux.

On the basis of the explicit computations of the different eras presented in the above sections, we make three key points:

1. An arrival time of 37 hr in the detector frame corresponds to a radial distance from the EMBH travelled by the ABM pulse of 3.94×10^{17} cm in the laboratory frame (see Tab. I).

2. It is likely that a few stars are present within that radius as members of a cluster. It has become evident from observations of dense clusters of star-forming regions that a stellar average density of typically 10^2pc^{-3} (Beck

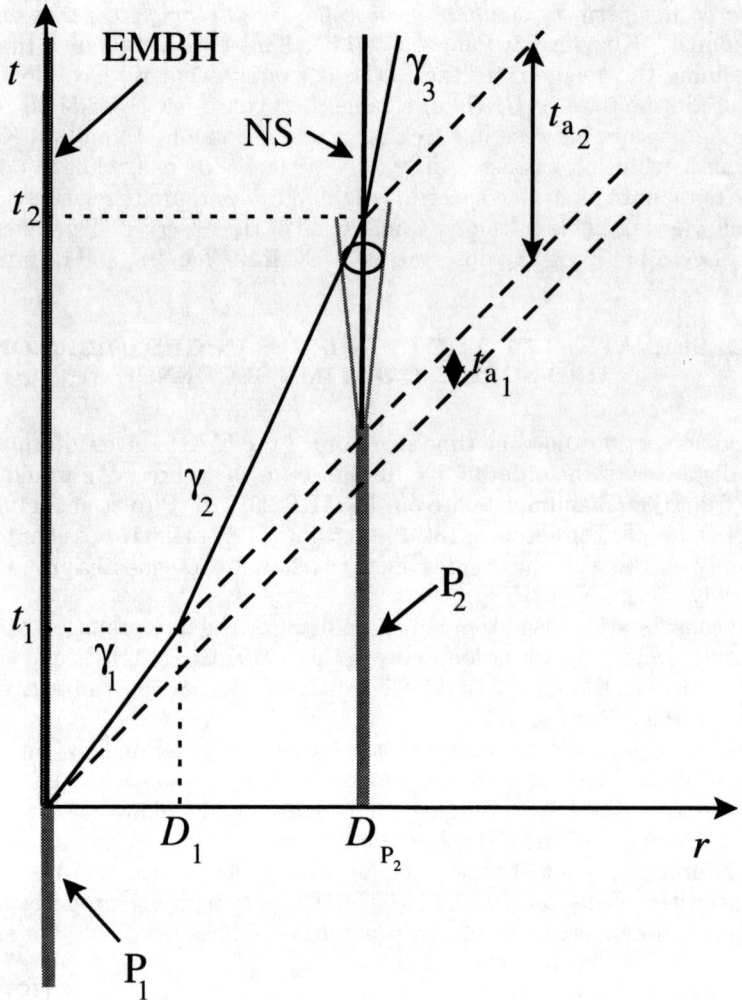

Figure 45: A qualitative simplified space-time diagram (in arbitrary units) illustrating the GSTS paradigm. The EMBH, originating from the gravitational collapse of a massive GRB-progenitor star P_1, and the massive supernova-progenitor star P_2-neutron star (P_2-NS) system, separated by a radial distance D_{P_2}, are assumed to be at rest in in the laboratory frame. Their worldlines are represented by two parallel vertical lines. The supernova shell moving at $0.1c$ generated by the P_2-NS transition is represented by the dotted line cone. The solid line represents the motion of the pulse, as if it would move with an "effective" constant gamma factor γ_1 during the eras reaching the condition of transparency. Similarly, another "effective" constant gamma factor $\gamma_2 < \gamma_1$ applies during era IV up to the collision with the P_2-NS system. A third "effective" constant gamma factor $\gamma_3 < \gamma_2$ occurs during era V after the collision as the nonrelativistic regime of expansion is reached. The dashed lines at 45 degrees represent signals propagating at speed of light.

et al. [8]) should be expected. There is also the distinct possibility for this case and other systems that the stars P_1 and P_2 are members of a binary system.

3. The possible observations at different wavelengths of the supernova crucially depend on the relative intensities between the GRB and the supernova as well as on the value of the distance and the redshift of the source. In the present case of GRB 991216, the expected optical and radio emission from the supernova are many orders of magnitude smaller than the GRB intensity. The opposite situation will be encountered in GRB 980425 (Ruffini et al. [151]).

In order to reach an intuitive understanding of these complex computations we present a schematic very simplified diagram (not to scale) in Fig. 45.

We now describe the sequence of events and the specific data corresponding to the GSTS paradigm:

1. The two stars P_1 and P_2 are separated by a distance $D_{P_2} = 3.94 \times 10^{17}$ cm in the laboratory frame, see Fig. 45. Both stars are at rest in the inertial laboratory frame. At laboratory time $t = 0$ and at comoving time

$\tau = 0$, the gravitational collapse of the GRB-progenitor star P_1 occurs, and the initial emission of gravitational radiation or a neutrino burst from the event then synchronizes this event with the arrival times $t_a = 0$ at the supernova-progenitor star P_2 and $t_a^d = 0$ for the distant observer at rest with the detector. The electromagnetic radiation emitted by the gravitational collapse process is instead practically zero, due to the optical thickness of the material at this stage (Bianco et al. [12], see Tab. I).

2. From Tab. I, at laboratory time $t_1 = 6.48 \times 10^3\,s$ and at a distance from the EMBH of $D_1 = 1.94 \times 10^{14}\,cm$, the condition of transparency for the PEMB pulse is reached and the P-GRB is emitted (see section IX). This time is recorded in arrival time at the detector $t_{a_1}^d = 8.41 \times 10^{-2}\,s$, and, at P_2, at $t_{a_1} = 4.20 \times 10^{-2}\,s$. The fact that the PEMB pulse in an arrival time of $8.41 \times 10^{-2}\,s$ covers a distance of $1.94 \times 10^{14}\,cm$ gives rise to an apparent "superluminal" effect. This apparent paradox can be straightforwardly explained by introducing an "effective" gamma factor, see Ruffini et al. [145].

3. At laboratory time $t = 1.73 \times 10^6\,s$ and at a distance from the EMBH of $5.18 \times 10^{16}\,cm$ in the laboratory frame, the peak of the E-APE is reached which is recorded at the arrival time $t_a = 9.93\,s$ at P_2 and $t_a^d = 19.87\,s$ at the detector. This also gives rise to an apparent "superluminal" effect.

4. At a distance $D_{P_2} = 3.94 \times 10^{17}\,cm$, the two bursts described in the above points 2) and 3) collide with the supernova-progenitor star P_2 at arrival times $t_{a_1} = 4.20 \times 10^{-2}\,s$ and $t_a = 9.93\,s$ respectively. They can then induce the supernova explosion of the massive star P_2.

5. The associated supernova shell expands with velocity $0.1c$.

6. The expanding supernova shell is reached by the ABM pulse generating the afterglow with a delay of $t_{a_2} = 18.5\,hr$ in arrival time following the arrival of the P-GRB and the E-APE. This time delay coincides with the interval of laboratory time separating the two events, since the P_2 is at rest in the inertial laboratory frame (see Ruffini et al. [145]). The ABM pulse has travelled in the laboratory frame a distance $D_{P_2} - D_1 \simeq D_{P_2} = 3.94 \times 10^{17}\,cm$ in a laboratory time $t_2 - t_1 \simeq t_2 = 1.32 \times 10^7\,s$ (neglecting the supernova expansion).

The collision of the pulse with the supernova shell occurs at $\gamma \simeq 4.0$. By this time the supernova shell has reached a dimension of $1.997 \times 10^{14}\,cm$, which is consistent with the observations from the Chandra satellite.

In these considerations on GRB 991216 the supernova remnant has been assumed to be close to but not exactly along the line of sight extending from the EMBH to the distant observer. If such an alignment should exist for other GRBs, it would lead to an observation of iron absorption lines as well as to an increase in the radiation observed in the afterglow corresponding to the crossing of the supernova shell by the ABM pulse. In fact, as the ABM pulse engulfs the baryonic matter of the remnant, above and beyond the normal interstellar medium baryonic matter, the conservation of energy and momentum implies that a larger amount of internal energy is available and radiated in the process (see section XIII). This increased energy-momentum loss will generally affect the slope of the afterglow decay, approaching more rapidly a nonrelativistic expansion phase (details are given in section XVIII).

It is quite clear that as soon as the relativistic transformations of the RSTT paradigm are duly taken into account, the sequence of events between the supernova and the GRB occurrences are exactly the opposite of the one postulated in the so-called "supranova" scenario (Vietri & Stella [180, 181], Vietri et al. [182]). This can be considered a very appropriate pedagogical example of how classical nonrelativistic applied to ultrarelativistic regimes can indeed subvert the very causal relation between events.

If we now turn to the possibility of dynamically implementing the scenario, there are at least three different possibilities:

1. Particularly attractive is the possibility that a massive star P_2 has rapidly evolved during its thermonuclear evolution to a white dwarf (see e.g. Chandrasekhar [25]). It it then sufficient that the P-GRB and the E-APE implode the star sufficiently as to reach a central density above the critical density for the ignition of thermonuclear burning. Consequently, the explosion of the star P_2 occurs, and a significant fraction of a solar mass of iron is generated. These configurations are currently generally considered precursors of some type I supernovae (see e.g. Filippenko [51] and references therein).

2. Alternatively, the massive star P_2 can have evolved to the condition of being close to the point of gravitational collapse, having developed the formation of an iron-silicon core, type II supernovae. The above transfer of energy momentum from the P-GRB and the E-APE may enhance the capture of the electrons on the iron nuclei and consequently decrease the Fermi energy of the core, leading to the onset of gravitational instability (see e.g. Bethe [11] p. 270 and followings). Since the time for the final evolution of a massive star with an iron-silicon core is short, this event requires a well tuned coincidence.

3. The pressure wave may trigger massive and instantaneous nuclear burning process, with corresponding changes in the chemical composition of the star, leading to the collapse.

The GSTS paradigm has been applied to the case of the GRB 980425 - SN1998bw which, with a red shift of 0.0083, is one of the closest and weaker GRBs observed. In this case, the radio and the optical emission of the supernova is distinctively observed. For this particular case, the EMBH appears to have a significantly lower value of the parameter ξ and the validity of the GSTS paradigm presented here is confirmed (see Ruffini et al. [151]).

XXVI. GENERAL CONSIDERATIONS ON THE EMBH FORMATION

Before concluding let us consider the problem of the EMBH formation. Such a problem has been debated for many years since the earliest discussions in 1970 in Princeton and has been finally clarified and addressed in general terms to justify the plausibility of the hypothesis in Ruffini [138]. There has been a basic change of paradigm. All the considerations on the electric charge of stars were traditionally directed, following the classical work by Shvartsman [173] all the way to the fundamental book by Punsly [125], to the presence of a net charge on the star surface in a steady state condition. The star can be endowed of rotation and magnetic field and surrounded by plasma, like in the case of Goldreich & Julian [69], or, in the case of absence of both magnetic field and rotation, the electrostatic processes can be related to the depth of the gravitational well, like in the treatment of Shvartsman [173]. However, in neither cases it is possible to reach the condition of the overcritical field needed for pair creation nor have the condition of no baryonic contamination discussed in sections III, VII and essential for the dyadosphere formation. The basic conceptual point is that GRBs are maybe the most violent transient phenomenon occurring in the universe and so the condition for the dyadosphere creation have to be searched in a transient phenomenon. The solution is related to the most transient phenomenon occurring in the life of a star: the process of gravitational collapse.

Having acquired such a fundamental understanding, the next step is to estimate the amount of polarization needed in order to reach the fully relativistic condition

$$\frac{Q}{M\sqrt{G}} = 1 \ . \tag{182}$$

Recalling that the charge to mass ratio of a proton is $q_p/\left(m_p\sqrt{G}\right) = 1.1 \times 10^{18}$, it is enough to have an excess of one quantum of charge every 10^{18} nucleons in the core of the collapsing star to obtain an extreme EMBH after the occurrence of the gravitational collapse. Physically this means that we are dealing with a process of charge segregation between the core and the outer part of the star which has the opposite sign of net charge in order to enforce the overall charge neutrality condition. We here emphasize the name "charge segregation" instead of the name "charge separation" in order to contrast a very mild charge surplus created in different part of the star, keeping the overall charge neutrality, from the much more extreme condition of charge separation in which all the charges of the atomic component of the star are separated. It is indeed reassuring that such a core, endowed with charge segregation, is indeed stable with respect to the Fermi-Chandrasekhar criteria for the stability of self-gravitating stars duly extended from the magnetic to the electric case: the electric energy of such a core is consistently smaller than its gravitational energy (see Boccaletti et al. [21]).

Such a condition of charge segregation between the core and the oppositely charged star surface layer can be reached under a very large number of physical conditions. We consider, for simplicity, one of the oldest example: the one of a star endowed with both a magnetic field and rotation. It is proved that a typical magnetic field expected for the ISM is $B_\circ \sim 10^{-5}\,G$ (Ferrière [50]). We further assume, consistently with the data which we have acquired and verified in the present article (see sections XIII, XIX), that also in the galaxy where GRB 991216 occurred the ISM has an average density of $n_{ism} = 1\,proton/cm^3$. From this value of density we have that an ISM cloud with mass $M \sim 10 M_\odot$ occupies a sphere of radius $R_\circ \sim 1.4 \times 10^{19}\,cm$. If this sphere collapse to a star with radius $R = R_\odot$, from the flux conservation we obtain that it is enough for this star to rotate with the most reasonable angular speed

$$\Omega \sim \frac{\xi M c \sqrt{G}}{R_\odot R_\circ^2 B_\circ} \tag{183}$$

to conclude that the progenitor star core is endowed of a charge to mass ratio equal to ξ. In the extreme case of Eq.(182) we have $\xi = 1$ and so the angular speed is $\Omega \sim 1.1 \times 10^{-3}\,rad/s$ — i.e. one round in $1.5\,hr$ — and correspondingly we have smaller Ω values for $\xi < 1$ (see Boccaletti et al. [21]). Clearly the overall neutrality is guaranteed by the oppositely charged baryonic matter which is the one measured by the B parameter in the EMBH model (see sections VIII–IX). The smallness of the B value clearly points to the absence of an extended envelope of the progenitor star.

The formation process of such an electromagnetised progenitor star will be clearly affected by the presence of differential rotation, the consequent amplification of the magnetic field and a variety of magnetohydrodynamical problems which will affect somewhat the simplicity of the heuristic Eq.(183). Similarly the process of gravitational collapse of such a progenitor star endowed with rotation will lead to complex phenomena of "gravitationally induced electromagnetic radiation" (Johnston et al. [80]) and of "electromagnetically induced gravitational radiation" (Johnston et al. [81]) which will tend to reduce both the eccentricity and the angular velocity of the collapsing core. The general outcome of gravitational collapse will be a Kerr-Newmann spacetime. It is interesting that such a general case will break the degeneracy in (μ, ξ) described in section X (see Ruffini et al. [152]). In this article we have addressed the much simpler case of a solution in which $(cL)/(GM^2) \ll 1$ and the treatment can be well approximated by a collapse described by a Reissner-Nordström geometry.

In addition to this scenario, based on the role of magnetic field and rotation, we are as well pursuing the possible generation of the charge segregation by quantum effects at the surface of the Fermi semi-degenerate core. In this framework, it is particularly interesting to consider the purely electric analog of the Chandrasekhar & Fermi [26] paper on the gravitational stability of self-gravitating magnetized stars. The stability condition, based on the virial theorem, is simply that the Coulomb energy of the inner core of a charged star should be smaller or equal than the gravitational energy of the star (Boccaletti et al. [21]). Previous to the collapse, the gravitational energy can be much smaller than the rest energy of the star and be amplified during the process of gravitational collapse reaching overcritical intensity of the electric field (see Fig. 46 and Ruffini [138]). It is interesting that the Chandrasekhar-Fermi inequality just leads to an extreme Reissner-Nordström solution.

In both these cases the Reissner-Nordström geometry appears indeed to be the relevant model for GRB 991216 as discussed in the previous sections. We shall return to non spherical configuration in forthcoming publications and/or when requested by observational evidence (see Ruffini et al. [152]).

XXVII. SOME PROPAEDEUTIC ANALYSIS FOR THE DYNAMICAL FORMATION OF THE EMBH

While the formation in time of the dyadosphere is the fundamental phenomena we are interested in, we can get an insight on the issue of gravitational collapse of an electrically charged star core studying in details a simplified model, namely a thin shell of charged dust. In De la Cruz & Israel [34], Israel [79] it is shown that the problem of a collapsing charged shell in general relativity can be reduced to a set of ordinary differential equations. We reconsider here the following relativistic system: a spherical shell of electrically charged dust which is moving radially in the Reissner-Nordström background of an already formed nonrotating EMBH of mass M_1 and charge Q_1, with $Q_1 \leq M_1$.

The world surface spanned by the shell divides the space-time into two regions: an internal one $\mathcal{M}_-$ and an external one $\mathcal{M}_+$. The line element in Schwarzschild like coordinate is (Cherubini et al. [27])

$$ds^2 = \begin{cases} -f_+ dt_+^2 + f_+^{-1} dr^2 + r^2 d\Omega^2 & \text{in } \mathcal{M}_+ \\ -f_- dt_-^2 + f_-^{-1} dr^2 + r^2 d\Omega^2 & \text{in } \mathcal{M}_- \end{cases}, \qquad (184)$$

where $f_+ = 1 - \frac{2M}{r} + \frac{Q^2}{r^2}$, $f_- = 1 - \frac{2M_1}{r} + \frac{Q_1^2}{r^2}$ and t_- and t_+ are the Schwarzschild-like time coordinates in $\mathcal{M}_-$ and $\mathcal{M}_+$ respectively. M is the total mass-energy of the system formed by the shell and the EMBH, measured by an observer at rest at infinity and $Q = Q_0 + Q_1$ is the total charge: sum of the charge Q_0 of the shell and the charge Q_1 of the internal EMBH.

Indicating by R the radius of the shell and by $T_\pm$ its time coordinate, the equations of motion of the shell become (Ruffini & Vitagliano [155])

$$\left(\frac{dR}{d\tau}\right)^2 = \frac{1}{M_0^2}\left(M - M_1 + \frac{M_0^2}{2R} - \frac{Q_0^2}{2R} - \frac{Q_1 Q_0}{R}\right)^2 - f_-(R)$$

$$= \frac{1}{M_0^2}\left(M - M_1 - \frac{M_0^2}{2R} - \frac{Q_0^2}{2R} - \frac{Q_1 Q_0}{R}\right)^2 - f_+(R), \qquad (185)$$

$$\frac{dT_\pm}{d\tau} = \frac{1}{M_0 f_\pm(R)}\left(M - M_1 \mp \frac{M_0^2}{2R} - \frac{Q_0^2}{2R} - \frac{Q_1 Q_0}{R}\right), \qquad (186)$$

where M_0 is the rest mass of the shell and τ is its proper time. Eqs.(185,186) (together with Eq.(184)) completely describe a 5-parameter (M, Q, M_1, Q_1, M_0) family of solutions of the Einstein-Maxwell equations. Note that Eqs.(185,186) imply that

$$M - M_1 - \frac{Q_0^2}{2R} - \frac{Q_1 Q_0}{R} > 0 \qquad (187)$$

holds for $R > M + \sqrt{M^2 - Q^2}$ if $Q < M$ and for $R > M_1 + \sqrt{M_1^2 - Q_1^2}$ if $Q > M$.

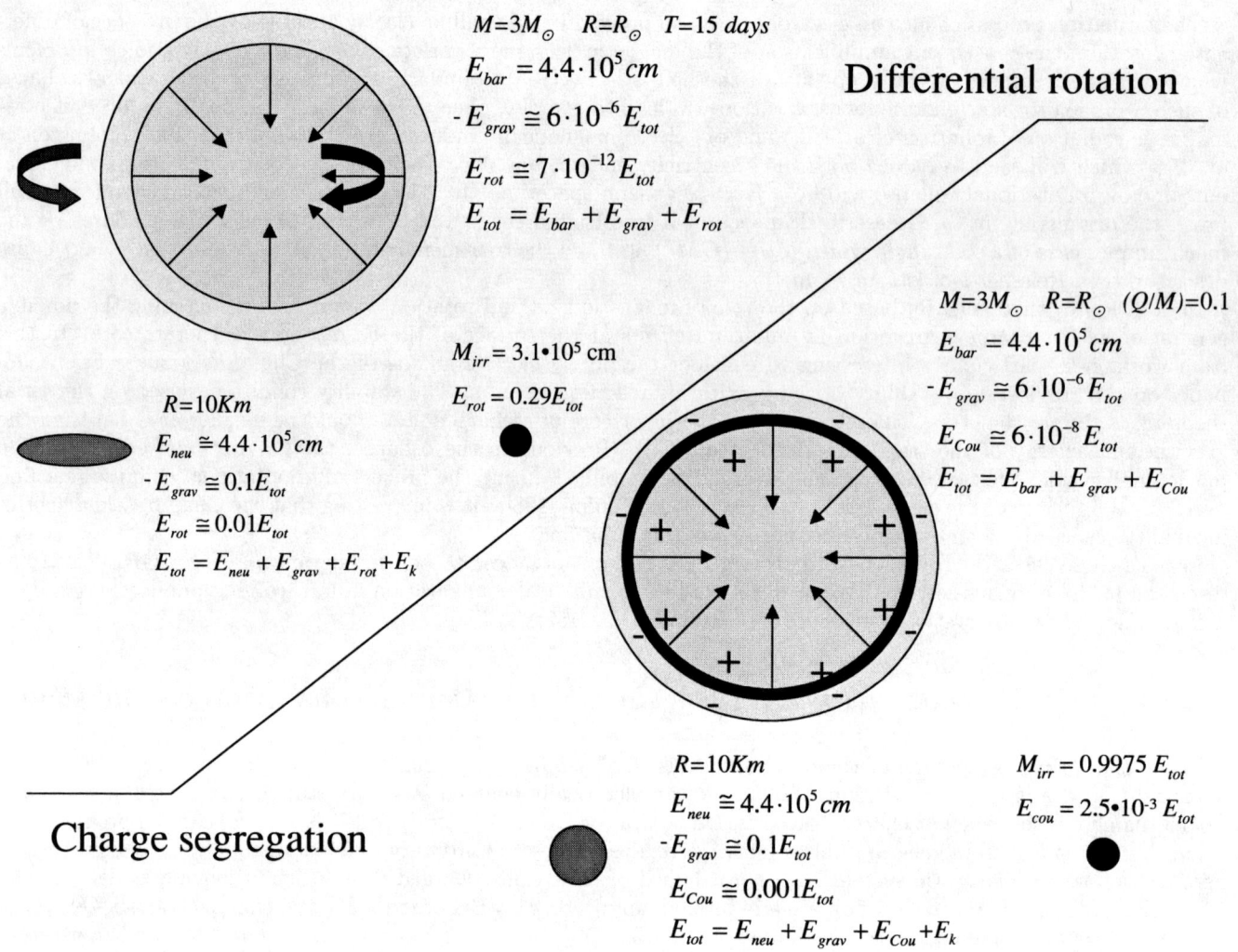

Figure 46: Quantitative description of the gravitational collapse to a neutron star and to a black hole of the core of a rotating progenitor rotating. The core is estimated to have a mass equal to $3M_\odot$, to have an initial radius $r = r_\odot$ and a rotation period of 15 days. Although the initial rotational energy is of the order of 10^{-11} of the total energy, the total rotational energy, in principle extractable, of the rotating black hole can be as high as of the order of 29%. On the lower-right side the same considerations are applied to the case of a neutral star formed by a core oppositely charged from its outermost envelope. The core is expected to have a mass of $3M_\odot$, a radius equal to $r_\odot$ and electromagnetic energy $Q/M = 0.1$. Although the initial Coulomb energy is only $\sim 10^{-7}$ of the total energy, which is in turn hundred times smaller than the gravitational energy, the final Coulomb energy can be as high as 2.5×10^{-3} of the total energy. In both cases, the amplification of the rotational energy and of the Coulomb energy, which indeed are the only two extractable forms of energy from a black hole, is due to the process of gravitational collapse.

For astrophysical applications (Ruffini et al. [158]) the trajectory of the shell $R = R(T_+)$ is obtained as a function of the time coordinate T_+ relative to the space-time region $\mathcal{M}_+$. In the following we drop the + index from T_+. From Eqs.(185,186) we have

$$\frac{dR}{dT} = \frac{dR}{d\tau}\frac{d\tau}{dT} = \pm \frac{F}{\Omega}\sqrt{\Omega^2 - F}, \tag{188}$$

where

$$F \equiv f_+(R) = 1 - \frac{2M}{R} + \frac{Q^2}{R^2}, \tag{189}$$

$$\Omega \equiv \Gamma - \frac{M_0^2 + Q^2 - Q_1^2}{2M_0 R}, \tag{190}$$

$$\Gamma \equiv \frac{M - M_1}{M_0}. \tag{191}$$

Since we are interested in an imploding shell, only the minus sign case in (188) will be studied. We can give the following physical interpretation of Γ. If $M - M_1 \geq M_0$, Γ coincides with the Lorentz γ factor of the imploding shell at infinity; from Eq.(188) it satisfies

$$\Gamma = \frac{1}{\sqrt{1-\left(\frac{dR}{dT}\right)^2_{R=\infty}}} \geq 1. \tag{192}$$

When $M - M_1 < M_0$ then there is a *turning point* R^*, defined by $\frac{dR}{dT}\big|_{R=R^*} = 0$. In this case Γ coincides with the "effective potential" at R^*:

$$\Gamma = \sqrt{f_-(R^*)} + M_0^{-1}\left(-\frac{M_0^2}{2R^*} + \frac{Q_0^2}{2R^*} + \frac{Q_1 Q_0}{R^*}\right) \leq 1. \tag{193}$$

The solution of the differential equation (188) is given by:

$$\int dT = -\int \frac{\Omega}{F\sqrt{\Omega^2 - F}} dR. \tag{194}$$

The functional form of the integral (194) crucially depends on the degree of the polynomial $P(R) = R^2(\Omega^2 - F)$, which is generically two, but in special cases has lower values. We therefore distinguish the following cases:

1. $M = M_0 + M_1$; $Q_1 = M_1$; $Q = M$: $P(R)$ is equal to 0, we simply have

$$R(T) = \text{const}. \tag{195}$$

2. $M = M_0 + M_1$; $M^2 - Q^2 = M_1^2 - Q_1^2$; $Q \neq M$: $P(R)$ is a constant, we have

$$T = \text{const} + \frac{1}{2\sqrt{M^2 - Q^2}}\left[(R + 2M)R \right.$$
$$\left. + r_+^2 \log\left(\frac{R-r_+}{M}\right) + r_-^2 \log\left(\frac{R-r_-}{M}\right)\right]. \tag{196}$$

3. $M = M_0 + M_1$; $M^2 - Q^2 \neq M_1^2 - Q_1^2$: $P(R)$ is a first order polynomial and

$$T = \text{const} + 2R\sqrt{\Omega^2 - F}\left[\frac{M_0 R}{3(M^2 - Q^2 - M_1^2 + Q_1^2)}\right.$$
$$\left. + \frac{(M_0^2 + Q^2 - Q_1^2)^2 - 9MM_0(M_0^2 + Q^2 - Q_1^2) + 12M^2 M_0^2 + 2Q^2 M_0^2}{3(M^2 - Q^2 - M_1^2 + Q_1^2)^2}\right]$$
$$- \frac{1}{\sqrt{M^2 - Q^2}}\left[r_+^2 \operatorname{arctanh}\left(\frac{R}{r_+}\frac{\sqrt{\Omega^2 - F}}{\Omega_+}\right)\right.$$
$$\left. - r_-^2 \operatorname{arctanh}\left(\frac{R}{r_-}\frac{\sqrt{\Omega^2 - F}}{\Omega_-}\right)\right], \tag{197}$$

where $\Omega_\pm \equiv \Omega(r_\pm)$.

4. $M \neq M_0 + M_1$: $P(R)$ is a second order polynomial and

$$T = \text{const} - \frac{1}{2\sqrt{M^2 - Q^2}}\left\{\frac{2\Gamma\sqrt{M^2 - Q^2}}{\Gamma^2 - 1} R\sqrt{\Omega^2 - F}\right.$$
$$+ r_+^2 \log\left[\frac{R\sqrt{\Omega^2 - F}}{R - r_+} + \frac{R^2(\Omega^2 - F) + r_+^2 \Omega_+^2 - (\Gamma^2 - 1)(R - r_+)^2}{2(R - r_+)R\sqrt{\Omega^2 - F}}\right]$$
$$- r_-^2 \log\left[\frac{R\sqrt{\Omega^2 - F}}{R - r_-} + \frac{R^2(\Omega^2 - F) + r_-^2 \Omega_-^2 - (\Gamma^2 - 1)(R - r_-)^2}{2(R - r_-)R\sqrt{\Omega^2 - F}}\right]$$
$$- \frac{[2MM_0(2\Gamma^3 - 3\Gamma) + M_0^2 + Q^2 - Q_1^2]\sqrt{M^2 - Q^2}}{M_0(\Gamma^2 - 1)^{3/2}} \log\left[\frac{R\sqrt{\Omega^2 - F}}{M}\right.$$
$$\left.\left.+ \frac{2M_0(\Gamma^2 - 1)R - (M_0^2 + Q^2 - Q_1^2)\Gamma + 2M_0 M}{2M_0 M \sqrt{\Gamma^2 - 1}}\right]\right\}. \tag{198}$$

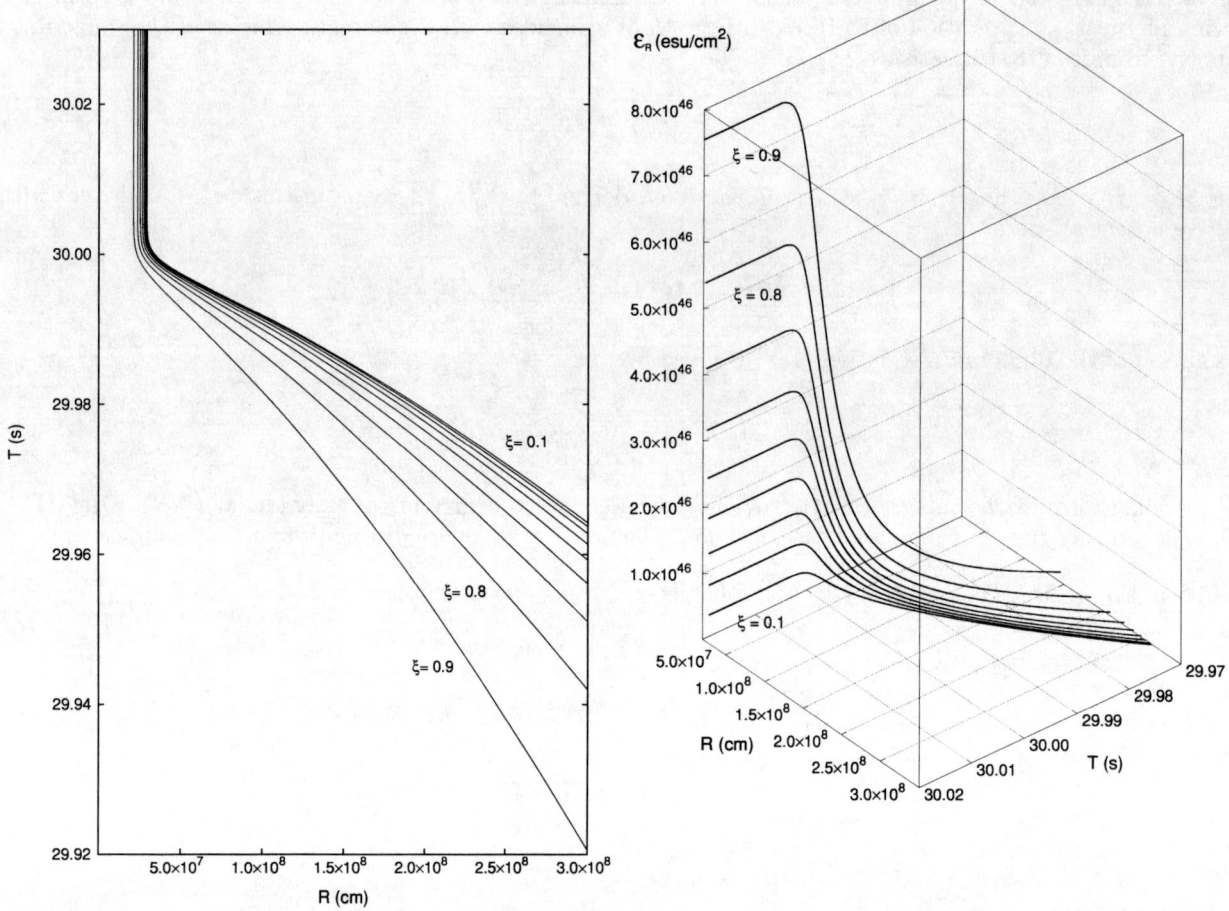

Figure 47: **Left)** Collapse curves in the plane (T, R) for $M = 20M_\odot$ and for different values of the parameter ξ. The asymptotic behavior is the clear manifestation of general relativistic effects as the horizon of the EMBH is approached. **Right)** Electric field behaviour at the surface of the shell for $M = 20M_\odot$ and for different values of the parameter ξ. The asymptotic behavior is the clear manifestation of general relativistic effects as the horizon of the EMBH is approached.

Of particular interest is the time varying electric field $\mathcal{E}_R = \frac{Q}{R^2}$ on the external surface of the shell. In order to study the variability of $\mathcal{E}_R$ with time it is useful to consider in the tridimensional space of parameters $(R, T, \mathcal{E}_R)$ the parametric curve $\mathcal{C} : \left(R = \lambda, \quad T = T(\lambda), \quad \mathcal{E}_R = \frac{Q}{\lambda^2} \right)$. In astrophysical applications (Ruffini et al. [158]) we are specially interested in the family of solutions such that $\frac{dR}{dT}$ is 0 when $R = \infty$ which implies that $\Gamma = 1$. In Fig. 47 we plot the collapse curves in the plane (T, R) for different values of the parameter $\xi \equiv \frac{Q}{M}$, $0 < \xi < 1$. The initial data (T_0, R_0) are chosen so that the integration constant in equation (197) is equal to 0. In all the cases we can follow the details of the approach to the horizon which is reached in an infinite Schwarzschild time coordinate. In Fig. 47 we plot the parametric curves $\mathcal{C}$ in the space $(R, T, \mathcal{E}_R)$ for different values of ξ. Again we can follow the exact asymptotic behavior of the curves $\mathcal{C}$, $\mathcal{E}_R$ reaching the asymptotic value $\frac{Q}{r_+^2}$. The detailed knowledge of this asymptotic behavior is of great relevance for the observational properties of the EMBH formation (see e.g. Ruffini & Vitagliano [155]).

In the case of a shell falling in a flat background ($M_1 = Q_1 = 0$) Eq.(185) reduces to

$$\left(\tfrac{dR}{d\tau}\right)^2 = \tfrac{1}{M_0^2} \left(M + \tfrac{M_0^2}{2R} - \tfrac{Q^2}{2R} \right)^2 - 1. \tag{199}$$

Introducing the total radial momentum $P \equiv M_0 u^r = M_0 \frac{dR}{d\tau}$ of the shell, we can express the kinetic energy of the shell as measured by static observers in $\mathcal{M}_-$ as $T \equiv -M_0 u_\mu \xi_-^\mu - M_0 = \sqrt{P^2 + M_0^2} - M_0$. Then from equation (199) we have

$$M = -\frac{M_0^2}{2R} + \frac{Q^2}{2R} + \sqrt{P^2 + M_0^2} = M_0 + T - \frac{M_0^2}{2R} + \frac{Q^2}{2R}. \quad (200)$$

where we choose the positive root solution due to the constraint (187). Eq.(200) is the *mass formula* of the shell, which depends on the time-dependent radial coordinate R and kinetic energy T. If $M \geq Q$, an EMBH is formed and we have

$$M = M_0 + T_+ - \frac{M_0^2}{2r_+} + \frac{Q^2}{2r_+}, \quad (201)$$

where $T_+ \equiv T(r_+)$ and $r_+ = M + \sqrt{M^2 - Q^2}$ is the radius of external horizon of the EMBH. We know from the Christodoulou-Ruffini EMBH mass formula that

$$M = M_{\text{irr}} + \frac{Q^2}{2r_+}, \quad (202)$$

so it follows that

$$M_{\text{irr}} = M_0 - \frac{M_0^2}{2r_+} + T_+, \quad (203)$$

namely that M_{irr} is the sum of only three contributions: the rest mass M_0, the gravitational potential energy and the kinetic energy of the rest mass evaluated at the horizon. M_{irr} is independent of the electromagnetic energy, a fact noticed by Bekenstein (Bekenstein [9]). We have taken one further step here by identifying the independent physical contributions to M_{irr}.

Next we consider the physical interpretation of the electromagnetic term $\frac{Q^2}{2R}$, which can be obtained by evaluating the conserved Killing integral

$$\int_{\Sigma_t^+} \xi_+^\mu T_{\mu\nu}^{(\text{em})} d\Sigma^\nu = \int_R^\infty r^2 dr \int_0^1 d\cos\theta \int_0^{2\pi} d\phi \, T^{(\text{em})}{}^0{}_0$$
$$= \frac{Q^2}{2R}, \quad (204)$$

where Σ_t^+ is the space-like hypersurface in $\mathcal{M}_+$ described by the equation $t_+ = t = \text{const}$, with $d\Sigma^\nu$ as its surface element vector and where $T_{\mu\nu}^{(\text{em})} = -\frac{1}{4\pi}\left(F_\mu{}^\rho F_{\rho\nu} + \frac{1}{4} g_{\mu\nu} F^{\rho\sigma} F_{\rho\sigma}\right)$ is the energy-momentum tensor of the electromagnetic field. The quantity in Eq.(204) differs from the purely electromagnetic energy

$$\int_{\Sigma_t^+} n_+^\mu T_{\mu\nu}^{(\text{em})} d\Sigma^\nu = \frac{1}{2} \int_R^\infty dr \sqrt{g_{rr}} \frac{Q^2}{r^2},$$

where $n_+^\mu = f_+^{-1/2} \xi_+^\mu$ is the unit normal to the integration hypersurface and $g_{rr} = f_+$. This is similar to the analogous situation for the total energy of a static spherical star of energy density ϵ within a radius R, $m(R) = 4\pi \int_0^R dr \, r^2 \epsilon$, which differs from the pure matter energy $m_p(R) = 4\pi \int_0^R dr \sqrt{g_{rr}} r^2 \epsilon$ by the gravitational energy (see Misner, Thorne, & Wheeler [102]). Therefore the term $\frac{Q^2}{2R}$ in the mass formula (200) is the *total* energy of the electromagnetic field and includes its own gravitational binding energy. This energy is stored throughout the region Σ_t^+, extending from R to infinity.

We now turn to the problem of extracting the electromagnetic energy from an EMBH see (see Christodoulou & Ruffini [29]). We can distinguish between two conceptually physically different processes, depending on whether the electric field strength $\mathcal{E} = \frac{Q}{r^2}$ is smaller or greater than the critical value $\mathcal{E}_c = \frac{m_e^2 c^3}{e\hbar}$. Here m_e and e are the mass and the charge of the electron. As already mentioned in this paper an electric field $\mathcal{E} > \mathcal{E}_c$ polarizes the vacuum creating electron-positron pairs (see Heisenberg & Euler [77]). The maximum value $\mathcal{E}_+ = \frac{Q}{r_+^2}$ of the electric field around an EMBH is reached at the horizon. We then have the following:

1. For $\mathcal{E}_+ < \mathcal{E}_c$ the leading energy extraction mechanism consists of a sequence of discrete elementary decay processes of a particle into two oppositely charged particles. The condition $\mathcal{E}_+ < \mathcal{E}_c$ implies

$$\xi \equiv \frac{Q}{\sqrt{G}M}$$
$$\lesssim \begin{cases} \frac{GM/c^2}{\lambda_C} \frac{\sqrt{G}m_e}{e} \sim 10^{-6} \frac{M}{M_\odot} & \text{if } \frac{M}{M_\odot} \leq 10^6 \\ 1 & \text{if } \frac{M}{M_\odot} > 10^6 \end{cases}, \quad (205)$$

where λ_C is the Compton wavelength of the electron. Denardo & Ruffini [35] and Denardo et al. [36] have defined as the *effective ergosphere* the region around an EMBH where the energy extraction processes occur. This region extends from the horizon r_+ up to a radius

$$r_{\text{Eerg}} = \frac{GM}{c^2} \left[1 + \sqrt{1 - \xi^2 \left(1 - \frac{e^2}{Gm_e^2}\right)}\right]$$
$$\simeq \frac{e}{m_e} \frac{Q}{c^2} . \qquad (206)$$

The energy extraction occurs in a finite number N_{PD} of such discrete elementary processes, each one corresponding to a decrease of the EMBH charge. We have

$$N_{\text{PD}} \simeq \frac{Q}{e} . \qquad (207)$$

Since the total extracted energy is (see Eq. (202)) $E^{\text{tot}} = \frac{Q^2}{2r_+}$, we obtain for the mean energy per accelerated particle $\langle E \rangle_{\text{PD}} = \frac{E^{\text{tot}}}{N_{\text{PD}}}$

$$\langle E \rangle_{\text{PD}} = \frac{Qe}{2r_+} = \frac{1}{2} \frac{\xi}{1+\sqrt{1-\xi^2}} \frac{e}{\sqrt{G}m_e} m_e c^2 \simeq \frac{1}{2} \xi \frac{e}{\sqrt{G}m_e} m_e c^2, \qquad (208)$$

which gives

$$\langle E \rangle_{\text{PD}} \lesssim \begin{cases} \left(\frac{M}{M_\odot}\right) \times 10^{21} eV & \text{if } \frac{M}{M_\odot} \leq 10^6 \\ 10^{27} eV & \text{if } \frac{M}{M_\odot} > 10^6 \end{cases} . \qquad (209)$$

One of the crucial aspects of the energy extraction process from an EMBH is its back reaction on the irreducible mass expressed in Christodoulou & Ruffini [29]. Although the energy extraction processes can occur in the entire effective ergosphere defined by Eq. (206), only the limiting processes occurring on the horizon with zero kinetic energy can reach the maximum efficiency while approaching the condition of total reversibility (see Fig. 2 in Christodoulou & Ruffini [29] for details). The farther from the horizon that a decay occurs, the more it increases the irreducible mass and loses efficiency. Only in the complete reversibility limit (Christodoulou & Ruffini [29]) can the energy extraction process from an extreme EMBH reach the upper value of 50% of the total EMBH energy.

2. For $\mathcal{E}_+ \geq \mathcal{E}_c$ the leading extraction process is a *collective* process based on an electron-positron plasma generated by the vacuum polarization, (see Fig. 4) as discussed in section III The condition $\mathcal{E}_+ \geq \mathcal{E}_c$ implies

$$\frac{GM/c^2}{\lambda_C} \left(\frac{e}{\sqrt{G}m_e}\right)^{-1} \simeq 2 \cdot 10^{-6} \frac{M}{M_\odot} \leq \xi \leq 1 . \qquad (210)$$

This vacuum polarization process can occur only for an EMBH with mass smaller than $2 \cdot 10^6 M_\odot$. The electron-positron pairs are now produced in the dyadosphere of the EMBH, (note that the dyadosphere is a subregion of the effective ergosphere) whose radius r_{ds} is given in Eq.(8). We have $r_{ds} \ll r_{\text{Eerg}}$. The number of particles created and the total energy stored in dyadosphere are given in Eqs.(10,12) respectively and we have approximately

$$N^\circ_{e^+e^-} \simeq \left(\frac{r_{ds}}{\lambda_C}\right) \frac{Q}{e} , \qquad (211)$$

$$E_{dya} \simeq \frac{Q^2}{2r_+} \qquad (212)$$

The mean energy per particle produced in the dyadosphere $\langle E \rangle_{\text{ds}} = \frac{E_{dya}}{N^\circ_{e^+e^-}}$ is then

$$\langle E \rangle_{\text{ds}} \simeq \frac{3}{8} \left(\frac{\lambda_C}{r_{ds}}\right) \frac{Qe}{r_+} , \qquad (213)$$

which can be also rewritten as

$$\langle E \rangle_{\text{ds}} \simeq \frac{1}{2} \left(\frac{r_{ds}}{r_+}\right) m_e c^2 \sim \sqrt{\frac{\xi}{M/M_\odot}} 10^5 keV . \qquad (214)$$

Such a process of vacuum polarization, occurring not at the horizon but in the extended dyadosphere region ($r_+ \leq r \leq r_{ds}$) around an EMBH, has been observed to reach the maximum efficiency limit of 50% of the total

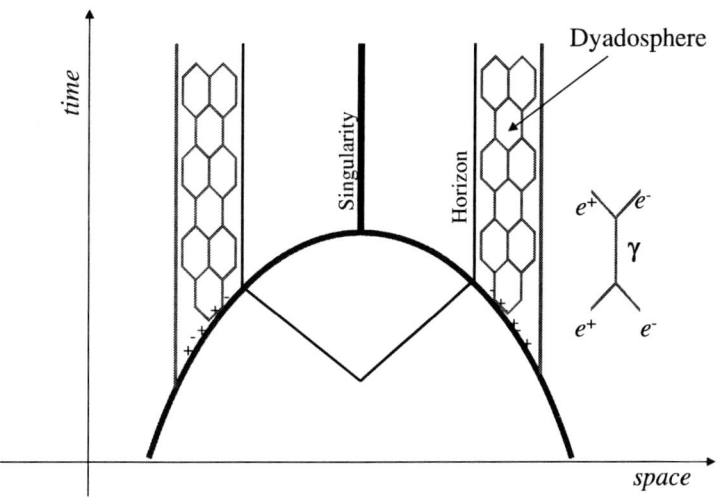

Figure 48: Space-time diagram of the collapse process leading to the formation of the dyadosphere. As the collapsing core crosses the dyadosphere radius the pair creation process starts, and the pairs thermalize in a neutral plasma configuration. Then also the horizon is crossed and the singularity is formed.

mass-energy of an extreme EMBH (see e.g. Preparata et al. [123]). The conceptual justification of this result follows from the present work: the e^+e^- creation process occurs at the expence of the Coulomb energy given by Eq. (204) and does not affect the irreducible mass given by Eq. (203), which indeed, as we have proved, does not depend of the electromagnetic energy. In this sense, $\delta M_{\rm irr} = 0$ and the transformation is fully reversible. This result will be further validated by the study of the dynamical formation of the dyadosphere, which we have obtained using the present work and Cherubini et al. [27] (see Ruffini et al. [158]).

Let us now compare and contrast these two processes. We have

$$r_{\rm Eerg} \simeq \left(\frac{r_{ds}}{\lambda_{\rm C}}\right) r \tag{215}$$

$$N_{\rm dya} \simeq \left(\frac{r_{ds}}{\lambda_{\rm C}}\right) N_{\rm PD}, \tag{216}$$

$$\langle E \rangle_{\rm dya} \simeq \left(\frac{\lambda_{\rm C}}{r_{ds}}\right) \langle E \rangle_{\rm PD}. \tag{217}$$

Moreover we see (Eqs. (209), (214)) that $\langle E \rangle_{\rm PD}$ is in the range of energies of UHECR, while for $\xi \sim 0.1$ and $M \sim 10 M_\odot$, $\langle E \rangle_{\rm ds}$ is in the gamma ray range. In other words, the discrete particle decay process involves a small number of particles with ultra high energies ($\sim 10^{21} eV$), while vacuum polarization involves a much larger number of particles with lower mean energies ($\sim 10 MeV$).

Having so established and clarified the basic conceptual processes of the energetic of the EMBH, we are now ready to approach, using the new analytic solution obtained, the dynamical process of vacuum polarization occurring during the formation of an EMBH as qualitatively represented in Fig. 48. The study of the dyadosphere dynamical formation as well as of the electron-positron plasma dynamical evolution will lead to the first possibility of directly observing the general relativistic effects approaching the EMBH horizon.

Before closing we would like to emphasize once more a basic point: all the considerations presented in the description of the preceding eras are based on the approximations in the description of the dyadosphere presented in section III. This treatment is very appropriate in estimating the general dependence of the energy of the P-GRB, the kinetic energy of the ABM pulse and consequently the intensity of the afterglow, as well as the overall time structure of the GRB and especially the time of the release of the P-GRB in respect to the moment of gravitational collapse and its relative intensity with respect to the afterglow. If, however, is addressed the issue of the detailed temporal structure of the P-GRB and its detailed spectral distribution, the above dynamical considerations on the dyadosphere formation are needed (see also Ruffini et al. [158]). In turn, this detailed analysis is needed if the general relativistic effects close to the horizon formation have to be followed. As expressed already in section. XII, all general relativistic quantum field theory effects are encoded in the fine structure of the P-GRB. As emphasized in section X, the only way to differentiate between solutions with same E_{dya} but different EMBH mass and charge is to observe the P-GRBs in the limit $B \to 0$, namely, to observe the short GRBs.

XXVIII. CONTRIBUTION OF THE EMBH MODEL TO THE BLACK HOLE THEORY

The aim of this section is to point out how the knowledge obtained from the EMBH model is of relevance also for the basic theory of black holes and further how very high precision verification of general relativistic effects in the very strong field near the formation of the horizon should be expected in the near future.

We shall first see how Eq.(203) for M_{irr},

$$M_{\text{irr}} = M_0 - \frac{M_0^2}{2r_+} + T_+, \tag{218}$$

leads to a deeper physical understanding of the role of the gravitational interaction in the maximum energy extraction process of an EMBH. This formula can also be of assistance in clarifying some long lasting epistemological issue on the role of general relativity, quantum theory and thermodynamics.

It is well known that if a spherically symmetric mass distribution without any electromagnetic structure undergoes free gravitational collapse, its total mass-energy M is conserved according to the Birkhoff theorem: the increase in the kinetic energy of implosion is balanced by the increase in the gravitational energy of the system. If one considers the possibility that part of the kinetic energy of implosion is extracted then the situation is very different: configurations of smaller mass-energy and greater density can be attained without violating Birkhoff theorem.

We illustrate our considerations with two examples: one has found confirmation from astrophysical observations, the other promises to be of relevance for gamma ray bursts (GRBs) (see Ruffini & Vitagliano [155]). Concerning the first example, it is well known from the work of Landau [85] that at the endpoint of thermonuclear evolution, the gravitational collapse of a spherically symmetric star can be stopped by the Fermi pressure of the degenerate electron gas (white dwarf). A configuration of equilibrium can be found all the way up to the critical number of particles

$$N_{\text{crit}} = 0.775 \frac{m_{Pl}^3}{m_0^3}, \tag{219}$$

where the factor 0.775 comes from the coefficient $\frac{3.098}{\mu^2}$ of the solution of the Lane-Emden equation with polytropic index $n=3$, and $m_{Pl} = \sqrt{\frac{\hbar c}{G}}$ is the Planck mass, m_0 is the nucleon mass and μ the average number of electrons per nucleon. As the kinetic energy of implosion is carried away by radiation the star settles down to a configuration of mass

$$M = N_{\text{crit}} m_0 - U, \tag{220}$$

where the gravitational binding energy U can be as high as $5.72 \times 10^{-4} N_{\text{crit}} m_0$.

Similarly Gamov (see Gamow & Critchfield [60]) has shown that a gravitational collapse process to still higher densities can be stopped by the Fermi pressure of the neutrons (neutron star) and Oppenheimer (Oppenheimer & Volkoff [105]) has shown that, if the effects of strong interactions are neglected, a configuration of equilibrium exists also in this case all the way up to a critical number of particles

$$N_{\text{crit}} = 0.398 \frac{m_{Pl}^3}{m_0^3}, \tag{221}$$

where the factor 0.398 comes now from the integration of the Tolman-Oppenheimer-Volkoff equation (see e.g. Harrison et al. [74]). If the kinetic energy of implosion is again carried away by radiation of photons or neutrinos and antineutrinos the final configuration is characterized by the formula (220) with $U \lesssim 2.48 \times 10^{-2} N_{\text{crit}} m_0$. These considerations and the existence of such large values of the gravitational binding energy have been at the heart of the explanation of astrophysical phenomena such as red-giant stars and supernovae: the corresponding measurements of the masses of neutron stars and white dwarfs have been carried out with unprecedented accuracy in binary systems (Gursky & Ruffini [71]).

From a theoretical physics point of view it is still an open question how far such a sequence can go: using causality nonviolating interactions, can one find a sequence of braking and energy extraction processes by which the density and the gravitational binding energy can increase indefinitely and the mass-energy of the collapsed object be reduced at will? This question can also be formulated in the mass-formula language of a black hole given in Christodoulou & Ruffini [29] (see also Ruffini & Vitagliano [155]): given a collapsing core of nucleons with a given rest mass-energy M_0, what is the minimum irreducible mass of the black hole which is formed?

Following Cherubini et al. [27] and Ruffini & Vitagliano [155], consider a spherical shell of rest mass M_0 collapsing in a flat space-time. In the neutral case the irreducible mass of the final black hole satisfies the equation (see Ruffini & Vitagliano [155])

$$M_{\text{irr}} = M = M_0 - \frac{M_0^2}{2r_+} + T_+, \tag{222}$$

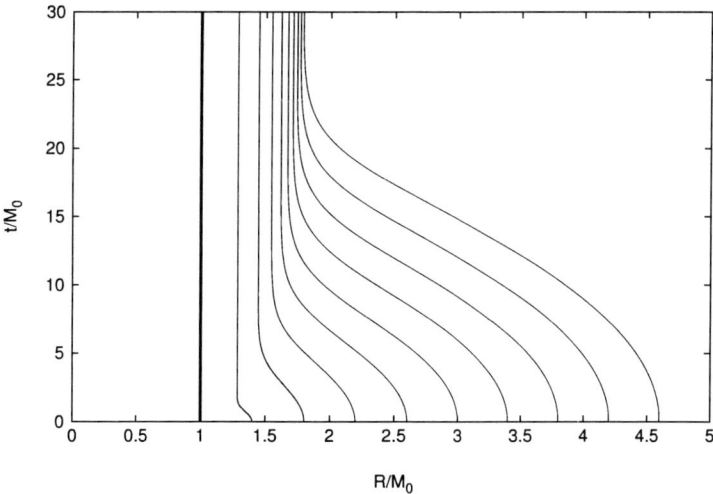

Figure 49: Collapse curves for neutral shells with rest mass M_0 starting at rest at selected radii R^* computed by using the exact solutions given in Cherubini et al. [27]. A different value of $M_{\rm irr}$ (and therefore of r_+) corresponds to each curve. The time parameter is the Schwarzschild time coordinate t and the asymptotic behaviour at the respective horizons is evident. The limiting configuration $M_{\rm irr} = \frac{M_0}{2}$ (solid line) corresponds to the case in which the shell is trapped, at the very beginning of its motion, by the formation of the horizon.

where M is the total energy of the collapsing shell and T_+ the kinetic energy at the horizon r_+. Recall that the area S of the horizon is (Christodoulou & Ruffini [29])

$$S = 4\pi r_+^2 = 16\pi M_{\rm irr}^2 \qquad (223)$$

where $r_+ = 2M_{\rm irr}$ is the horizon radius. The minimum irreducible mass $M_{\rm irr}^{\rm (min)}$ is obtained when the kinetic energy at the horizon T_+ is 0, that is when the entire kinetic energy T_+ has been extracted. We then obtain the simple result

$$M_{\rm irr}^{\rm (min)} = \frac{M_0}{2}. \qquad (224)$$

We conclude that in the gravitational collapse of a spherical shell of rest mass M_0 at rest at infinity (initial energy $M_{\rm i} = M_0$), an energy up to 50% of $M_0 c^2$ can in principle be extracted, by braking processes of the kinetic energy. In this limiting case the shell crosses the horizon with $T_+ = 0$. The limit $\frac{M_0}{2}$ in the extractable kinetic energy can further increase if the collapsing shell is endowed with kinetic energy at infinity, since all that kinetic energy is in principle extractable.

In order to illustrate the physical reasons for this result, using the formulas of Cherubini et al. [27], we have represented in Fig. 49 the world lines of spherical shells of the same rest mass M_0, starting their gravitational collapse at rest at selected radii R^*. These initial conditions can be implemented by performing suitable braking of the collapsing shell and concurrent kinetic energy extraction processes at progressively smaller radii (see also Fig. 50). The reason for the existence of the minimum (224) in the black hole mass is the "self closure" occurring by the formation of a horizon in the initial configuration (thick line in Fig. 49).

Is the limit $M_{\rm irr} \to \frac{M_0}{2}$ actually attainable without violating causality? Let us consider a collapsing shell with charge Q. If $M \geq Q$ an EMBH is formed. As pointed out in Ruffini & Vitagliano [155] the irreducible mass of the final EMBH does not depend on the charge Q. Therefore Eqs. (222) and (224) still hold in the charged case with $r_+ = M + \sqrt{M^2 - Q^2}$. In Fig. 50 we consider the special case in which the shell is initially at rest at infinity, i.e. has initial energy $M_{\rm i} = M_0$, for three different values of the charge Q. We plot the initial energy M_i, the energy of the system when all the kinetic energy of implosion has been extracted as well as the sum of the rest mass energy and the gravitational binding energy $-\frac{M_0^2}{2R}$ of the system (here R is the radius of the shell). In the extreme case $Q = M_0$, the shell is in equilibrium at all radii (see Cherubini et al. [27]) and the kinetic energy is identically zero. In all three cases, the sum of the extractable kinetic energy T and the electromagnetic energy $\frac{Q^2}{2R}$ reaches 50% of the rest mass energy at the horizon, according to Eq. (224).

What is the role of the electromagnetic field here? If we consider the case of a charged shell with $Q \simeq M_0$, the electromagnetic repulsion implements the braking process and the extractable energy is entirely stored in the

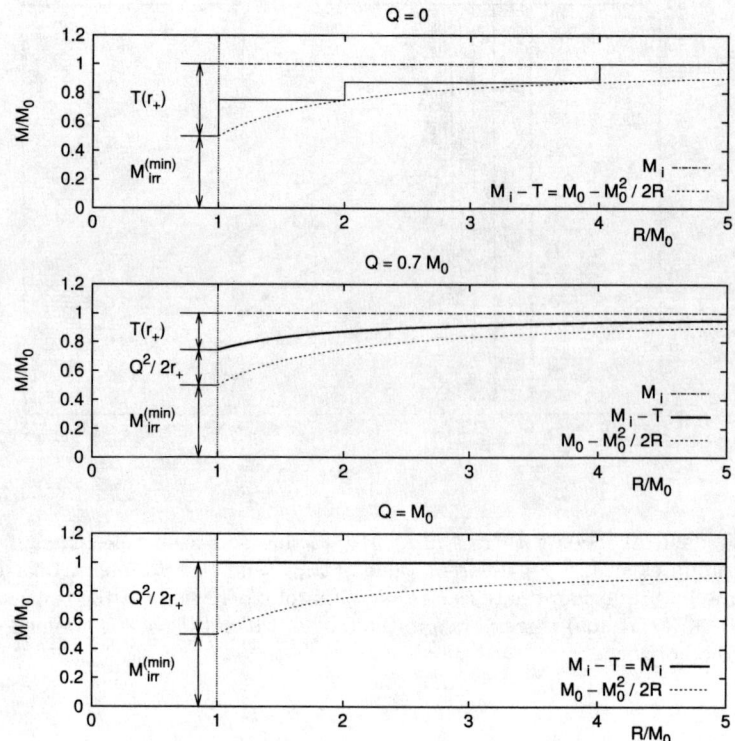

Figure 50: Energetics of a shell such that $M_i = M_0$, for selected values of the charge. In the first diagram $Q = 0$; the dashed line represents the total energy for a gravitational collapse without any braking process as a function of the radius R of the shell; the solid, stepwise line represents a collapse with suitable braking of the kinetc energy of implosion at selected radii; the dotted line represents the rest mass energy plus the gravitational binding energy. In the second and third diagram $Q/M_0 = 0.7$, $Q/M_0 = 1$ respectively; the dashed and the dotted lines have the same meaning as above; the solid lines represent the total energy minus the kinetic energy. The region between the solid line and the dotted line corresponds to the stored electromagnetic energy. The region between the dashed line and the solid line corresponds to the kinetic energy of collapse. In all the cases the sum of the kinetic energy and the electromagnetic energy at the horizon is 50% of M_0. Both the electromagnetic and the kinetic energy are extractable. It is most remarkable that the same underlying process occurs in the three cases: the role of the electromagnetic interaction is twofold: a) to reduce the kinetic energy of implosion by the Coulomb repulsion of the shell; b) to store such an energy in the region around the EMBH. The stored electromagnetic energy is extractable as shown in Ruffini & Vitagliano [155].

electromagnetic field surrounding the EMBH (see Ruffini & Vitagliano [155]). In Ruffini & Vitagliano [155] we have outlined two different processes of electromagnetic energy extraction. We emphasize here that the extraction of 50% of the mass-energy of an EMBH is not specifically linked to the electromagnetic field but depends on three factors: a) the increase of the gravitational energy during the collapse, b) the formation of a horizon, c) the reduction of the kinetic energy of implosion. Such conditions are naturally met during the formation of an extreme EMBH but are more general and can indeed occur in a variety of different situations, e.g. during the formation of a Schwarzschild black hole by a suitable extraction of the kinetic energy of implosion (see Fig. 49 and Fig. 50).

Now consider a test particle of mass m in the gravitational field of an already formed Schwarzschild black hole of mass M and go through such a sequence of braking and energy extraction processes. Kaplan (Kaplan [82]) found for the energy E of the particle as a function of the radius r

$$E = m\sqrt{1 - \frac{2M}{r}}. \tag{225}$$

It would appear from this formula that the entire energy of a particle could be extracted in the limit $r \to 2M$. Such 100% efficiency of energy extraction has often been quoted as evidence for incompatibility between General Relativity and the second principle of Thermodynamics (see Bekenstein [10] and references therein). J. Bekenstein and S. Hawking have gone as far as to consider General Relativity not to be a complete theory and to conclude that in order to avoid inconsistencies with thermodynamics, the theory should be implemented through a quantum description (Bekenstein [10], Hawking [75]). Einstein himself often expressed the opposite point of view (see e.g.

Dyson [44]).

The analytic treatment presented in Cherubini et al. [27] can clarify this fundamental issue. It allows to express the energy increase E of a black hole of mass M_1 through the accretion of a shell of mass M_0 starting its motion at rest at a radius R in the following formula which generalizes Eq. (225):

$$E \equiv M - M_1 = -\frac{M_0^2}{2R} + M_0\sqrt{1 - \frac{2M_1}{R}}, \qquad (226)$$

where $M = M_1 + E$ is clearly the mass-energy of the final black hole. This formula differs from the Kaplan formula (225) in three respects: a) it takes into account the increase of the horizon area due to the accretion of the shell; b) it shows the role of the gravitational self energy of the imploding shell; c) it expresses the combined effects of a) and b) in an exact closed formula.

The minimum value $E_{\min}$ of E is attained for the minimum value of the radius $R = 2M$: the horizon of the final black hole. This corresponds to the maximum efficiency of the energy extraction. We have

$$E_{\min} = -\frac{M_0^2}{4M} + M_0\sqrt{1 - \frac{M_1}{M}} = -\frac{M_0^2}{4(M_1+E_{\min})} + M_0\sqrt{1 - \frac{M_1}{M_1+E_{\min}}}, \qquad (227)$$

or solving the quadratic equation and choosing the positive solution for physical reasons

$$E_{\min} = \tfrac{1}{2}\left(\sqrt{M_1^2 + M_0^2} - M_1\right). \qquad (228)$$

The corresponding efficiency of energy extraction is

$$\eta_{\max} = \frac{M_0 - E_{\min}}{M_0} = 1 - \tfrac{1}{2}\frac{M_1}{M_0}\left(\sqrt{1 + \frac{M_0^2}{M_1^2}} - 1\right), \qquad (229)$$

which is strictly *smaller than* 100% for *any* given $M_0 \neq 0$. It is interesting that this analytic formula, in the limit $M_1 \ll M_0$, properly reproduces the result of equation (224), corresponding to an efficiency of 50%. In the opposite limit $M_1 \gg M_0$ we have

$$\eta_{\max} \simeq 1 - \tfrac{1}{4}\frac{M_0}{M_1}. \qquad (230)$$

Only for $M_0 \to 0$, Eq. (229) corresponds to an efficiency of 100% and correctly represents the limiting reversible transformations introduced in Christodoulou & Ruffini [29]. It seems that the difficulties of reconciling General Relativity and Thermodynamics are ascribable not to an incompleteness of General Relativity but to the use of the Kaplan formula in a regime in which it is not valid. The generalization of the above results to stationary black holes is being considered.

XXIX. CONCLUSIONS

The EMBH theory has been here applied for the first time to fit the experimental data of GRB 991216. This process has given us the opportunity to rethink the entire GRB process in an unitary description starting from the moment of gravitational collapse all the way up to the latest phases of the afterglow. We have identified the three fundamental actors of the GRB phenomenon in:

1. E_{dya}. Having reanalyzed in section III the physics of the dyadosphere we have pointed out in Fig. 17 that the same value of E_{dya} can be obtained from an entire family of (μ, ξ) parameters (i.e. E_{dya} is degenerate in (μ, ξ)). We have then shown in the reexamination of all the GRB eras that all the results depend only on the value of E_{dya} and not on the particular value of (μ, ξ) (see sections VIII,IX,XIII,XIV). The only exception to this occurs in the era I (see section VII) which is the only one relevant for short GRBs.

2. B. The crucial role played by the baryonic remnant of the progenitor star in determining the relative intensity ratio and the time delay between the P-GRB and the E-APE has been summarized already in the two Figs. 11–11 in the introduction.

3. ISM. The density n_{ism} of the interstellar medium and its inhomogeneities appears to have a fundamental role in the intensity and the temporal substructures of the E-APE and the afterglow.

The observational data agree with the predictions of the model on:

1. the intensity ratio, 1.58×10^{-2}, between the P-GRB and the E-APE, which strongly depends on the parameter B;

2. the absolute intensities for both the P-GRB and the E-APE, respectively 7.54×10^{51} erg and 4.75×10^{53};

3. the arrival time of the P-GRB and the peak of the E-APE, respectively 8.41×10^{-2} s and 19.87 s;

4. the power-law index n of the afterglow, predicted $n_{theo} = -1.6$ and observed $n_{obs} = -1.616 \pm 0.067$ (see sections XVIII, XIX);

5. the temporal structure of the E-APE and its correlation with the inhomogeneity in the ISM;

6. the spectral distribution of the X-ray and γ-ray emission.

Concerning the total energy of GRB 991216, $E_{dya} = 4.83 \times 10^{53}$ erg is found in the EMBH theory. This value is systematically larger than the ones quoted in the current literature by Panaitescu & Kumar [110] and by Halpern et al. [72] due to the fact that they respectively consider beaming angles of $3° - 4°$ and $6°$. These considerations have been shown to be untenable in section XIX. There is still a difference of $\sim 28\%$ between the total energy implied by the EMBH theory (4.83×10^{53} erg) and the value quoted by Halpern ($E_{dya} = 6.7 \times 10^{53}$ erg) in the case of spherical emission. We trust that this is a consequence of the underlying assumption of the spectral distribution of the radiation assumed by Halpern et al. [72] (see e.g. Frail et al. [54]), which should be reassessed on the ground of our theoretical results (see also Ruffini et al. [152]).

These results can certainly be considered the success of the EMBH theory.

Before closing, we like to stress how GRBs, if duly theoretically interpreted, can open a main avenue of inquiring on totally new physical and astrophysical regimes. This program is very likely one of the greatest computational efforts in physics and astrophysics and cannot be actuated using shortcuts.

From the point of view of fundamental physics new regimes are explored:

1. The process of energy extraction from black holes. It is interesting that the analysis of GRBs has promoted a new effort in developing new theoretical tools for approaching the dynamical phase of collapse as expressed in section. XXVII. These results have further clarified some basic issue related to the energy extraction process from black hole (see e.g. Ruffini [138]). It was already known from the definition of the ergosphere (see Ruffini & Wheeler [134]) that the rotational energy extraction process do occur in an extended region around the horizon of a black hole. The fortunate situation that the energy extraction process in GRBs occurs in a condition of almost perfect spherical symmetry have allowed us to focus on the second fundamental parameter of black holes, namely the electric energy. The spherical symmetry has allowed as well to develop some powerful theoretical tools (see section XXVII) which have allowed to reach a better understanding of the role of kinetic energy of implosion in the process of gravitational collapse, in the storage of electromagnetic energy in the region around black holes and to establish as well a new upper limit in the energy extraction process in the gravitational collapse up to 50% of the initial rest mass of the system (see section XXVIII). These results are of general validity and do transcend the work on the EMBH theory, although they are motivated by these researches. Interestingly this work, by giving a new expression for the efficiency of transforming gravitational energy into mechanical work (see section XXVIII), has opened up a new opportunity of debating the relation between general relativity, thermodynamics and quantum theory, which is certainly one of the most profound and important topic of research in the entire realm of fundamental physics.

2. The quantum and general relativistic effects of matter-antimatter creation near the black hole horizon. It is well known that one of the most important topics pursued in the last seventy years in physics has been the possibility, postulated by Sauter, Heisemberg, Euler, Schwinger to create matter-antimatter from the vacuum. In order to have the first experimental and observational evidence for this phenomenon, three major approaches are being followed:

   a) In central collisions of heavy ions near the Coulomb barrier, as first proposed in Gerštein & Y. B. Zel'dovich [63, 64] (see also Popov & Rozhdestvenskaya [120], Popov [121], Zel'dovich & Popov [191]). Efforts in experimentally implementing this idea at GSI were made since early 80's. Despite some apparently encouraging result (Schweppe et al. [170]), such efforts have failed so far due to the small contact time of the colliding ions (see e.g. Ahmad et al. [2], Bär et al. [7], Ganz et al. [61], Heinz et al. [76], Leinberger et al. [87]). Typically the electromagnetic energy involved in the collisions of heavy ions with impact parameter $l_1 \sim 10^{-12}$cm is $E_1 \sim 10^{-6}$erg.

   b) At the focus of an X-ray free electron laser (XFEL) (see Ringwald [131], Roberts et al. [132] and references therein). This idea will be possibly testable at DESY, where the XFEL is part of the design of the collider TESLA, as well as at SLAC, where the so-called Linac Coherent Light Source (LCLS) has been proposed. The electromagnetic energy at the focus of an XFEL is $E_2 \sim 10^6$erg concentrated in a region of linear extension

$l_2 \sim 10^{-8}$ cm (Ringwald [131]).

c) Around an electromagnetic black hole (EMBH) (Damour & Ruffini [32], Preparata et al. [122, 123]), giving rise to the observed phenomenon of GRBs (see e.g. Ruffini et al. [143, 144, 145, 147]). The electromagnetic energy of an EMBH of mass $M \sim 10 M_\odot$ and charge $Q \sim 0.1 M/\sqrt{G}$ is $E_3 \sim 10^{54}$ ergs and it is deposited in a region of linear extension $l_3 \sim 10^8$ m (Preparata et al. [123], Ruffini & Vitagliano [155]).

There is the very distinct possibility that in this race the success will be reached by the observations in relativistic astrophysics more than from the high energy experiments on the Earth. This will be certainly a splendid success which will be only second to the discovery of Helium first in the stars and then on the Earth! Quite apart from the discovery in itself, the detection of vacuum polarization in the astrophysical settings presents distinctively new physical phenomena as Ruffini et al. [159]. The very important topic to be covered in the forthcoming months is the study of the dynamical phase of gravitational collapse and to follow the effects of such process of vacuum polarization in the dynamical phase. It will be also important to follow the development of this process all the way to the emission of the P-GRB (Ruffini & Vitagliano [157]).

3. The physics of ultrarelativisitc shock waves with Lorentz gamma factor $\gamma > 100$. We are expecting much progress in this topic from the understanding of the instantaneous spectrum of GRBs. Some preliminary results along this line are presented in Ruffini et al. [149]. See also section XXIV.

From the point of view of astronomy and astrophysics also new regimes are explored:

1. The occurrence of gravitational collapse to a black hole from a critical mass core of mass $M \gtrsim 10 M_\odot$, which clearly differs from the values of the critical mass encountered in the study of stars "catalyzed at the endpoint of thermonuclear evolution" (white dwarfs and neutron stars).

2. The extremely high efficiency of the spherical collapse to a black hole, where almost 99.99% of the core mass collapses leaving negligible remnant. The EMBH theory offers an unprecedented tool in order to map with great accuracy all the matter distribution around the newly formed EMBH from the horizon all the way to the ISM. This concept was pioneered by Dermer & Mitman [41] who proposed to use GRB sources as "tomographic images of the density distributions of the medium surrounding the sources of GRBs". It is important to emphasize that the very precise reading of the matter distribution encoded in the data of the P-GRB, the E-APE and the afterglow in GRB 991216 is in marked disagreement with the matter distribution postulated by the "collapsar" scenario (see MacFadyen & Woosley [89], Paczyński [109], Woosley [189]). This conclusion is evidenced not only by the absence of beaming already mentioned above, but also for the paucity of the baryonic matter encountered by the PEM pulse in its way out from the EMBH. There is no evidence for the presence either of a baryonic disk component nor of a conspicuous baryonic remnant. We actually have $B = 3.0 \times 10^{-3}$. Unlike the case of formation of a neutron star, the mass of the remnant of the progenitor star is very small indeed. This mass, determined by B, is very accurately inferable from the relative intensity and temporal distance between the P-GRB and the E-APE (see above). In the present case we have $M_B \sim 8.1 \times 10^{-4} M_\odot$. The presence of the remnant is also important for guaranteeing the overall charge neutrality of the system formed by the oppositely charged collapsing core and the remnant. It has been pointed out in section XXVI that this condition of charge separation between the collapsing core and the remnant occurs only during the relevant part of the gravitational collapse process which, we recall, for a $10 M_\odot$ is of the order of 30 seconds.

3. The necessity of developing a fine tuning in the final phases of thermonuclear evolution of the stars, both for the star collapsing to the black hole and the surrounding ones, in order to explain the possible occurrence of the "induced gravitational collapse".

New regimes are as well encountered from the point of view of nature of GRBs:

1. The basic structure of GRBs is uniquely composed by a proper-GRB (P-GRB) and the afterglow. The most general GRB contains three different components: the P-GRB, the E-APE and the rest of the afterglow. The ratio between the P-GRB and the E-APE intensity and their temporal separation is a function of the B parameter (see Figs. 11–11). The best fit is obtained for $B = 3.0 \times 10^{-3}$ (see section XV). We recall that in the present case for $B < 2.5 \times 10^{-5}$ the energy of the P-GRB would be larger than the one of the E-APE and the energy of the dyadosphere would be mainly emitted in what have been called the "short bursts", while for $B > 2.5 \times 10^{-5}$ the energy of the E-APE would predominate and the energy of the dyadosphere would be mainly carried by the ABM pulse and emitted in the afterglow.

2. The long bursts are then simply explained as the peak of the afterglow (the E-APE) and their observed time variability is explained in terms of inhomogeneities in the interstellar medium (ISM). The difficulties encountered by *all* theoretical models, through the years, in order to explain the so called "long bursts" are resolved in a

drastic way (see section XVI). The so called "long bursts" are *not* bursts at all. They represent just the E-APE which was interpreted as a burst only due to the noise threshold in the BATSE observations (see Fig. 10). The E-APE is emitted at distances from the EMBH in the range $1.0 \times 10^{16} \sim 1.0 \times 10^{17}$ cm, see Tab. I, namely well outside the size of the progenitor star and already deep in interstellar space. The fact that the crossing of such distance, which is a typical dimension of an interstellar cloud, appears to occur in arrival time in only ~ 100 seconds is perfectly explained by the relativistic transformations encoded in the RSTT paradigm corresponding to a gamma factor between 100 and 300 (see section V and Tab. I). This effect would be interpreted within a classical and incorrect astronomical picture by a "superluminal" behaviour propagating at $\sim 3.6 \times 10^4 c$ (see Tab. I).

3. The short bursts are identified with the P-GRBs and the crucial information on general relativistic and vacuum polarization effects are encoded in their spectra and intensity time variability. In the limit $B \to 0$ the entire dyadosphere energy is emitted in the P-GRB. These events represents the "short bursts" class, for which the afterglow intensity is smaller than the P-GRB emission and below the actual observational limits (see section XII). It is interesting that the proposed differentiation between the "short bursts" and "long bursts" within the EMBH theory is merely due to the amount of baryonic matter in the remnant, described by the B parameter, and totally independent from the process of gravitational collapse which is clearly identical in both cases. This explains at once the recently found conclusion that the distribution of short and long GRBs have essentially the same characteristic peak luminosity (Schmidt [169]). Also the result expressed in Fig. 6 that the average temperature corresponding to the P-GRB emission does increase for decreasing values of the B parameter can explain the observed fact that the "short bursts", which are obtained in the limit $B \to 0$, are systematically harder than "long bursts" (Kouveliotou et al. [83]).

A new class of space missions to acquire information on such extreme new regimes are urgently needed. The detailed observations of the yet unexplored region in the range up to 10 seconds in Fig. 29 and the corresponding observations of the "short bursts" by a new class of space missions with higher sensitivity than the BATSE instrument appear to be of great importance. Such observations should allow to directly observe for the first time the general relativistic and extreme quantum field theory effects connected to the process of formation of the EMBH. It can be of some interest to explore the possibility of observing in these regimes the "gravitationally induced electromagnetic radiation" (Johnston et al. [80]) and the "electromagnetically induced gravitational radiation" (Johnston et al. [81]) phenomena as well as to explore the possibility of developing neutrino detectors. This will need further developments of the predictions of the EMBH theory in these general relativistic and ultra-high-energy particle phenomena.

---

[1] Abramowitz, M., Stegun, I.A., Editors, *Handbook of Mathematical Functions*, National Bureau of Standards, Washington, D.C., 1970
[2] Ahmad, I., et al., 1995, Phys. Rev. Lett., 75, 2658
[3] Amati, L. et al. 2000, Science, 290, 953
[4] Band, D., et al., 1993, ApJ, 413, 281
[5] BATSE GRB Light Curves, http://gammaray.msfc.nasa.gov/batse/grb/lightcurve/
[6] BATSE Rapid Burst Response, 1999, http://gammaray.msfc.nasa.gov/ kippen/batserbr/
[7] Bär, R., et al., 1995, Nucl. Phys. A, 583, 237
[8] Beck, S.C., Turner, J.L., Kovo, O., 2000, AJ, 120, 244
[9] Bekenstein, J.D., 1971, Phys. Rev. D, 4, 2185
[10] Bekenstein, J.D., 1973, Phys. Rev. D, 7, 2333
[11] Bethe, H.A., 1991, *The road from Los Alamos*, Touchstone Ed. by American Institute of Physics (New York)
[12] Bianco, C.L., Ruffini, R., Xue, S.-S., 2001a, A&A, 368, 377
[13] Bianco, C.L., Chardonnet, P., Ruffini, R., Xue, S.-S., 2002, in preparation
[14] Bini, D., Cherubini, C., Jantzen, R., Ruffini, R., 2002, Prog. Theor. Phys., 107, 967
[15] Biretta, J.A., Sparks, W.B., Macchetto, F., 1999, ApJ, 520, 621
[16] Blandford, R.D., McKee, C.F., 1976, Phys. Fluids, 19, 1130
[17] Blinnikov, S.I., Kozyreva, A.V., Panchenko, I.E., 1999, Astron. Rep., 43, 739 (preprint: astro-ph/9902378)
[18] Bloom, L. et al., 1999, Nature, 401, 453
[19] Böttcher, M., Dermer, C.D., 2000, ApJ, 532, 281
[20] Bloom, J.S., Kulkarni, S.R., Djorgowski, S.G., 2002, AJ, 123, 1111
[21] Boccaletti, D., Ohanian, H.C., Ruffini, R., 2003, in preparation
[22] Borozdin, K.N., Trudolyubov, S.P., 2003, ApJ, 583, L57
[23] Carter, B., 1997, Proceedings of the Eighth Marcel Grossmann Meeting on General Relativity, Eds. T. Piran, R. Ruffini, World Scientific, Singapore, p. 136

[24] Cavallo, G., Rees, M.J., 1978, MNRAS, 183, 359
[25] Chandrasekhar, S., *Why are the stars as they are*, in *Physics and Astrophysics of Neutron Stars and Black Holes*, R. Giacconi and R. Ruffini Eds., 1978, North Holland, Elsevier
[26] Chandrasekhar, S., Fermi, E., 1953, ApJ, 118, 116
[27] Cherubini, C., Ruffini, R., Vitagliano, L., 2002, Phys. Rev. B, 545, 226
[28] Chiang, J., Dermer, C.D., 1999, ApJ, 512, 699
[29] Christodoulou, D., Ruffini, R., 1971, Phys. Rev. D, 4, 3552
[30] Corbet, R., Smith, D.A., 2000, in Rossi2000: Astrophysics with the Rossi X-ray Timing Explorer, Greenbelt, USA
[31] Costa, E., 2002, invited talk in "IX Marcel Grossmann Meeting on General Relativity", V. Gurzadyan, R. Jantzen & R. Ruffini editors, World Scientific (Singapore)
[32] Damour, T., Ruffini, R., 1975, Phys. Rev. Lett., 35, 463
[33] Davies, M.B., Benz, W., Piran, T., Thielemann, F.K., 1994, ApJ, 431, 742
[34] De la Cruz, V., & Israel, W., 1967, Nuovo Cimento A, 51, 744.
[35] Denardo, G., Ruffini, R., 1973, Phys. Lett., 45B, 259
[36] Denardo, G., Hively, L., Ruffini, R., 1974, Phys. Lett., 50B, 270
[37] Dermer, C.D., 1998, ApJ, 501, L157
[38] Dermer, C.D., Chiang, J., 1998, in *Proceedings of High Energy Processes in Accreting Black Holes*, held Graftavallen, Sweden, June 29- July 4, 1998, J. Poutanen & R. Svensson editors (preprint: astro-ph/9810222)
[39] Dermer, C.D., Böttcher, M., Chiang J., 1999a, ApJ, 515, L49
[40] Dermer, C.D., Chiang, J., Böttcher, M., 1999b, ApJ, 513, 656
[41] Dermer, C.D., Mitman, K.E., 1999, ApJ, 513, L5
[42] Dermer, C.D., 2002, ApJ, 574, 65
[43] Drozdova, D.N., Panchenko I.E., 1997, A&A, 324, 17
[44] Dyson, F., 2002, communication at "Science and Ultimate Reality", Symposium in honour of J. A. Wheeler, Princeton
[45] Eichler, D., Livio, M., Piran, T., Schramm, D.N., 1989, Nature, 340, 126
[46] Einstein, A., 1905, Ann. Phys. (Germany), 17, 891
[47] Fenimore, E.E., Madras, C.D., Nayakshin, S., 1996, ApJ 473, 998
[48] Fenimore, E.E., 1999, ApJ, 518, 375
[49] Fenimore, E.E., Cooper C., Ramirez-Ruiz, E., Sumner, M.C., Yoshida, A., Namiki, M., 1999, ApJ, 512, 683
[50] Ferrière, K.M., 2001, Rev. Mod. Phys., 73, 1031
[51] Filippenko, A.V., 1997, Annu. Rev. Astron. Astrophys., 35, 309
[52] Fishman, G., Meegan, C., 1995, Annu. Rev. Astron. Astrophys., 33, 415
[53] Frail, D. et al., 2000, ApJ, 538, L129
[54] Frail, D. et al., 2001, ApJ, 562, L55
[55] Frontera, F., et al., 2000, ApJS, 127, 59
[56] Fryer, C.L., Woosley, S.E., Herant, M., Davies, M.B., 1999, ApJ, 520, 650
[57] Galama, T.J. et al., 1998a, IAU Circ., 6895
[58] Galama, T.J. et al., 1998b, Nature, 395, 670
[59] Galama, T.J. et al., 2000, ApJ, 536, 185
[60] Gamow, G., Critchfield, C.L., 1951, "Theory of Atomic Nucleus and Energy Sources", Clarendon Press, Oxford
[61] Ganz, R. et al., 1996, Phys. Lett. B, 389, 4
[62] Garnavich, P. et al., 2000, ApJ, 543, 61
[63] Gerštein, S.S., Zel'dovich, Y.B., 1969, Lett. Nuovo Cimento, 1, 835
[64] Gerštein, S.S., Zel'dovich, Y.B., 1970, Sov. Phys. JETP, 30, 358
[65] Giacconi, R., Ruffini, R. (Eds.), 1978, *Physics and Astrophysics of Neutron Stars and Black Holes*, North Holland, Elsevier
[66] Gibbons, G.W., 1975, Commun. Math. Phys., 44, 245
[67] Gold, T., 1968, Nature, 218, 731
[68] Gold, T., 1969, Nature, 221, 27
[69] Goldreich, P., Julian, W.H., 1969, ApJ, 157, 869
[70] Gou, L.J., Dai, Z.G., Huang, Y.F., Lu, T., 2001, A&A, 368, 464
[71] Gursky, H., Ruffini, R., editors, 1975, "Neutron Stars, Black Holes and binary X-Ray Sources", Reidel, Dordrecht
[72] Halpern, J.P., Uglesich, R., Mirabal, N., Kassin, S., Thorstensen, J., Keel, W.C., Diercks, A., Bloom, J.S., Harrison, F., Mattox, J., Eracleous, M., 2000, ApJ, 543, 697
[73] Halzen, F., "High energy neutrino astronomy" in "Weak Interactions and Neutrinos, Proceedings of the 17th International Workshop. Cape Town, South Africa", Edited by C. A. Dominguez and R. D. Viollier, World Scientific Publishers (Singapore, 2000), p.123
[74] Harrison, B.K., Thorne, K.S., Wakano, M., Wheeler, J.A., 1965, "Gravitation Theory and Gravitational Collapse", University of Chicago Press, Chicago.
[75] Hawking, S.W., Nature, 1974, 248, 30
[76] Heinz, S. et al., 1998, Eur. Phys. J. A, 1, 27
[77] Heisenberg, W., Euler, H., 1935, Zeits. Phys., 98, 714
[78] Hewish, A., Bell, S.J., Pilkington, J.D., Scott, P.F., Collins, R.A., 1968, Nature, 217, 709
[79] Israel, W., 1966, Nuovo Cimento B, 44, 1
[80] Johnston, M., Ruffini, R., Zerilli, F., 1973, Phys. Rev. Lett., 31, 1317

[81] Johnston, M., Ruffini, R., Zerilli, F., 1974, Phys. Lett., 49B, 185
[82] Kaplan, S.A., 1949, Zh. Eksp. & Teor. Fiz., 19, 951
[83] Kouveliotou, C., Meegan, C.A., Fishman, G.J., Bath, N.P., Briggs, M.S., Koshut, T.M., Paciesas, W.S., Pendleton, G.N., 1993, ApJ, 413, L101
[84] Kulkarni, S.R. et al., 1998 Nature, 395, 663
[85] Landau, L., 1932, Phys. Zeits. Sowj., 1, 285
[86] Landau, L.D., Lifshitz, E.M., "Course of Theoretical Physics - Volume 6: Fluid Mechanics", 2nd edition, paperback, 1995, Butterworth Heinemann, section 135, pag. 510
[87] Leinberger, U. et al., 1997, Phys. Lett. B, 394, 16
[88] Leinberger, U. et al., 1998, Eur. Phys. J. A, 1, 249
[89] MacFadyen, A.I., Woosley, S.E., 1999, ApJ, 524, 262
[90] Mao, S., Yi, I., 1994, ApJ, 424, L131
[91] Mészáros, P., Rees, M.J., 1992a, MNRAS, 257, 29p
[92] Mészáros, P., Rees, M.J., 1992b, ApJ, 397, 570
[93] Mészáros, P., Rees, M.J., 1993, ApJ, 405, 278
[94] Mészáros, P., Rees, M.J., 1997a, ApJ, 476, 232
[95] Mészáros, P., Rees, M.J., 1997b, ApJ, 482, L29
[96] Mészáros, P., Rees, M.J., Wijers, R.A.M.J., 1998, ApJ, 499, 301
[97] Mészáros, P., Rees, M.J., 2000, ApJ, 530, 292
[98] Mészáros, P., Rees, M.J., 2001, ApJ, 556, L37
[99] Mészáros, P., 2002, Annu. Rev. Astron. Asttroph., 40, 137
[100] Mirabel, I.F., Rodriguez, L.F., 1994, Nature, 371, 46
[101] Mirabel, I.F., Rodriguez, L.F., 1999, Annu. Rev. Astron. Asttroph., 37, 409
[102] Misner, C.W., Thorne, K.S., Wheeler J.A., *Gravitation*, San Francisco: Freeman (1973), chapter 23.
[103] Narayan, R., Paczyński, B., Piran, T., 1992, ApJ, 395, L83
[104] Norris, J.P., et al., 1986, ApJ, 301, 213
[105] Oppenheimer, J.R., Volkoff, G., 1939, Phys. Rev. D, 55, 374
[106] Paciesas, W.S. et al., 1999, ApJS, 122, 465
[107] Paczyński, B., 1991, Acta Astronomica, 41, 257
[108] Paczyński, B., Xu, G., 1994, ApJ, 427, 708
[109] Paczyński, B., 1998, ApJ, 494, L45
[110] Panaitescu, A., Kumar P., 2001, ApJ, 554, 667
[111] Panaitescu, A., Mészáros, P., Rees, M.J., 1998, ApJ, 503, 314
[112] Panaitescu, A., Mészáros, P., 1998a, ApJ, 492, 683
[113] Panaitescu, A., Mészáros, P., 1998b, ApJ, 501b, 772
[114] Panaitescu, A., Mészáros, P., 1999, ApJ, 526, 707
[115] Piran, T., 1999, Phys. Rep. 314, 575
[116] Piran, T., 2001, talk at 2000 Texas Meeting (preprint: astro-ph/0104134)
[117] Piro, L. et al., 1998, GCN 155
[118] Piro, L. et al., 1999, ApJ, 514, L73
[119] Piro, L. et al., 2000, Science, 290, 955
[120] Popov, V.S., Rozhdestvenskaya, T.I., 1971, JETP Lett., 14, 177
[121] Popov, V.S., 1972, Sov. J. Nucl. Phys., 14, 257
[122] Preparata, G., Ruffini, R., Xue, S.-S., 2003, in *Proceedings of the VII Italian-Korean meeting*, ed. Journal of the Korean Physical Society, in press (preprint: astro-ph/0204080)
[123] Preparata, G., Ruffini, R., Xue, S.-S., 1998b, A&A, 338, L87
[124] Preparata, G., Ruffini, R., Xue, S.-S., 2001, in *Proceedings of the III ICRA Network Workshop and VI Italo-Korean Meeting*, C.Cherubini & R. Ruffini editors, SIF
[125] Punsly, B., "Black Hole Gravitohydromagnetics", Springer, 2001
[126] Rees M.J., 1966, Nature 211, 468
[127] Rees, M.J., Mészáros, P., 1994, ApJ, 430, L93
[128] Rhoads, J.E., 1997a, ApJ, 487, L1
[129] Rhoads, J.E., 1997b, in AIP Conf. Proc. 428, Gamma-Ray Bursts: Fourth Huntsville Symposium, ed. C. Meegan, R. Preece & T. Koshut (New York: AIP), 699
[130] Rhoads, J.E., 1999, ApJ, 525, 737
[131] Ringwald, A., 2001, Phys. Lett. B, 510, 107
[132] Roberts, C.D., Schmidt, S.M., Vinnik, D.V., 2002, Phys. Rev. Lett., 89, 153901
[133] Rol, E. et al., 1999, GCN Circ. 491 (http://gcn.gsfc.nasa.gov/gcn/gcn3/491.gcn3)
[134] Ruffini, R., Wheeler, J.A., 1971, a chapter in "The Significance of Space Research for Fundamental Physics", A.F. Moore, V. Hardy, eds., European Space Research Organization (ESRO) book No. SP-52, Paris, updated in Rees, M., Ruffini, R., Wheeler, J.A., 1974, "Black Holes, Gravitational Waves and Cosmology", Gordon and Breach Science Publisher, New York, London, Paris
[135] Ruffini, R., in *Physics and Astrophysics of Neutron Stars and Black Holes*, Giacconi, R., Ruffini, R., Ed. and coauthors, North Holland, Amsterdam, 1978

[136] Ruffini, R., 1998, in "Black Holes and High Energy Astrophysics", Proceedings of the 49th Yamada Conference Ed. H. Sato and N. Sugiyama, Universal Ac. Press, Tokyo, 1998
[137] Ruffini, R., in "Fluctuating Paths and Fields - Dedicated to Hagen Kleinert on the Occasion of His 60th Birthday", Eds. W. Janke, A. Pelster, H.-J. Schmidt, and M. Bachmann, World Scientific, Singapore, 2001, p. 771
[138] Ruffini R., 2002, in *Proceedings of the Ninth Marcel Grossmann Meeting on General Relativity*, Gurzadyan V.G., Jantzen R.T. & Ruffini R. editors, World Scientific (Singapore)
[139] Ruffini R., 2003, in *Proceedings of the 34th COSPAR Scientific Assembly - Houston, TX, USA, 2002*, Pian, E., Masetti, N., Piro, L., editors, Elsevier, in press.
[140] Ruffini, R. & Wheeler, J.A., 1971, Physics Today, 24,(1), 30
[141] Ruffini, R., Salmonson, J.D., Wilson, J.R., Xue, S.S., 1999, A&A, 350, 334, A&AS, 138, 511
[142] Ruffini, R., Salmonson, J.D., Wilson, J.R., Xue, S.S., 2000, A&A, 359, 855
[143] Ruffini, R., Bianco, C.L., Chardonnet, P., Fraschetti, F., Xue, S.-S., 2001a, ApJ, 555, L107
[144] Ruffini, R., Bianco, C.L., Chardonnet, P., Fraschetti, F., Xue, S.-S., 2001b, ApJ, 555, L113
[145] Ruffini, R., Bianco, C.L., Chardonnet, P., Fraschetti, F., Xue, S.-S., 2001c, ApJ, 555, L117
[146] Ruffini, R., Bianco, C.L., Chardonnet, P., Fraschetti, F., Xue, S.-S., 2001e, Nuovo Cimento, 116, 99
[147] Ruffini, R., Bianco, C.L., Chardonnet, P., Fraschetti, F., Xue, S.-S., 2002, ApJ, 581, L19
[148] Ruffini, R., Bianco, C.L., Chardonnet, P., Fraschetti, F., Xue, S.-S., 2003a, Int. Journ. Mod. Phys. D, to appear
[149] Ruffini, R., Bianco, C.L., Chardonnet, P., Fraschetti, F., Xue, S.-S., 2003b, ApJ, submitted to
[150] Ruffini, R., Bianco, C.L., Chardonnet, P., Fraschetti, F., Xue, S.-S., 2003c, in preparation
[151] Ruffini, R., Bianco, C.L., Chardonnet, P., Fraschetti, F., Xue, S.-S., 2003d, in preparation
[152] Ruffini, R., Bianco, C.L., Chardonnet, P., Fraschetti, F., Xue, S.-S., 2003e, A&A, submitted to
[153] Ruffini, R., Bianco, C.L., Chardonnet, P., Fraschetti, F., Xue, S.-S., 2003f, in preparation
[154] Ruffini, R., Bianco, C.L., Chardonnet, P., Fraschetti, F., Xue, S.-S., 2003g, in preparation
[155] Ruffini, R., Vitagliano, L., 2002a, Phys. Lett. B, 545, 233
[156] Ruffini, R., Vitagliano, L., 2003a, Int. Journ. Mod. Phys. D, 12, 121
[157] Ruffini, R., Vitagliano, L., 2003b, in preparation
[158] Ruffini, R., Vitagliano, L., Xue, S.-S., 2003h, in preparation
[159] Ruffini, R., Vitagliano, L., Xue, S.-S., 2003i, in preparation
[160] Sagar, R. et al., 2000, Bull Astron. Soc. India, 28, 15
[161] Salmonson, J.D., Wilson, J.R., Mathews, G.J, 2001, ApJ, 553, 471
[162] Sari, R., 1997, ApJ, 489, L37
[163] Sari, R., 1998, ApJ, 494, L49
[164] Sari, R., Piran, T., 1997, ApJ, 485, 270
[165] Sari, R., Piran, T., Narayan, R., 1998, ApJ, 497, L17
[166] Sari, R., Piran, T., Halpern, J.P., 1999, ApJ, 519, L17
[167] Sari, R., Piran, T., 1999, ApJ, 520, 641
[168] Schaefer, B., 2000, GCN Circ. 517 (http://gcn.gsfc.nasa.gov/gcn/gcn3/517.gcn3)
[169] Schmidt, M., 2001, ApJ, 559, L79
[170] Schweppe, J., et al., 1983, Phys. Rev. Lett., 51, 2261
[171] Schwinger, J., 1951, Phys. Rev., 98, 714
[172] Shemi, A., Piran, T., 1990, ApJ, 365, L55
[173] Shvartsman, V.F., 1097, Soviet Physics JETP, 33, 475
[174] Strong, I.B., in "Neutron Stars, Black Holes and Binary X-Ray Sources", Gursky, H. and Ruffini, R., editors, D. Reidel Publishing Company, 1975
[175] Takeshima, T., Markwardt, C., Marshall, F., Giblin, T., Kippen, R.M., 1999, GCN Circ. 478 (http://gcn.gsfc.nasa.gov/gcn/gcn3/478.gcn3)
[176] Taub, A.H., 1948, Phys. Rev., 74, 328
[177] van Dick, R.S., Schwinberg, P.B., Dehmelt, H.G., 1977, Phys. Rev. Lett., 38, 310
[178] van Paradijs, J., Kouveliotou, C., Wijers, R.A.M.J., 2000, Annu. Rev. Astron. Astrophys., 38, 379
[179] Vietri, M., 1997, ApJ, 478, L9
[180] Vietri, M., Stella, L., 1998, ApJ, 507, L45
[181] Vietri, M., Stella, L., 1999, ApJ, 527, L43
[182] Vietri, M., Perola, C., Piro, L., Stella, L., 1999, MNRAS 308, L29
[183] Watson, D., Reeves, J.N., Osborne, J.P., Tedds, J.A., O'Brien, P.T., Tomas, L., Ehle, M., 2002a, A&A, 395, L41
[184] Watson, D., Reeves, J.N., Osborne, J., O'Brien, P.T., Pounds, K.A., Tedds, J.A., Santos-Lleó, M., Ehle, M., 2002b, A&A, 393, L1
[185] Waxman, E., 1997, ApJ, 491, L19
[186] Wilson, J.R., Mathews, G.J., Marronetti, P., 1996, Phys. Rev. D, 54, 1317
[187] Wilson, J.R., Salmonson, J.D., Mathews, G.J., 1997, in Gamma-Ray Bursts: 4th Huntsville Symposium, ed. C. A. Meegan, R. D. Preece, T. M. Koshut (American Institute of Physics)
[188] Wilson, J.R., Salmonson, J.D., Mathews, G.J., 1998, in 2nd Oak Ridge Symposium on Atomic and Nuclear Astrophysics (IOP Publishing Ltd).
[189] Woosley, S.E., 1993, ApJ, 405, 273
[190] Zaumen, W.T., 1974, Nature, 247, 530
[191] Zel'dovich, Y.B., Popov, V.S., 1972, Sov. Phys. Usp., 14, 673
[192] Zel'dovich, Ya.B. & Rayzer, Yu.R., 1966, *Physics of shock waves and high-temperature hydrodynamic phenomena*, Ed. by Wallace D. Hayes and Ronald F. Probstein, Academic press (New York and London)

Chaotic Phenomena in Astrophysics and Cosmology

V.G.Gurzadyan

*ICRA, Department of Physics, University of Rome "La Sapienza", Rome,
Italy and Department of Theoretical Physics, Yerevan Physics Institute, Yerevan, Armenia*

Chaos is a typical property of many-dimensional nonlinear systems and is revealed in a number of problems of astrophysics and cosmology. Particularly, chaos made to revise the two-hundred year old views on the evolution of Solar system, while the theory of interstellar matter, dynamics of stellar systems cannot be considered neglecting the chaotic effects. The lectures notes cover the following topics: dynamics of the Solar system, relaxation of galaxies and star clusters, the substructure of galaxy clusters, hyperbolicity in the Wheeler-DeWitt superspace and the stability of cosmological solutions. Thus we aimed to cover as broad topics as possible, at the same time showing the diversity of approaches and mathematical tools. For pedagogical reasons, the techniques such as the estimation Kolmogorov-Sinai entropy, the hyperbolicity in pseudo-Riemannian spaces are described in some detail, so that they can be applied in various problems.

I. INTRODUCTION

Chaos is a typical property of many-dimensional nonlinear systems. Its role it revealed in various problems of astrophysics and cosmology. Chaos made to revise the two-hundred year old views on the evolution of Solar system. Theory of interstellar matter, dynamics of star clusters and galaxies at present cannot be considered without chaotic effects.

Astronomical topics themselves had remarkable impact on the development of chaotic dynamics. The Henon-Heiles system, one of the first systems with revealed chaotic properties, was proposed for the study of the motion of a star in a galactic potential. Much earlier, Poincare's classical work on the foundations of the theory of dynamical systems emerged from the problem of small perturbations in the planetary dynamics.

In the present lectures I will discuss only several astrophysical and cosmological problems. The choice of the problems is determined with the aim, first, to cover as broad topics as possible, second, to show the diversity of approaches and mathematical tools. I will start from planetary dynamics, moving to galactic dynamics, to cosmology, and to the instability in the Wheeler-DeWitt superspace. For pedagogical reasons I will describe the techniques, such as the estimation Kolmogorov-Sinai entropy, the dealing with hyperbolicity in pseudo-Riemannian spaces, so that

they can be applied for any other problems. Obviously, numerous other problems, methods and results remain out of these lectures, most of them, however, can be traced from the references; for chaos see[1–3], for applications of our interest see[4–7].

I will start from a brief review of the elements of theory of dynamical systems, to introduce the main concepts used in the subsequent chapters.

II. ELEMENTS OF ERGODIC THEORY

A. Dynamical systems

Ergodic theory is the metric theory of dynamical systems, i.e. which deals with spaces for which a measure is defined but not a metric.

In the following brief account of elements of smooth ergodic theory, I will concentrate on the classification of dynamical systems by the degree of their statistical properties; for details see[8–10].

The key concept is obviously, that of the dynamical system. Initially the dynamical systems were understood as mechanical systems, however later that term was generalized for variety of physical systems of non-mechanical origin. Cosmological solutions of Einstein equations can be considered as such examples.

Dynamical system $(M, \mathcal{B}, \mu, T)$ is considered defined if M is a smooth manifold, $\mathcal{B}$ is a σ–algebra of measurable sets on M, and μ is a complete measure on $\mathcal{B}$, and T^t is a one-parameter group of diffeomorphisms defined by vector field **v**

$$\mathbf{v}(x) = \frac{dT^t x}{dt}. \tag{1}$$

One-parameter groups are called flows, by the term borrowed from hydrodynamics. The apparent abstractness of the definition implies quite general and natural properties for physical systems.

B. Classification of dynamical systems, mixing, relaxation

Flows are called *ergodic*, if for any measurable invariant set A

$$T^t A = A = T^{-t} A, \qquad (2)$$

its measure μ takes only the values

$$\mu(A) = \begin{cases} 0 \\ 1 \end{cases}. \qquad (3)$$

One can show that for measure-preserving ergodic flows the time-average almost everywhere equals the phase space average

$$\int_M f d\mu = \lim_{t \to \infty} \frac{1}{t} \int_0^t f(T^{-\tau}(x)) d\tau.$$

In physical literature this property is often considered as a definition of an ergodic system since it is enough and sufficient. The property of ergodicity is one of rare definitions of smooth ergodic theory which can be generalized also for spaces with infinite measure. Ergodicity, however, is a weak statistical property and therefore is less important for actual physical problems.

The far more importance for statistical physics of another property, mixing, has been established firstly by Gibbs. Ergodic theory provides definitions for mixing of various degrees.

Weak mixing is indicated by the condition for $\forall f, g \in L^2$

$$\lim_{t \to \infty} \frac{1}{t} \int_0^t \left[\int_M f(T^{-\tau} x) g d\mu - \int_M f d\mu \int_M g d\mu \right]^2 d\tau = 0. \qquad (4)$$

The 'weakness' of the property of weak mixing can be seen from the following limit

$$\lim_{t \to \infty} \frac{1}{t} \int_0^t | \mu(T^{-\tau} A \cap B) - \mu(A)\mu(B) | d\tau = 0, \qquad (5)$$

implying that $T^t A$ becomes independent of the set B only if some parts of the trajectory are not taken into account.

Note the absence of the factor '1/t' and hence increase in the convergence rate in the definition of the property of *mixing*

$$\lim_{t \to \infty} \int_M f(T^t x) g d\mu = \int_M f d\mu \int_M g d\mu. \qquad (6)$$

Analogically the property of *m-fold mixing* for m functions is generalized as follows

$$\lim_{t_1, \ldots, t_m \to \infty} \int_M f_0 f_1(T^{t_1} x) \ldots f_m(T^{t_1 + \ldots + t_m} x) d\mu = \prod_{i=0}^m \int_M f_i d\mu. \qquad (7)$$

These properties describe systems with increasing statistical properties in the sense that, systems with mixing possess the property of weak mixing, and those with n-fold mixing include also that of mixing and weak mixing but not vice versa.

Systems with mixing are evidently also ergodic ones. However, for the systems with mixing, as opposed to ergodic ones, a set $A \in \mathcal{B}$ evolves in such a way (preserving its measure and connection) that the measure of the part which intersects the set $B \in \mathcal{B}$ tends in time to be proportional to the measure of B

$$\lim_{t \to \infty} \frac{\mu(T^t A \bigcap B)}{\mu(A)} = \mu(B). \qquad (8)$$

Compare this limit with the following one for ergodic systems

$$\lim_{t \to \infty} \frac{1}{t} \int_0^t \mu(T^{-\tau} A \cap B) d\tau = \mu(A)\mu(B), \qquad (9)$$

The latter limit is said to converge *in the Cesaro sense*, while the limit for the mixing case is ordinary converging. In other words, for ergodic systems the initial fluctuations tend to zero only in the time-average, for mixing systems their absolute value decreases as well.

Hence the property of mixing guarantees the existence of a final state of measure μ

$$\mu_t(A) = \mu_0(T^t A), \ A \in \mathcal{B}$$

to which the system tends smoothly, so that

$$\lim_{t \to \infty} \int_M f d\mu_t = \int_M f d\mu. \tag{10}$$

In physical terminology the final state is called *equilibrium*, the process of tending to that state is the *relaxation*.

Even stronger statistical property, K-mixing, possess K-systems (Kolmogorov systems) systems for which the following limit

$$\lim_{t \to \infty} \sup |\mu(A \cap B) - \mu(A)\mu(B)| = 0 \tag{11}$$

exists, where the upper limit is taken for the smallest σ-algebra containing $T^{t_0} A$ for $t < t_0$. Kolmogorov systems possess n-fold mixing of arbitrary n.

Strongest statistical properties are possessed by hyperbolic (Axiom-A), Anosov, and Bernoulli systems.[49]

We will define Anosov systems[12] which by their statistical properties are equivalent (isomorphic) to Bernoulli shifts.

The flow f^t is of Anosov type if for its all trajectories $\{f^t\}$ there exist subspaces $E^s(f^t(x)), E^u(f^t(x))$ of the tangential space $TM_{f^t(x)}$, and numbers $C > 0, \lambda > 0$, such that

$$TM_{f^t(x)} = E^s(f^t(x)) \oplus E^u(f^t(x)),$$
$$df^\tau E^s(f^t(x)) = E^s(f^{t+\tau}(x)), \qquad df^\tau E^u(f^t(x)) = E^u(f^{t+\tau}(x)),$$

and for all $t > 0$, one has

$$\| df^t v \| \leq C e^{-\lambda t} \| v \|, \qquad v \in E^s;$$
$$\| df^t v \| \geq C^{-1} e^{\lambda t} \| v \|, \qquad v \in E^u.$$

The subspaces E^s and E^u are stable (converging) and unstable (expanding) subspaces.

For physical systems this definition implies exponential instability at each point of the phase trajectory and at any small perturbation. Anosov systems are subclass of hyperbolic systems and they possess an important property of structural stability. Roughly it means that the perturbed systems possess the property of the unperturbed one, i.e. the strong instability acts towards preserving of the properties of the system. Though the conditions of Anosov systems never or almost never are satisfied for real physical systems, nevertheless it appears that the structural stability can be peculiar to certain types of instable physical systems.

Geodesic flow on a compact manifold with negative constant curvature is an example of an Anosov system, and was studied long ago by Hadamard, Hopf and Hedlund. Their works had inspired Krylov[13] to apply those ideas to physical systems.

If the systems with mixing can tend to equilibrium by any law (e.g. polynomial), hyperbolic systems tend to that state exponentially.

C. Kolmogorov-Sinai entropy

The problem of distinguishing different features of dynamical systems, and the formulation of corresponding characterizing criteria is a central one in ergodic theory. The main efforts in this direction were concentrated on the study of spectral properties of dynamical systems, until in 1958 Kolmogorov[14] discovered a new metric invariant, the entropy.

Consider the entropy of a splitting ξ_i of the measurable manifold M

$$H(\xi) = \sum_{i=1}^{d} \mu(\xi_i) \ln(\xi_i), \tag{12}$$

where $\xi_i \in \mathcal{B}$ and

$$\xi_i \bigcap \xi_j = \emptyset \quad \text{if } i \neq j, \tag{13}$$

$$\bigcup_{i=1}^{d} \xi_i = M. \tag{14}$$

Then the Kolmogorov-Sinai (KS) entropy h is the limit

$$h(f) = \sup \lim_{n \to \infty} \frac{1}{n} H(\xi^n), \tag{15}$$

where

$$\xi^n = \bigvee_{j=0}^{n-1} f^{-j}\xi,$$

and the upper limit is taken over all measurable splittings.

Dynamical systems with positive KS-entropy $h > 0$ are usually called *chaotic*, while those with $h = 0$ are called *regular* ones. In particular, Anosov and Kolmogorov systems, which are typical systems with mixing, have positive KS-entropy $h > 0$, while most of the ergodic ones have $h = 0$. Therefore, the ergodic systems are not considered as chaotic according to this definition.

For the above mentioned geodesic flows on spaces with constant negative curvature $R < 0$, the KS-entropy equals

$$h = \sqrt{-R}. \tag{16}$$

The KS-entropy is related to the Lyapunov characteristic exponents λ_i via the Pesin formula

$$h(f) = \int_M \sum_{\lambda_i(x) > 0} \lambda_i(x) d\mu(x); \tag{17}$$

we see that a system with at least one non-zero Lyapunov exponent has positive KS-entropy. The use of Lyapunov exponents for many-dimensional systems is not always well defined, nevertheless it was efficiently applied for stellar systems[15, 16].

Finally let us mention another important characteristic of dynamical systems, the correlation function, defined by

$$b_{g,g'}(t) = \int_M g(f^t x) g'(x) d\mu - \int_M g(x) g'(x) d\mu. \tag{18}$$

Although at present estimates of the correlation functions exist only for a few dynamical systems (including numerical results on some billiards), for Anosov systems it has been shown that the correlation functions decay exponentially, i.e., $\exists \alpha_{g,g'}, \beta, t > 0$ so that

$$|b_{g,g'}(t)| \leq \alpha_{g,g'} \exp(-\beta t), \tag{19}$$

where

$$\beta \simeq h(f).$$

III. CHAOTIC SOLAR SYSTEM

Results obtained in recent decades have revealed the crucial role of chaotic effects in planetary dynamics. For detailed reviews I refer to [17–19], where various evidences of chaos, particularly in the asteroid belt, in the motion of comets, are discussed, along with the methods of overlapping resonances, estimations of Lyapunov exponents, Wisdom-Holman symplectic mapping and other techniques used in those studies.

Before considering the stability of the Solar system, let us formulate the two key theorems, Poincare's and Kolmogorov's, which were crucial in the efforts on this long-standing problem.

N-dimensional system is considered as integrable if its first integrals $I_1, ..., I_N$ in involution are known, i.e. their Poisson brackets are zero. As follows from the Liouville theorem, if the set of levels

$$M_I = \{I_j(x) = I_j^0, j = 1, ..., N\}$$

is compact and connected, then it is diffeomorphic to N-dimensional torus

$$T^N = \{(\theta_1, ..., \theta_N), mod\, 2\pi\},$$

and the Hamiltonian system performs a conditional-periodic motion on M_I. Poincare theorem states that *for a system with perturbed Hamiltonian*

$$H(I, \varphi, \epsilon) = H_0(I) + \epsilon H_1(I, \varphi, \epsilon), \tag{20}$$

where I, φ are action-angle coordinates, at small $\epsilon > 0$ no other integral exists besides the one of energy $H = const$, if H_0 fulfills the nondegeneracy condition,

$$det|\partial\omega/\partial I| \neq 0, \tag{21}$$

i.e. the functional independence of the frequencies $\omega = \partial H_0/\partial I$ *of the torus over which the conditional-periodic winding is performed.*

Though this theorem does not specify the behavior of the trajectories of the system on the energy hypersurface, up to 1950s it was widely believed that such perturbed systems have to be chaotic.

Kolmogorov's theorem[20] of 1954, the main theorem of Kolmogorov-Arnold-Moser theory, however, showed that at certain conditions the perturbed Hamiltonian systems can remain stable.

It states:

If the system (20) satisfies the nondegeneracy condition (21) and H_1 is an analytic function, then at enough small $\epsilon > 0$ most of non-resonant tori, i.e. tori with rationally independent frequencies satisfying the condition

$$\sum n\omega_k \neq 0, \tag{22}$$

do not disappear and the measure of the complement of their union set $\mu(M) \to 0$ *at* $\epsilon \to 0$.

KAM-theory was initially considered as supporting the views on the stability of the Solar system, though it says nothing about the limiting value of the perturbation ϵ.

However, though the level of direct applicability of the KAM theory for the Solar system remains unclear, it turned out that, the joint application of both theoretical and numerical methods at present computer's possibilities is rather efficient.

The frequency map technique developed by Laskar [21–25] is based on the approach of KAM theory. This method enabled numerical treatment of long-term planetary evolution in terms of a perturbed Hamiltonian system using the idea that, if a quasi-periodic function is given numerically on the complex domain, then it is possible to approximate it via a quasi-periodic function with an accuracy higher than that given by standard Fourier series. Namely, the quasi-periodic function is represented over a finite time interval as a finite number of terms [22]

$$z_j(t) = z_0 e^{i\nu_j t} + \sum_{k=1}^{N} a_{m_k} e^{i<m_k, \nu>t}. \tag{23}$$

Then the frequencies and complex amplitudes are computed via an iterative procedure. For example, the first frequency is determined by the maximum amplitude of $\phi(\sigma)$

$$\phi(\sigma) = <1/2\pi \int_{-T}^{T} f(t) e^{i\sigma t} \chi(t) dt>, \tag{24}$$

where $\chi(t)$ is an even-weight function.

Numerical integration with a time step of 500 years over the time span of about 200 million years reveals that the inner planets of the Solar system are chaotic, due to the presence of two secular resonances, one due to Mars and Earth at $\theta = 2(g_4 - g_3) - (s_4 - s_3)$ and another due to Mercury, Venus and Jupiter at $\sigma = (g_1 - g_5) - (s_1 - s_2)$, where g_i and s_i are the frequencies of the perihelions and nodes, respectively.

Laskar's calculations revealed *the chaotic behavior of Mercury's orbit with eccentricity variations up to 0.05, its overlapping with the orbit of Venus and with inevitable escape of Mercury from its orbit*. Chaotic behavior was discovered also for the obliquities of the planets, particularly for Mars, varying from 0 to 60 degrees. Obviously, this

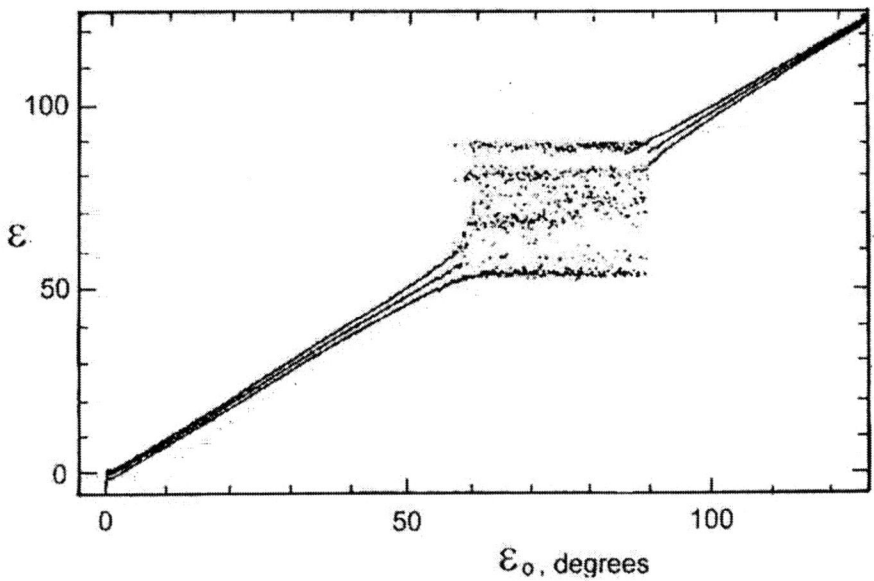

FIG. 1: The maximum, mean and minimum obliquity variations of the Earth depending on the initial obliquity ϵ_0. Note the chaotic zone at $\epsilon_0 = 60° - 90°$ and the stability for its other values. The present obliquity of the Earth is within the stability zone due to the presence of the Moon, and transfers into the chaotic zone at the absence of the Moon (reprinted from Ref.[24], with permission from Nature).

fact has to be taken into account while studying the past evolution of the atmosphere of Mars. *Chaotic behavior of the obliquity of the Earth would range even wider, within 0 and 85 degrees, with all dramatic consequences for the climate of the Earth, however, only at the absence of the Moon. The Moon therefore, is damping the obliquity variations up to 1.3 degrees, thus stabilizing the Earth's climate[24].*

We see that, the chaotic effects are not only able to influence essentially the dynamics of the Solar system but even the Earth's climate.

IV. GALACTIC DYNAMICS

A. N-body gravitating systems and geodesic flows

Many properties of statistical mechanics of globular clusters and galaxies can be studied considering a N-body gravitating system described by Hamiltonian

$$H(p,r) = \sum_{a=1}^{N}\sum_{i=1}^{3} \frac{p_{(a,i)}^2}{2m_a} + U(r), \qquad (25)$$

$$U(r) = -\sum_{a<b} \frac{Gm_a m_b}{|r_{ab}|}, \qquad (26)$$

$$r_{ab} = r_a - r_b. \qquad (27)$$

We will use a well known method existing in classical mechanics, the Maupertuis principle[8], enabling one to represent a Hamiltonian system as a geodesic flow on some Riemannian manifold. In physical problems this approach was firstly used by Krylov[13] and in stellar dynamics in[26, 27].

By means of the Maupertuis principle the Hamiltonian equations

$$\dot{r}^\mu = \frac{\partial H}{\partial p_\mu}, \qquad \dot{p}_\mu = -\frac{\partial H}{\partial r^\mu}, \qquad (28)$$

are reduced to the geodesic equation

$$\nabla_u u = 0 \qquad (29)$$

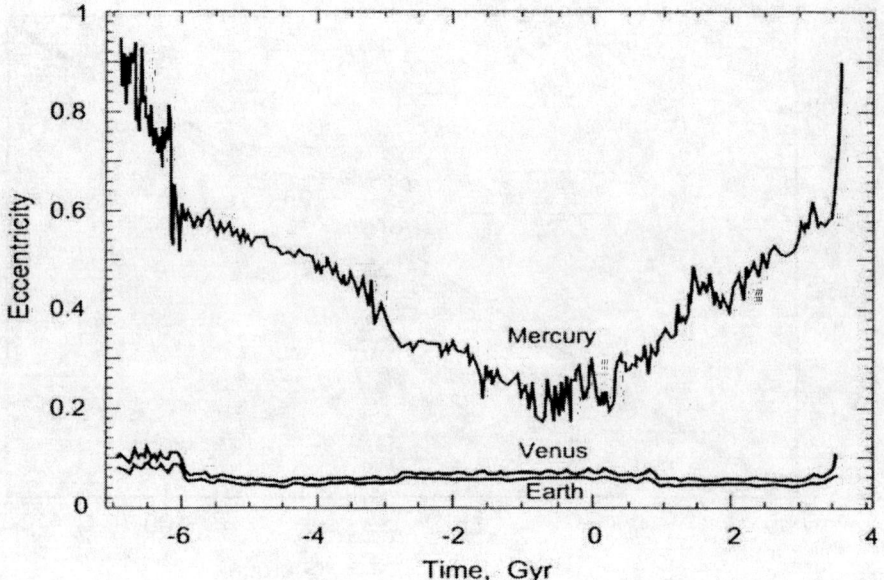

FIG. 2: The variation of the eccentricities of Mercury, Venus and Earth by time. The chaotic variations are suppressed for the Earth and Venus, but are significant for Mercury and will lead to its escape from the Solar system within a period about 3.5 Gyr (Laskar 1994).

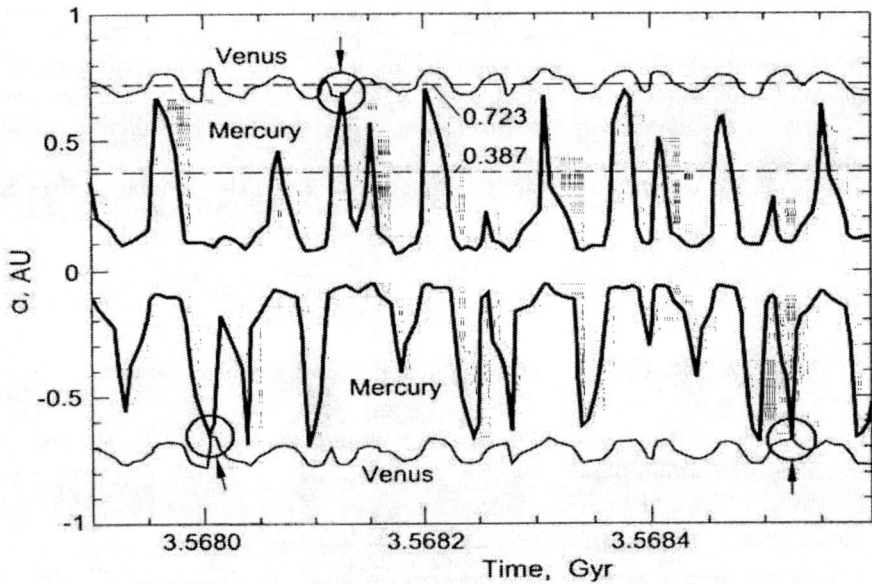

FIG. 3: The overlapping of the major semi-axes of orbits of Mercury and Venus due to chaotic effects (Laskar 1994).

on the region of configurational space

$$M = \{W = E - V(r) > 0\}$$

with the Riemannian metric

$$ds^2 = [E - V(r_{1,1}, ..., r_{N,3})] \sum_{a=1}^{N} \sum_{i=1}^{3} (dr_{a,i})^2, \tag{30}$$

where E is the total energy of the system. The condition of conservation of the total energy of the system

$$H(p,q) = E$$

is equivalent to the condition on the velocity associated with the geodesic

$$\| u \| = 1,$$

while the affine parameter along the geodesic is determined by

$$ds = \sqrt{2(E - V(r))}dt. \tag{31}$$

The statistical properties of the geodesic flow are determined from the Jacobi equation

$$\nabla_u \nabla_u n + Riem(n, u)u = 0. \tag{32}$$

For a vector field satisfying the orthogonality condition

$$<n, u> = 0,$$

the Jacobi equation can be written in the form

$$\frac{d^2 \| n \|^2}{ds^2} = -2K_{u,n} \| n \|^2 + 2 \| \nabla_u n \|^2, \tag{33}$$

where $K_{u,n}$ is the two-dimensional curvature

$$K_{u,n} = \frac{<Riem(n,u)u, n>}{\| u \|^2 \| n \|^2 - <u,n>^2} = \frac{<Riem(n,u)u, n>}{\| n \|^2}. \tag{34}$$

Jacobi equation has the solution

$$\| n(s) \| \geq \frac{1}{2} \| n(0) \| \exp(\sqrt{-2k}s), \qquad s > 0 \tag{35}$$

if

$$k = \max_{u,n}\{K_{u,n}\} < 0. \tag{36}$$

Then, the geodesic flow is an Anosov system.

Thus the negativity of the two-dimensional curvature the criterion of the instability of the geodesic flow.

B. Chaos in spherical systems

So, we have to calculate the two-dimensional curvature $K_{u,n}$ for the Hamiltonian of the gravitating N-body system. The Riemann curvature has the form[26, 27]

$$Riem_{\mu\lambda\nu\rho} = -\frac{1}{2W}[g_{\mu\nu}W_{\lambda\rho} + g_{\lambda\rho}W_{\mu\nu} - g_{\mu\rho}W_{\lambda\nu} - g_{\nu\rho}W_{\mu\lambda}] \tag{37}$$

$$-\frac{3}{4W^2}[(g_{\mu\lambda}W_\nu - g_{\mu\nu}W_\lambda)W_\rho + (g_{\nu\rho}W_\lambda - g_{\lambda\rho}W_\nu)W_\mu] \tag{38}$$

$$-\frac{1}{4W^2}[g_{\mu\nu}g_{\lambda\rho} - g_{\mu\lambda}g_{\nu\rho}] \| \partial W \|^2, \tag{39}$$

where

$$g_{\mu\nu} = W\delta_{\mu\nu},$$

$$W_\mu = \frac{\partial W}{\partial r^\mu}, \qquad W_{\mu\nu} = \frac{\partial^2 W}{\partial r^\mu \partial r^\nu}.$$

The analysis shows that $K_{u,n}$ is sign-indefinite, which means that no universal function

$$\tau : (\text{all } N\text{-body systems}) \to R_+$$

exists and hence no unique relaxation time scale can exist for all N-body gravitating systems.

For spherical systems, however, the two-dimensional curvature at large N limit is shown to be strongly negative, since is determined by the scalar curvature R

$$K_{u,n} = \frac{R}{3N(3N-1)} \qquad (40)$$

which is negative for $N > 2$

$$R = -\frac{3N(3N-1)}{W^3}\left(\frac{1}{4} - \frac{1}{2N}\right)(\nabla W)^2 - (3N-1)\frac{\Delta W}{W^2} < 0. \qquad (41)$$

Based on the mixing properties of dynamical systems described in previous sections, one can define the *relaxation time* for spherical systems. The explicit formula for that relaxation time taking into account the nonlinear interaction of all N bodies of the system has been estimated in[26, 27] (see also[28]). For real stellar systems its value is shorter than the two-body relaxation time scale, but longer than the dynamical (crossing) time. The non-compactness of the phase space is not a crucial difficulty at the current conditions of the stellar systems.

The disk gravitating systems can be studied using the Lie algebra of all vector fields with zero divergence[29] on the two-dimensional torus T^2, since the kinetic energy of the element of the moving fluid induces a right–invariant Riemannian metric on $SDiff(T^2)$. The principle of least action, which determines the motion of an incompressible fluid in terms of the geodesics of this metric, plays the role of the Maupertuis principle. One can show that, though the motion in disk galaxies is exponentially instable, the velocity field remains constant, so one cannot speak about a relaxation in the same sense as for spherical systems[30].

C. Relative chaos in stellar systems

The approach described above enables to consider instability of various configurations of stellar systems. Average the Jacobi equation over the geodesic deviation vector

$$\frac{d^2 z}{ds^2} = \frac{1}{3N} r_u(s) + <\parallel \nabla_u n \parallel^2 >, \qquad (42)$$

where

$$n = z\hat{n}, \qquad \parallel \hat{n} \parallel^2 = 1,$$

and $r_u(s)$ denotes the Ricci curvature in the direction of the velocity of the geodesic (*Ric* is the Ricci tensor)

$$r_u(s) = \frac{Ric(u,u)}{u^2} = \sum_{\mu=1}^{3N-1} K_{n_\mu, u}(s), \qquad (n_\mu \perp n_\nu, n_\mu \perp u). \qquad (43)$$

The Ricci tensor has the expression

$$Ric_{\lambda\rho} = -\frac{1}{2W}[\Delta W g_{\lambda\rho} + (3N-2)W_{\lambda\rho}]$$
$$+ \frac{3}{4W^2}[(3N-2)W_\lambda W_\rho] - [\frac{3}{4W^2} - \frac{(3N-1)}{4W^2}]g_{\lambda\rho} \parallel dW \parallel^2 .$$

Then the criterion of relative instability is[31]:
the more unstable of two systems is the one with smaller negative r

$$r = \frac{1}{3N} \inf_{0 \le s \le s_*}[r_u(s)], \qquad r < 0 \qquad (44)$$

within a given interval $0 \le s \le s_*$, i.e., this system should be unstable with a higher probability in the same interval.

Numerical exploitation of the Ricci criterion of relative instability for the different models of stellar systems has shown that, e.g., a spherical system with a central mass is more unstable than a homogeneous one, spherical systems are more instable than disk-like ones, etc [31–33].

Thus, the chaotic effects are governing the relaxation and hence the evolution of globular clusters and elliptical galaxies and are determining various properties of galaxies of other morphological types.

V. GALAXY CLUSTERS: SUBSTRUCTURE AND BULK FLOWS

N-body gravitational systems, as we saw above, are exponentially instable systems. This fact gives a key to the possibility of reconstruction of certain properties of based on the limited observational information, which usually includes the 2D coordinates and 1D (line-of-sight) velocities and the magnitudes of the galaxies. We will show how one can reconstruct the hierarchical substructure and the bulk regular flows of the subgroups in the clusters of galaxies[4, 34].

The developed S-tree technique is based on the geometrical methods of theory of dynamical systems discussed in the previous sections, namely, on the introduction of the concept *degree of boundness* of N particles.

Consider two particles, so that $x_1(t)$ and $x_2(t)$, $t \in (-T, T)$ are their trajectories when their interaction is taken into account, and $y_1(t)$ and $y_2(t)$, when the interaction is "switched off".

It is easy to see that the deviation of trajectories within certain time interval

$$m = \max_{i=1,2} \mathcal{M}(x_i(\cdot) - y_i(\cdot)), \tag{45}$$

can be taken as a measure of the degree of boundness with respect to a local norm

$$\mathcal{M}(x(\cdot)) = \sup_{t \in (-T,T)} \{|x(t)|, |\dot{x}(t)|\}. \tag{46}$$

Consider balls of radius r at each point of trajectories of the two interacting particles x_i. The union of those balls

$$C_i(r) = \bigcup_{t \in (-T,T)} B_{x_i(t)}(r), \ i = 1, 2 \ .$$

of such minimal radius m which contains all trajectories of the particles will denote the free corresponding particles.

Two particles are considered to be ρ-bound for $\rho > 0$ if $m \leq \rho$.

This is easily generalized to any finite number of particles. N particles labeled by the set of integers $\mathcal{A} = \{1, \ldots, N\}$ form a ρ-bound cluster if the distance between the corresponding trajectories of the system of interacting particles and free ones is less than the maximal deviation of all of the particles:

$$m = \max_{a \in \mathcal{A}} \mathcal{N}(x_a(\cdot) - y_a(\cdot)) \leq \rho \ . \tag{47}$$

One can then define the boundness function so that for the given local norm $\mathcal{N}$

$$\mathcal{P}_Z(Y) = \max_{a \in Y} \mathcal{N}(z_a(\cdot) - y_a(\cdot)), \tag{48}$$

where $z_a(t), y_a(t)$ are the solutions of corresponding equations for some time interval $(-T, T)$. In other words the boundness of $\mathcal{Y}$ in $\mathcal{Z}$ is the maximum deviation of the trajectories of its particles taking into account only internal interactions compared to the situation when interactions with particles in $\mathcal{Z}$ are also included. The goal is to split $\mathcal{A}$ into ρ–subsystems, i.e., to obtain the map Σ for this choice of boundness function.

The definition of a ρ–bound cluster given above can be reformulated now as a set of corresponding particles A being a connected subgraph of the graph Γ (equivalent the matrix) so that there is no other connected subgraph B including A: $A \subset B$. If one defines P as follows

$$\mathcal{P}_X Y = \max_{\substack{y \in Y \\ z \in X \setminus Y}} \{D_{yz}\} \ , \tag{49}$$

the problem of the search of a ρ-bound cluster is reduced to that of a connected Γ-graph, i.e. to tree-diagrams.

The algorithm of the construction of tree-diagrams based on the estimation of the two-dimensional curvature as containing information both on the coordinates and velocities of the all particles, is developed and applied to various clusters of galaxies. As a result subgroups of galaxies, 'galaxy associations', of specific dynamical properties are detected in the studied Abell clusters[35], triggering later observations by the provided lists of galaxies (see, e.g.[37]). The S-tree method, together with certain general assumptions on the velocity distribution function of galaxies, can be efficiently used also for the determination of the bulk velocities of the subgroups.

The nonlinearity of the Newtonian interaction, thus, is providing a clue to the reconstruction of the substructure and internal dynamics of clusters of galaxies.

VI. GENERAL RELATIVITY AND COSMOLOGY

The impossibility of the direct application of results of theory of dynamical systems developed for Riemannian spaces is the main difficulty arising while studying the chaos in General Relativity and cosmology where one deals with pseudo-Riemannian spaces. Therefore, first, one has to reformulate the concepts described in previous sections and applied for astrophysical Newtonian systems, for the case of pseudo-Riemannian spaces. We will provide the reformulation of the property of hyperbolicity and the covariant definition of the Lyapunov exponents as given in[38], which are basic concepts for the study of chaos, and then, using them, will consider the stability of cosmological solutions, particularly, of inflationary ones.

I will not discuss the mixmaster models which had essentially provoked the studies on chaos in cosmology, since they are covered in Kirillov's lectures at VIII Brazilian School of Cosmology and Gravitation[39]. For the further progress on those models in the context of Non-Abelian gauge, string theories and pre-Big-Bang scenarios I refer to reviews[40–42].

A. Hyperbolicity in pseudo-Riemannian spaces

Consider a geodesic flow on M, i.e. a group of mappings $\{S^t\}$ of a space $T^\lambda M$

$$T^\lambda M = \{(x,u); x \in T_x M, g(u,u) = ||u||^2 = \lambda\}, \lambda = 0, \pm 1.$$

Each mapping performs a shift of a linear element $\xi = (x, u)$ along the geodesic on distance t.

Let $\gamma(t)$ be a geodesic on M passing by a point $x \in M$, and $\{E_a\}$ is a fixed n-dimensional basis on $T_x M$. Transferring $\{E_a\}$ parallel along $\gamma(t)$, i.e. getting a basis at every t, one has a Fermi basis on $T_x M$.

Each vector $X \in T_\gamma M$ can be represented via Fermi basis

$$X(t) = X^a(t) E_a. \tag{50}$$

with the E-norm

$$||X||_E^2 = \Sigma (X_a)^2$$

for basis $\{E_a\}$.

Let $\{E_{a'}\}$ be another basis. Then a non-singular matrix $\Phi_a^{b'}$ exists, such that

$$E_a = \Sigma \Phi_a^{b'} E_{b'}.$$

Since both $\{E_a\}$ and $\{E_{a'}\}$ are Fermi bases, the latter relation has to be satisfied also for constant $\Phi_a^{b'}$.

Then

$$X^{b'}(t) = \Sigma \Phi_a^{b'} X^a(t). \tag{51}$$

In view of non-singularity of $\Phi_a^{b'}$, we can write

$$C \Sigma (X_a)^2 \leq \Sigma (X_{b'})^2 \leq C^{-1} \Sigma (X_a)^2$$

or

$$C ||X||_E^2 \leq ||X||_E^2 \leq C^{-1} ||X||_E^2,$$

where C is a positive constant.

Definition of hyperbolicity. Geodesic $\gamma_x(t) = S^t(\xi), ||\dot\gamma_x(t)||^2 = \lambda$ is λ-hyperbolic, if there exist subspaces $W^s(S^t(\xi))$ and $W^u(S^t(\xi))$ and $W^0(S^t(\xi))$ of the tangent space $T_{S^t(\xi)} T^\lambda M$ and numbers $A \neq 0, 0 < \mu < 1$, such that

$$T_{S^t(\xi)} T^\lambda M = W^s(S^t(\xi)) \oplus W^u(S^t(\xi)) \oplus W^0(S^t(\xi)),$$

$$dS^\tau W^s(S^t(\xi)) = W^s(S^{t+\tau}(\xi)), dS^\tau W^u(S^t(\xi)) = W^u(S^{t+\tau}(\xi)),$$

where $W^0(S^t(\xi))$ is a 1D space defined by the flow vector.

For each $t, \tau > 0$ and for a certain basis $\{E_a\}$ we have

$$||dS^\tau v||_E^2 \leq A^2 \mu^{2\tau} ||v||_E^2, v \in W^s(S^t(\xi)),$$

$$||dS^\tau v||_E^2 \geq A^{-2} \mu^{-2\tau} ||v||_E^2, v \in W^u(S^t(\xi)),$$

where

$$||v||_E^2 = ||d\pi_\lambda v||_E^2 + ||Kv||_E^2, v \in TT^\lambda M, \pi_\lambda : TT^\lambda M \to TM,$$

and K is the mapping of connection ∇.

The definition is $\{E_a\}$ invariant.

Definition Geodesic flow is λ−hyperbolic, if its all geodesics are λ-hyperbolic.

Jacobi field is defined along the geodesic determined by the Jacobi equation

$$\nabla_u \nabla_u Y + R(u,v)Y = 0. \tag{52}$$

Correspond now to each vector $v \in TT^\lambda M$ a solution $Y(t)$ of Jacobi equation with initial conditions

$$Y_v(0) = d\pi v, \nabla_u Y_v(0) = kv.$$

The resulting mapping

$$f : v \to Y_v(t)$$

is an isomorphism and

$$d\pi dS^t(v) = Y_v(t), K dS^t v = \nabla_v Y_v(t).$$

¿From the Jacobi equation we have

$$||dS^t v||_E^2 = ||Y_v(t)||_E^2 + ||\nabla_u Y_u(t)||_E^2. \tag{53}$$

the latter equation and the Jacobi one enable to check the hyperbolicity condition.

Definition. Lyapunov characteristic exponent for maximal geodesic γ and vector v is defined as

$$\chi(\gamma, v) = \lim_{t \to \infty} \sup \frac{ln||dS^t_\gamma v||_E^2}{2t}. \tag{54}$$

Definition. Geodesic γ is stable if for any $\epsilon > 0, \exists \delta(\epsilon > 0$ such that from $||v||_E^2 < \delta$ follows the condition $||dS^t_\gamma v||^2 < \epsilon$ for any t. Otherwise γ is unstable. The latter two definitions are also basis-invariant.

Let us now define a convenient basis.

For arbitrary geodesic $\gamma(t)$ we choose the following orthonormal basis at point $\gamma(t)$

$$E_0 = \gamma(0) = u, E_1, ..., E_{n-1};$$

$$g(E_a, E_b) = \begin{pmatrix} -1 & 0 & 0 & 0 & 0 \\ 0 & 1 & 0 & 0 & 0 \\ 0 & 0 & . & 0 & 0 \\ 0 & 0 & 0 & . & 0 \\ 0 & 0 & 0 & 0 & 1 \end{pmatrix} \tag{55}$$

where E_a is a dual basis.

If the following conditions are satisfied

$$\nabla_u E_a = 0 = \nabla_u E^b,$$

then the basis on $T^0 M$ can be defined as

$$E_0 = u,$$

$$g(E_a, E_b) = \begin{pmatrix} 0 & -1 & 0 & 0 & 0 \\ -1 & 0 & 0 & 0 & 0 \\ 0 & 0 & . & 0 & 0 \\ 0 & 0 & 0 & . & 0 \\ 0 & 0 & 0 & 0 & 1 \end{pmatrix} \qquad (56)$$

and on $T^1 M$

$$E_0 = u,$$

$$g(E_a, E_b) = \begin{pmatrix} 1 & 0 & 0 & 0 & 0 \\ 0 & 1 & 0 & 0 & 0 \\ 0 & 0 & . & 0 & 0 \\ 0 & 0 & 0 & . & 0 \\ 0 & 0 & 0 & 0 & 1 \end{pmatrix} \qquad (57)$$

For the vector field

$$Y(t) = Z^a(t) E_a$$

the Jacobi equation can be written in the form

$$\ddot{Z}^a(t) + K^a_b(t) Z^b(t) = 0, \qquad (58)$$

where

$$K^a_b = <E^a, R(u, E_b)u> = R^a_{bcd} u^c u^d.$$

For the above defined basis on $T^0 M$ we have

$$\dot{Z} = 1.$$

which means that none of geodesic flows can be 0-hyperbolic.

These definitions for spaces of Lorentzian signature (-,+,...,+), can be generalized for the signatures (-,,-, +,,+).

The covariant definition of hyperbolicity and Lyapunov exponents given above enable the consideration of the stability problem of cosmological solutions.

B. The ADM principle and geodesic flows in Wheeler-DeWitt superspace

The problem of the stability of cosmological solutions is a problem of stability in Wheeler-DeWitt superspace. We will consider this problem using the method of geodesic flows[43]. So, first, we have to define the Hamiltonian system, then reduce it to a flow of geodesics using the definition of the hyperbolicity given in the previous section.

Arnowitt-Deser-Misner (ADM) method provides the scheme of the sought Hamiltonian formulation, assuming as given the 3-geometries of the initial and final Cauchy hypersurfaces. We will consider locally isotropic and homogeneous cosmological models with scalar field when the metric can be given as

$$h_{ij} = \sigma^2 \bar{h}_{ij} \qquad (59)$$

where

$$\sigma^2 = \frac{4\pi G}{3} [\int \bar{h}^{1/2} d^3 x]^{-1}, \; \bar{h} = \det \bar{h}_{ij}$$

Consider the Lagrangian of the scalar field

$$L_\phi = -g^{-1/2} [\phi_{,a} \phi_{,b} g^{ab} + V(\phi)] \qquad (60)$$

and the action

$$I = \int p_\alpha d\alpha + p_\chi d\chi - N H_{\text{ADM}} dt$$

where the ADM Hamiltonian is

$$H_{ADM} = \frac{1}{2}e^{-3\alpha}[-p_\alpha^2 + p_\chi^2] + e^{3\alpha}[U(x) - \frac{k}{2}e^{-2\alpha}], \qquad (61)$$

where

$$\alpha = \ln a, \chi = \sigma\phi, U(x) = \frac{4\pi G\sigma^2}{3}V(\phi).$$

As usual, the variation with respect the lapse function N leads to the condition

$$H_{ADM} = 0.$$

To reduce the Hamiltonian system to the geodesic flow let us split the hypersurface into the following regions

$$W^+\{x|V(x) > 0\}, W^-\{x|V(x) < 0\},$$

so that, if the metric in region W^- is Riemannian, then

$$g_{ab}v^a v^b = -2V > 0, d\tau = (g_{ab}u^a u^b/-2V)^{-1/2} ds \qquad (62)$$

and we can write the variation

$$extI|_{H=0} = ext \int (-2V)^{1/2}(g_{ab}u^a u^b)^{1/2} ds = ext \int (G_{ab}u^a u^b)^{1/2} ds.$$

Here $G_{ab} = -Vg_{ab}$ is also a Riemannian metric.

Choosing the affine parameter s in order to satisfy the condition

$$||u||^2 = G_{ab}u^a u^b = 1 \qquad (63)$$

we have

$$G_{ab}u^a u^b = -V g_{ab} v^a v^b (d\tau/ds)^2 = 2V^2 (d\tau/ds)^2.$$

Reparameterizing the affine parameter

$$ds = 2^{1/2}(-V)d\tau,$$

we arrive at the flow of geodesics in the region W^-

$$H = 1/2 g^{ab} p_a p_b + V \to \{G_{ab} = -Vg_{ab}, ds = 2^{1/2}(-V)d\tau, ||u||^2 = 1\}. \qquad (64)$$

As regards for the region W^+, the classical system cannot end up in it.

Now, if the metric g is pseudo-Riemannian, following the same scheme we end up with the geodesic flow

$$H \to \{|V|g_{ab}, 2^{1/2}|V|d\tau, -signV\}. \qquad (65)$$

Thus, we reduced the ADM Hamiltonian system to a geodesic flow on a pseudo-Riemannian manifold.

To study the stability of the geodesic flow we have to proceed from the Jacobi equation, which has the form

$$\frac{d^2 z^i}{d\tau^2} + \gamma(z)\frac{dz^i}{d\tau} + \omega_j^i(z) = 0, \qquad (66)$$

where

$$\gamma = -\frac{d}{d\tau}\ln|V|, \omega_j^i = 2V^2 K_j^i, i, j, = 1, 2...k-1.$$

Using the variables

$$z^i = AY^i$$

we arrive to the equation

$$\ddot{Y}^i + 2(\dot{A}/A + \gamma)\dot{Y}^i + [(\ddot{A}/A + \gamma\dot{A}/A))\delta^i_j + \omega^i_j]Y^j = 0 \qquad (67)$$

where

$$\dot{A}/A = -1/2\gamma, A = |V|^{1/2}.$$

Particularly for the case $k = 0$ we have the simplified expressions

$$G = e^{6\alpha}|U|, |V| = e^{3\alpha}|U|,$$

$$\gamma(\tau) = -3\frac{d\alpha}{d\tau} - \frac{1}{U}\frac{dU}{d\tau}$$

$$K(\tau) = -\frac{1}{2}\frac{\Box lnG}{G} = -\frac{1}{2}\frac{ln\Box|V|}{e^{6\alpha}|U|},$$

$$\omega(\tau) = U\Box ln|U|, \Box = -\frac{\partial^2}{\partial\alpha^2} + \frac{\partial^2}{\partial\chi^2}.$$

C. Stability of inflationary solutions

Consider the scalar field

$$U = \lambda\chi^n/n \qquad (68)$$

and the conditions

$$\dot{H} << H^2, \dot{\chi}^2 << U, |\dot{\chi}/\chi| << H$$

and $\chi >> 1$ at $\alpha \to \infty$. Then we have

$$\dot{\gamma} = -3\dot{H}, \frac{1}{2}\dot{\gamma} + \frac{\gamma^2}{4} \simeq \frac{9}{4}H^2,$$

and the Jacobi equation

$$\ddot{Y}^i - [-\frac{nU}{\chi^2} + (\frac{3}{2}H)^2]Y = 0. \qquad (69)$$

¿From the Einstein equation

$$\dot{H} + 3H^2 = 6U,$$

and in view of the condition $\dot{H} << H^2$ we have $H^2 \simeq 2U$. If $\chi >> (2n/3)^{1/2}$, then $nU/\chi^2 << 9/4H^2$, and we have the Jacobi equation in a simple form

$$\ddot{Y} - (9/4\dot{\alpha}^2)Y = 0. \qquad (70)$$

¿From its solution we obtain

$$\delta^2 = z^2 + \dot{z}^2 \simeq const\, U + const\, \frac{U'^4}{U^2} \simeq const\, \chi^n + const\, \chi^{2n-4}. \qquad (71)$$

We see that δ decreases for any $n > 2$ and therefore we have Lyapunov stability of the inflationary solutions. The last formula enables also to obtain the law of the decay of perturbations at various n. For example at $n = 4$ we have exponential decay of perturbations, and the larger is n, the more stable the solution is.

Thus, we showed how one can deal with the stability problem in pseudo-Riemannian spaces, and particularly, study the stability of inflationary solutions.

VII. INSTABILITY IN SUPERSPACE

How typical is the given cosmological solution? This is a basic question posed since the early days of the study of the Einstein equations. In the context of later developments, particularly in quantum cosmology, the question can be reformulated in the form: to what degree are the minisuperspace models typical in superspace given the huge extrapolations involved?

The study of dynamics in the Wheeler-DeWitt superspace, more precisely, the properties of geodesic flows in superspace can provide a more general view on how typical the minisuperspace models are. The problem, however, is far more difficult than the one posed in conventional hyperbolicity theory, since one deals both with infinite dimensional and pseudo-Riemannian manifolds.

However, as we saw above, hyperbolicity can be defined for such manifolds and now we will consider the case of homogeneous cosmological models [44].

The metric of Wheeler-DeWitt superspace is

$$G^{ijkl} = \frac{1}{2}[\gamma^{ik}\gamma^{jl} + \gamma^{il}\gamma^{jk} - 2\gamma^{ij}\gamma^{kl}], \tag{72}$$

where

$$\gamma = det\gamma_{ij}, i,j,k,l = 1,...,n.$$

Then one can derive

$$G^{ijkl}d\gamma_{ij}d\gamma_{jl} = -d\xi^2 + \frac{n\xi^2}{16(n-1)}tr(\gamma^{-1}\partial\gamma/\partial\xi^A\gamma^{-1}\partial\gamma/\partial\xi^B)d\xi^A d\xi^B \tag{73}$$

where

$$\xi = 4((n-1)/n)^{1/2}\gamma^{1/4}, \ A, B = 1, ..., n(n+1)/2 - 1,$$

Then, moving to the subspace of superspace

$$\bar{W} = (\gamma_{ij}; \gamma_{ij} \in W, \gamma = 1)$$

with a metric induced by the metric of the superspace W

$$G_{AB}d\xi^A d\xi^B = tr(\gamma^{-1}\partial\gamma/\partial\xi^A\gamma^{-1}\partial\gamma/\partial\xi^B)d\xi^A d\xi^B, \tag{74}$$

the existence of non-zero Lyapunov numbers

$$\Sigma_i \omega_i = -n/4 < 0$$

can be shown for the solutions of the Jacobi equation

$$z_i^a(t) = x_i^a \exp[(\pm(-\omega_i)^{1/2}t]. \tag{75}$$

This implies the exponential instability of the geodesic flow in that subspace of the superspace.

For models with a scalar field Armen Kocharyan[45] was able to show that the instability is exponential if:

1. Gravitational and matter fields vary quickly with respect the potential;

2. The Universe undergoes inflation in a local domain.

The smaller are the dimension and the number of scalar fields, the stronger is the instability.

These results lead to the following general conclusion:

The quantized system in a finite-dimensional submanifold is not typical in superspace due to the existence of virtual perturbations along the frozen directions which are unstable.

This implies that minisuperspace models cannot be considered as fair approximations of superspace models.

VIII. CONCLUSION

I stop here. As we saw, chaos is an inevitable ingredient of the Universe, but it needs particular efforts to deal with. The spectrum of the problems and methods will obviously grow further. I conclude with mentioning the use of Kolmogorov complexity (algorithmic information) for the study of properties of Cosmic Background Radiation and relations of the latter with thermodynamical and cosmological time arrows.[46–48]

IX. ACKNOWLEDGEMENTS

I greatly acknowledge my collaboration with A.A.Kocharyan and G.K.Savvidy and discussions with D.V.Anosov, V.I. Arnold and Ya.B.Pesin on the topics discussed in these lectures.

---

[1] Sagdeev R.Z., Usikov D.A., Zaslavsky G.M., *Nonlinear Physics*, Harwood, 1988.
[2] Ott E., *Chaos in Dynamical Systems*, Cambridge University Press, 1993.
[3] Zaslavsky G.M., *Physics of Chaos in Hamiltonian Dynamics*, Imperial College Press, 1998.
[4] Gurzadyan V.G., Kocharyan A.A., *Paradigms of the Large-Scale Universe*, Gordon and Breach, 1994.
[5] Gurzadyan V.G., Pfenniger D. (Eds.), *Ergodic Concepts in Stellar Dynamics*, Springer. 1994.
[6] Hobill D., Burd A., Coley A. (Eds.), *Deterministic Chaos in General Relativity*, Plenum, 1994.
[7] Gurzadyan V.G., Ruffini R. (Eds.), *The Chaotic Universe*, World Scientific, 2000.
[8] Arnold V.I., *Mathematical Methods of Classical Mechanics*, Springer, 1989.
[9] *Dynamical Systems. Modern Problems in Mathematics*, vols.1,2, Ed.Ya.G.Sinai, Springer, 1989.
[10] Katok A., Hassenblatt B., *Introduction to the Modern Theory of Dynamical Systems*, Cambridge University Press, 1996.
[11] Smale S., Finding a Horseshoe on the Beaces of Rio, in: *The Chaos Avant-Garde: Memories of the Early Days of Chaos Theory*, World Scientific, 2000.
[12] Anosov D.V., *Geodesic Flows on Closed Riemannian Spaces with Negative Curvature*, Comm. MIAN, vol.90, 1967.
[13] Krylov N.S., *Studies on Foundation of Statistical Mechanics*, Publ. AN SSSR, Leningrad, 1950.
[14] Kolmogorov A.N., Doklady AN SSSR, 119, 861, 1958.
[15] Pfenniger D., A&A,165, 74, 1986.
[16] Pfenniger D., in[5]
[17] Murray C.D., Dermott S.F., *Solar System Dynamics*, Cambridge University Press, 1999.
[18] Lecar M. et al, Ann.Rev.Astron.Astroph. 39, 581, 2001.
[19] Morbidelli A., *Modern Celestial Mechanics*, Taylor and Francis, 2002.
[20] Kolmogorov A.N., Doklady AN SSSR, 98, 527, 1954.
[21] Laskar J., Nature, 338, 237, 1989.
[22] Laskar J., Physica D, 67, 257, 1993.
[23] Laskar J., Robutel P., Nature, 361, 608, 1993.
[24] Laskar J., Joutel F., Robutel P., Nature, 361, 615, 1993.
[25] Laskar J., A&A, 287, L9, 1994.
[26] Gurzadyan V.G., Savvidy G.K., Doklady AN SSSR, 277, 69, 1984.
[27] Gurzadyan V.G., Savvidy G.K., A&A, 160, 203, 1986.
[28] Lang K.R., *Astrophysical Formulae*, vol.II,p.95, Springer, 1999.
[29] Arnold V.I., Annales de l'Institute Fourier, XVI, 319, 1966.
[30] Gurzadyan V.G., Kocharyan A.A., A&A, 205, 93, 1988.
[31] Gurzadyan V.G., Kocharyan A.A., Ap&SS, 135, 307, 1987.
[32] El-Zant A.A., A&A, 326, 113, 1997.
[33] El-Zant A.A., Gurzadyan V.G., Physica D, 122, 241, 1998.
[34] Bekarian K.M., Melkonian A.A., Complex Systems, 11, 323, 1997.
[35] Gurzadyan V.G., Mazure A., MNRAS, 295, 177, 1998.
[36] Bekarian K.M., Ph.D Thesis, Yerevan State University, 2001.
[37] Mario-Franch A., Aparicio A., ApJ, 568, 174, 2002.
[38] Gurzadyan V.G., Kocharyan A.A., YerPhI-920(71), 1986.
[39] Kirillov A.A., in: *Cosmology and Gravitation*, Ed.M.Novello, Editions Frontieres, 1996.
[40] Belinski V. in[7]
[41] Matinyan S.G., in: *Proc.IX Marcel Grossmann meeting*, World Scientific, 2002; gr-qc/0010054.
[42] Damour T., hep-th/0204017, 2002.
[43] Gurzadyan V.G., Kocharyan A.A., Sov.Phys-JETP, 66, 651, 1988.
[44] Gurzadyan V.G., Kocharyan A.A., Mod.Phys.Lett. 2A, 921. 1987.
[45] Kocharyan A.A., Comm.Math.Phys., 143, 27, 1991.
[46] Gurzadyan V.G., Europhys.Lett. 46, 114, 1999.
[47] Gurzadyan V.G., Ade P.A.R., de Bernardis P., Bianco C.L., Bock J.J., Boscaleri A., Crill B.P., De Troia G., Ganga K., Giacometti M., Hivon E., Hristov V.V., Kashin A.L., Lange A.E., Masi S., Mauskopf P.D., Montroy T., Natoli P., Netterfield C.B., Pascale E., Piacentini F., Polenta G., Ruhl J., astro-ph/0210021, 2002.
[48] Allahverdyan A.E., Gurzadyan V.G. J.Phys.A, 35, 7243, 2002.
[49] As mentions Smale,[11] he had found the horseshoe transformation, the classic example of Axiom-A systems in Brazil, on Rio beaches.

String and M-theory Cosmology*

E. J. Copeland

Centre for Theoretical Physics, CPES, University of Sussex, Brighton BN1 9QJ, United Kingdom

In these lectures we review recent advances in string cosmology. Starting with the Dilaton-Moduli Cosmology (known also as the Pre Big Bang), we go on to include the effects of axion fields and address the thorny issue of the Graceful Exit in String Cosmology. This is followed by a review of density perturbations arising in string cosmology and we finish with a brief introduction to the impact moving of five branes on the Dilaton-Moduli cosmological solutions.

I. INTRODUCTION

String theory, and its most recent incarnation, that of M-theory, has been accepted by many as the most likely candidate theory to unify the forces of nature as it includes General Relativity in a consistent quantum theory. If it is to play such a pivotal role in particle physics, it should also include in it all of cosmology. It should provide the initial conditions for the Universe, perhaps even explain away the singularity associated with the standard big bang. It should also provide a mechanism for explaining the observed density fluctuations, perhaps by providing the inflaton field or some other mechanism which would lead to inflation. Should the observations survive the test of time, string theory should be able to provide a mechanism to explain the current accelerated expansion of the Universe. In other words, even though it is strictly a theory which can unify gravity with the other forces in the very early Universe, for consistency, as a theory of everything it will have a great deal more to explain. In this article, we will introduce some of the developments that have occurred in string cosmology over the past decade or so, initially basing the discussion on an analyse of the low energy limit of string theory, and then later extending it to include branes arising in Heterotic M-theory.

II. DILATON-MODULI COSMOLOGY (PRE-BIG BANG)

Strings live in 4+d spacetime dimensions, with the extra d dimensions being compactified. For homogeneous, four–dimensional cosmologies, where all fields are uniform on the surfaces of homogeneity, we can consider the compactification of the $(4+d)$-dimensional theory on an isotropic d-torus. The radius, or 'breathing mode' of the internal space, is then parameterized by a modulus field, β, and determines the volume of the internal dimensions. We can then assume that the $(4+d)$-dimensional metric is of the form

$$ds^2 = -dt^2 + g_{ij}dx^i dx^j + e^{\sqrt{2/d}\beta}\delta_{ab}dX^a dX^b \tag{1}$$

where indices run from $(i,j) = (1,2,3)$ and $(a,b) = (4,\ldots,3+d)$ and δ_{ab} is the d-dimensional Kronecker delta. The modulus field β is normalized in such a way that it becomes minimally coupled to gravity in the Einstein frame.

The low energy action that is commonly used as a starting point for string cosmology is the four dimensional effective Neveu-Schwarz- Neveu-Schwarz (NS-NS) action given by:

$$S_* = \int d^4x \sqrt{|g|} e^{-\varphi}\left[R + (\nabla\varphi)^2 - \frac{1}{2}(\nabla\beta)^2 - \frac{1}{2}e^{2\varphi}(\nabla\sigma)^2\right], \tag{2}$$

where φ is the effective dilaton in four dimensions, and σ is the pseudo–scalar axion field which is dual to the fundamental NS–NS three–form field strength present in string theory, the duality being given by

$$H^{\mu\nu\lambda} = \epsilon^{\mu\nu\lambda\kappa} e^{\varphi} \nabla_\kappa \sigma. \tag{3}$$

The dimensionally reduced action (2) may be viewed as the prototype action for string cosmology because it contains many of the key features common to more general actions. Cosmological solutions to these actions have been extensively discussed in the literature – for a review see [1]. Some of them play a central role in the pre–big bang

---

* Based on Lectures given at Xth Brazilian School of Cosmology and Gravitation, Rio de Janeiro 29 July – 9 August 2002.

inflationary scenario, first proposed by Veneziano [2, 3]. An important point can be seen immediately in (2) where there is a non-trivial coupling of the dilaton to the axion field, a coupling which will play a key role later on when we are investigating the density perturbations arising in this scenario.

All homogeneous and isotropic external four–dimensional spacetimes can be described by the Friedmann-Robertson-Walker (FRW) metric. The general line element in the string frame can be written as

$$ds_4^2 = a^2(\eta)\left\{-d\eta^2 + d\Omega_\kappa^2\right\}, \tag{4}$$

where $a(\eta)$ is the scale factor of the universe, η is the conformal time and $d\Omega_\kappa^2$ is the line element on a 3-space with constant curvature κ:

$$d\Omega_\kappa^2 = d\psi^2 + \left(\frac{\sin\sqrt{\kappa}\psi}{\sqrt{\kappa}}\right)^2 \left(d\theta^2 + \sin^2\theta d\varphi^2\right) \tag{5}$$

To be compatible with a homogeneous and isotropic metric, all fields, including the pseudo–scalar axion field, must be spatially homogeneous.

The models with vanishing form fields, but time-dependent dilaton and moduli fields, are known as *dilaton-moduli-vacuum* solutions. In the Einstein–frame, these solutions may be interpreted as FRW cosmologies for a stiff perfect fluid, where the speed of sound equals the speed of light. The dilaton and moduli fields behave collectively as a massless, minimally coupled scalar field, and the scale factor in the Einstein frame is given by

$$\tilde{a} = \tilde{a}_*\sqrt{\frac{\tau}{1+\kappa\tau^2}} \tag{6}$$

where $\tilde{a} \equiv e^{-\varphi/2}a$, $\tilde{a}_*$ is a constant and we have defined a new time variable:

$$\tau \equiv \begin{cases} \kappa^{-1/2}|\tan(\kappa^{1/2}\eta)| & \text{for } \kappa > 0 \\ |\eta| & \text{for } \kappa = 0 \\ |\kappa|^{-1/2}|\tanh(|\kappa|^{1/2}\eta)| & \text{for } \kappa < 0 \end{cases}. \tag{7}$$

The time coordinate τ diverges at both early and late times in models which have $\kappa \geq 0$, but $\tau \to |\kappa|^{-1/2}$ in negatively curved models. There is a curvature singularity at $\eta = 0$ with $\tilde{a} = 0$ and the model expands away from it for $\eta > 0$ or collapses towards it for $\eta < 0$. The expanding, closed models recollapse at $\eta = \pm\pi/2$ and there are no bouncing solutions in this frame.

The corresponding string frame scale factor, dilaton and modulus fields are given by the 'rolling radii' solutions [4]

$$a = a_*\sqrt{\frac{\tau^{1+\sqrt{3}\cos\xi_*}}{1+\kappa\tau^2}}, \tag{8}$$

$$e^\varphi = e^{\varphi_*}\tau^{\sqrt{3}\cos\xi_*}, \tag{9}$$

$$e^\beta = e^{\beta_*}\tau^{\sqrt{3}\sin\xi_*} \tag{10}$$

The integration constant ξ_* determines the rate of change of the effective dilaton relative to the volume of the internal dimensions. Figures 1 and 2 show the dilaton-vacuum solutions in flat FRW models when stable compactification has occurred, so that the volume of the internal space is fixed, with ξ_* mod $\pi = 0$.

The solutions just presented have a scale factor duality which when applied simultaneously with time reversal implies that the Hubble expansion parameter $H \equiv d(\ln a)/dt$ remains invariant, $H(-t) \to H(t)$, whilst its first derivative changes sign, $\dot{H}(-t) \to -\dot{H}(t)$. A decelerating, post–big bang solution – characterized by $\dot{a} > 0$, $\ddot{a} < 0$ and $\dot{H} < 0$ – is mapped onto a pre–big bang phase of inflationary expansion, since $\ddot{a}/a = \dot{H} + H^2 > 0$. The Hubble radius H^{-1} decreases with increasing time and the expansion is therefore super-inflationary. Thus, the pre-big bang cosmology ($\kappa = 0$ case in Eqns. (8–10)) is one that has a period of super-inflation driven simply by the kinetic energy of the dilaton and moduli fields [2, 3]. This is related by duality to the usual FRW post–big bang phase. The two branches are separated by a curvature singularity, however, and it is not clear how the transition between the pre– and post–big bang phases might proceed. This will be the focus of attention in section five.

The solution for a flat ($\kappa = 0$) FRW universe corresponds to the well–known monotonic power-law, or 'rolling radii', solutions. For $\cos\xi_* < -1/\sqrt{3}$ there is accelerated expansion, i.e., inflation, in the string frame for $\eta < 0$ and $e^\varphi \to 0$ as $t \to -\infty$, corresponding to the weak coupling regime. The expansion is an example of 'pole–law' inflation [5, 6].

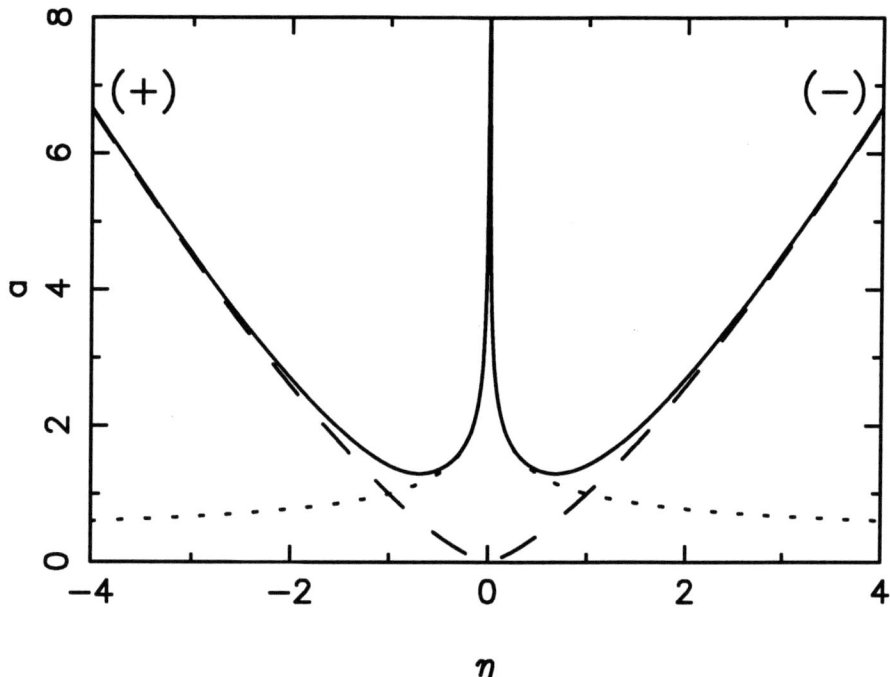

FIG. 1: String frame scale factor, a, as a function of conformal time, η, for flat $\kappa = 0$ FRW cosmology in dilaton-vacuum solution in Eq. (8) with $\xi_* = 0$ (dashed-line), $\xi_* = \pi$ (dotted line) and dilaton-axion solution in Eq. (13) with $r = \sqrt{3}$ (solid line). The $(+)$ and $(-)$ branches are defined in the text.

The solutions have semi-infinite proper lifetimes. Those starting from a singularity at $t = 0$ for $t \geq 0$ are denoted as the $(-)$ branch in Ref. [7], while those which approach a singularity at $t = 0$ for $t \leq 0$ are referred to as the $(+)$ branch (see figures 1–2). These $(+/-)$ branches do *not* refer to the choice of sign for $\cos \xi_*$. On either the $(+)$ or $(-)$ branches of the dilaton-moduli-vacuum cosmologies we have a one-parameter family of solutions corresponding to the choice of ξ_*, which determines whether e^φ goes to zero or infinity as $t \to 0$. These solutions become singular as the conformally invariant time parameter $\eta \equiv \int dt/a(t) \to 0$ and there is no way of naively connecting the two branches based simply on these solutions [7].

In the Einstein frame, where the dilaton field is minimally coupled to gravity, the scale factor given in Eq. (6), becomes

$$\tilde{a} = \tilde{a}_* |\eta|^{1/2} \tag{11}$$

As $\eta \to 0$ on the $(+)$ branch, the universe is collapsing with $\tilde{a} \to 0$, and the comoving Hubble length $|d(\ln \tilde{a})/d\eta|^{-1} = 2|\eta|$ is decreasing with time. Thus, in both frames there is inflation taking place in the sense that a given comoving scale, which starts arbitrarily far within the Hubble radius in either conformal frame as $\eta \to -\infty$, inevitably becomes larger than the Hubble radius in that frame as $\eta \to 0$. The significance of this is that it means that perturbations can be produced in the dilaton, graviton and other matter fields on scales much larger than the present Hubble radius from quantum fluctuations in flat spacetime at earlier times – this is a vital property of any inflationary scenario.

For completeness, it is worth mentioning that these solutions can be extended to include a time-dependent axion field, $\sigma(t)$, by exploiting the $SL(2, R)$ S-duality invariance of the four–dimensional, NS-NS action [4]. We now turn our attention to this fascinating case.

III. DILATON-MODULI-AXION COSMOLOGIES

The cosmologies containing a non–trivial axion field can be generated immediately due to the global $SL(2, R)$ symmetry of the action (2). The resultant solutions are [4]:

$$e^\varphi = \frac{e^{\varphi_*}}{2} \left\{ \left(\frac{\tau}{\tau_*}\right)^{-r} + \left(\frac{\tau}{\tau_*}\right)^r \right\}, \tag{12}$$

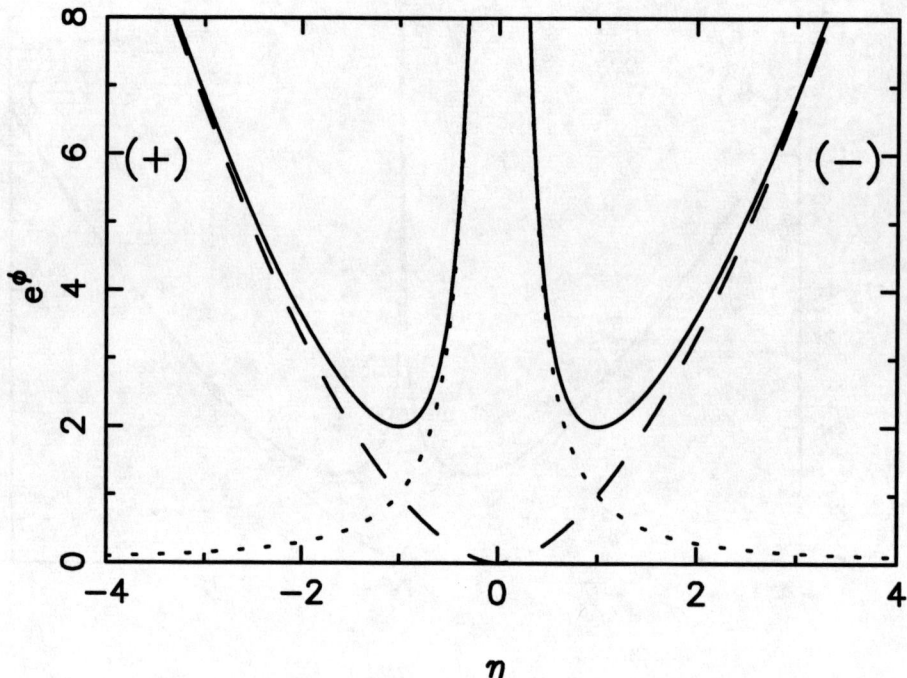

FIG. 2: Dilaton, e^φ, as a function of conformal time, η, for flat $\kappa = 0$ FRW cosmology in dilaton-vacuum solution in Eq. (9) with $\xi_* = 0$ (dashed-line), $\xi_* = \pi$ (dotted line) and dilaton-axion solution in Eq. (12) with $r = \sqrt{3}$ (solid line).

$$a^2 = \frac{a_*^2}{2(1+\kappa\tau^2)}\left\{\left(\frac{\tau}{\tau_*}\right)^{1-r} + \left(\frac{\tau}{\tau_*}\right)^{1+r}\right\}, \qquad (13)$$

$$e^\beta = e^{\beta_*}\tau^s, \qquad (14)$$

$$\sigma = \sigma_* \pm e^{-\varphi_*}\left\{\frac{(\tau/\tau_*)^{-r} - (\tau/\tau_*)^r}{(\tau/\tau_*)^{-r} + (\tau/\tau_*)^r}\right\}, \qquad (15)$$

where the exponents are related via $r^2 + s^2 = 3$, and without loss of generality we may take $r \geq 0$.

In all cases, the dynamics of the axion field places a *lower* bound on the value of the dilaton field, $\varphi \geq \varphi_*$. In so doing, the axion smoothly interpolates between two dilaton–moduli–vacuum solutions, where its dynamical influence asymptotically becomes negligible. The effects of time–dependent axion solutions for the scale-factor and dilaton are plotted in Figure2 1 and 2 for the flat FRW model when the modulus field is trivial ($s = 0$). When the internal space is static, it is seen that the string frame scale factors exhibit a bounce. However we still have a curvature singularity in the Einstein frame as $\tau \to 0$. The actual time-dependent axion solutions is shown in Figure 3.

The spatially flat solutions reduce to the power law, dilaton–moduli–vacuum solution given in Eqs. (8–10) at early and late times. When $\eta \to \pm\infty$ the solution approaches the vacuum solution with $\sqrt{3}\cos\xi_* = +r$, while as $\eta \to 0$ the solution approaches the $\sqrt{3}\cos\xi_* = -r$ solution. Thus, the axion solution interpolates between two vacuum solutions related by an S-duality transformation $\varphi \to -\varphi$. When the internal space is static the scale factor in the string frame is of the form $a \propto t^{1/\sqrt{3}}$ as $\eta \to \pm\infty$, while as $\eta \to 0$ the solution becomes $a \propto t^{-1/\sqrt{3}}$. These two vacuum solutions are thus related by a scale factor duality that inverts the spatial volume of the universe. This asymptotic approach to dilaton–moduli–vacuum solutions at early and late times will lead to a particularly simple form for the semi-classical perturbation spectra that is independent of the intermediate evolution. However, there is a down side to these solutions from the standpoint of pre big bang cosmologies. As $\eta \to \pm\infty$ and as $\eta \to 0$ the solution approaches the strong coupling regime where $e^\varphi \to \infty$. Thus there is no weak coupling limit, the axion interpolates between two strong coupling vacuum solutions. We will shortly see how a similar affect arises when we include a moving brane in the dilaton-moduli picture, as it too mimics the behaviour of a non-minimally coupled axion field.

The overall dynamical effect of the axion field is negligible except near $\tau \approx \tau_*$, when it leads to a bounce in the dilaton field. Within the context of M–theory cosmology, the radius of the eleventh dimension is related to the dilaton by $r_{11} \propto e^{\varphi/3}$ when the modulus field is fixed. This bound on the dilaton may therefore be reinterpreted as a lower bound on the size of the eleventh dimension.

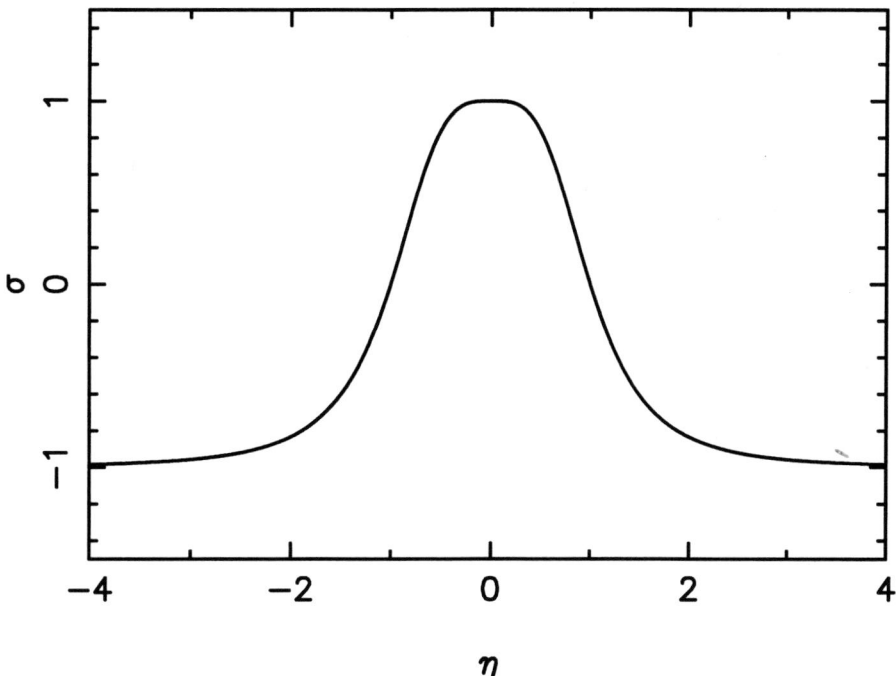

FIG. 3: Axion, σ, as a function of conformal time, η, for flat $\kappa = 0$ FRW cosmology in dilaton-axion solution in Eq. (15) with $r = \sqrt{3}$ (solid line).

IV. FINE TUNING ISSUES

The question over the viability of the initial conditions required in the pre Big Bang scenario has been a cause for many an argument both in print and in person. Since both $\dot{H}$ and $\dot{\varphi}$ are positive in the pre–big bang phase, the initial values for these parameters must be *very small*. This raises a number of important issues concerning fine–tuning in the pre–big bang scenario [8–14]. There needs to be enough inflation in a homogeneous patch in order to solve the horizon and flatness problems which means that the dilaton driven inflation must survive for a sufficiently long period of time. This is not as trivial as it may appear, however, since the period of inflation is limited by a number of factors.

The fundamental postulate of the scenario is that the initial data for inflation lies well within the perturbative regime of string theory, where the curvature and coupling are very small [3]. Inflation then proceeds for sufficiently homogeneous initial conditions [12, 13], where time derivatives are dominant with respect to spatial gradients, and the universe evolves into a high curvature and strongly–coupled regime. Thus, the pre–big bang initial state should correspond to a cold, empty and flat vacuum state. Initial the universe would have been huge relative to the quantum scale and hence should have been well described by classical solutions to the string effective action. This should be compared to the initial state which describes the standard hot big bang, namely a dense, hot, and highly curved region of spacetime. This is quite a contrast and a primary goal of pre–big bang cosmology must be to develop a mechanism for smoothly connecting these two regions, since we believe that the standard big bang model provides a very good representation of the current evolution of the universe.

Our present observable universe appears very nearly homogeneous on sufficiently large scales. In the standard, hot big bang model, it corresponded to a region at the Planck time that was 10^{30} times larger than the horizon size, $l_{\rm Pl}$. This may be viewed as an initial condition in the big bang model or as a final condition for inflation. It implies that the comoving Hubble radius, $1/(aH)$, must decrease during inflation by a factor of at least 10^{30} if the horizon problem is to be solved. For a power law expansion, this implies that

$$\left|\frac{\eta_f}{\eta_i}\right| \leq 10^{-30} \quad (16)$$

where subscripts i and f denote values at the onset and end of inflation, respectively.

In the pre–big bang scenario, Eq. (9) implies that the dilaton grows as $e^{\varphi} \propto |\eta|^{-\sqrt{3}}$, and since at the start of the post–big bang epoch, the string coupling, $g_s = e^{\varphi/2}$, should be of order unity, the bound (16) implies that the initial value of the string coupling is strongly constrained, $g_{s,i} \leq 10^{-26}$. Turner and Weinberg interpret this constraint as a severe fine–tuning problem in the scenario, because inflation in the string frame can be delayed by the effects of spatial curvature [8]. It was shown by Clancy, Lidsey and Tavakol that the bounds are further tightened when spatial anisotropy is introduced, actually preventing pre–big bang inflation from occurring [9]. Moreover, as we have seen the dynamics of the NS–NS axion field also places a lower bound on the allowed range of values that the string coupling may take [4]. In the standard inflationary scenario, where the expansion is quasi–exponential, the Hubble radius is approximately constant and $a \propto (-\eta)^{-1}$. Thus, the homogeneous region grows by a factor of $|\eta_i/\eta_f|$ as inflation proceeds. During a pre–big bang epoch, however, $a \propto (-\eta)^{-1/1+\sqrt{3}}$ and the increase in the size of a homogeneous region is reduced by a factor of at least $10^{30\sqrt{3}/(1+\sqrt{3})} \approx 10^{19}$ relative to that of the standard inflation scenario. This implies that the initial size of the homogeneous region should exceed 10^{19} in string units if pre–big bang inflation is to be successful in solving the problems of the big bang model [2, 10]. The occurrence of such a large number was cited by Kaloper, Linde and Bousso as a serious problem of the pre–big bang scenario, because it implies that the universe must already have been large and smooth by the time inflation began [10].

On the other hand, Gasperini has emphasized that the initial homogeneous region of the pre–big bang universe is not larger than the horizon even though it is large relative to the string/Planck scale [15]. The question that then arises when discussing the naturalness, or otherwise, of the above initial conditions is what is the basic unit of length that should be employed. At present, this question has not been addressed in detail.

Veneziano and collaborators conjectured that pre–big bang inflation generically evolves out of an initial state that approaches the Milne universe in the semi–infinite past, $t \to -\infty$ [12, 13]. The Milne universe may be mapped onto the future (or past) light cone of the origin of Minkowski spacetime and therefore corresponds to a non–standard representation of the string perturbative vacuum. The proposal was that the Milne background represents an early time attractor, with a large measure in the space of initial data. If so, this would provide strong justification for the postulate that inflation begins in the weak coupling and curvature regimes and would render the pre-big bang assumptions regarding the initial states as 'natural'. However, Clancy *et al.* took a critical look at this conjecture and argued that the Milne universe is an unlikely past attractor for the pre–big bang scenario [16]. They suggested that plane wave backgrounds represent a more generic initial state for the universe [9]. Buonanno, Damour and Veneziano have subsequently proposed that the initial state of the pre–big bang universe should correspond to an ensemble of gravitational and dilatonic waves [14]. They refer to this as the state of 'asymptotic past triviality'. When viewed in the Einstein frame these waves undergo collapse when certain conditions are satisfied. In the string frame, these gravitationally unstable areas expand into homogeneous regions on large scales.

To conclude this Section, it is clear that the question of initial conditions in the pre–big bang scenario is currently unresolved. We turn our attention now to another unresolved problem for the scenario – the Graceful Exit.

V. THE GRACEFUL EXIT

We have seen how in the pre Big Bang scenario, the Universe expands from a weak coupling, low curvature regime in the infinite past, enters a period of inflation driven by the kinetic energy associated with the massless fields present, before approaching the strong coupling regime as the string scale is reached. There is then a branch change to a new class of solutions, corresponding to a post big bang decelerating Friedman-Robertson-Walker era. In such a scenario, the Universe appears to emerge because of the gravitational instability of the generic string vacua – a very appealing picture, the weak coupling, low curvature regime is a natural starting point to use the low energy string effective action. However, how is the branch change achieved without hitting the inevitable looking curvature singularity associated with the strong coupling regime? The simplest version of the evolution of the Universe in the pre-big bang scenario inevitably leads to a period characterised by an unbounded curvature. The current philosophy is to include higher-order corrections to the string effective action. These include both classical finite size effects of the strings (α' corrections arising in higher order derivatives), and quantum string loop corrections (g_s corrections). The list of authors who have worked in this area is too great to mention here, for a detailed list see [1, 17]. A series of key papers were written by Brustein and Madden, in which they demonstrated that it is possible to include such terms and successfully have an exit from one branch to the other [18, 19]. More recently this approach has been generalised by including combinations of classical and quantum corrections [20]. Brustein and Madden [18, 19] made use of the result that classical corrections can stabilize a high curvature string phase while the evolution is still in the weakly coupled regime[21]. The crucial new ingredient that they added was the inclusion of terms of the type that may result from quantum corrections to the string effective action and which induce violation of the null energy condition (NEC – The Null Energy Condition is satisfied if $\rho + p \geq 0$, where ρ and p represent the effective energy density and pressure of the additional sources). Such extra terms mean that evolution towards a decelerated FRW phase is possible. Of course this violation of the null energy condition can not continue indefinitely, and eventually it needs to be turned off in order to stabilise the dilaton at a fixed value, perhaps by capture in a potential minimum or by radiation production – another problem for string theory!

The analysis of [18] resulted in a set of necessary conditions on the evolution in terms of the Hubble parameters H_S in the string frame, H_E in the Einstein frame and the dilaton φ, where they are related by $H_E = e^{\varphi/2}(H_S - \frac{1}{2}\dot\varphi)$. The conditions were:

- Initial conditions of a (+) branch and $H_S, \dot\varphi > 0$ require $H_E < 0$.

- A branch change from (+) to (−) has to occur while $H_E < 0$.

- A successful escape and exit completion requires NEC violation accompanied by a bounce in the Einstein frame after the branch change has occurred, ending up with $H_E > 0$.

- Further evolution is required to bring about a radiation dominated era in which the dilaton effectively decouples from the "matter" sources.

In the work of [19], employing both types of string inspired corrections, the authors made use of the known fact [21] that α' corrections created an attractive fixed point for a wide range of initial conditions which stabilized the evolution in a high curvature regime with linearly growing dilaton. This then caused the evolution to undergo a branch change, all of this occurring for small values of the dilaton (weak coupling), so the quantum corrections could be ignored. However, the linearly growing dilaton meant that the quantum corrections eventually become important. Brustein and Madden employed these to induce NEC violation and allow the universe to escape the fixed point and complete the transition to a decelerated FRW evolution. As an explicit example in [20] we consider a string theory motivated example where we include a number of higher derivative α' terms. Our starting point is the minimal 4−dimensional string effective action:

$$\Gamma^{(0)} = \frac{1}{\alpha'}\int d^4x \sqrt{-g}\mathcal{L}^{(0)} = \frac{1}{\alpha'}\int d^4x \sqrt{-g} e^{-\varphi}\left\{R + (\partial_\mu\varphi)^2\right\}. \tag{17}$$

By low-energy tree-level effective action, we mean that the string is propagating in a background of small curvature and the fields are weakly coupled. However, the evolution from the pre-big bang era to the present is understood to be characterised by a regime of growing couplings and curvature. This means that the Universe will have to evolve through a phase when the field equations of this effective action are no longer valid. Hence, the low-energy dynamical description has to be supplemented by corrections in order to reliably describe the transition regime.

The finite size of the string will have an impact on the evolution of the scale factor when the curvature of the Universe reaches a critical level, corresponding to the string length scale $\lambda_S \sim \sqrt{\alpha'}$ (fixed in the string frame), and such corrections are expected to stabilise the growth of the curvature into a de-Sitter like regime of constant curvature and linearly growing dilaton [21]. Eventually the dilaton will play a major role, and since the loop expansion is governed by powers of the string coupling parameter $g_S = e^\varphi$, these quantum corrections will modify dramatically the evolution when we reach the strong coupling region [18, 19]. This should correspond to the stage when the Universe completes a smooth transition to the post-big bang branch, characterised by a fixed value of the dilaton and a decelerating FRW expansion. One of the unresolved issues of the transition concerns whether or not the actual exit takes place at large coupling, $e^\varphi \geq 1$. If it occurred whilst the coupling was still small, then we would be happy to use the perturbative corrections we are adopting.

The type of corrections we consider involve truncations of the classical action at order α'. The most general form for a correction to the string action up to fourth-order in derivatives has been presented in refs [22, 23]:

$$\begin{aligned}\Gamma^{(C)} &= \frac{1}{\alpha'}\int d^4x \sqrt{-g}\mathcal{L}^{(C)} \\ &= -\frac{1}{4}\int d^4x \sqrt{-g} e^{-\varphi}\Big\{aR_{GB}^2 + b\nabla^2\varphi(\partial_\mu\varphi)^2 \\ &\quad + c\{R^{\mu\nu} - \frac{1}{2}g^{\mu\nu}R\}\partial_\mu\varphi\partial_\nu\varphi + d(\partial_\mu\varphi)^4\Big\}.\end{aligned} \tag{18}$$

$R_{GB}^2 = R_{\mu\nu\lambda\rho}R^{\mu\nu\lambda\rho} - 4R_{\mu\nu}R^{\mu\nu} + R^2$ is the Gauss-Bonnet combination which guarantees the absence of higher derivatives. In fixing the different parameters in this action we require that it reproduces the usual string scattering amplitudes [24]. This constrains the coefficient of $R_{\mu\nu\lambda\rho}^2$ with the result that the pre-factor for the Gauss-Bonnet term has to be $a = -1$. But the Lagrangian can still be shifted by field redefinitions which preserve the on-shell amplitudes, leaving the three remaining coefficients of the classical correction satisfying the constraint $2(b+c) + d = -16a$.

There is as yet no definitive calculation of the full loop expansion of string theory. This is of course a big problem if we want to try and include quantum effects in analysing the graceful exit issue. The best we can do, is to propose plausible terms that we hope are representative of the actual terms that will eventually make up the loop corrections.

We believe that the string coupling g_S actually controls the importance of string-loop corrections, so as a first approximation to the loop corrections we multiply each term of the classical correction by a suitable power of the string coupling [18, 19]. When loop corrections are included, we then have an effective Lagrangian given by

$$\mathcal{L} = \mathcal{L}^{(0)} + \mathcal{L}^{(C)} + Ae^{\varphi}\mathcal{L}^{(C)} + Be^{2\varphi}\mathcal{L}^{(C)}, \tag{19}$$

where $\mathcal{L}^{(0)}$ is given in Eq. (17) and $\mathcal{L}^{(C)}$ given in Eq. (18). The constant parameters A and B actually control the onset of the loop corrections.

Not surprisingly the field equations need to be solved numerically, but this can be done and the solutions are very encouraging as they show there exists a large class of parameters for which successful graceful exits are obtained [20]. For example the natural setting $b = c = 0$ leads to the well-known form which has given rise to most of the studies on corrections to the low-energy picture. In references [18, 21], the authors demonstrated that this minimal classical correction regularises the singular behaviour of the low-energy pre-big bang scenario. It drives the evolution to a fixed point of bounded curvature with a linearly growing dilaton (the star in Figure 4 – which agrees with the results of [18, 21]), suggesting that quantum loop corrections -known to allow a violation of the null energy condition $(p + \rho < 0)$- would permit the crossing of the Einstein bounce to the FRW decelerated expansion in the post-big bang era. Indeed, the addition of loop corrections leads to a $(-)$ FRW-branch as pictured in Figure 4. However, we still have to freeze the growth of the dilaton. Following [18], we introduce by hand a particle creation term of the form $\Gamma_\varphi \dot\varphi$, where Γ_φ is the decay width of the φ particle, in the equation of motion of the dilaton field and then coupling it to a fluid with the equation of state of radiation in such a way as to conserve energy overall. This allows us to stabilise the dilaton in the post-big bang era with a decreasing Hubble rate, similar to the usual radiation dominated FRW cosmology. We should point out though, that although it is possible to have a successful exit, it is not so easy to ensure that the exit takes place in a weakly coupled regime, and typically we found that as the exit was approached $\varphi_\text{final} \sim 0.1 - -0.3$. Thus it is fair to say that although great progress has been made on the question of Graceful Exit in string cosmology, it remains a problem in search of the full solution. It is a fascinating problem, and not surprisingly alternative prescriptions which aim to address this issue have recently been proposed, involving colliding branes [25] and Cyclic universes [27]. We now turn our attention to the observational consequences of string cosmology, in particular the generation of the observed cosmic microwave background radiation.

VI. DENSITY PERTURBATIONS IN STRING COSMOLOGY

We have to consider inhomogeneous perturbations that may be generated due to vacuum fluctuations, and follow the formalism pioneered by Mukhanov and collaborators [28, 29]. During a period of accelerated expansion the comoving Hubble length, $|d(\ln a)/d\eta|^{-1}$, decreases and vacuum fluctuations which are assumed to start in the flat-spacetime vacuum state may be stretched up to exponentially large scales. The precise form of the spectrum depends on the expansion of the homogeneous background and the couplings between the fields. The comoving Hubble length, $|d(\ln \tilde{a})/d\eta|^{-1} = 2|\eta|$, does indeed decrease in the Einstein frame during the contracting phase when $\eta < 0$. Because the dilaton, moduli fields and graviton are minimally coupled to this metric, this ensures that small-scale vacuum fluctuations will eventually be stretched beyond the comoving Hubble scale during this epoch.

As we remarked earlier, the axion field is taken to be a constant in the classical pre-big bang solutions. However, even when the background axion field is set to a constant, there will inevitably be quantum fluctuations in this field. We will see that these fluctuations can not be neglected and, moreover, that they are vital if the pre-big bang scenario is to have any chance of generating the observed density perturbations.

In the Einstein frame, the first-order perturbed line element can be written as

$$d\tilde{s}^2 = \tilde{a}^2(\eta)\left\{-(1+2\widetilde{A})d\eta^2 + 2\widetilde{B}_{,i}d\eta dx^i + [\delta_{ij} + h_{ij}]dx^i dx^j\right\}, \tag{20}$$

where $\widetilde{A}$ and $\widetilde{B}$ are scalar perturbations and h_{ij} is a tensor perturbation.

A. Scalar metric perturbations

First of all we consider the evolution of linear metric perturbations about the four-dimensional spatially flat dilaton-moduli-vacuum solutions given in Eqs. (8–10). Considering a single Fourier mode, with comoving wavenumber k, the perturbed Einstein equations yield the evolution equation

$$\widetilde{A}'' + 2\tilde{h}\widetilde{A}' + k^2\widetilde{A} = 0, \tag{21}$$

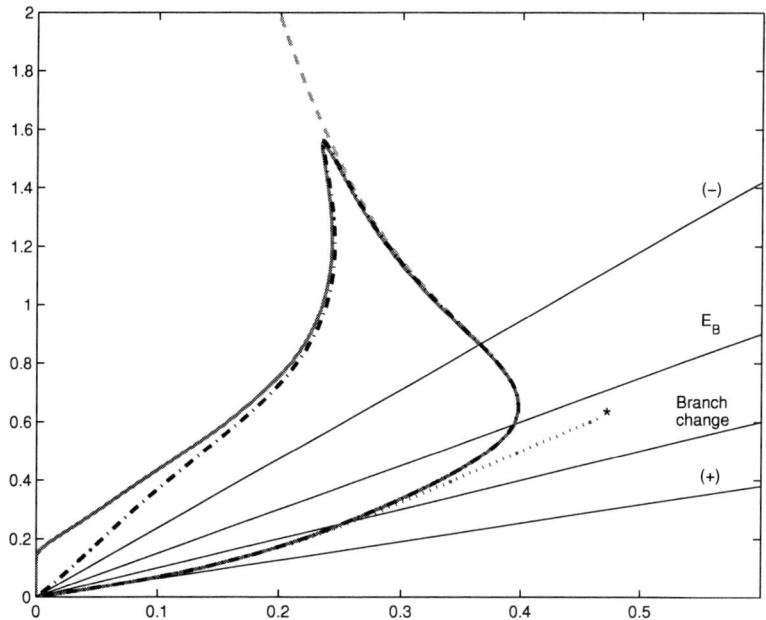

FIG. 4: Hubble expansion in the S-frame as a function of the dilaton for a successful exit with $a = -1$, $b = c = 0$ and $d = 16$. The y-axis corresponds to H, and the x-axis to $2\dot\varphi/3$. The initial conditions for the simulations have been set with respect to the lowest-order analytical solutions at $t_S = -1000$. The straight black lines describe the bounds quoted in Section II. The dotted magenta line shows the impact of the classical correction due to the finite size of the string. A ∗ denotes the fixed point. The contribution of the one-loop expansion is traced with a dashed cyan line ($A = 4$). The dash-dotted blue line represents the incorporation of the two-loop correction without the Gauss-Bonnet combination ($B = -0.1$). Finally, the green plain line introduces radiation with $\Gamma_\varphi = 0.08$ and stabilises the dilaton.

plus the constraint

$$\widetilde{A} = -(\widetilde{B}' + 2\widetilde{h}\widetilde{B}), \qquad (22)$$

where $\widetilde{h}$ is the Hubble parameter in the Einstein frame derived from Eq. (11), and $\widetilde{A}' \equiv \frac{d\widetilde{A}}{d\eta}$. In the spatially flat gauge we have the simplification that the evolution equation for the scalar metric perturbation, Eq. (21), is independent of the evolution of the different massless scalar fields (dilaton, axion and moduli), although they will still be related by the constraint

$$\widetilde{A} = \frac{\varphi'}{4\widetilde{h}}\delta\varphi + \frac{\beta'}{4\widetilde{h}}\delta\beta, \qquad (23)$$

where $\delta\varphi$ and $\delta\beta$ are the perturbations in φ and β respectively. To first-order, the metric perturbation, $\widetilde{A}$, is determined solely by the dilaton and moduli field perturbations, although its evolution is dependent only upon the Einstein frame scale factor, $\widetilde{a}(\eta)$, given by Eq. (11), which in turn is determined solely by the stiff fluid equation of state for the homogeneous fields in the Einstein frame.

One of the most useful quantities we can calculate is the curvature perturbation on uniform energy density hypersurfaces (as $k\eta \to 0$). It is commonly denoted by ζ [30] and in the Einstein frame, we obtain

$$\zeta = \frac{\widetilde{A}}{3}, \qquad (24)$$

in any dilaton–moduli–vacuum or dilaton–moduli–axion cosmology [31, 32].

The significance of ζ is that in an expanding universe it becomes constant on scales much larger than the Hubble scale ($|k\eta| \ll 1$) for purely adiabatic perturbations. In single-field inflation models this allows one to compute the density perturbation at late times, during the matter or radiation dominated eras, by equating ζ at "re-entry" ($k = \widetilde{a}\widetilde{H}$) with that at horizon crossing during inflation. To calculate ζ, hence the density perturbations induced in the pre-big bang scenario we can either use the vacuum fluctuations for the canonically normalised field at early times/small scales (as $k\eta \to -\infty$) or use the amplitude of the scalar field perturbation spectra to normalise the solution for $\widetilde{A}$. This yields, (after some work), the curvature perturbation spectrum on large scales/late times (as $k\eta \to 0$):

$$\mathcal{P}_\zeta = \frac{8}{\pi^2} l_{\rm Pl}^2 \widetilde{H}^2 (-k\eta)^3 [\ln(-k\eta)]^2 , \qquad (25)$$

where $l_{\rm Pl}$ is the Planck length in the Einstein frame and remains fixed throughout. The scalar metric perturbations become large on superhorizon scales ($|k\eta| < 1$) only near the Planck era, $\widetilde{H}^2 \sim l_{\rm Pl}^{-2}$.

The spectral index of the curvature perturbation spectrum is conventionally given as [33]

$$n \equiv 1 + \frac{d \ln \mathcal{P}_\zeta}{d \ln k} \qquad (26)$$

where $n = 1$ corresponds to the classic Harrison-Zel'dovich spectrum for adiabatic density perturbations favoured by most models of structure formation in our universe. By contrast the pre–big bang era leads to a spectrum of curvature perturbations with $n = 4$. Such a steeply tilted spectrum of metric perturbations implies that there would be effectively no primordial metric perturbations on large (super-galactic) scales in our present universe if the post-Big bang era began close to the Planck scale. Fortunately, as we shall see later, the presence of the axion field could provide an alternative spectrum of perturbations more suitable as a source of large-scale structure. The pre-big bang scenario is not so straightforward as in the single field inflation case, because the full low-energy string effective action possesses many fields which can lead to non-adiabatic perturbations. This implies that density perturbations at late times may not be simply related to ζ alone, but may also be dependent upon fluctuations in other fields.

B. Tensor metric perturbations

The gravitational wave perturbations, h_{ij}, are both gauge and conformally invariant. They decouple from the scalar perturbations in the Einstein frame to give a simple evolution equation for each Fourier mode

$$h_k'' + 2\widetilde{h} h_k' + k^2 h_k = 0 . \qquad (27)$$

This is exactly the same as the equation of motion for the scalar perturbation given in Eq. (21) and has the same growing mode in the long wavelength ($|k\eta| \to 0$) limit given by Eq. (25). The spectrum depends solely on the dynamics of the scale factor in the Einstein frame given in Eq. (11), which remains the same regardless of the time-dependence of the different dilaton, moduli or axion fields. It leads to a spectrum of primordial gravitational waves steeply growing on short scales, with a spectral index $n_T = 3$ [3], in contrast to conventional inflation models which require $n_T < 0$ [33]. The graviton spectrum appears to be a robust and distinctive prediction of any pre-big bang type evolution based on the low-energy string effective action, although recently in the non-singular model of section 5, we have demonstrated how passing through the string phase could lead to a slight shift in the tilt closer to $n_T \sim 2$ [34]

C. Dilaton–Moduli–Axion Perturbation Spectra

We will now consider inhomogeneous linear perturbations in the fields about a homogeneous background given by [32, 35]

$$\varphi = \varphi(\eta) + \delta\varphi(\mathbf{x},\eta), \quad \sigma = \sigma(\eta) + \delta\sigma(\mathbf{x},\eta), \quad \beta = \beta(\eta) + \delta\beta(\mathbf{x},\eta) . \qquad (28)$$

The perturbations can be re-expressed as a Fourier series in terms of Fourier modes with comoving wavenumber k. Considering the production of dilaton, moduli and axion perturbations during a pre-big bang evolution where the background axion field is constant, $\sigma' = 0$, the evolution of the homogeneous background fields are given in Eqs. (9–10). The dilaton and moduli fields both evolve as minimally coupled massless fields in the Einstein frame. In particular, the dilaton perturbations are decoupled from the axion perturbations and the equations of motion in the spatially flat gauge become

$$\delta\varphi'' + 2\widetilde{h}\delta\varphi' + k^2 \delta\varphi = 0, \qquad (29)$$
$$\delta\beta'' + 2\widetilde{h}\delta\beta' + k^2 \delta\beta = 0, \qquad (30)$$
$$\delta\sigma'' + 2\widetilde{h}\delta\sigma' + k^2 \delta\sigma = -2\varphi' \delta\sigma' , \qquad (31)$$

Note that these evolution equations for the scalar field perturbations defined in the spatially flat gauge are automatically decoupled from the metric perturbations, although as we have said they are still related to the scalar metric perturbation, $\widetilde{A}$ through Eq. (23).

On the (+) branch, i.e., when $\eta < 0$, we can normalise modes at early times, $\eta \to -\infty$, where all the modes are far inside the Hubble scale, $k \gg |\eta|^{-1}$, and can be assumed to be in the flat-spacetime vacuum. Whereas in conventional inflation where we have to assume that this result for a quantum field in a classical background holds at the Planck scale, in this case the normalisation is done in the zero-curvature limit in the infinite past. Just as in conventional inflation, this produces perturbations on scales far outside the horizon, $k \ll |\eta|^{-1}$, at late times, $\eta \to 0^-$.

Conversely, the solution for the (−) branch with $\eta > 0$ is dependent upon the initial state of modes far outside the horizon, $k \ll |\eta|^{-1}$, at early times where $\eta \to 0$. The role of a period of inflation, or of the pre-big bang (+) branch, is precisely to set up this initial state which otherwise appears as a mysterious initial condition in the conventional (non-inflationary) big bang model.

The power spectrum for perturbations is commonly denoted by

$$\mathcal{P}_{\delta x} \equiv \frac{k^3}{2\pi^2} |\delta x|^2, \tag{32}$$

and thus for modes far outside the horizon ($k\eta \to 0$) we have

$$\mathcal{P}_{\delta\varphi} = \frac{32}{\pi^2} l_{\text{Pl}}^2 \widetilde{H}^2 (-k\eta)^3 [\ln(-k\eta)]^2, \tag{33}$$

$$\mathcal{P}_{\delta\beta} = \frac{32}{\pi^2} l_{\text{Pl}}^2 \widetilde{H}^2 (-k\eta)^3 [\ln(-k\eta)]^2, \tag{34}$$

where $\widetilde{H} \equiv \widetilde{a}'/\widetilde{a}^2 = 1/(2\widetilde{a}\eta)$ is the Hubble rate in the Einstein frame. The amplitude of the perturbations grows towards small scales, but only becomes large for modes outside the horizon ($|k\eta| < 1$) when $\widetilde{H}^2 \sim l_{\text{Pl}}^{-2}$, i.e., the Planck scale in the Einstein frame. The spectral tilt of the perturbation spectra is given by

$$n - 1 \equiv \Delta n_x = \frac{d \ln \mathcal{P}_{\delta x}}{d \ln k} \tag{35}$$

which from Eqs. (33) and (34) gives $\Delta n_\varphi = \Delta n_\beta = 3$ (where we neglect the logarithmic dependence). This of course is the same steep blue spectra we obtained earlier for the metric perturbations, which of course is far from the observed near H-Z scale invariant spectrum. We have recently examined the case of the evolution of the field perturbations in the non-singular cosmologies of section five and as with the metric-perturbation case, amongst a number of new features that emerge there is a slight shift produced in the spectral index [36].

While the dilaton and moduli fields evolve as massless minimally coupled scalar fields in the Einstein frame, the axion field's kinetic term still has a non-minimal coupling to the dilaton field. This is evident in the equation of motion, Eq. (31), for the axion field perturbations $\delta\sigma$. The non-minimal coupling of the axion to the dilaton leads to a significantly different evolution to that of the dilaton and moduli perturbations.

After some algebra, we find that the late time evolution in this case is logarithmic with respect to $-k\eta$, (for $\mu \neq 0$)

$$\mathcal{P}_{\delta\sigma} = 64\pi l_{\text{Pl}}^2 C^2(\mu) \left(\frac{e^{-\varphi}\widetilde{H}}{2\pi}\right)^2 (-k\eta)^{3-2\mu}, \tag{36}$$

where $\mu \equiv |\sqrt{3}\cos\xi_*|$ and the numerical coefficient

$$C(\mu) \equiv \frac{2^\mu \Gamma(\mu)}{2^{3/2}\Gamma(3/2)}, \tag{37}$$

approaches unity for $\mu \to 3/2$.

The key result is that the spectral index can differ significantly from the steep blue spectra obtained for the dilaton and moduli fields that are minimally coupled in the Einstein frame. The spectral index for the axion perturbations is given by [32, 35]

$$\Delta n_\sigma = 3 - 2\sqrt{3}|\cos\xi_*| \tag{38}$$

and depends crucially upon the evolution of the dilaton, parameterised by the value of the integration constant ξ_*. The spectrum becomes scale-invariant as $\sqrt{3}|\cos\xi_*| \to 3/2$, which if we return to the higher-dimensional underlying theory corresponds to a fixed dilaton field in ten-dimensions. The lowest possible value of the spectral tilt Δn_σ is $3 - 2\sqrt{3} \simeq -0.46$ which is obtained when stable compactification has occurred and the moduli field β is fixed. The more rapidly the internal dimensions evolve, the steeper the resulting axion spectrum until for $\cos \xi_* = 0$ we have $\Delta n_\sigma = 3$ just like the dilaton and moduli spectra.

When the background axion field is constant these perturbations, unlike the dilaton or moduli perturbations, do not affect the scalar metric perturbations. Axion fluctuations correspond to isocurvature perturbations to first-order. However, if the axion field does affect the energy density of the universe at later times (for instance, by acquiring a mass) then the spectrum of density perturbations need not have a steeply tilted blue spectrum such as that exhibited by the dilaton or moduli perturbations. Rather, it could have a nearly scale-invariant spectrum as required for large-scale structure formation. Such an exciting possibility has received a great deal of attention recently, notably in [37-41], and could be a source for the 'curvaton' field recently introduced by Lyth and Wands as a way of converting isocurvature into adiabatic perturbations [42]. Time will tell if the axion has any role to play in cosmological density perturbations although already it is beginning to look as the curvaton route is an interesting one to follow in this context [43, 44].

VII. SMOKING GUNS?

Are there any distinctive features that we should be looking out for which would act as an indicator that the early Universe underwent a period of kinetic driven inflation? We have already mentioned the possibility of observing the presence of axion fluctuations in the cosmic microwave background anisotropies. Some of the other smoking guns include:

- The spectrum of primordial gravitational waves steeply growing on short scales, with a spectral index $n_T = 3$, although of no interest on large scales, such a spectrum could be observed by the next generation of gravitational wave detectors such as the Laser Interferometric Gravitational Wave Observatory (LIGO) if they are on the right scale [34, 45, 46]. The current frequency of these waves depends on the cosmological model, and in general we would require either an intermediate epoch of stringy inflation, or a low re-heating temperature at the start of the post-big bang era [47] to place the peak of the gravitational wave spectrum at the right scale. Nonetheless, the possible production of high amplitude gravitational waves on detector scales in the pre-big bang scenario is in marked contrast to conventional inflation models in which the Hubble parameter decreases during inflation.

- Because the scalar and tensor metric perturbations obey the same evolution equation, their amplitude is directly related. The amplitude of gravitational waves with a given wavelength is commonly described in terms of their energy density at the present epoch. For the simplest pre-big bang models this is given in terms of the amplitude of the scalar perturbations as

$$\Omega_{\text{gw}} = \frac{2}{z_{\text{eq}}} \mathcal{P}_\zeta \qquad (39)$$

where $z_{\text{eq}} = 24000\Omega_o h^2$ is the red-shift of matter-radiation equality. The advanced LIGO configuration will be sensitive to $\Omega_{\text{gw}} \approx 10^{-9}$ over a range of scales around 100Hz. However, the maximum amplitude of gravitational waves on these scales is constrained by limits on the amplitude of primordial scalar metric perturbations on the same scale [47]. In particular, if the fractional over-density when a scalar mode re-enters the horizon during the radiation dominated era is greater than about 1/3, then that horizon volume is liable to collapse to form a black hole with a lifetime of the order the Hubble time and this would be evaporating today! If we find PBH's and gravitational waves together then this would indeed be an exciting result for string cosmology!

- Evidence of a primordial magnetic field could have an interpretation in terms of string cosmology. In string theory the dilaton is automatically coupled to the electromagnetic field strength, for example in the heterotic string effective action the photon field Lagrangian is of the form $\mathcal{L} = e^{-\varphi} F_{\mu\nu} F^{\mu\nu}$, where the field strength is derived from the vector potential, $F_{\mu\nu} = \nabla_{[\mu} A_{\nu]}$.

Now in an isotropic FRW cosmology the magnetic field must vanish to zeroth-order, and thus the vector field perturbations are gauge-invariant and we can neglect the metric back-reaction to first-order. In the radiation gauge ($A^0 = 0$, $A^i_{|i} = 0$) then the field perturbations can be treated as vector perturbations on the spatial hypersurfaces. The field perturbation A_i turns out to have a clear unique dependence on the dilaton field. In fact the time dependence of the dilaton (rather than the scale factor) leads to particle production during the pre-big bang from an initial vacuum state [48-50]. Using the pre-big bang solutions given in Eqs. (8)-(10), we find that the associated Power spectrum of the gauge fields have a minimum tilt for the spectral index for $\xi_* = 0$

when $\mu = (1+\sqrt{3})/2$ with a spectral tilt $\Delta n_{em} = 4 - \sqrt{3} \approx 2.3$. This is still strongly tilted towards smaller scales, which currently is too steep to be observably acceptable.

VIII. DILATON-MODULI COSMOLOGY INCLUDING A MOVING FIVE BRANE.

We turn our attention in this final section to M-theory, and in particular to cosmological solutions of four-dimensional effective heterotic M-theory with a moving five-brane, evolving dilaton and T modulus [51]. It turns out that the five-brane generates a transition between two asymptotic rolling-radii solutions, in a manner analogous to the case of the NS-NS axion discussed in section three. Moreover, the five-brane motion generally drives the solutions towards strong coupling asymptotically. The analogous solutions to those presented in the pre-big-bang involves a negative-time branch solution which ends in a brane collision accompanied by a small-instanton transition. Such an exact solution should be of interest bearing in mind the recent excitement that has been generated over the Ekpyrotic Universe scenario, which involves solving for the collision of two branes [25, 26].

The four-dimensional low-energy effective theory we will be using is related to the underlying heterotic M-theory. Of particular importance for the interpretation of the results is the relation to heterotic M-theory in five dimensions, obtained from the 11-dimensional theory by compactification on a Calabi-Yau three-fold. This five-dimensional theory provides an explicit realisation of a brane-world. The compactification of 11 dimensional Horava-Witten theory, that is 11-dimensional supergravity on the orbifold $S^1/Z_2 \times M_{10}$, to five dimensions on a Calabi-Yau three fold, leads to the appearance of extra three-branes in the five-dimensional effective theory. Unlike the "boundary" three-branes which are stuck to the orbifold fix points, however, these three-branes are free to move in the orbifold direction, and this leads to a fascinating new cosmology.

Our starting point is the four dimensional action

$$S = -\frac{1}{2\kappa_P^2} \int d^4x \sqrt{-g} \left[\frac{1}{2} R + \frac{1}{4}(\nabla\varphi)^2 + \frac{3}{4}(\nabla\beta)^2 + \frac{q_5}{2} e^{(\beta-\varphi)} (\nabla z)^2 \right], \qquad (40)$$

where φ is the effective dilaton in four dimensions, β is the size of the orbifold, z is the modulus representing the position of the five brane and satisfies $0 < z < 1$, and q_5 is the five brane charge. Due to the non-trivial kinetic term for z, solutions with exactly constant φ or β do not exist as soon as the five-brane moves. Therefore, the evolution of all three fields is linked and (except for setting z = const) cannot be truncated consistently any further. Looking for cosmological solutions for simplicity, we assume the three-dimensional spatial space to be flat. Our Ansatz then reads

$$ds^2 = -e^{2\nu} d\tau^2 + e^{2\alpha} d\mathbf{x}^2,$$

$$\varphi = \varphi(\tau), \qquad \alpha = \alpha(\tau), \qquad \beta = \beta(\tau), \qquad z = z(\tau).$$

The cosmological solutions are given by [51]

$$\alpha = \frac{1}{3} \ln \left| \frac{t-t_0}{T} \right| + \alpha_0 \qquad (41)$$

$$\beta = p_{\beta,i} \ln \left| \frac{t-t_0}{T} \right| + (p_{\beta,f} - p_{\beta,i}) \ln \left(\left| \frac{t-t_0}{T} \right|^{-\delta} + 1 \right)^{-\frac{1}{\delta}} + \beta_0 \qquad (42)$$

$$\varphi = p_{\varphi,i} \ln \left| \frac{t-t_0}{T} \right| + (p_{\varphi,f} - p_{\varphi,i}) \ln \left(\left| \frac{t-t_0}{T} \right|^{-\delta} + 1 \right)^{-\frac{1}{\delta}} + \varphi_0 \qquad (43)$$

$$z = d \left(1 + \left| \frac{T}{t-t_0} \right|^{-\delta} \right)^{-1} + z_0. \qquad (44)$$

where t is the proper time, the time-scales t_0 and T are arbitrary constants as are the constants d and z_0 which parameterise the motion of the five-brane. For $-\infty < t < t_0$ we are in the positive branch of the solutions and for $t_0 < t < \infty$ we are in the negative branch.

We see that both expansion powers for the scale factor α are given by $1/3$, a fact which is expected in the Einstein frame. The initial and final expansion powers for β and φ are less trivial and are subject to the constraint

$$3p_{\beta,n}^2 + p_{\varphi,n}^2 = \frac{4}{3} \qquad (45)$$

for $n = i, f$. These are mapped into one another by

$$\begin{pmatrix} p_{\beta,f} \\ p_{\varphi,f} \end{pmatrix} = P \begin{pmatrix} p_{\beta,i} \\ p_{\varphi,i} \end{pmatrix}, \qquad P = \frac{1}{2} \begin{pmatrix} 1 & 1 \\ 3 & -1 \end{pmatrix}. \tag{46}$$

This map is its own inverse, that is $P^2 = 1$, which is a simple consequence of time reversal symmetry. The power δ is explicitly given by

$$\delta = p_{\beta,i} - p_{\varphi,i}. \tag{47}$$

For $\delta < 0$ we are in the negative branch and for $\delta > 0$ we are in the positive time branch. Finally, we have

$$\varphi_0 - \beta_0 = \ln\left(\frac{2q_5 d^2}{3}\right). \tag{48}$$

The solutions have the following interpretation: at early times, the system starts in the rolling radii solution characterised by the initial expansion powers p_i while the five-brane is practically at rest. When the time approaches $|t - t_0| \sim |T|$ the five-brane starts to move significantly which leads to an intermediate period with a more complicated evolution of the system. Then, after a finite comoving time, in the late asymptotic region, the five-brane comes to a rest and the scale factors evolve according to another rolling radii solution with final expansion powers p_f. Hence the five-brane generates a transition from one rolling radii solution into another one. While there are perfectly viable rolling radii solutions which become weakly coupled in at least one of the asymptotic regions, the presence of a moving five-brane always leads to strong coupling asymptotically, a phenomenon similar to what we observed in the dilaton-moduli-axion dynamics (see Figure 2).

These general results can be illustrated by an explicit example. Focusing on the negative-time branch and considering the solutions with an approximately static orbifold at early time, Figure 5 shows the evolution of β and φ, whereas Figure 6 shows the evolution of the dynamical brane.

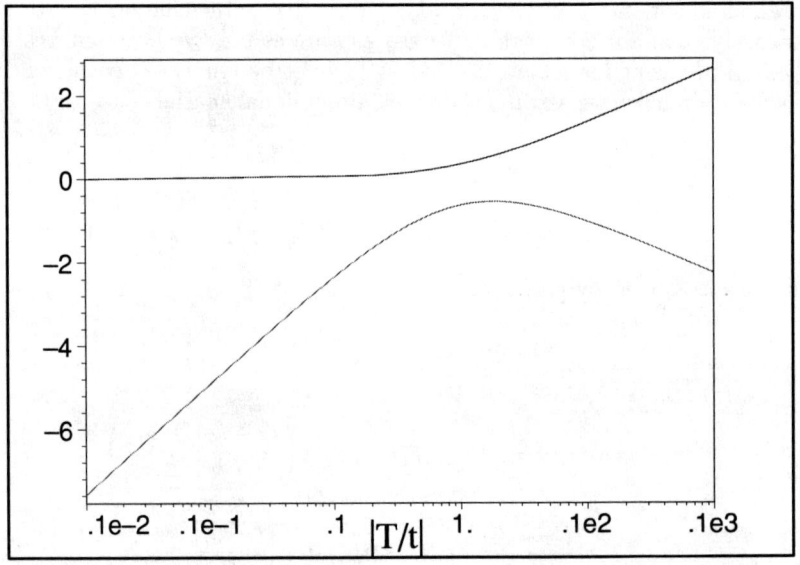

FIG. 5: Time-behaviour of β (upper curve) and φ (lower curve).

At early times, $|t - t_0| \gg |T|$, the evolution is basically of power-law type with powers p_i, because at early time the five-brane is effectively frozen at $z \simeq d + z_0$ and does not contribute a substantial amount of kinetic energy. This changes dramatically once we approach the time $|t - t_0| \sim |T|$. In a transition period around this time, the brane moves from its original position by a total distance d and ends up at $z \simeq z_0$. At the same time, this changes the behaviour of the moduli β and φ until, at late time $|t| \ll |T|$, they correspond to another rolling radii solution with powers controlled by p_f. Concretely, the orbifold size described by β turns from being approximately constant at early time to expanding at late time, while the Calabi-Yau size controlled by φ undergoes a transition from expansion to contraction. We also find that as with the axion case discussed earlier, the solution runs into strong coupling in both asymptotic regions $t - t_0 \to -\infty$ and $t - t_0 \to 0$ which illustrates our general result.

In Fig. 6 we have shown a particular case which leads to brane collision. The five-brane is initially located at $d + z_0 \simeq 0.9$ and moves a total distance of $d = 1.5$ colliding with the boundary at $z = 0$ at the time $|t - t_0|/|T| \simeq 1$.

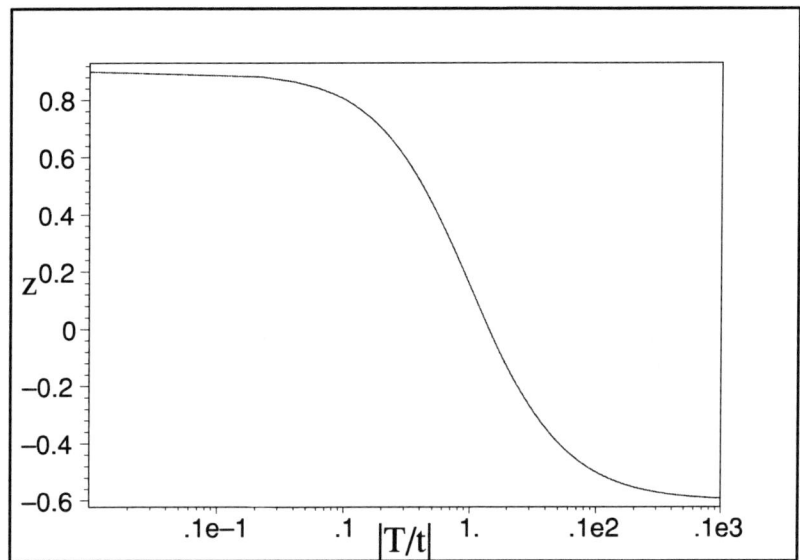

FIG. 6: Time-behaviour of the five-brane position modulus z for the example specified in the text. The boundaries are located at $z = 0, 1$ and the five-brane collides with the $z = 0$ boundary at $|t/T| \simeq 1$.

This represents an explicit example of a negative-time branch solution which ends in a small-instanton brane-collision. Solving for these systems has only just the begun, but already interesting features have emerged including a new mechanism for baryogenesis arising from the collision of two branes [52], and a more detailed understanding of the vacuum transitions associated with brane collisions [53].

IX. SUMMARY

In this article we have addressed a number of issues relating to string cosmology. We have seen how rolling radii solutions associated with the low energy string action lead to new inflationary solutions, and how the inclusion of the axion field perturbations can generate scale invariant density fluctuations, although they are primarily isocurvature in nature. The thorny issues of initial conditions and Graceful Exit facing the pre Big Bang scenario have been discussed and possible resolutions proposed. Observational features of string cosmology today have been discussed including gravitational wave detection and anisotropies in the cosmic microwave background. Finally, we have related these solutions to the exciting new solutions arising in M-theory cosmology, and showed how a moving five brane could act in a manner similar to the axion field in the pre Big Bang case. This is an exciting time for string and M-theory cosmology, the subject is developing at a very fast rate, and no doubt there will be new breakthroughs emerging over the next few years. Hopefully out of these we will be in a position to address a number of the issues I have raised in this article, as well as other key ones such as stabilising the dilaton and explaining the current observation of an accelerating Universe.

There are a number of areas that I have not had time to discuss, string statistical mechanics, multi brane models, warped geometries. Hopefully it is clear from what I have said so far though, that in the words of the 'Spontaneous Harmony Breakers' who sang in the conference banquet, when doing string cosmology you will need to "think with lots of fantasy, think a new world".

Acknowledgments

I am very grateful to Professor Mario Novello for inviting me to this beautiful summer school, and to all the participants for helping to make it such a memorable experience for me, not only through the physics, but through the games of soccer in which England tried to compete with the South American skills of Brazil!

---

[1] J. H. Lidsey, D. Wands and E. J. Copeland, Phys. Rep. **337**, 343 (2000)
[2] G. Veneziano, Phys. Lett. **B265**, 287 (1991)
[3] M. Gasperini and G. Veneziano. Astropart. Phys. **1**, 317 (1993)
[4] E. J. Copeland, A. Lahiri, and D. Wands, Phys. Rev. **D50**, 4868 (1994)
[5] M. D. Pollock and D. Sahdev, Phys. Lett. **B222**, 12 (1989)
[6] J. J. Levin and K. Freese, Phys. Rev. **D47**, 4282 (1993)
[7] R. Brustein and G. Veneziano, Phys. Lett. **B329**, 429 (1994)
[8] M. S. Turner and E. J. Weinberg, Phys. Rev. **D56**, 4604 (1997)
[9] D. Clancy, J. E. Lidsey, and R. Tavakol, Phys. Rev. **D58**, 044017 (1998)
[10] N. Kaloper, A. D. Linde, and R. Bousso, Phys. Rev. **D59**, 043508 (1999)
[11] J. Maharana, E. Onofri, and G. Veneziano, J. High Energy Phys. **01**, 004 (1998)
[12] G. Veneziano, Phys. Lett. **B406**, 297 (1997)
[13] A. Buonanno, K. A. Meissner, C. Ungarelli, and G. Veneziano, Phys. Rev. **D57**, 2543 (1998)
[14] A. Buonanno, T. Damour, and G. Veneziano, Nucl. Phys. **B543**, 275 (1999)
[15] M. Gasperini, Phys. Rev. **D61** 087301 (2000)
[16] D. Clancy, J. E. Lidsey, and R. Tavakol, Phys. Rev. **D59**, 063511 (1999)
[17] M. Gasperini's web page, http://www.to.infn.it/ gasperin/
[18] R. Brustein and R. Madden, Phys. Lett. **B410**, 110 (1997) R. Brustein and R. Madden, Phys. Lett. **B410**, 110 (1997)
[19] R. Brustein and R. Madden, Phys. Rev. **D57**, 712 (1998) R. Brustein and R. Madden, Phys. Rev. **D57**, 712 (1998).
[20] C. Cartier, E. J. Copeland and R. Madden, JHEP **0001**, 035 (2000)
[21] M. Gasperini, M. Maggiore, and G. Veneziano, Nucl. Phys. **B494**, 315 (1997)
[22] K.A. Meissner, Phys. Lett. **B392**, 110 (1997).
[23] N. Kaloper and K.A. Meissner, Phys. Rev. **D56**, 7940 (1997)
[24] R.R. Metsaev and A.A. Tseytlin, Nucl. Phys. **B293**, 385 (1987)
[25] J. Khoury, B. A. Ovrut, P. J. Steinhardt and N. Turok, Phys. Rev. **D64**, 123522 (2001) and hep-th/0108187
[26] R. Kallosh, L. Kofman and A. Linde, Phys. Rev. **D64**, 123523 (2001)
[27] P. J. Steinhardt and N. Turok, hep-th/0111030 and hep-th/0111098
[28] V. F. Mukhanov, Sov. Phys. JETP **68**, 1297 (1988)
[29] V. F. Mukhanov, H. A. Feldman, and R. H. Brandenberger, Phys. Rep. **215**, 203 (1992)
[30] J. M. Bardeen, P. J. Steinhardt, and M. S. Turner, Phys. Rev. **D28**, 679 (1983)
[31] R. Brustein, M. Gasperini, M. Giovannini, V. F. Mukhanov, and G. Veneziano, Phys. Rev. D **51**, 6744 (1995)
[32] E. J. Copeland, R. Easther, and D. Wands, Phys. Rev. **D56**, 874 (1997)
[33] A. R. Liddle and D. H. Lyth, Phys. Rep. **231**, 1 (1993)
[34] C. Cartier, E. J. Copeland and M. Gasperini, Nucl. Phys. **B607**, 406 (2001)
[35] E. J. Copeland, J. E. Lidsey, and D. Wands, Nucl. Phys. **B506**, 407 (1997)
[36] C. Cartier, J. Hwang and E. J. Copeland, Phys. Rev. **D64**, 103504 (2001)
[37] R. Durrer, M. Gasperini, M. Sakellariadou, and G. Veneziano, Phys. Lett. **B436**, 66 (1998)
[38] R. Durrer, M. Gasperini, M. Sakellariadou, and G. Veneziano, Phys. Rev. **D59**, 043511 (1999)
[39] A. Melchiorri, F. Vernizzi, R. Durrer and G. Veneziano, Phys. Rev. Lett. **83**, 4464 (1999)
[40] F. Vernizzi, A. Melchiorri and R. Durrer, Phys. Rev. **D63**, 063501 (2001).
[41] K. Enqvist and M. S. Sloth, hep-ph/0109214
[42] D. Lyth and D. Wands, Phys. Lett. **B524**, 5 (2002)
[43] V. Bozza, M. Gasperini, M. Giovannini and G. Veneziano, Phys. Lett. B **543**, 14 (2002)
[44] V. Bozza, M. Gasperini, M. Giovannini and G. Veneziano, arXiv:hep-ph/0212112.
[45] B. Allen and R. Brustein, Phys. Rev. **D55**, 3260 (1997)
[46] M. Maggiore, Phys. Rept. **331**, 283 (2000)
[47] E. J. Copeland, A. R. Liddle, J. E. Lidsey, and D. Wands, Phys. Rev. **D58**, 063508 (1998)
[48] D. Lemoine and M. Lemoine, Phys. Rev. **D52**, 1955 (1995)
[49] M. Gasperini, M. Giovannini, and G. Veneziano, Phys. Rev. Lett. **75**, 3796 (1995)
[50] M. Gasperini, M. Giovannini, and G. Veneziano, Phys. Rev. **D52**, 6651 (1995)
[51] E.J. Copeland, J. Gray and A. Lukas, Phys. Rev. **D64**, 126003 (2001)
[52] M. Bastero-Gill, E.J. Copeland, J. Gray, A. Lukas, M. Plumacher, Phys. Rev. **D66**, 066005 (2002)
[53] N. Antunes, E.J. Copeland, M. Hindmarsh and A. Lukas, hep-th/0208219

Canonical quantization of general relativity: the last 18 years in a nutshell

Jorge Pullin

Department of Physics and Astronomy, Louisiana State University, 202 Nicholson Hall, Baton Rouge, LA 70803

This is a summary of the lectures presented at the Xth Brazilian school on cosmology and gravitation. The style of the text is that of a lightly written descriptive summary of ideas with almost no formulas, with pointers to the literature. We hope this style can encourage new people to take a look into these results. We discuss the variables that Ashtekar introduced 18 years ago that gave rise to new momentum in this field, the loop representation, spin networks, measures in the space of connections modulo gauge transformations, the Hamiltonian constraint, application to cosmology and the connection with potentially observable effects in gamma-ray bursts and conclude with a discussion of consistent discretizations of general relativity on the lattice.

I. INTRODUCTION: WHERE IS STRING THEORY IN THESE LECTURES?

The words "quantum gravity" have become associated by physicists over the years with "difficult problem". Attempts to quantize general relativity started almost immediately after quantum mechanics was establishes (see the article by Rovelli [1] for a concise history of the field). Among the people involved are the most stellar names in physics. Indeed, one should expect problems when attempting to apply the rules of quantum mechanics to general relativity. Although general relativity was developed before quantum mechanics, the latter was introduced in the context of Newtonian physics. Already the incorporation of special relativity required some effort and in fact, the introduction of quantum field theory, which in many ways is an extension of quantum mechanics. General relativity however, is a much more radical revision of physics than special relativity. It is a theory of space-time itself as opposed to a theory of entities living in a spacetime. Quantum mechanics was firmly based on the latter viewpoint. This key element separates general relativity from almost all other physics theories. The invariance of the theory under coordinate redefinitions, which is more clearly viewed as invariance of the variables under diffeomorphisms, is not present in any other significant physical theory. In fact, we have only learnt how to apply the rules of quantum mechanics to theories invariant under diffeomorphisms relatively recently [2]. And the theories in question, like BF theory or Chern–Simons theory, are remarkably simpler than general relativity. These theories are only superficially field theories, in that they are described in terms of fields, but the true degrees of freedom of the theories are finite in number. They are, in fact, mechanical systems instead of field theories. Solutions of the equations of motion are given by fields that are "trivial", the only non-trivialities coming from possible topological features of the manifolds the theories live on. Once one realizes this fact, it should not be surprising that their quantization becomes relatively straightforward.

There is a strong sociological element involved in quantum gravity as well. After the many successes of quantum field theory following World War II, it could only be expected that the application of the same powerful techniques to general relativity should finally conquer the problem. But this was not so. The application of perturbation theory to general relativity taught many interesting lessons, consolidating the ideas of gauge and ghosts. But it ultimately appeared to fail. General relativity appears to be perturbatively non-renormalizable. The practitioners of quantum field theory became so discouraged by this fact, that they adopted the point of view that general relativity should be abandoned as a physical theory. It is not that the successes of the theory explaining the classical world are in question. It is the fundamental nature of the theory. In this point of view general relativity would play the role of an effective theory. The Lagrangian and field equations of general relativity should be viewed as, for instance, those of the Navier–Stokes theory of fluids. A highly successful and useful theory, but not one that anyone would care to quantize, for instance, to describe a quantum fluid. Just like in the case of quantum fluids one quantizes a theory underlying the Navier–Stokes one, one would quantize a theory underlying general relativity. A theory that reproduces general relativity only in certain regimes, but is richer in other regimes. Another analogy that comes to mind is the Fermi theory of weak interactions, non-renormalizable and just a low energy manifestation of a richer theory, the theory of weak interactions. This point of view led to the development of supersymmetric theories, supergravity, Kaluza-Klein theories, superstrings and M-theory.

Is this point of view the only one? Strictly speaking, the answer is yes. General relativity only describes gravity, and therefore a richer theory should come into play in order to have a unified picture of all interactions, so general relativity indeed should be the limit of a larger theory. But even if one ignores all other interactions, are we completely sure that general relativity cannot be quantized? This appears as an academic question. After all, if we know we need a larger theory, why bother with determining if general relativity can be quantized? The reason this question is, in the view of some people, not of purely academic interest, is that general relativity has features that we would all desire in a unified description of nature. Most notably, the fact that space-time is dynamical and invariant under diffeomorphisms. In a sense, general relativity is perhaps the simplest theory incorporating these features with non-trivial content. Lessons learnt from attempting to quantize it should therefore be very valuable at the time of quantizing a richer theory. These are the reasons propelling a small but non-trivial (see [3] for some statistics) minority of physicists to study the quantization of general relativity.

But isn't the fact that the theory is non-renormalizable an indictment of this program? How could one quantize such a theory? To understand this we need to separate an intrinsic question (is the theory quantizable or not) from a procedural question (can we quantize it using perturbation theory). These two questions are different. In fact, we know of examples of theories that admit a quantum description and that we do not know how to treat perturbatively. De Witt's group [4] studied sigma models that have this feature. But more striking is the example of general relativity in $2+1$ dimensions. In dimensions lower than four, the Einstein equations just state that the metric is flat. General relativity in such a situation is only an apparent field theory, since the only solution to the equations is "constant". One can have degrees of freedom if the topology of the manifold is non-trivial. Yet, when perturbative quantization of general relativity in $2+1$ dimensions was attempted, the theory appeared to be non-renormalizable more or less in the same way the four dimensional version was. It was only when Witten [5] noticed that one should be able to perform a non-perturbative quantization, and carried it out, that people realized one could find ways to treat the theory perturbatively [6].

The general relativity in $2+1$ dimensions example exhibits in a dramatic way the pitfalls expecting anyone attempting to quantize these kinds of theories. The lesson is that the symmetries of these theories are far more elaborate than usual. In the case of $2+1$ gravity, the symmetry group is so large that the theory is rendered trivial by it. Quantization schemes that do not take this into account, fail. In $2+1$ dimensions, unravelling the symmetry was easy because one has full control of the theory. The general exact solution of the equations of motion is not only known in closed form, but a good intuitive handle on its meaning is available. Nothing like this occurs in $3+1$ dimensions. Learning how to gain a comparable handle is the task at hand. It is obviously a difficult task. We will never have the general solution of the Einstein equations in four dimensions in closed form. That may prevent us from ever getting the kind of intuitive handle on the theory that is needed in order to quantize it.

The non-perturbative quantization of gravity was pioneered by DeWitt in the 60's, following the early efforts of Dirac and Bergmann. An immediate problem that was encountered is that the kind of variables in terms of which gravity is usually described, is very different from the ones used in the successful quantum field theories of particle physics. In the latter, the fundamental variable is a connection. This made many of the techniques that had been developed for handling particle physics theory not applicable to general relativity. A change in this situation occurred when Ashtekar introduced a formulation of general relativity in terms of a connection that had a very elegant and simple canonical structure. In fact, the theory resembled a Yang–Mills theory and opened the possibility of introducing the techniques so successfully used in that context to general relativity. These lectures will give glimpses onto some of the results that have arisen ever since.

II. ORGANIZATION AND COVERAGE

These notes are based on lectures. Due to the finite lecture time (further compressed by a two day plane delay!) and lack of expertise of the author in some areas it was not possible to cover many topics. A big, broad topic I missed is spin foam approaches to the path integral. This will be covered soon by a forthcoming review paper by Perez [7]. I will not discuss the beautiful results on black hole entropy. These are of great importance, since they are precise calculations that do not shy away from taking into account the full dynamics of the theory. The paper by Ashtekar et al. [8] has references to all the early literature. Very recent work by Varadarajan [9] and others [10], showing a connection between the loop representation and more traditional Fock pictures and the work of Thiemann's group [11] on semi-classical states could not be covered. The beauty of these results requires a level of detail that was not possible in the format of the lectures. Finally, although the notes attempt to guide a newcomer to the literature, they have not been prepared carefully enough as to attempt to be a comprehensive review. Loll [12], Rovelli [13] and Thiemann [14]Living Reviews articles covering lattice approaches, loop quantization and canonical quantum gravity respectively. Carlip [15] has a superb review on quantum gravity in general, giving the essentials of all approaches. For a lighter reading, Smolin's [16] recent book covers several aspects of quantum gravity. Detailed discussion of the early results are found in Ashtekar's books [17], other topics can also be seen in the book we wrote with Gambini [18]. Baez and Muniain have an introductory book to knot theory with applications to gravity [19].

III. CANONICAL QUANTIZATION

Canonical quantization is the oldest and most conservative approach to quantization. It demands one to control the theory well, not allowing to bypass several detailed questions, namely what is the space of states of the theory, what is the inner product, what is observable. Every physicist has performed a few canonical quantizations in courses on quantum mechanics. To canonically quantize one roughly follows the following steps: a) one picks a Poisson algebra of classical quantities that is large enough to span the physics of interest (in ordinary mechanics q and p, for instance); b) one represents these quantities as operators acting on a space of wavefunctions and the Poisson algebra as an algebra of commutators; c) if the theory has constraints, that is, quantities that vanish identically classically, one has to impose that they vanish quantum mechanically as operators; d) an inner product has to be introduced on the space of wavefunctions that are annihilated by the constraints; e) Predictions for the expectation values of observables (quantities that have vanishing Poisson brackets with the constraints) can be worked out. Notice that several of these steps involve choices. For instance, there is no unique way to choose an inner product, or to choose a certain set of classical quantities to be promoted to operators.

For general relativity one has to start by casting the theory in a canonical form. This was done by Dirac and Bergmann in the 50's and 60's (for a more modern discussion see [17]) . The fundamental canonical variable is the metric of a spatial surface q_{ab} (people normally use q instead of g to avoid confusion with the space-time metric). Its canonically conjugate momenta is usually denoted $\tilde{\pi}^{ab}$ and the tilde denotes that it is a density, as momenta usually are. The momenta are closely related to the extrinsic curvature, which is also closely related to the time derivative of the three metric. The theory has four constraints. These are relationships among the variables at a given instant of time. Three of them form a vector and are called the "vector" or diffeomorphism constraint. When one has constraints in canonical theories, it is due to the presence of symmetries. The diffeomorphism constraint can be shown to be associated with the invariance of general relativity under spatial diffeomorphisms. The remaining constraint is associated with the invariance of general relativity under diffeomorphisms off the spatial surface. It is usually called the scalar or Hamiltonian constraint. Unfortunately, the canonical treatment breaks the symmetry between space and time in general relativity and the resulting algebra of constraints is not the algebra of four diffeomorphisms, the Hamiltonian constraint is singled out and behaves differently. The algebra of constraints is closed in the sense that Poisson brackets of constraints are proportional to constraints, but the Poisson bracket of two Hamiltonian constraints is proportional to a diffeomorphism constraint through a function of the canonical variables. This we will see, will cause difficulties at the time of quantization.

If one performs a Legendre transform one finds that the Hamiltonian of general relativity is just a combination of the above mentioned constraints. That is, the Hamiltonian of the theory vanishes. This is due to the fact that the notion of time introduced in order to set up the canonical theory is a fiducial, arbitrarily introduced time. The canonical formalism "knows" that relativity does not really single out a preferred time and responds back by saying that the Hamiltonian associated with any artificial time vanishes. Physical time can only be retrieved in the canonical theory through an elaborate process with many difficulties (Kuchař's [20] article contains a detailed discussion of the "problem of time")

One can attempt a canonical quantization by considering wavefunctions of the metric $\psi(q)$. One can represent the metric as a multiplicative operator and its canonically conjugate momentum as a functional derivative. One can then attempt to promote the constraints to operatorial equations. The diffeomorphism constraint, which is linear in the momenta, is relatively simple to implement. It implies that the wavefunctions are really only functions of the diffeomorphism invariant content of q and not of q itself. This is natural and elegant, but is also problematic: we do not know how to code in a simple way the diffeomorphism invariant information of q. Therefore the solution to constraint presented is natural but also quite formal; we cannot write it or handle it in an explicit way. A worse situation arises when one considers the Hamiltonian constraint. The latter is a non-polynomial function of the canonical variables that requires regularization. Most regularizations used in particle physics depend on the presence of a background (c-number) metric, which we do not have available in quantum gravity. No satisfactory treatment of the Hamiltonian constraint in this context has ever been found.

Worse, the lack of control on the space of functions considered also implies that we do not know any kind of useful inner product to be introduced.

Finally, in the last step we were supposed to compute expectation values of the observables (see [21] for references on the observables problem) of the theory. Observables have to be quantities that have vanishing Poisson brackets with the constraints. This implies they are invariants under the symmetries of the theory and that as quantum expressions they will act upon the space of physical states (solutions to the constraints) in such a way as to keep us within that space. Unfortunately no such quantities are known for general relativity in a generic situation. If one reduces the theory by introducing additional symmetries, sometimes observables can be found. For instance in cosmological models or if the space-times considered are asymptotically flat. What is happening here is that since the Hamiltonian of the theory is a combination of constraints, finding observables is tantamount to finding the constants of motion of the theory. But the constants of motion can be seen as re-expressing the initial conditions for a given solution of the equations of motion as functions of phase space. This, of course, requires solving the equations of motion, something we cannot do for general relativity in closed form, unless we have symmetries present. This has suggested the possibility that the problem could perhaps be tackled in an approximate form. Progress has been recently made on this issue in the new variable context as well [22].

IV. THE NEW VARIABLES

The introduction of Ashtekar's new variables generated the new momentum that has invigorated the field in the last 18 years. For pedagogical reasons, the new variables are best introduced in a two stage process. The first stage is to use, instead of the metric of space as a fundamental variable, a set of triads $\tilde{E}^a_i$ (the tilde denotes a density weight, introduced for convenience, a is a spatial index and i labels the three frame fields). People had considered using the triad as a canonical variable. The description closest to the notation used in these days is given by Barbero [23]. Extra constraints arise since the theory is now invariant under frame field transformations (rotations) as well. The Hamiltonian is still a complicated non-polynomial function of the canonical variables. So the introduction of triads per se is not too helpful. The real breakthrough was the realization by Ashtekar that the Sen [24] connection could be used as a canonically conjugate momentum to the triad. Usually the canonically conjugate momentum considered for the triad is proportional to the extrinsic curvature. The Sen connection adds a piece given by the spin connection of the triad. Actually the sum of these two terms can be done while multiplying the extrinsic curvature times a constant (this constant is called the Immirzi parameter), yielding a one-parameter family of possible canonically conjugate variables (all members of the family are related by a canonical transformation). We call this the generalized Sen connection. If one rewrites the constraints in terms of the triads and the generalized Sen connection several things happen. The set of constraints introduced due to the symmetry of the theory under triad rotations now takes the form of "divergence of the triad equal zero". The divergence is taken with respect to the generalized Sen connection. If one writes the connection as A and the triads as E the resulting equation is exactly a Gauss law $D_a \tilde{E}^a_i = 0$, like the one present in $SO(3)$ Yang-Mills theory (the $SO(3)$ arises due to the symmetry under rotations of the triads) . The diffeomorphism constraint takes a form that resembles a Poynting vector $\tilde{E}^a_i F^i_{ab} = 0$. This is nice, since the latter is clearly associated with the momentum of the fields and fields without net linear momentum are the only ones invariant under diffeomorphisms. Finally the Hamiltonian constraint still is a complicated non-Polynomial function of the variables. However, if one chooses the Immirzi parameter equal to the imaginary unit (this is if one considers a Lorentzian signature space-time, for an Euclidean signature the Immirzi parameter should be chosen equal to one), the non-polymialities cancel out. One is left with a Hamiltonian constraint that takes a simple, polynomial (in fact at most quadratic) form in terms of the canonical variables $\tilde{E}^a_i \tilde{E}^b_j F^i_{ab} = 0$.

Another appealing aspect is that written in terms of these new variables, general relativity appears as a Yang-Mills theory with a set of extra constraints (and with a different Hamiltonian). This opened the possibility of introducing in general relativity tools that were used in Yang-Mills theory for its quantization. One of these tools is the use of loops.

V. LOOPS AND THE LOOP REPRESENTATION

We could now attempt a canonical quantization of the theory we just discussed. One could pick wavefunctions that are functionals of the connection $\psi[A]$ just like in Yang-Mills theory and promote the connection and the triad to canonically conjugate operators. Notice that this is already quite a departure from the traditional quantization where one took functionals of the metric. One now has to impose the constraints as operator equations. The Gauss law as an operator just demands that the wavefunctions be $SO(3)$ invariant (gauge invariant in the Yang-Mills language).

An interesting set of gauge invariant functionals of the connection is given by considering the trace of the holonomy of the connection along a loop.

$$W_\gamma(A) = \mathrm{Tr}\left(P \exp \oint dy^a A_a\right). \tag{1}$$

These quantities are called "Wilson loops". A very attractive feature is that these quantities constitute a basis for all gauge invariant functions [25]. That is, given any gauge invariant function of a connection it can in principle be expressed as a linear combination of Wilson loops based on different loops. The coefficients of this expansion therefore contain all the gauge invariant information of the wavefunction. Therefore whenever we think of a gauge invariant functional of a connection $\psi[A]$ one can alternatively think of the coefficients $\psi(\gamma)$ of its expansion in the Wilson loop basis (γ is a loop). Representing the functions in this way is what is known as the "loop representation". In this representation wavefunctions are functions of loops and operators are geometric operators that act on loops. Such representation was first proposed for Yang–Mills theory by Gambini and Trias [26] and for general relativity in terms of the new variables by Rovelli and Smolin [27].

A caveat is that the basis provided by Wilson loops is really an overcomplete basis. The coefficients in the expansion $\psi(\gamma)$ are therefore constrained by certain relations, known as the Mandelstam identities [28]. For a function of a loop to be admissible as a wavefunction in the loop representation, it should satisfy these identities. This can be challenging to achieve. As we will see a solution to this problem was eventually found in terms of spin networks.

Since one is automatically considering gauge invariant functions only when one works in the loop representation, Gauss law identically vanishes. We are therefore left with the diffeomorphism and Hamiltonian constraints only. The beauty of the loop representation lies in the natural action the diffeomorphism constraint acquires in this representation. The diffeomorphism constraint acts on wavefunctions by shifting infinitesimally the loop. Therefore it is immediate to solve the diffeomorphism constraint. One simply has to consider wavefunctions of loops such that they are invariant under deformations of the loops. Such functions are studied by the branch of mathematics called knot theory and are known as knot invariants. We therefore see that we can solve the diffeomorphism constraint in terms of a set of functions on which there is a lot of mathematical knowledge.

A further surprise that the loop representation yielded was that it appeared to also help in solving the Hamiltonian constraint [29]. In retrospect, this result appears as of quite limited importance, but it provided a quite significant boost of interest in the subject at the time it was found, so we will review it here. Let us go back briefly to the connection representation. Suppose we want to promote the Hamiltonian constraint to a quantum operator. We choose a factor ordering such that the triads are to the right. One needs to regularize the operator. Let us choose a simple minded point splitting, putting the two triads and the curvature at slightly different points. The triads operating on the Wilson loop produce a result that is proportional to the tangent to the loop at the point where they act (one can see this simply by noting that it is the only vector present at that point). That means that the two functional derivatives, viewed as a tensor, produce as a result a symmetric tensor, since the result is proportional to a vector times itself. This tensor is contracted with F_{ab}, which is antisymmetric. Therefore the result vanishes. Notice that this result does not depend on the details of the regularization. This result was first noticed by Jacobson and Smolin [29]. One caveat is that for it to be true, the loops in question have to be smooth. If a loop is not smooth, it contains points where there is more than one tangent and the constraint is not automatically zero, since the two functional derivatives could be proportional to different vectors and do not yield a symmetric tensor anymore.

If we now go back to the loop representation, the previous result suggests that if we consider knot invariants $\psi(\gamma)$ that have support on smooth loops only, one would appear to solve all the constraints of quantum gravity. This result by Rovelli and Smolin [27] generated a lot of excitement. But there are problems in attempting to solve the constraints in such a way. First of all, generic knot invariants with support on smooth loops only fail to satisfy the Mandelstam identities. This can be fixed by using spin networks as we will see later. But more importantly, smooth loops appear to be too simple to carry interesting physics. As we will see, operators like the volume of space vanish identically unless one has intersections. Therefore this space of solutions of the constraints is very likely a "degenerate" subspace that is not large enough to do meaningful physics. It was however, of great historical importance.

Another early result of interest in the loop representation was based on an observation due to Kodama [30]. The observation is that if one considers the exponential of the integral over space of the Chern–Simons form built from the Sen connection $\exp(kS_{CS}) = \exp\left(k\left[\text{Tr}\int A \wedge \partial A + 2/3 A \wedge A \wedge A\right]\right)$, it automatically solves the Hamiltonian constraint in the connection representation if one introduces a cosmological constant. The cosmological constant produces an extra terms in the Hamiltonian of the form $\Lambda/6 \tilde{E}_i^a \tilde{E}_j^b \tilde{E}_k^c \epsilon^{ijk} \epsilon_{abc}$. To see that the Kodama state solves the constraint we only need to know that when one acts with $\delta/\delta A_a^i$ on it, one gets something proportional to $\epsilon^{abc} F_{bc}^i$; this one can see through a calculation. To put it in simpler terms, for this state "$\tilde{E} \sim \tilde{B}$," that is the triad is proportional to the magnetic field built from the Sen connection. Therefore the two terms in the Hamiltonian constraint, which can be schematically written as "$\hat{E}\hat{E}\hat{B} + \Lambda/6 \hat{E}\hat{E}\hat{E}$" can be made to cancel each other by a choice of the constant k in the Kodama state.

The Kodama state has been found to have connection, in the cosmological context, with the Hartle–Hawking and Vilenkin vacua [30]. Of interest as well is its expression in the loop representation. To transform this state to the loop representation we wish to find its expansion in the basis of Wilson loops, that is, the coefficients,

$$\psi(\gamma) = \int DA \exp(kS_{CS}) W_\gamma(A) \qquad (2)$$

This integral has been extensively studied in a different context, that of Chern–Simons theory [31]. That is, consider a theory whose action is S_{CS}. Then the integral can be viewed as the computation of the expectation value of the Wilson loop in a Chern–Simons theory. The result is a function of a loop that is diffeomorphism invariant, that is, it is a knot invariant. This invariant is the Kauffman bracket, which is closely related to the Jones polynomial, a knot invariant of great interest in the mathematical literature. Since this invariant is the transform of a state that is annihilated by the Hamiltonian constraint in the connection representation, it should be annihilated by the Hamiltonian constraint in the loop representation. This has been checked explicitly [32]. It is remarkable that this invariant from the mathematical literature, which was developed in completely independent fashion from any idea related to the gravitational field, manages to solve the quantum Einstein equations.

VI. FORMAL DEVELOPMENTS

As we discussed in the previous section, the early years after the introduction of the new variables and the loop representation were ripe with intriguing and promising results, that appeared to be a significant step forward in the construction of a canonical theory of quantum gravity. However, many of the results were of formal nature only. There was not a good control on the space of states that one was operating on. Here we will mention some formal developments that took place in the mid 90's that helped gain better control on the calculations being performed.

A. Spin networks

As we discussed before, Wilson loops are an overcomplete basis for gauge invariant functions. They are constrained by a set of nonlinear identities called Mandelstam identities. The simplest such identity comes from the following identity of $SU(2)$ matrices in the fundamental representation $(\text{Tr}A)(\text{Tr}B) = \text{Tr}AB + \text{Tr}AB^{-1}$. In terms of Wilson loops this would read $W_\gamma W_\eta = W_{\gamma \circ \eta} + W_{\gamma \circ \eta^{-1}}$ with $\circ$ denoting loop composition. Now, it is clear that these identities stem from the fact that we are working in the fundamental representation of $SU(2)$ to construct the Wilson loops. One does not need to do so. Consider a diagram like the one in the figure.

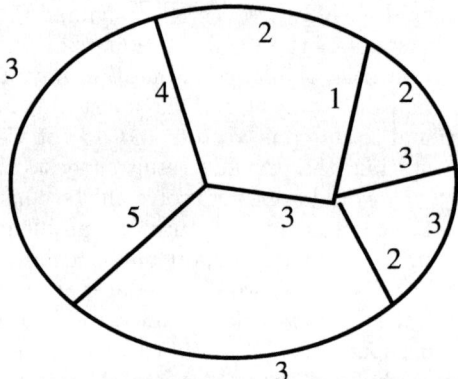

FIG. 1: A spin network.

One could construct holonomies along all the links in the diagram, in principle, with different representations of $SU(2)$ on each link (the representations are labeled by an integer). At each of the vertices one would use invariant tensors in the group to "tie up" the holonomies in such a way as to have at the end of the process a gauge invariant quantity. Such a quantity is a natural generalization of the Wilson loop. The diagrams like the one in the figure are called spin networks. Since the invariants so constructed do not depend on any particular representation, there are no relations between them as when we were working in the fundamental representation only. Therefore the Mandelstam identities are automatically solved. This was first realized by Rovelli and Smolin [34]. Spin networks also allow to do calculations in a natural and simple way, as we will see in the following sections.

B. Measures of integration

Calculations like the ones needed to transform states to the loop representation require a measure of integration in the space of connections modulo gauge transformations. Such integrations are also of interest to construct inner products. These are really functional integrals in spaces of infinite dimensionality. There is little experience on how to construct such integrals. Ashtekar and Lewandowski [33] among others have pioneered the construction of measures of integration in such spaces. They begin by choosing a given set of functions that will be integrable. The functions chosen are "cylindrical functions". These are functionals on the infinite dimensional space that really only depend on certain "directions" or "projections" of the space against a set of Schwarz test functions. The projection is achieved through the use of Wilson loops, or even more easily, spin networks. It might appear that cylindrical functions are too simple to be able to capture interesting physics. But for the case of a scalar field, for instance, the Fock measure can be constructed with cylindrical functions. In the case of spin networks the resulting measure of integration is really simple, it just states that spin networks are actually an orthogonal basis, ie, $<s_1|s_2> = \delta_{s_1,s_2}$ where the delta on the right hand side is one if the spin networks are equal or zero otherwise. The measures constructed are naturally diffeomorphism invariant. If one considers the class of spin networks related by diffeomorphisms with s_1 and the class related with s_2 one can construct an inner product on the classes simply by demanding that their inner product be zero if no member of the s_1 class coincides with at least one member of the s_2 class. The subject of measures of integration has several mathematical subtleties. Physicists can get a very readable succinct account in the paper by Ashtekar, Marolf and Mourao [35].

C. Areas and volumes

Most quantities of physical interest will involve products of the fundamental fields and therefore will require regularization. The latter is a non-trivial subject in quantum gravity. Most regularization procedures used in particle physics require the use of metric information. In particle physics the metric is a c-number. But this is not the case in gravity. If one insists in using regularization procedures that involve the metric, one should consider it an operator. This can complicate quite a bit the task of regularizing expressions. Alternatively, one can introduce an external c-number metric into the formalism, and hope that after one regularizes, there is no trace left of this artificial element in the construction. It is of interest to notice that some operators of (limited) physical interest can indeed be computed that are well defined in spite of the use of regularizations. These operators represent the area of a surface and the volume of a region of space. At first it appears that these operators will be difficult to regularize. The classical expression for the area of a surface Σ is $A(\Sigma) = \int \sqrt{\tilde{E}_i^a \tilde{E}_i^b n_a n_b} d^2x$ where n_a is the normal to the surface. The presence of the square root might at first suggest that regularization will be problematic. However, partitioning the area in small elements of area one can quickly see that for the quantity inside the square root spin network states are actually eigenstates. The end result for main portion of the spectrum of the area operator is

$$\hat{A}(\Sigma)|s> = \sum_L \sqrt{J_L(J_L+1)} \ell_P^2 |s> \quad (3)$$

where the sum is over all links of the spin network that pierce the area and J_L is the valence of the link L. We see that areas are quantized in terms of the Planck length squared ℓ_P^2. This was first noticed by Rovelli and Smolin [36]. Later Ashtekar and Lewandowski [37] did a comprehensive analysis of the spectrum of the area operator. The quantization of the area reveals another surprise. Most people expected areas to be quantized, but the expectation was that the spectrum would be equally spaced, i.e, $n\ell_P^2$. It is not. This has consequences. Bekenstein and Mukhanov [38] have shown that assuming that the spectrum of the area is equally spaced has serious implications for the validity of the thermal spectra of black holes. Rovelli and collaborators [39] have shown that these problems can be solved by considering the correct spectrum.

For the volume operator results are quite similar. The volume operator acting on a spin network gives a nonzero result if within the region considered there are intersections of valence equal or larger than four. One gets a picture of spin networks in which each links carries "quanta of area" and each intersection of valence four or larger carries "quanta of volume". Both operators are finite and well defined without reference to any background metric structure in spite of the fact that they had to be regularized. Several calculations of the spectrum of the volume can be found in [40].

VII. PHYSICAL PREDICTIONS: GAMMA RAY BURSTS

It is clear that one cannot really discuss any physics emerging from quantum gravity until one has dealt with the Hamiltonian constraint. Attempting to do so would be equivalent to trying to do physics after handling two of the three components of Gauss' law in $SU(2)$ Yang–Mills theories. One can attempt to do some calculations "at the kinematical level" (i.e. ignoring the constraints) in the hope that some of the basic features of the calculations will persist when the constraints are enforced, but this is not guaranteed. It is important to preface the discussion of this section with these caveats since they very much apply to what we will discuss.

It took everyone by surprise when Amelino-Camelia and collaborators [41] at CERN argued that in the detection of gamma ray bursts one could find traces of quantum gravity phenomenology. For years it had been common lore that quantum gravity required energies so high that it could only have relevant effects in the big bang or inside black holes. The possibility of detecting quantum gravity effects via gamma ray bursts goes as follows: the gamma rays that arrive on Earth have travelled a very long distance, since gamma ray bursts are expected to be cosmological. That distance appears even larger if one measures in terms of the number of wavelengths of a gamma ray. If one assumes that when a wave propagates through the "quantum foam" each wavelength gets disturbed by an amount of the order of the Planck length, then the smallness of this number can be compensated by the huge number of wavelengths involved in traveling from the burst to Earth. If one inputs numbers it turns out that detectable dispersions (differences in times of arrival) of 0.01 seconds in gamma rays that differ by $300 keV$ as those detected by the BATSE experiment imply that quantum gravity has to happen at energies larger than $10^{16} GeV$ in order not to be visible. This is only three orders of magnitude away from the Planck scale! This led to a lot of interest in these observations.

Within loop quantum gravity some calculations have been performed to attempt to estimate these effects, at a kinematical level [42]. The calculations require a number of simplificatory assumptions. Otherwise one would have to deal with Einstein-Maxwell theory and work out a semiclassical limit. In general the assumptions have been that the electromagnetic field has been treated classically and only the Maxwell part of the Hamiltonian constraint is considered. One finds modified Maxwell equations that imply that there is birefringence in the propagation of waves. Similar results can be found for neutrinos [43]. The birefringence for photons has been severely constrained experimentally [44]. A much more careful recalculation of the effects done recently confirms several of the general features of the original calculations [11].

The main problem with these predictions is that in order to have a non-vanishing effect at the lower order in terms of the energy of the gamma rays, one needs to introduce rather unnatural assumptions in terms of the quantum state considered (otherwise one could not generate a birefringence, which is tantamount to a parity violation). If one does not make these assumptions, then the effects only arise at the next order in E/E_{Planck} and are completely undetectable. In terms of the original work of Amelino Camelia et al. they postulated a non-standard dispersion relation of the form

$$cp^2 = E^2 \left(1 + \alpha \frac{E}{E_{Planck}}\right) + O\left(\frac{E^2}{E_{Planck}^2}\right) \tag{4}$$

and the effects would be observable if α is non-vanishing. A non-vanishing α implies a fractional power in a dispersion relation, which is unusual.

More importantly, all these calculations are implying that one is violating Lorentz invariance. This is a huge step to take. There is significant discussion of the implications in the current literature (see [45] and references therein).

VIII. THIEMANN'S HAMILTONIAN CONSTRAINT

One of the initial encouragements that the new variables introduced was that the Hamiltonian constraint appears as a polynomial function of the fundamental variables. This suggested that one could perhaps promote it to a quantum operator and several attempts to regularize it were carried out. However, there is an obvious fundamental flaw in attempting this. The Hamiltonian constraint is quadratic in the triads. The triads are densities of weight one, meaning that the constraint is a density of weight two. More precisely, the version of the constraint that is nice and polynomial is a density of weight two. One could turn it into a density of weight one by dividing by the determinant of the metric, but then the resulting operator would be complicated and non-polynomial. Why is it a problem that it is a density of weight two? Suppose we wished to promote it to an operator in the loop or spin network representation. What could such an operator be? We have at our disposal a manifold, and a set of lines in it. We have available a density of weight one, the Dirac delta, which is naturally defined on any manifold. But we do not have a density of weight two. And we cannot multiply Dirac deltas. Therefore if one found a regularization of the doubly densitized Hamiltonian constraint, what has to be happening is that one provided the extra density weight via the regulator. And therefore the imprint of the regulator will not disappear upon regularization.

All these difficulties were bridged when Thiemann [46] discovered how to handle the single-densitized Hamiltonian constraint. The expression for the constraint is,

$$\tilde{H} = \frac{\tilde{E}_i^a \tilde{E}_j^b F_{ab}^k \epsilon^{ijk}}{\sqrt{\tilde{E}_i^a \tilde{E}_j^b \tilde{E}_k^c \epsilon^{ijk} \epsilon_{abc}}}, \tag{5}$$

and Thiemann noticed that

$$\frac{\tilde{E}_i^a \tilde{E}_j^b \epsilon^{ijk}}{\sqrt{\tilde{E}_i^a \tilde{E}_j^b \tilde{E}_k^c \epsilon^{ijk}\epsilon_{abc}}} = 2\{A_a, V\} \tag{6}$$

where V is the volume of the three manifold. The Hamiltonian constraint can therefore be written as,

$$H(N) = \int d^3x N(x) \text{Tr}(F_{ab}\{A_c, V\})\epsilon^{abc}. \tag{7}$$

When we first discussed the Hamiltonian constraint with the new variables, we noted that it was important that we take the Immirzi parameter to be the imaginary unit. That made certain non-polynomial terms disappear, but at the price of making the variables complex. Thiemann noted that through a similar use of identities as the one we discussed, these non-polynomial terms could also be reexpressed in terms of Poisson brackets. Therefore there is no need anymore of taking the Immirzi parameter to be imaginary and from now on one can work with variables that are completely real.

Thiemann proposed a quantization for the above mentioned Hamiltonian constraint. The procedure consists in introducing a lattice. He chooses an irregular lattice (tetrahedral). In terms of this lattice, he approximates the expression for the classical Hamiltonian constraint using holonomies. Omitting many details, the idea is that the "F_{ab}" term is represented by a closed loop going around a triangle on one of the faces of the elementary tetrahedron and the "A_c" is represented by a line holonomy that is retraced to recover gauge invariance. The classical Hamiltonian constraint discretized on the lattice is therefore only a function of holonomies and the volume of the manifold. The attractive aspect of this is that both holonomies and the volume of the manifold can be represented by well defined finite operators in the spin network representation. Therefore producing a well defined, finite Hamiltonian constraint is tantamount to "putting hats on the classical expression" since all the ingredients can be naturally quantized without divergences!

There are a couple of caveats that need to be noted. The "F_{ab}" can be constructed by many different kinds of elementary loops. As long as they shrink to a point when the lattice is refined they all will represent properly the curvature. This indicates that there is therefore huge ambiguity in how to define the operator. An additional ambiguity is the valence of the holonomy that represents the curvature [47]. Moreover, a crucial element for the Hamiltonian to be well defined is that it act on diffeomorphism invariant states. On such states the details of how the holonomy that represents the curvature is placed with respect to the spin network are immaterial. This, in turn, ensures that the resulting quantum theory is consistent. If one acts with two Hamiltonian constraints, the two loops added are indistinguishable from each other and therefore if one acts in the opposite order the final result is the same. The Hamiltonian constraint therefore commutes with itself. Now, the classical Poisson algebra of constraints stated that the Poisson bracket of two Hamiltonians should be proportional to a diffeomorphism. If one promotes this to a quantum operatorial expression and acts on diffeomorphism invariant states, the right hand side will give zero since they are annihilated by diffeomorphisms. Therefore the commutator of two Hamiltonians should vanish as well [48].

Thiemann goes on to show that similar constructions can be carried out for general relativity coupled to fundamental matter fields: Yang–Mills, Higgs, Fermions [49]. This achievement is quite remarkable. We are in the presence of the first finite, well defined, anomaly-free, non-trivial theory of quantum gravity ever presented. The theory fulfills the promise of acting as a "natural regulator of matter fields" in the sense that no divergences are present when the theory is coupled to matter.

Is quantum gravity finally achieved? The answer is still not known. What has been found is a theory (more precisely infinitely many theories due to the ambiguities) that are well defined. Although this is no small feat in this context, We do not know if any of these theories contains the correct physics of gravity. This will only be confirmed or contradicted when a semiclassical approximation is worked out so we can make contact with more familiar results. Active investigations along these lines are being pursued by Thiemann and collaborators [11] and Ashtekar and collaborators [10].

There are some aspects of Thiemann's construction that appear somewhat troubling. The same construction can be worked out in $2+1$ dimensional gravity [50]. If one studies the solutions of the quantum Hamiltonian constraint one finds many more states than the ones allowed by Witten's theory. However, if one demands that they be normalized with the inner product we discussed, the Witten sector is all that is left. This can be seen as a positive result (after all, we get the correct theory) or as a negative one (the correct theory only is recovered after choosing carefully an inner product). It appears that in $3+1$ dimensions Thiemann's Hamiltonian also admits too many solutions. The fact that the constraint algebra can only be recovered on diffeomorphism invariant states, where it is only Abelian, is also troubling. Though again, there is no genuine interest in states that are not diffeomorphism invariant. Other worries were expressed in a paper by Smolin [51]. The general consensus at the moment is that there appear to be worries that the Hamiltonian does not capture the correct physics, but no one can make a theorem out of the worries to prove that Thiemann's Hamiltonian is wrong. The verdict will come when further explorations of the semiclassical approximation are worked out.

IX. BOJOWALD'S COSMOLOGIES

The idea of exploring quantum gravity effects in the simplified context of cosmological models has held appeal over the years. Yet, due to the lack of a theory of quantum gravity, the approach taken was rather bizarre. People would consider general relativity, then reduce the classical theory to only cosmological metrics (which in the case of homogeneous cosmologies reduces the equations to ordinary differential equations, losing the field theoretical nature of general relativity). The resulting theory was then quantized and some interpretations were attempted. The main criticism that was levied against this kind of investigations is that "imposing a symmetry then quantizing" does not have to agree with "a sector of the quantum theory with a given symmetry". That is, there is no guarantee that what one sees in quantum cosmology will appear at all when one gets a handle of the full theory and studies cosmological situations.

With Thiemann's introduction of viable theories for quantum gravity, it therefore became ubiquitous to attempt to study quantum cosmology "properly". That is, study the sector of the full quantum theory of gravity that approximates homogeneous cosmologies. This is what Bojowald [52] set out to do. He finds that in isotropic and homogeneous quantum cosmology states reduce to "spin networks with only one link". This is understandable since everything happens "at a single point" in a homogeneous model. The presence of the link is needed to make sense of the operators involving connections (one needs more than a point to have a notion of a connection!). The quantum states therefore are labeled by an integer $|n>$ which corresponds to the valence of the single link of the spin network. Bojowald constructs a version of Thiemann's Hamiltonian acting on these states. He also finds a well defined version of the volume operator.

One of the long held beliefs in quantum cosmology is that quantum effects will eliminate the big bang singularity. In Bojowald's case this is actually realized in practice. Considering the case of a flat Robertson–Walker metric $ds^2 = -dt^2 + a(t)^2 dx^2$, he finds that he can find a finite, well defined expression for the operator $1/a(t)$. This is done through similar identities as those that led to equation (6). Since the resulting operator is finite, it suggests that the singularity can be avoided. Remarkably, if one analyzes the relationship of $a(t)$ with the volume operator (which should be $1/a(t) = V^{-1/3}$ such relationship does hold quantum mechanically when the universe is "large". But when the universe becomes of the size of a few Planck volumes, the relationship is broken, the volume goes to zero but $1/a$ remains finite [53]. The avoidance of the singularity can be implemented concretely in this approach through discrete equations of motion that actually never become singular. And the theory can be coupled to various matter sources without introducing singular behaviors through the use of the defined $1/a(t)$ to implement the couplings.

Bojowald goes on to introduce a notion of time for these cosmologies. Since everything is discrete, his notion of time is discrete too. The evolution equations are recursion relations and he shows that for large universes they reproduce the results of usual Wheeler-DeWitt-based quantum cosmology [54]. And these evolution equations also allow to evolve non-singularly through the point where one expects the singularity classically. Remarkably, even an argument for inflation being generated by quantum gravity can be found in this context [55].

The fact that the cosmological reduction of Thiemann's Hamiltonian appears to give the correct physics of quantum cosmology is considered by some as an indication that the right physics of gravity is contained. One should be aware of the fact, however, that Bojowald's construction implies a limiting procedure. Quantum states peaked on homogeneous cosmologies are really distributions and therefore one needs to extend the operators defined for other states to them. This extension is non-trivial and there might be ambiguities in it that allow to "correct" things in order to get the right physics. Although this is not what Bojowald set out to do, it might have accidentally happened. Moreover, part of the worries about Thiemann's Hamiltonian have to do with the constraint algebra. In homogeneous cosmologies, since everything takes place "at a point" there is only one Hamiltonian constraint that therefore is obviously Abelian, which agrees with Thiemann's general result.

Nevertheless, it is striking that detailed attractive predictions in the cosmological context can be extracted from the proposed Hamiltonian constraint.

X. CONSISTENT DISCRETIZATIONS: A NEW FRAMEWORK?

When we discussed Thiemann's Hamiltonian constraint we mentioned that he started from a given classical theory in the continuum and introduced a lattice to discretize the Hamiltonian constraint. The lattice Hamiltonian is then promoted to a quantum operator naturally. The idea of using lattices to regularize gravity is not new. The novelty is the use of the recently acquired knowledge about well defined operators and states. Lattice approaches, however, are plagued by difficulties to which Thiemann's approach may not be immune. The difficulty has to do with the fact that in the case of general relativity lattice regularizations breaks the symmetry of the theory under diffeomorphisms (on the contrary, in the Yang–Mills case, lattice gauge theory has the advantage of providing a gauge invariant regularization). The theories one gets on the lattice therefore have a considerably different structure than the theory they attempt to approximate in the continuum. It is the personal impression of the author that at the time of quantizing the discrete theories, one needs to take their structure seriously.

In particular, most discrete approximations that one constructs for general relativity, end up being inconsistent (their equations do not admit a single solution). This is well known, for instance, in numerical relativity. When one wishes to integrate the Einstein equations on a computer, they are approximated by finite difference equations. Whereas in the continuum theory if one solves the constraint equations initially, the evolution equations guarantee the constraints will hold at all time, this is not true of the discrete equations. There is therefore no way to satisfy simultaneously the constraint equations (at all times) and the evolution equations. Most people in numerical relativity use "free evolution" that is, they accept that they will fail to satisfy the constraints at later times as part of the numerical error of the solution they incur.

In quantization the last argument does not work. If a theory is inconsistent, there is little sense in attempting a quantization. Most attempts to lattice quantum gravity suffer from this problem. For instance, when one discretizes the Hamiltonian constraint, the discrete constraints fail to close an algebra (this is reasonable, since algebras are associated with infinitesimal symmetries and nothing can be infinitesimal in a discrete theory). If the constraints do not close an algebra one can generate further constraints by taking Poisson brackets. If one is not careful one ends up with too many constraints and the theory has no solutions. This is the canonical manifestation of the inconsistency.

These kind of problems are very basic and are even present in very simple systems. For instance, if one considers a Newtonian particle and discretizes Newton's equations, it is a well known fact that energy fails to be conserved. In fact astronomers who wish to follow planetary motion have known this for a long time and construct special discretizations of Newton's laws that automatically conserve energy and angular momentum. Our goal is to find something similar in the gravitational context, that is, a discretization scheme that automatically preserves the constraints.

T.D. Lee [56] proposed a way to fix the problems of the Newtonian particle that can be easily translated to the gravitational case. The idea consists in enforcing the constraints through a suitable choice of Lagrange multipliers. In the case of general relativity, one chooses the lapse and shift in such a way that in the next step the constraints are satisfied. This has no counterpart in the continuum theory. One has four constraints to enforce and four quantities to solve for (the equations to be solved are coupled algebraic non-linear equations).

We have recently worked out [57] the canonical theory for such consistent discretizations and applied it to Yang–Mills and BF theory and presented the prescription for the gravitational case. In this approach, since the constraints are automatically satisfied, most of the conceptually hard problems that arose due to attempting to impose the constraints disappear. Quantum gravity become a conceptually clear yet computationally challenging problem. There are many attractive features in this approach. Since the initial data fixes the lapse and the shift, and quantum mechanically one generically has a superposition of initial data, one automatically has a superposition of discretizations. The goal of "averaging out" over all discretizations is implemented naturally. We have applied these ideas to a cosmological model. Classically one finds that if one runs the model backwards, unless one fine tunes the initial data, no singularity is present. This is natural, generically the singular point will not fall on the lattice. When one quantizes however, the fact that the singularity classically only occurs for a set of measure zero in the possible initial data implies that the singularity is not present. We see a remarkable agreement with Bojowald's prediction, although the details and motivation are different.

Much more will have to be explored to see if the consistent discretizations are a viable route for quantization. In particular we have little experience with the complicated non-linear equations that fix the lapse and the shift. What if they quickly generate negative lapses or singularities? It is imperative that experience be gained, particularly in midi-superspace examples where there are field theoretic degrees of freedom. Even these cases are computationally challenging given the complexity of the algebraic non-linear coupled equations that determine the lapse and the shift.

XI. SUMMARY

The last 18 years have seen a renaissance of canonical quantum gravity. The field has been brought to a complete new level in terms of mathematical sophistication and possibilities for discussing physical consequences and applications. It is becoming more evident by the day that not only it is not clear that general relativity has a problem at the time of its quantization, but that actually the quantization of Einstein's theory has a lot to teach us about physics. This physics might be of interest just in the context of pure general relativity or in the context of proposed unified theories of all interactions, like string theory.

XII. ACKNOWLEDGMENTS

I am grateful to A. Pérez, M. Bojowald, J. Lewandowski, T. Thiemann for comments on the manuscript. I also wish to thank the organizers of the Brazilian school for the invitation to participate. This paper was completed at the Erwin Schrödinger Institute in Vienna. My work in this field has for many years been in collaboration with Rodolfo Gambini. This work was supported by grants NSF-PHY0090091, funds from the Horace Hearne Jr. Institute for Theoretical Physics and the Fulbright Commission in Montevideo.

---

[1] C. Rovelli, "Notes for a brief history of quantum gravity," arXiv:gr-qc/0006061.
[2] See D. Birmingham, M. Blau, M. Rakowski and G. Thompson, Phys. Rept. **209**, 129 (1991) for a review.
[3] C. Rovelli, "Strings, loops and others: A critical survey of the present approaches to quantum gravity, arXiv:gr-qc/9803024.
[4] J. de Lyra et al., Phys. Rev. D **46** (1992) 2538.
[5] E. Witten, Nucl. Phys. B **311**, 46 (1988).
[6] S. Deser, J. G. McCarthy and Z. Yang, Phys. Lett. B **222**, 61 (1989).
[7] A. Perez, Class. Quan. Grav. (to appear).
[8] A. Ashtekar, J. C. Baez and K. Krasnov, Adv. Theor. Math. Phys. **4**, 1 (2000) [arXiv:gr-qc/0005126].
[9] M. Varadarajan, Phys. Rev. D **64**, 104003 (2001);**66**, 024017 (2002) [arXiv:gr-qc/0104051,0204067];
[10] A. Ashtekar and J. Lewandowski, Class. Quant. Grav. **18**, L117 (2001); A. Ashtekar, S. Fairhurst and J. L. Willis, "Quantum gravity, shadow states, and quantum mechanics," arXiv:gr-qc/0207106.
[11] T. Thiemann, "Complexifier coherent states for quantum general relativity," arXiv:gr-qc/0206037; H. Sahlmann and T. Thiemann, "Towards the QFT on curved spacetime limit of QGR. I: A general scheme," arXiv:gr-qc/0207030; "II: A concrete implementation," arXiv:gr-qc/0207031; H. Sahlmann, T. Thiemann and O. Winkler, Nucl. Phys. B **606**, 401 (2001) [arXiv:gr-qc/0102038] and references therein.
[12] R. Loll, Living Rev. Rel. **1**, 13 (1998) [arXiv:gr-qc/9805049].
[13] C. Rovelli, Living Rev. Rel. **1**, 1 (1998) [arXiv:gr-qc/9710008].
[14] T. Thiemann, arXiv:gr-qc/0110034.
[15] S. Carlip, Rept. Prog. Phys. **64**, 885 (2001) [arXiv:gr-qc/0108040].
[16] L. Smolin, "Three roads to quantum gravity", London, UK: Weidenfeld & Nicolson (2000).
[17] A. Ashtekar (Notes prepared in collaboration with R. Tate), "Lectures on non-perturbative canonical gravity", Advanced Series in Astrophysics and Cosmology Vol. 6, World Scientific, Singapore (1991); "New perspectives in canonical quantum gravity", Bibliopolis (Naples, Italy) (1988).
[18] R. Gambini, J. Pullin, "Loops, knots, gauge theories and quantum gravity", Cambridge University Press (1996).
[19] J. Baez and J. P. Muniain, *Singapore, Singapore: World Scientific (1994) 465 p. (Series on knots and everything, 4)*.
[20] K. Kuchař, "Time and interpretations in quantum gravity", in " Proceedings of the 4th Canadian Conference on General Relativity and Relativistic Astrophysics: University of Winnipeg, 16-18 May, 1991", G. Kunstatter, D.E. Vincent, J.G. Williams (Editors), World Scientific, Singapore (1992).
[21] C. Rovelli, Phys. Rev. D **65**, 044017 (2002) [arXiv:gr-qc/0110003].
[22] R. Gambini and J. Pullin, Phys. Rev. Lett. **85**, 5272 (2000) [arXiv:gr-qc/0008031]; Class. Quant. Grav. **17**, 4515 (2000) [arXiv:gr-qc/0008032].
[23] J. F. Barbero, Phys. Rev. D **51**, 5507 (1995) [arXiv:gr-qc/9410014].
[24] A. Sen, Phys. Lett. B **119** (1982) 89.
[25] R. Giles, Phys. Rev. **D24**, 2160 (1981).
[26] R. Gambini, A. Trias, Nucl. Phys. **B278**, 436 (1986).
[27] C. Rovelli, L. Smolin, Phys. Rev. Lett. **61**, 1155 (1988); Nucl. Phys. **B331**, 80 (1990).
[28] S. Mandelstam, Phys. Rev. **D19**, 2391 (1979).
[29] T. Jacobson, L. Smolin, Nucl. Phys. **B299**, 295 (1988).
[30] H. Kodama, Phys. Rev. **D42**, 2548 (1990).
[31] E. Witten, Commun. Math. Phys **121**, 351 (1989).
[32] B. Brügmann, R. Gambini, J. Pullin, Phys. Rev. Lett. **68**, 431 (1992).
[33] A. Ashtekar and J. Lewandowski, J. Math. Phys. **36**, 2170 (1995) [arXiv:gr-qc/9411046]; also in "Quantum Gravity and Knots", ed. by J. Baez, Oxford Univ. Press (1993).
[34] C. Rovelli and L. Smolin, Phys. Rev. D **52**, 5743 (1995) [arXiv:gr-qc/9505006].
[35] A. Ashtekar, D. Marolf and J. Mourao, "Integration On The Space Of Connections Modulo Gauge Transformations," arXiv:gr-qc/9403042.
[36] C. Rovelli and L. Smolin, Nucl. Phys. B **442**, 593 (1995) [Erratum-ibid. B **456**, 753 (1995)] [arXiv:gr-qc/9411005].
[37] A. Ashtekar and J. Lewandowski, Class. Quant. Grav. **14**, A55 (1997) [arXiv:gr-qc/9602046]; Adv. Theor. Math. Phys. **1**, 388 (1998) [arXiv:gr-qc/9711031].
[38] J. D. Bekenstein and V. F. Mukhanov, Phys. Lett. B **360**, 7 (1995) [arXiv:gr-qc/9505012].
[39] M. Barreira, M. Carfora and C. Rovelli, Gen. Rel. Grav. **28**, 1293 (1996) [arXiv:gr-qc/9603064].

[40] Adv. Theor. Math. Phys. **1**, 388 (1998) [arXiv:gr-qc/9711031]; T. Thiemann, J. Math. Phys. **39**, 3347 (1998) [arXiv:gr-qc/9606091]; R. Loll, Nucl. Phys. B **460**, 143 (1996) [arXiv:gr-qc/9511030]; R. Loll, Phys. Rev. Lett. **75**, 3048 (1995) [arXiv:gr-qc/9506014]; J. Lewandowski, Class. Quant. Grav. **14**, 71 (1997) [arXiv:gr-qc/9602035].

[41] G. Amelino-Camelia, J. R. Ellis, N. E. Mavromatos, D. V. Nanopoulos and S. Sarkar, Nature **393**, 763 (1998) [arXiv:astro-ph/9712103].

[42] R. Gambini and J. Pullin, Phys. Rev. D **59**, 124021 (1999) [arXiv:gr-qc/9809038]; J. Alfaro, H. A. Morales-Tecotl and L. F. Urrutia, Phys. Rev. D **65**, 103509 (2002) [arXiv:hep-th/0108061].

[43] J. Alfaro, H. A. Morales-Tecotl and L. F. Urrutia, Phys. Rev. Lett. **84**, 2318 (2000) [arXiv:gr-qc/9909079].

[44] R. J. Gleiser and C. N. Kozameh, Phys. Rev. D **64**, 083007 (2001) [arXiv:gr-qc/0102093].

[45] J. Magueijo and L. Smolin, arXiv:gr-qc/0207085.

[46] T. Thiemann, Class. Quant. Grav. **15**, 839 (1998) [arXiv:gr-qc/9606089].

[47] M. Gaul and C. Rovelli, Class. Quant. Grav. **18**, 1593 (2001) [arXiv:gr-qc/0011106].

[48] T. Thiemann, Class. Quant. Grav. **15**, 1207 (1998) [arXiv:gr-qc/9705017].

[49] T. Thiemann, Class. Quant. Grav. **15**, 1281 (1998) [arXiv:gr-qc/9705019].

[50] T. Thiemann, Class. Quant. Grav. **15**, 1249 (1998) [arXiv:gr-qc/9705018].

[51] L. Smolin, "The classical limit and the form of the Hamiltonian constraint in non-perturbative quantum general relativity," arXiv:gr-qc/9609034.

[52] M. Bojowald, Class. Quant. Grav. **17**, 1489 (2000) [arXiv:gr-qc/9910103]; Class. Quant. Grav. **17**, 1509 (2000) [arXiv:gr-qc/9910104]; Class. Quant. Grav. **18**, 1055 (2001) [arXiv:gr-qc/0008052];

[53] M. Bojowald, Phys. Rev. Lett. **86**, 5227 (2001) [arXiv:gr-qc/0102069].

[54] M. Bojowald, Class. Quant. Grav. **18**, L109 (2001) [arXiv:gr-qc/0105113].

[55] M. Bojowald, arXiv:gr-qc/0206054.

[56] T. D. Lee, in "How far are we from the gauge forces" Antonino Zichichi, ed. Plenum Press, (1985)

[57] R. Gambini and J. Pullin, arXiv:gr-qc/0206055; C. Di Bartolo, R. Gambini and J. Pullin, arXiv:gr-qc/0205123.

The Strong-Coupling Expansion and the Singularities of the Perturbative Expansion

N. F. Svaiter

Centro Brasileiro de Pesquisas Fisicas,
Rua Dr.Xavier Sigaud 150,
22290-180 Rio de Janeiro, RJ Brazil

We discuss the strong-coupling expansion in the $(\lambda\varphi^4)_d$ theory, quantum electrodynamics and also in different scalar models in a d-dimensional Euclidean space, and the singularities of this perturbative expansion assuming the ultra-local approximation. We analyse the analytic structure of the zero-dimensional generating functions in the complex coupling constant plane for the case of $(\lambda\varphi^4)_d$ and also quantum electrodynamics. We find a superposition between a branch point and essential singularity at $g_0 = 0$ in the scalar model and also in quantum electrodynamics. Further, still using the strong-coupling expansion as a formal representation for the generating functional of complete Schwinger functions given by $Z(h)$, we present an interacting field theory model that we call the Sinh-Gordon model ($V(g_1, g_2; \varphi) = g_1(\cosh(g_2\,\varphi) - 1)$) where the ultra-local generating functional given by $Q_0(h)$ coincides with the generating functional of complete Schwinger functions. In other words, the ultra-local approximation is exact, i.e. $Z(h) \equiv Q_0(h)$. As a consequence, it is possible to calculate the vacuum energy per unit-volume. From the distribution of the the zeros of the modified Bessel function of third kind it is possible to show that the model has a finite vacuum energy per unit-volume. Finally we sketch how it is possible to obtain a renormalized generating functional for all Schwinger functions, going beyond the ultra-local approximation.

I. INTRODUCTION

The purpose of this lecture is twofold. First we would like to review the strong-coupling expansion in theories which are infrared free in a four dimensional spacetime. Second we wish to study the analytic structure of the ultra-local generating functional in the complex coupling constant plane in $(\lambda\varphi^4)_d$ theory and also quantum electrodynamics. This work is a natural extension of the program developed by Klauder and others [1] [2] that have been studying the ultra-local generating functional in different scalar infrared free models.

In quantum field theory, the perturbative renormalization approach is an algoritm, where first we control all the ultraviolet divergences of a theory using a procedure to obtain well defined expressions for each Feynman diagram. Second, we have to implement a renormalization prescription where the divergent part of each Feynman diagram is canceled by a suitable counterterm. For complete reviews of this program see for example Ref. [3] or Ref. [4]. Concerning the first step, there are different ways to transform the Feynman diagrams in well defined finite quantities. The most simple way is to modify the field theory at short distances by introducing a sharp cut-off in momentum space or a more elaborated version as the Pauli-Villars regularization [5]. In this case we simply modify the propagator for large momentum. We can also replace the continuum Euclidean space by a hypercubic lattice, with lattice space a. This is called the lattice regularization method. Finally, a more convenient regularization procedure is the dimensional regularization [6] [7] [8] [9], which is particulary well suited to deal with gauge theories. To implement the renormalization procedure, using dimensional regularization, for example we use the fact that the ultraviolet divergences of Feynman diagrams appear as poles of some function defined in a complex plane. The next step in the perturbative renormalization is to cancel the principal part of the Laurent series of the analytic regularized expressions introducing counterterms in the theory. There are different ways to disregard the divergent part of each Feynman diagram. For example we can use the minimal subtraction scheme (MS) where the counterterms just cancel the principal part of the Laurent series of the analytic regularized expressions, or any different renormalization scheme. The arbitrariness of the method must be cured by the renormalization group equations [10] [11] [12]. In this framework, the field theory models are classified as perturbatively super-renormalizables, renormalizables and non-renormalizables. The fundamental difference between a perturbatively renormalizable and a non-renormalizable field theory model is the fact that in a non-renormalizable theory the usual renormalization procedure we use to remove the infinities that arise in the usual perturbative expansion introduces infinite new "empirical" parameters in the theory. This comes from the fact that we need to specify the finite part of each counterterm. Consequently, in the infinite cut-off limit, the predictibility or the physical consistence of non-renormalizable theories is missing. See for instance the discussion developed in Ref. [13]. Nevertheless, there are some models where it is possible to construct a physically sensible version of the theory. A well known example is the Gross-Neveu model, where the theory is not renormalizable in the usual sense, however in the $\frac{1}{N}$ expansion is renormalizable for $d < 4$ [14] [15].

In the case of perturbatively renormalizable theories, although the renormalization procedure can be implemented in a mathematical consistent way, it is still not clear how the renormalized perturbative series can be summed up in different models [16] [17] [18]. There are many results showing that the perturbative series that we obtain in different perturbatively renormalizable theories in $d=4$, does not converge for any value of the coupling constants of the interacting theories. Well known theories with such kind of problems are scalar models with a $\lambda\varphi^4$ self-interaction and also quantum electrodynamics. If someone try to perform a partial resummation of the perturbative series in both theories, the Landau poles appear [19] [20] [21]. To circumvent the problem of non-hermitian Hamiltonians in the infinite cut-off limit, in a four dimensional spacetime, the $(\lambda\varphi^4)_4$ model, the $O(N)_4$ model and also quantum electrodynamics must be trivial. The problem of the singularities in the three and four-point connected functions for quantum electrodynamics and $(\lambda\varphi^4)_4$ respectively, for positive values of the coupling constants is related to the fact that in these theories the high frequency fluctuations are more strongly coupled than the lower frequencies ones. In others words, both theories are infrared-free in a four dimensional spacetime. We found a completely distinct behavior in a theory where the high frequency fluctuations are more weakly coupled than the lower frequencies ones. In this situation, at least the zero charge problem does not appear.

Therefore, a new step in the development of quantum field theories was given by the construction of non-abelian gauge field theories and the discovery of the asymptotic freedom [22] [23] [24], after the construction of the renormalization group equations. From the renormalization group equations a important classification of different field theory models arises. The models are either asymptotically free or IR (infrared) stables. As we discussed, in the renormalization group approach, the triviality of $(\lambda\varphi^4)_4$ model, the $O(N)_4$ model and also quantum electrodynamics in a four dimensional spacetime is a reflection of the absence of a non-trivial ultraviolet stable fixed point in the Callan-Symanzik β-function. Moreover, the unification of the statistical mechanics and some models in quantum field theory was achieved in progressive steps. A long time ago Schwinger introduced the idea of Euclidean fields, where the classical action must be continued to Euclidean time. Symanzik [25] constructed the Euclidean functional integral where the vacuum persistence functional defined in Minkowski spacetime becomes a statistical mechanics average of classical fields weighted by a Boltzmann probability. At the same time, a deeper insight into our understanding of the renormalization procedure in different models in field theory was given by the study of the critical phenomena and the Wilson version of the renormalization group equations [26]. Further, Osterwalder and Schrader [27] proved that for scalar theories, the Euclidean Green's functions or the Schwinger functions, which are the moments of the Boltzman measure, are equivalent to the Minkowski Green's functions. A new step in our understanding of the limitation of the perturbative approach in quantum field theories was achieved by Aizenman [28] and also Frohlich [29]. These authors proved that for the case of $(\lambda\varphi^4)_d$ theory, using the lattice regularization with nearest neighborhood realization of the Laplacian leads to a trivial theory in the continuum limit for $d\geq 5$. With additional assumptions, for $d=4$ it is also possible to obtain the triviality of the theory.

It is important to point out that some authors claim that these results are odd, since for $d=4$ the renormalized perturbative series is non-trivial and for $d\geq 5$ the theory is perturbatively non-renormalizable. For example, Klauder sustains that the triviality of $(\lambda\varphi^4)_d$ for $d\geq 4$ is still an open problem [30]. Making use of the correspondence principle, this author has been emphasizing that the quantization of a non-trivial classical theory can not be a non-interacting quantum theory. Furthermore, he claims that an alternative regularization procedure can gives a non-trivial theory in the infinite cut-off limit. Some results going in this direction have been obtained by Gallavotti and Rivasseau [31]. These authors discussed the $(\lambda\varphi^4)_d$ theory with more general scalar regularized theories where the realization of the Laplacian is not restricted to the nearest neighbours and also the presence of antiferromagnetic couplings. They suggested that the ultraviolet limit of such lattice regularized field theories is not a Gaussian field theory, consequently it is possible to obtain a non-trivial ultraviolet limit. We shall emphasize that the ultraviolet behavior of the $(\lambda\varphi^4)_d$ model for $d\geq 4$ is an strong coupling problem, since in the weak coupling perturbative expansion the high orders terms of the perturbative series are dominant in the large cut-off limit, as has been discussed by many authors. See for instance the discussion in Ref. [32].

In the case of a strong-coupling regime of a theory, we can perform a "new" perturbative expansion by using two different approaches. The first is to define a new parameter in order to implement a "new" weak coupled perturbative expansion. For example, the $\frac{1}{N}$ expansion [33], where N is the number of the components of the field in some isotopic space (the dimension of the order parameter), is a realization of this approach. Of course we are still using the standard perturbative scheme, performing a perturbative expansion with respect to the anharmonic terms of the theory. The second approach is based on the following idea: in a formal representation for the generating functional of complete Schwinger functions of the theory we treat the off-diagonal terms of the Gaussian factor as a perturbation about the remaining terms of the functional integral. This approach has been called in the literature the strong-coupling expansion.

The purpose of this lecture is to discuss the strong-coupling expansion in some infrared stable models, and investigate the analytic structure of the generating functional of complete Schwinger functions in the complex coupling constant planes. We are studying the singularities of the strong-coupling perturbative expansion in different models investigating the analytic structure of the zero-dimensional generating function in the complex coupling constant planes. We present such study for the case of $(\lambda\varphi^4)_d$ and also quantum electrodynamics.

We would like to mention briefly that the study of the analytic structure of theories in the complex coupling constant plane has been used by many authors in quantum field theory. It is well known that the behavior of the standard perturbative series in powers of the coupling constant at large order is related to the analytic structure of the partition function in a neighborhoodhood of the origin in the complex coupling constant plane. For example, Bender and Wu [34] studied the anharmonic oscillator an pointed out that there is a relation between the nth Rayleigh-Schrodinger coefficients and the lifetime of the unstable states of a negative coupling constant anharmonic oscillator.

The organization of the lecture is as follows: In section II we discuss the standard weak and the strong-coupling expansions for the $(\lambda\varphi^4)_d$ model. Section III we perform the study of the zero-dimensional $\lambda\varphi^4$ model in the absence of sources. In section IV we discuss the strong-coupling expansion in quantum electrodynamics also in absence of fermionic sources and the analysis of the singularities of the zero-dimensional generating function in the complex coupling constant plane is performed. In section V we introduce sources to briefly analyse different scalar models. In section VI we sketch how it is possible to go beyond the ultra-local approximation. Finally, section VII contains our conclusions. Throughout this lecture we use $\hbar = c = 1$.

II. THE WEAK AND THE STRONG COUPLING PERTURBATIVE EXPANSION FOR SCALAR $(\lambda\varphi^4)_d$ MODEL

Let us consider a neutral scalar field with a $(\lambda\varphi^4)$ self-interaction, defined in a d-dimensional Minkowski spacetime. The vacuum persistence functional in Minkowski spacetime that we call the generating functional of all vacuum expectation value of time-ordered products of the theory has a Euclidean conterpart defining the generating functional of complete Schwinger functions. Actually, the $(\lambda\varphi^4)_d$ Euclidean theory is defined by these Euclidean Green's functions. The Euclidean generating functional $Z(h)$ is formally defined by the following functional integral:

$$Z(h) = \int [d\varphi] \, \exp\left(-S_0 - S_I + \int d^d x \, h(x)\varphi(x)\right), \quad (1)$$

where the familiar action that describes a free scalar field is given by

$$S_0(\varphi) = \int d^d x \left(\frac{1}{2}(\partial\varphi)^2 + \frac{1}{2}m_0^2\varphi^2\right), \quad (2)$$

and the interacting part, defined by the non-Gaussian contribution is given by the following term in the action:

$$S_I(\varphi) = \int d^d x \frac{g_0}{4!} \varphi^4(x). \quad (3)$$

In the Eq.(1) $[d\varphi]$ is the appropriate measure, formally given by $[d\varphi] = \prod_x d\varphi(x)$, g_0 and m_0^2 are respectivelly the bare coupling constant and mass squared. It is convenient to introduce an arbitrary parameter μ with mass dimension to define a dimensionless coupling constant $g_0 = \mu^{4-d}\lambda$. Finally, $h(x)$ is a smooth function that we introduce to generate the Schwinger functions of the theory by functional derivatives. The conventional procedure, that it is called the weak-coupling expansion is to perform a perturbative expansion with respect to the non-Gaussian terms of the action. As a consequence of this formal expansion all the n-point unrenormalized Schwinger functions are expressed in a powers series of the bare coupling constant g_0. Let us summarize how to perform the weak-coupling expansion in the $(\lambda\varphi^4)_d$ theory. Therefore let us first consider the Gaussian functional integral $Z_0(h)$ defined by

$$Z_0(h) = \mathcal{N} \int [d\varphi] \, \exp\left(-\frac{1}{2}\varphi K \varphi + h\varphi\right), \quad (4)$$

where we are using the compact notation of Zinn-Justin [35], and each of the terms in Eq.(4) are given respectively by

$$\varphi K \varphi = \int d^d x \int d^d y \, \varphi(x) K(m_0; x, y) \varphi(y), \qquad h\varphi = \int d^d x \, \varphi(x) h(x). \quad (5)$$

The symmetric kernel is defined by $K(m_0; x, y) = (-\Delta + m_0^2)\delta^d(x-y)$ and Δ denotes the Laplacian in R^d. As usual the normalization factor is defined using that $Z_0(h)|_{h=0} = 1$. Therefore we have $\mathcal{N} = \left[det(\frac{-\Delta+m_0^2}{2\pi})\right]^{\frac{1}{2}}$ but in the following we are absorbing this normalization factor in the functional measure. It is convenient to introduce the inverse kernel $G_0(m_0; x - y)$ by the following identity, where we are using translational invariance:

$$\int d^d z\, G_0(m_0; x - z) K(m_0; z - y) = \delta^d(x - y). \tag{6}$$

Since Eq.(4) involves Gaussian functional integrals, it is possible to show that $Z_0(h)$ can be expressed in terms of the inverse kernel $G_0(m_0; x - y)$ i.e., in terms of the free two-point Schwinger function. Simple manipulations performing only Gaussian integrals allow us to write the following identity:

$$\int [d\varphi]\, \exp\left(-S_0 + \int d^d x\, h(x)\varphi(x)\right) = \exp\left(\frac{1}{2} \int d^d x \int d^d y\, h(x) G_0(m_0; x - y) h(y)\right). \tag{7}$$

This construction is fundamental to define the weak-coupling perturbative expansion. Using Eq.(1), Eq.(2) and Eq.(7) we are able to write the generator functional of all unrenormalized Schwinger functions $Z(h)$ as

$$Z(h) = \exp\left(-\int d^d x\, \mathcal{L}_\mathcal{I}(\frac{\delta}{\delta h})\right) \exp\left(\frac{1}{2} \int d^d x \int d^d y\, h(x) G_0(m_0; x - y) h(y)\right), \tag{8}$$

where $\mathcal{L}_\mathcal{I}$ is defined by the non-Gaussian contribution to the action, and in the conventional perturbative expansion the generating functional of complete Schwinger functions is formally given by the following infinite series:

$$Z(h) = \sum_{n=0}^{\infty} \frac{(-1)^n}{n!} \left(\int d^d x\, \mathcal{L}_\mathcal{I}(\frac{\delta}{\delta h})\right)^n \exp\left(\frac{1}{2} \int d^d x \int d^d y\, h(x) G_0(m_0; x - y) h(y)\right). \tag{9}$$

To generate all the n-point Schwinger functions we have only to perform a suitable number of functional differentiations in $Z(h)$ with respect to the source $h(x)$ and set the source to zero in the end. Thus we have that the renormalized n-point Schwinger functions are defined by

$$G_n(x_1, x_2, .., x_n) = Z^{-1}(h = 0) \left[\frac{\delta}{\delta h(x_1)} \cdots \frac{\delta}{\delta h(x_n)} Z(h)\right]|_{h=0}. \tag{10}$$

This general method can be used to derive the weak-coupling perturbative expansion in different theories. Observe that it is possible to generalize this formalism including the product of composite sources [36] [37] [38], but in this paper we limit ourselves to models withouth composite operators. To generate only the connected diagrams $G_n^{(c)}(x_1, .., x_n)$ let us consider the "free energy", or the generating functional of the connected Green's functions, defined by the following object $F(h) = \ln Z(h)$. Thus the generating functional of connected Schwinger functions has the following functional Taylor expansion

$$F(h) = \sum_{n=1}^{\infty} \frac{1}{n!} \int \prod_{k=1}^{n} d^d x_k \prod_{k=1}^{n} h(x_k)\, G_n^{(c)}(x_1, .., x_n). \tag{11}$$

Note that in the perturbative expansion we are expanding in powers of the unrenormalized coupling constant. From the above discussions we see that we have a formal expansion of the generating functional of all Schwinger functions $Z(h)$ in powers of the coupling constant g_0 defined by $f(g) = \sum_{k=0}^{\infty} f_k\, g^k$, where $A_k(h)$ are unrenormalizable perturbative coefficients.

Let us briefly discuss the condition of renormalizability of scalar theories. As for example, for any $(\varphi^2)^\delta$ self-interaction, we have that the inequality $\delta \leq \frac{d}{d-2}$ is the condition of the renormalizability or not of the theory, that can be obtained by the Sobolev inequality [39]

$$\left[\int |\varphi(x)|^{2\delta} d^d x\right]^{\frac{1}{\delta}} \leq c_d \left[\int \left((\partial \varphi)^2 + m_0^2\, \varphi^2\right) d^d x\right], \tag{12}$$

that assert that for $d \geq 3$ and for non-zero m_0^2 if $\delta \leq \frac{d}{d-2}$ there exists a constant that depends only on the dimension of the Euclidean space d, that satisfies the above equation. It is clear that the condition of renormalizability is saturated by $\delta = \frac{d}{d-2}$. After a regularization and renormalization procedure it is possible to show that any physically measurable quantity $f(g)$ can be expanded in power series defined by $f(g) = \sum_{k=0}^{\infty} f_k\, g^k$, where g is the renormalized coupling constant and f_k are perturbative coefficients.

As we discussed, the perturbative renormalization is not enough to obtain well defined physical quantities, since in general the series that we obtain from perturbatively renormalizable theories is divergent. For example for $P(\varphi)_2$ the renormalized perturbative series for any connected Schwinger function that can be obtained by a Wick ordering is divergent [40]. For the $(\lambda\varphi^4)_3$ model the same divergent behavior was founded by de Calan and Rivasseau [41]. Although in general the series that we obtain from perturbatively renormalizable theories is divergent, they must be an asymptotic expansion of the solutions of our theories. In other words, in a specific theory even though the series diverges, a finite number of terms of the series is still a good approximation of the function in question. Since from a general asymptotic series it is not possible to determine which function the asymptotic series is describing, because two different functions can have the same asymptotic series, we have to use some powerful method to obtaining the solutions of our theory. The Borel summability is the method that allow us to obtain the solutions of the theory from the asymptotic series.

Let us discuss with more details the asymptotic expansion of a function and the Borel summability. Suppose a function $f(z)$ defined in the extended complex plane. The series of the form $\sum_{n=0}^{\infty} a_n z^{-n}$ which need not converge for any value of z, is called the asymptotic expansion or representation of the function $f(z)$ which is defined for every sufficient large z if we define (i) $S_N(z) = \sum_{n=1}^{N} a_n z^{-n}$ and $(ii) R_N(z) = z^N |f(z) - S_N(z)|$, and for every fixed N, we have $lim_{|z| \to \infty} R_N(z) = 0$. It is clear that it is possible to consider the series of the form $\sum_{n=0}^{\infty} a_n z^n$. Again, this series is the asymptotic representation of $f(z)$, i.e. $f(z) \sim \sum_{n=0}^{\infty} a_n z^n$ if for a small z we still have (i) and in (ii) we have $R_N(z) = z^{-N}|f(z) - S_N(z)|$ and for every fixed N we have $lim_{|z| \to 0} R_N(z) = 0$.

From the above definitions there are two main questions. The first is the question whether a function under consideration possesses an asymptotic expansion, which we call the expansion problem. There is also the question of how the function is to be found, which is represented by a given asymptotic expansion, that we call the summation problem. Note that any function can have only one asymptotic expansion, or we can show that the function in question has no asymptotic expansion. Suppose $f(x) = \exp(-x)$, $x > 0$. It is clear that all the coefficients of the asymptotic expansion are zero since $x^k \exp(-x) \to 0$ for every $k \geq 0$ when $x \to \infty$. This simple result shows that different functions may have the same asymptotic expansion. For example if $h(z)$ has an asymptotic representation it is clear that $h(z) + \exp(-z)$ or $h(z) + a \exp(-bx)$ for $b > 0$ have the same asymptotic representation. Now, suppose a function $f(z)$ which has the asymptotic expansion defined by a divergent series. Thus we have

$$f(z) \sim \sum_{k=0}^{\infty} f_k z^k. \tag{13}$$

Let us define the Borel transform of $f(z)$ that we call $B_f(z)$ and it is given by

$$B_f(z) = \sum_{k=0}^{\infty} \frac{1}{k!} f_k z^k. \tag{14}$$

The key point if that the Borel transform of the series given by $B_f(z)$ may converge even if the series is divergent. It is clear that $f(z)$ can be obtained from $B_f(z)$ by the inverse Borel transform given by

$$f(z) = \int_0^{\infty} exp(-t) \, B_f(zt) \, dt, \tag{15}$$

and this construction is an indispensable tool to recover a function from its asymptotic expansion in quantum field theory. Thus let us repeat this construction for the n-point Schwinger function of any renormalizable theory. If the perturbative series does not converge and must be an asymptotic expansion for the solutions of our theory, the Borel summability method can be used to recover the solutions of our theory. In the standard perturbative expansion we express the 2n-point renormalized Schwinger functions as the following power series for the renormalized coupling constant g:

$$G_{2n}(x_1, x_2, .., x_{2n}) \sim \sum_{k=0}^{\infty} g^k \, G_{2n}^{(k)}(x_1, x_2, .., x_{2n}). \tag{16}$$

Let us define the Borel transform of the n-point Schwinger function by

$$G_{2n}(\tau; x_1, x_2, .., x_{2n}) = \sum_{k=0}^{\infty} \frac{\tau^k}{k!} G_{2n}^{(k)}(x_1, x_2, .., x_{2n}), \tag{17}$$

and it is clear that from the inverse Borel transform we have

$$G_{2n}(g; x_1, x_2, .., x_{2n}) = \int_0^\infty \exp(-\frac{\tau}{g}) G_{2n}(\tau; x_1, x_2, .., x_{2n}). \tag{18}$$

The series wich defines $G_{2n}(\tau; x_1, x_2, .., x_{2n})$ has a much better convergence than the original series of $G_{2n}(x_1, x_2, .., x_{2n})$. If there are no singularities in the positive axis of the Borel transform of perturbative series, generated by instantons or renormalons, the Borel summability is a powerful way to extract results from a divergent series [42] [43].

Let us assume that we are studying the strong-coupling regime of a theory, as for example the ultraviolet limit of a non-asymptoticaly free theory. In this situation an alternative perturbation expansion is mandatory. Consequently we now turn to discuss the alternative expansion that has been called the strong-coupling expansion [44] [45] [46] [47] [48]. The fundamental idea is to treat the off-diagonal terms of the Gaussian factor as a perturbation about the remaining terms in the integral, instead of perform a perturbative expansion with respect to the anharmonic terms of the theory.

Let us suppose a Euclidean space compact with or withouth a boundary, and suppose some eliptic semi-positive and self-adjoint differential operator D acting on scalar functions on the manifold. The basic example is $D = -\Delta$ or $D = -\Delta + m^2$. It is clear that the kernel $K(m_0; x - y)$ is given by $K(m_0; x - y) = D \delta^d(x - y)$. Let us assume first the self-interacting $(\lambda\varphi^4)_d$ model. It is clear that by treating the off-diagonal terms of the Gaussian factor as a perturbation about the remaining terms in the integral, we have the following formal expression for the generating functional of complete Schwinger functions $Z(h)$ or the Schwinger functional:

$$Z(h) = \exp\left(-\frac{1}{2}\int d^d x \int d^d y \frac{\delta}{\delta h(x)} K(m_0; x-y) \frac{\delta}{\delta h(y)}\right) Q_0(h), \tag{19}$$

where the functional integral $Q_0(h)$ is given by

$$Q_0(h) = \mathcal{N} \int [d\varphi] \exp\left(\int d^d x \left(-\frac{g_0}{4!}\varphi^4(x) + h(x)\varphi(x)\right)\right), \tag{20}$$

and $\mathcal{N}$ is a normalization that can be found using that $Q_0(h)|_{h=0} = 1$. As we discussed, using translational invariance the kernel $K(m_0; x - y)$ is defined by $K(m_0; x - y) = (-\Delta + m_0^2) \delta^d(x - y)$. At this point it is convenient to introduce a small imaginary contribution in $h(x)$, consequently $h(x) = \Re(h) + i\Im(h)$. Although the functional integral $Q_0(h)$ is not a product of Gaussian integrals, can be view formaly as an infinite product of ordinary integrals, one for each point of the d-dimensional Euclidean space. The fundamental problem of the strong-coupling expansion, as we will see is how to construct non-Gaussian measures to define the Schwinger functional. It is clear that it is possible to obtain a formal expansion of $Z(h)$ as a perturbative series in the form

$$Z(h) = Z^{(0)}(h) + Z^{(1)}(h) + ..., \tag{21}$$

where the two first terms of the series are given respectively by $Z^{(0)}(h) = Q_0(h)$, and

$$Z^{(1)}(h) = \left(-\frac{1}{2}\int d^d x \int d^d y \frac{\delta}{\delta h(x)} K(m_0; x-y) \frac{\delta}{\delta h(y)}\right) Q_0(h), \tag{22}$$

and so on. The main difference from the the standard perturbative expansion is that we have an expansion of the generating functional of complete Schwinger functions in inverse powers of the coupling constant. For the case of the self-interacting scalar field we are perturbing around the static ultra-local model [49] [50] [51] [52], where different points of the Euclidean space are decoupled since the gradient terms are dropped.

This representation of $Z(h)$, defined by the unrenormalized perturbative series can be truncated and this situation we call the order of the approximation. For example, if $Z(h) \equiv Q_0(h)$ we call it the ultra-local approximation or the zero-order approximation. The next order approximation where we link every two points in the Euclidean space we call it the first-order approximation since $Z(h) \equiv Z^{(0)}(h) + Z^{(1)}(h)$. Assuming that $h(x) \equiv h$, we have that the Schwinger generating functional in the first-order approximation is defined by:

$$Z(h) = Q_0(h) - \frac{1}{2}\frac{\partial^2}{\partial h^2} \int d^d x \int d^d y \, K(m_0; x-y) Q_0(h). \tag{23}$$

It is not difficult to present an expression for $Z^{(1)}(h)$. Let us first use the Fourier representation for the two-point Schwinger function $G_0(m_0; x - y)$ and the kernel $K(m_0; x - y)$. Thus we have

$$G_0(m_0; x-y) = \frac{1}{(2\pi)^d} \int d^d q \, \tilde{G}(m_0; q) \exp(iq(x-y)), \tag{24}$$

and

$$K(m_0; x-y) = \frac{1}{(2\pi)^d} \int d^d q \, \tilde{K}(m_0; q) \exp(iq(x-y)), \tag{25}$$

where $\tilde{G}(m_0; q) = (q^2+m_0^2)^{-1}$. Using Eq.(6), Eq.(24) and Eq.(25) we have that $\tilde{G}(m_0; q) \tilde{K}(m_0; q) = 1$. Consequently $\tilde{K}(m_0; q) = (q^2 + m_0^2)$. The next step is to introduce the analytic regularized two-point function $G_0(m_0, \alpha; x-y)$ defined by

$$G_0(m_0, \alpha; x-y) = \frac{1}{(2\pi)^d} \int d^d q \, \frac{\exp(iq(x-y))}{(q^2+m_0^2)^\alpha}, \tag{26}$$

where α is a complex regulator parameter. Thus we have that the kernel $K(m_0; x-y)$ is defined by $K(m_0; x-y) = G_0(m_0, \alpha; x-y)|_{\alpha=-1}$. Thus it is possible to write that $Z^{(1)}(h) = F_1(h) + F_2(h)$ where

$$F_1(h) = -\frac{1}{2} m_0^2 \frac{\partial^2}{\partial h^2} \int d^d x \int d^d y \, \delta^d(x-y) Q_0(h) \tag{27}$$

and

$$F_2(h) = -\frac{1}{2} \frac{\partial^2}{\partial h^2} \int d^d x \int d^d y \int \frac{d^d q}{(2\pi)^d} q^2 \exp(iq(x-y)) Q_0(h). \tag{28}$$

Although $F_1(h)$ and $F_2(h)$ are ill defined expressions, in the section VI we will show how it is possible to regularize and renormalize $Z^{(1)}(h)$. As we stressed, although the strong-coupling expansion is very inconvenient for pratical calculations in the continuum R^d Euclidean space, it is very natural in the lattice. The technical problems that we have to deal in the strong-coupling expansion in the continuum Euclidean space are the following: first, we have to define non-Gaussian measures for functionals, and second, we have to show how it is possible to regularize and renormalize the generating functional going beyond the ultra-local approximation. In this lecture we are not interested in regularizing the complete series of the strong-coupling perturbative expansion. We are interested to study the analytic structure of the generating functional of the Schwinger functions in the complex coupling constant plane. We will show that the singularity that we will find in the zero-dimensional generating function of the $\lambda\varphi^4$ model is a superposition between an essential and a branch point singularity. For the case of quantum electrodynamics we found also a combination between a branch point and a essential singularity in the complex coupling constant plane.

Using the fact that we have invariance of the functional integral with respect to the choice of the quadratic part, let us consider the modified-strong coupling expansion. The new generating functional of the complete Schwinger functions can be defined by the following perturbative series:

$$Z(h) = \exp\left(-\frac{1}{2} \int d^d x \int d^d y \frac{\delta}{\delta h(x)} K(m_0, \sigma; x-y) \frac{\delta}{\delta h(y)}\right) Q_0(\sigma; h), \tag{29}$$

where the "new" ultra-local functional integral is given by

$$Q_0(\sigma; h) = \mathcal{N} \int [d\varphi] \exp\left(\int d^d x \left(-\frac{\sigma}{2} m_0^2 \varphi^2(x) - \frac{g_0}{4!} \varphi^4(x) + h(x)\varphi(x)\right)\right). \tag{30}$$

Note that we split the quadratic part of the functional integral proportional to the mass squared in two parts: in the off-diagonal terms of the Gaussian factor and in the ultra-local generating functional. The "modified" kernel is defined by $K(m_0, \sigma; x-y) = (-\Delta + (1-\sigma)m_0^2)\delta^d(x-y)$, where σ is a complex parameter defined in the closed region $0 \leq Re(\sigma) \leq 1$. This choice is based in our ability to choose σ in the closed region only to simplify our calculations.

Before study the zero-dimensional model let us briefly discuss the ultra-local model in the context of the standard perturbative expansion, where the functional generator of complete Schwinger functions are given by Eq.(29) and Eq.(30), respectively. We are following the discussion developed in Ref. [2]. If we compare the Fourier representation for the tree-level two-point Schwinger functions of the ultra-local model with of the Gaussian model we find that the absence of the gradient term makes the model perturbatively non-renormalizable in $d=4$. This is do to the fact that the ultra-local two-point function $G(m_0; x-y)$ is proportional to the delta function. As was discussed by Klauder and others, the distributional properties of the fields that appears in the functional integral is atenuated by the gradient term. To see this fact let us study the $d=4$ case. If we compare with the usual loop expansion of the theory with the gradient term included we have the following. In the one-loop approximation only two graphs, i.e the two and the four-point functions are divergent and the others are ultraviolet finite ($d=4$) and it is clear that for any

d there is a k such that all the graphs $k(d)$ are ultraviolet finite diagrams. For the case of the ultra-local model all the contributions of the one-loop graphs are ultraviolet divergent for any d. Consequently in the one-loop approximation we have to introduce counterterms of the kind $\varphi^6(x)$, $\varphi^8(x)$, etc and the theory is perturbatively non-renormalizable in the standard perturbative approach.

In the next section we are interested in studying the analytic structure of the zero-dimensional $\lambda\varphi^4$ theory. In the formal representation for the Schwinger functional $Z(h)$ appear divergences coming from differents origins. The first kind is related to the infinite volume and continuum hypotesis for the Euclidean space and can be controlled by the introduction of a box and a regulator function with a renormalization procedure. The second kind of divergences is related to the functional form of the non-Gaussian part of the action. The study of the ultra-local model in different field theories can easily show us the structure of the singularities for the renormalized perturbative series and also the structure of the singularities for the n-point Schwinger functions in the complex coupling constant plane. In other words, we are interested in showinghow are the singularities of the Green's functions obtained from the generating functional of the complete Schwinger functions $Z(h)$. How is the analytic structure of the renormalized Schwinger functions in the complex coupling constant plane?

III. ANALYSIS OF THE ANALYTIC STRUCTURE OF THE ULTRA-LOCAL MODEL

The aim of this section is to analyze the analytic structure of the zero-dimensional $\lambda\varphi^4$ model in the complex g_0 coupling constant plane. As we discussed before, the first term of the strong coupling expansion of $Z(h)$ is exactly the ultra-local model [50] [51] [53] [54], also called the static independent value model. Since we can interpret the ultra-local model as an infinite product of ordinary integrals, let us introduce a Euclidean lattice and analyse the generating function defined in each point of the Euclidean lattice given by

$$z(m_0, g_0; h) = \frac{1}{\sqrt{2\pi}} \int_{-\infty}^{\infty} d\varphi \, \exp(-\frac{1}{2}m_0^2 \varphi^2 - \frac{g_0}{4!}\varphi^4 + h\varphi), \tag{31}$$

where for simplicity we are assuming $\sigma = 1$. The generating function in the absence of external source will be defined by $z(m_0, g_0; h)|_{h=0} \equiv z_0(m_0, g_0)$. Consequently, let us analyse the following integral with a quartic probability distribution in which the zero-dimensional partition function $z_0(m_0, g_0)$ given by

$$z_0(m_0, g_0) = \frac{1}{\sqrt{2\pi}} \int_{-\infty}^{\infty} d\varphi \, \exp(-\frac{1}{2}m_0^2 \varphi^2 - \frac{g_0}{4!}\varphi^4). \tag{32}$$

Although the integrand of the above equation is integrable in the interval $[0, \infty)$ and the exponential power series is convergent everywhere, it is not uniformly convergent, consequently the series can not be integrated term by term. Let us define $z^{(1)}(m_0, g_0)$ and this quantity is the asymptotic expansion for the solution given by $z_0(m_0, g_0)$. Thus we have $z_0(m_0, g_0) \sim z^{(1)}(m_0, g_0)$ and choosing $m_0^2 = 1$ it is not difficult to show that

$$z^{(1)}(m_0, g_0)|_{m_0^2=1} = \sum_{k=0}^{\infty} (-g_0)^k c_k, \tag{33}$$

where the coefficients c_k are given by $c_k = \frac{(4k-1)!!}{(4!)^k k!}$. The $z^{(1)}(m_0, g_0)$ series does not converge for any value of g_0, i.e., must be the asymptotic expansion for $z_0(m_0, g_0)$. The asymptotic expansion for the zero-dimensional partition function given by $z^{(1)}(m_0, g_0)$ has the contribution from the vacuum diagrams [55] [56], and each coefficient c_k is given by the sum of symmetry factors over all diagrams of order k. In the other hand, it is easy to verify that the integral given by Eq.(32) can be solved exactly and the result is given by

$$z_0(m_0, g_0) = (\frac{3}{2g_0})^{\frac{3}{4}} m_0^2 \Psi(\frac{3}{4}, \frac{3}{2}; \frac{3m_0^4}{2g_0}), \tag{34}$$

where $\Psi(a, c; z)$ is is the confluent hypergeometric function of second kind [57]. Since we are interested to study the analytic structure of $z_0(m_0, g_0)$ in the complex plane, let us use the principle of analytic continuation [58]. We have a branch cut for $|arg\, z| = \pi$. The confluent hypergeometric function of second kind $\Psi(a, c; z)$ is a many value function of z, and we shall consider that the function can be analytic continue to the whole complex z plane except for a branch cut for $|arg\, z| = \pi$. So we have to consider its principal branch in the plane cut along the negative real axis. The analytic continuation corresponds to the definition for $z_0(m_0, g_0)$ in the whole coupling constant complex plane except for a branch cut for $|arg\, z| = \pi$. Although the theory can be defined by analytic continuation, this theory probably violates the Osterwalder-Schrader axioms, which ensure that it is possible to obtain the Wightman functions from the Schwinger functions defined in Euclidean space using the Osterwalder-Schraeder reconstruction theorem. As we discussed, the generating function in the absence of external sources was defined by $z_0(m_0, g_0)$ and we call it as the zero-dimensional partition function. This situation defines a genuine stocastic process, since the Schwinger functions of the theory are the moments of a non-Gaussian probability distribution. In the case of the analytic extended zero-dimensional generating function we are losing the positive probability interpretation.

In the next section we will discuss the strong-coupling expansion for quantum electrodynamics and shows the existence of a combination between branch point and a essential singularity in the complex coupling constant plane.

IV. THE STRONG-COUPLING EXPANSION AND THE ULTRA-LOCAL APPROXIMATION IN QUANTUM ELECTRODYNAMICS

It is not difficult to study quantum electrodynamics also in the context of the strong-coupling expansion. Although in quantum electrodynamics does not exist arguments like the Osterwalder and Schraeder reconstruction theorem that it was obtained for scalar theories, we assume that such a theory defined in a Minkowski spacetime can be extended to the Euclidean formulation. Accepting this point, let us investigate the strong-coupling regime of the theory. The treatment is standard and the idea is the same as for the scalar theory, treating the kinetic terms of the theory as a perturbation and solving the self-interacting part exactly. The only point that we have to call the attention of the reader is that we have to introduce a mass term for the photon field. Here we are following the approach developed by Cooper and Kenway [47] and Itzykson, Parisi and Zuber [59]. The generating functional of all Schwinger functions for quantum electrodynamics defined in a d-dimensional Euclidean space is given by

$$Z(\bar{\eta}, \eta, J_\mu) = \int d\bar{\psi} \, d\psi \, dA_\mu \exp\left(-\int d^d x \, \mathcal{L}(A_\mu, \psi, \bar{\psi}) + sources - terms\right), \tag{35}$$

where $A_\mu(x)$ and $\psi(x)$ are respectively the gauge and fermion field and $\mathcal{L}(A_\mu, \psi, \bar{\psi})$ is given by

$$\mathcal{L} = \frac{1}{4}(F_{\mu\nu})^2 + m^2 A_\mu^2 + \bar{\psi}(\gamma_\mu \partial_\mu + ie\gamma_\mu A_\mu)\bar{\psi}. \tag{36}$$

Note that it is necessary to introduce a photon mass in the same way that we did for the case of the scalar theory in order to regulate the photon part of the strong-coupling expansion. Although the introduction of the mass term eliminates the necessity of a gauge fixing term, it is useful to work in a general gauge.

The Euclidean form of the generating functional of complete Schwinger functions in a d-dimensional space in a covariant gauge is given by

$$Z(\bar{\eta}, \eta, J_\mu) = K_A K_\psi \int d\bar{\psi} \, d\psi \, dA_\mu \exp\left(-\int d^d x \left(\frac{1}{2}m^2 A^2 + ie\bar{\psi}\gamma_\mu A_\mu \psi + souce - terms\right)\right) \tag{37}$$

where

$$K_A = \exp\left(\frac{1}{2} \int d^d x \int d^d y \frac{\delta}{\delta J_\mu(x)} D^{-1}_{\mu\nu}(x,y) \frac{\delta}{\delta J_\mu(y)}\right), \tag{38}$$

and also

$$K_\psi = \exp\left(-\int d^d x \int d^d y \frac{\delta}{\delta \eta(x)} S^{-1}(x,y) \frac{\delta}{\delta \bar{\eta}(y)}\right). \tag{39}$$

In a covariant gauge we have $D^{-1}_{\mu\nu}(\alpha; x-y) = \left(\delta_{\mu\nu}\Delta - (1-\frac{1}{\alpha})\partial_\mu\partial_\nu\right)\delta^d(x-y)$ and also $S^{-1}(x,y) = \gamma_\mu \partial_\mu \delta^d(x-y)$. As we have been discussing, the generating functional defined by the remaining functional integral is a product of one-dimensional integrals in each point of the Euclidean space. Consequently, let us study the zero-dimensional generating function of quantum electrodynamics defined by $z(\eta, \bar{\eta}, j)$, where we are choosing $m^2 = 1$ [60]. Thus in each point of the Euclidean space we have the following generating function:

$$z(\eta, \bar{\eta}, j) = \frac{1}{\pi} \int dA \, d\psi \, d\bar{\psi} \exp\left(-\frac{A^2}{2} - \bar{\psi}(1-eA)\psi + \bar{\psi}\eta + \bar{\eta}\psi + jA\right), \tag{40}$$

where $\bar{\psi}$ and ψ are complex conjugate c-numbers variables. Integrating over the $\bar{\psi}$ and ψ c-numbers variables it is easy to show that the generating function given by Eq.(40) can be written as

$$z(\eta, \bar{\eta}, j) = \int \frac{dA}{(1-eA)} \exp\left(-\frac{A^2}{2} + \bar{\eta}\eta(1-eA) + jA\right). \tag{41}$$

Let us first study the generating functional $z(\eta, \bar{\eta}, j)$ before the use of the Furry's theorem and let us call this generating function at zero external sources as $I(v)$, were $v = -\frac{1}{e}$. Thus we have

$$I(v) = v \int_0^\infty \frac{da}{(a+v)} \exp(-\frac{a^2}{2}). \tag{42}$$

Following Stieljes [61] it is possible to show that if a function $F(x)$ is defined by

$$F(x) = \int_0^\infty \frac{f(u)}{(x+u)} du, \tag{43}$$

the following series $\sum_{n=1}^\infty a_n x^{-n}$ is an asymptotic expansion for the integral in which the coefficients a_n are given by $(-1)^{n-1} a_n = \int_0^\infty f(u) u^{n-1} du$, where $n = 1, 2, ...$

Thus generating function at zero external sources $I(z)$ possesses an asymptoptic expansion

$$I(v) \sim \sum_{n=1}^\infty a_n v^{-n+1}, \tag{44}$$

where the coefficients are given by $(-1)^{n-1} a_n = \int_0^\infty dx\, x^{n-1} \exp(-\frac{1}{2}x^2)$. Using the fact that quantum electrodynamics is charge conjugation invariant, and making use of the Furry's theorem, we have that the generating function of quantum electrodynamics must be given by

$$z(\eta, \bar{\eta}, j) = \int \frac{dA}{(1-e^2A^2)^{\frac{1}{2}}} \exp\left(-\frac{A^2}{2} + \bar{\eta}\eta(1-eA)^{-1} + jA\right). \tag{45}$$

Note that the photon propagator is given by $G = <A^2>$ and the electron propagator is $S = <\frac{1}{1-e^2A^2}>$ where the average is over the measure $\frac{dA}{(1-e^2A^2)^{\frac{1}{2}}} \exp(-\frac{A^2}{2})$.

Let us proceed in the study of the generating function for the zero-dimensional quantum electrodynamics with j, $\bar{\eta}$ and η being the c-number sources. In order to obtain some information about the structure of the singularities of the generating function, and also the generating functional of the Schwinger functions in the complex coupling constant plane, let first try to solve the integral defined by Eq.(45). It is clear that the integral that defines $z(\eta, \bar{\eta}, j)$ is meaningful only for e^2 negative. To simplify our discussion let us firt assume that the c-number sources $\bar{\eta}$ and η are zero i.e. $\bar{\eta} = \eta = 0$. In this particular situation we have

$$z(\eta, \bar{\eta}, j)|_{\eta=\bar{\eta}=0} = \int \frac{dA}{(1-e^2A^2)^{\frac{1}{2}}} \exp\left(-\frac{A^2}{2} + jA\right). \tag{46}$$

The integrand of Eq.(46) has two branch points at $A = \frac{1}{e}$ and $A = -\frac{1}{e}$. To perform the integral, let us first make the replacement $e \to ie$. In this case the branch points appear in the imaginary axis of the A complex plane and after we impose that our function $z_0(ie; j)$ is defined only for $A > 0$ we have

$$z_0(ie; j) = \int_0^\infty \frac{dA}{(1+e^2A^2)^{\frac{1}{2}}} \exp\left(-\frac{A^2}{2} + jA\right). \tag{47}$$

Even after this "improvement", it is very difficult to express the integral in terms of known functions. Nevertheless an asymptoptic expansion for small j can be found. Using a Taylor expansion for $z_0(ie; j)$ near $j = 0$ we have

$$z_0(ie; j) = z_0(ie; j)|_{j=0} + j \frac{\partial z_0(ie; j)}{\partial j}\Big|_{j=0} + ... \tag{48}$$

Let us calculate first $z_0(ie; j)|_{j=0}$. We have

$$z_0(ie; j)|_{j=0} = \int \frac{dA}{(1+e^2A^2)^{\frac{1}{2}}} \exp(-\frac{A^2}{2}). \tag{49}$$

Thus, it is not difficult to show that $z_0(ie; j)|_{j=0}$ is given by

$$z_0(ie; j)|_{j=0} = \frac{1}{2e} \exp\left(\frac{1}{4e^2}\right) K_0\left(\frac{1}{4e^2}\right), \tag{50}$$

where $K_0(z)$ is the Macdonald funcntion of zero order. It is not difficult to perform the integral that appears in the second term of the Taylor expansion. A simple calculation gives

$$\frac{\partial z_0(ie; j)}{\partial j}\bigg|_{j=0} = \sqrt{\frac{\pi}{2e^2}} \exp\left(\frac{1}{2e^2}\right)\left(1 - \Phi\left(\sqrt{\frac{1}{2e^2}}\right)\right) \tag{51}$$

where $\Phi(x)$ is the Fresnel integral defined by

$$\Phi(x) = \frac{2}{\sqrt{\pi}} \int_0^x dt \, \exp(-t^2). \tag{52}$$

Using Eq.(48), Eq.(50) and Eq.(56) we have

$$z_0(ie; j) = \frac{1}{2e}\left(\exp\left(\frac{1}{4e^2}\right) K_0\left(\frac{1}{4e^2}\right) + j\sqrt{\frac{\pi}{2e^2}} \exp\left(\frac{1}{2e^2}\right)\left(1 - \Phi\left(\sqrt{\frac{1}{2e^2}}\right)\right)\right) + .. \tag{53}$$

It is clear that there is a combination of branch point and a essential singularity in the complex coupling constant plane in the zero-dimensional model. There are two different ways to interpret our results. The first one is to assume that the zero-dimensional generating function is an analytic function defined in the extended complex plane. Since the ultra-local generating functional emerges from the strong-coupling expansion we have to interpret the zero-dimensional generating function as a Laurent expansion of this object at large e^2, i.e., $e^2 \to \infty$. The second is to try to extend the validity of the expansion and assume that it is possible to obtain results even for small coupling constant. In this case we call the zero-dimensional generating function as the Laurent expansion in the neighboorhood of the origin of the complex coupling constant plane.

In the next section we will analyse more general scalar models with polynomial and non-polynomial interactions in the context of the strong-coupling expansion.

V. THE ULTRA-LOCAL APPROXIMATION IN SCALAR MODELS

As we discussed many times in this lecture, in the infinite cut-off limit of an infrared free theory, the usual perturbative expansion where we assume that the non-Gaussian contribution is a perturbation of the corresponding free theory can not be implemented and an alternative scheme must be used. The strong coupling expansion is alternative perturbative expansion that is suitable for treat this situation.

It is remarkable that in the strong coupling expansion different theories can be treated in the same way, since we factor out the free part of the Lagrangian density and evaluate the remaining non-Gaussian contribution in a closed form. From this discussion we see that the "new" expansion can be performed for any polynomial or non-polynomial interaction $V(g_i; \varphi)$, where $g_i, i = 1, 2, ..n$ are the coupling constants of the model. In a different context, for the study of non-polynomial scalar models at finite temperature in the one-loop approximation, see for instance Ref. [62]. Thus the generating functional is given by

$$Z(h) = \left(1 - \frac{1}{2}\int d^d x \int d^d y \frac{\delta}{\delta h(x)} K(m_0, \sigma; x-y) \frac{\delta}{\delta h(y)} + ...\right) Q_0(\sigma; h), \tag{54}$$

where the ultra-local generating functional $Q_0(\sigma; h)$ is defined by the following functional integral:

$$Q_0(\sigma; h) = \mathcal{N} \int [d\varphi] \exp\left(\int d^d x \left(-\frac{\sigma}{2} m_0^2 \varphi^2(x) - V(g_i; \varphi) + h(x)\varphi(x)\right)\right), \tag{55}$$

and $\mathcal{N}$ is the normalization factor. In this section we would like to consider the ultra-local approximation ($Z(h) = Q_0(\sigma; h)$) for different scalar models. Although this approximation is not satisfatory, the aim our our investigation is more concentrated in the general structure of the singularities in the complex coupling constant plane than on specific numerical calculations. Consequently let us study the ultra-local generating functional $Q_0(\sigma; h)$ in details. This generating functional is a mean zero Gaussian functional integral and using the fact that the fields defined in each point of the Euclidean space are statistically independent we are able to write

$$Q_0(\sigma; h) = \exp\left(-\int d^d x \, L(\sigma; h(x))\right). \tag{56}$$

The formulas given by Eq.(55) and Eq.(56) are fundamental for our study. Let us see how it is possible to extract some informations from different models. We shall consider two different situations: first when $V(\varphi) = V(-\varphi)$) and second when $V(\varphi) \neq V(-\varphi)$). We limit ourselves to models with only one component. The generalization of our investigations to models with more than one component does not present any difficulty. For a discussion of the strong-coupling expansion in the $O(N)$ model, see for instance [45].

There are two questions that we would like to ask at this point: the first is related to the structure of the singularities in the complex coupling constant plane in different models, and the second is in which circumstances the ultra-local approximation is exact? We are interested to study four different scalar models. The first one is the usual $(\lambda\varphi^4)_d$ model with a interaction Lagrangian defined by

$$V_I(g_0;\varphi) = \frac{g_0}{4!}\varphi^4(x). \tag{57}$$

The diference between the treatment that we are giving in this section from the previous ones is that we are introducing the source in order to generate the Schwinger functions. The second model is only a straightforward generalization of the previous model where p is an positive integer and it is defined by

$$V_{II}(g_0;\varphi) = g_0\varphi^p(x). \tag{58}$$

Of course, p must be even. Although the model is non-renormalizable in a four dimensional Euclidean space for $p > 4$ using the weak coupling expansion, in the strong coupling approach the divergences of the model are similar to the $(\lambda\varphi^4)_d$ model. The third model is also a generalization for the previous one and the interaction Lagrangian is given by

$$V_{III}(\beta,\gamma;\varphi) = \beta\varphi^p(x) + \gamma\varphi^{-p}(x). \tag{59}$$

Finally we would like to discuss the Sinh-Gordon model defined by

$$V_{IV}(\beta,\gamma;\varphi) = \beta(\cosh\gamma\varphi(x) - 1), \tag{60}$$

where β and γ are bare parameters. Before starting to discuss each model we would like to discuss a general procedure for dealing with functional integrals. The general method for dealing with the ultra-local generating functional is to consider the transition from the continuum formulation to a lattice obtaining in this way a regularized form for the generating functional. Thus, in the continuum formulation we have that the ultra-local generating functional is given by

$$Q_0(\sigma;h) = \exp\left(\frac{1}{2}\delta^d(0)\int d^dx\, ln\frac{z(\sigma;h)}{z_0(\sigma)}\right). \tag{61}$$

Replacing the Euclidean space R^d by a hyper-cubic lattice of spacing a we have that

$$Q_0(\sigma;h) = \exp\left(\frac{1}{a^d}\int d^dx\, ln\frac{z(\sigma;h)}{z_0(\sigma)}\right), \tag{62}$$

which can be written in the following form

$$Q_0(\sigma;h) = \exp\left(\frac{1}{2}\sum_i \ln\frac{z_i(\sigma;h)}{z_{0i}(\sigma)}\right). \tag{63}$$

Consequently we have that the "free-energy" in the ultra-local approximation is given by $W_0 = \ln Q_0$ and it is defined by

$$W_0(\sigma;h) = \frac{1}{2}\sum_i \ln\frac{z_i(\sigma;h)}{z_{0i}(\sigma)}. \tag{64}$$

We will analyse this expression in the end of this section. Thus assuming a lattice it is possible to show that $Q_0(h)$ is given by

$$Q_0(\sigma;h) = \mathcal{N}\exp\left(\frac{1}{a^d}\int d^dx\, \ln z(\sigma;h(x))\right), \tag{65}$$

where the zero-dimensional generating function is given by

$$z(m_0, g_i, \sigma; h) = \frac{1}{\sqrt{2\pi}} \int_{-\infty}^{\infty} d\varphi \exp\left(a^d(-\frac{\sigma}{2}m_0^2\varphi^2 - V(g_i; \varphi) + h\varphi)\right), \qquad (66)$$

and the normalization is given by $Q_0(\sigma; h)|_{h=0} = 1$. For simplicity we are choosing $\sigma = 1$ and $a = 1$, to analyse the models above.

Let us start with the model that we briefly discussed in section III, characterized by the interaction Lagrangian $V_I(g_0; \varphi)$ defined by Eq.(57). Thus we have that the zero-dimensional generating function $z_1(m_0, g_0; h)$ is given by

$$z_1(m_0, g_0; h) = \sqrt{\frac{2}{\pi}} \int_0^\infty d\varphi \, \exp(-\frac{1}{2}m_0^2 \varphi^2 - \frac{g_0}{4!}\varphi^4) \cosh h\varphi. \qquad (67)$$

As we discussed it is possible to find $z_1(m_0, g_0; h)|_{h=0}$ in a closed form, nevertheless it is not possible to express $z_1(m_0, g_0; h)$ in terms of known functions. If we try to expand $\exp(h\varphi)$ in power series, and in order to solve the resulting integrals, we interchange the summation and the integration we thus have problems. The power series is uniformly convergent only if $|h\varphi| < 1$. Consequently, $z^{(1)}(m_0, g_0; h)$ is the asymptotic expansion of $z_1(m_0, g_0; h)$ and write that $z^{(1)}(m_0, g_0; h) \sim z_1(m_0, g_0; h)$. Thus we have

$$z^{(1)}(m_0, g_0; h) = \sum_{k=0}^{\infty} h^{2k} f_k(m_0, g_0), \qquad (68)$$

where the coefficients f_k are given by

$$f_k(m_0, g_0) = \frac{(-1)^k}{\sqrt{2\pi}} \frac{2^{k+1}}{2k!} (\frac{\partial}{\partial m_0^2})^k z_0(m_0, g_0), \qquad (69)$$

where $z_0(m_0, g_0)$ is the generating function in the absence of sources and it is given by Eq.(34). Simple substitution gives us the asymptotic representation for $z_1(m_0, g_0; h)$ in terms of derivatives of the confluent hypergeometric function of second kind. Note that we are able to simplify our calculations choosing $\sigma = 0$, but this situation will be a particular case of the next studied model. Let us study the singularities of $z^{(1)}(m_0, g_0; h)$ in the complex coupling constant for $0 < |g_0| < \infty$. Simple manipulations gives

$$z^{(1)}(m_0, g_0; h) = (\frac{3}{2g_0})^{\frac{3}{4}} \left(\sqrt{\frac{2}{\pi}} m_0^2 \Psi(\frac{3}{4}, \frac{3}{2}; \frac{3m_0^4}{2g_0}) + \sum_{k=1}^{\infty} h^{2k} c_k (\frac{\partial}{\partial m_0^2})^k \Psi(\frac{3}{4}, \frac{3}{2}; \frac{3m_0^4}{2g_0}) \right), \qquad (70)$$

where the coefficients c_k are given by $c_k = \frac{(-1)^k}{\sqrt{2\pi}} \frac{2^{k+1}}{2k!}$. Using the fact that the derivatives of the confluent hypergeometric functions of second kind are given by

$$\frac{d^n}{dz^n} \Psi(\alpha, \gamma; z) = (-1)^n (\alpha)_n \Psi(\alpha + n, \gamma + n; z), \qquad (71)$$

where the coefficients $(\alpha)_k$ are defined by

$$(\alpha)_0 = 1, \ldots (\alpha)_k = \frac{\Gamma(\alpha + k)}{\Gamma(\alpha)} = \alpha(\alpha + 1)\ldots(\alpha + k - 1), \qquad (72)$$

for $k = 1, 2, ,$. We have the following behavior. In the series representation for $z^{(1)}(m_0, g_0; h)$ the principal part has an infinite number of terms consequently it has an essential singularity at $g_0 = 0$. Note also the presence of a a branch point at $g_0 = 0$ and also at $g_0 = \infty$ in the complex coupling constant plane.

The second model that we are interested to discuss has been extensivelly studied by Klauder [49] [50] is given by the interaction Lagrangian $V_{II}(g_0; \varphi) = g_0 \varphi^p$, where p is an even natural number. Although in the conventional perturbative expansion this model is non-renormalizable, in the strong-perturbative expansion the model is included in the same class of the $\lambda \varphi^4$ model, and if we are able to extract finite results from one we are able to extract from the other. For this model it is interesting to assume that $\sigma = 0$ and also $a = 1$. Thus we have that the zero-dimensional generating function $z_2(g_0; h)$ associated with this model is given by

$$z_2(g_0; h) = \sqrt{\frac{2}{\pi}} \int_0^\infty d\varphi \, \exp(-g_0 \varphi^p) \cosh h\varphi. \qquad (73)$$

Since it is not possible to express $z_2(g_0; h)$ in terms of known function we have to use the same method that we have been using. Let us express the zero-dimensional generating fuction $z_2(g_0; h)$ in the absence of sources in a closed form and expand $\cosh(h\varphi)$ in power series, and in order to solve the resulting integrals we interchange the summation and the integration. Again $z^{(2)}(g_0; h)$ that we obtained after use the power expansion is the asymptotic expansion of $z_2(g_0; h)$ and write that $z^{(2)}(g_0; h) \sim z_2(g_0; h)$. It is clear that using that

$$\int_0^\infty dx\, x^{\nu-1} \exp(-\mu x^p) = \frac{1}{p} \mu^{-\frac{\nu}{p}} \Gamma(\frac{\nu}{p}), \tag{74}$$

for $\mu > 0$, $\nu > 0$ and $p > 0$, a very simple calculation shows that the zero-dimensional generating function $z_2(g_0; h)$ in the absence of sources is given by

$$z_2(g_0; h)|_{h=0} = \frac{1}{p}\sqrt{\frac{2}{\pi}} (\frac{1}{g_0})^{\frac{1}{p}} \Gamma(\frac{1}{p}). \tag{75}$$

From the above discussion we have that the asymptotic representation of the generating function $z_2(g_0; h)$ is given by $z^{(2)}(g_0; h)$ where

$$z^{(2)}(g_0, h) = \sum_{k=0}^\infty h^{2k} c(p,k) (\frac{1}{g_0})^{\frac{2k+1}{p}}, \tag{76}$$

and the coefficients $c(p,k)$ are given by

$$c(p,k) = \frac{1}{p}\sqrt{\frac{2}{\pi}} \frac{1}{(2k)!} \Gamma\left(\frac{2k+1}{p}\right). \tag{77}$$

Until now we have been considered only polynomials models. However we are not simple interested in such kind of models. A interesting model that we would like to consider is given by the following idealized interaction Lagrangian given by Eq.(59). Note that in this model we have a suppression of the configurations fluctuations around $\varphi = 0$. For p even the model has two minima. It is not difficult to show that the model has a power series representation. The equilibrium values are given by $\varphi_0 = (\frac{\gamma}{\beta})^{\frac{1}{2p}}$ and $\varphi_0 = -(\frac{\gamma}{\beta})^{\frac{1}{2p}}$. Let us choose the case where $\varphi_0 > 0$ and define a new field $\phi(x) = (\varphi(x) - \varphi_0)$. Using the binomial expansion and its generalization

$$(1+x)^\alpha = \sum_{k=0}^\infty \binom{\alpha}{k} x^k, \ |x| < 1. \tag{78}$$

where the coefficients of the expansion are given by

$$\binom{\alpha}{0} = 1, \ \binom{\alpha}{k} = \frac{\alpha(\alpha-1)...(\alpha-k+1)}{k!}, \ for\ k \geq 1, \tag{79}$$

we have

$$V_{III}(\gamma,\beta;\phi) = \sum_{k=0}^p c(p,k) \phi(x)^k + \gamma \varphi_0^{-p-k} \sum_{k=p+1}^\infty \binom{-p}{k} \phi(x)^k. \tag{80}$$

The coefficients $c(p,k)$ are given by

$$c(p,k) = \left(\beta \varphi_0^{-p-k} \binom{p}{k} + \gamma \varphi_0^{-p-k} \binom{-p}{k}\right). \tag{81}$$

Note that the generalization of the binomial series is valid for any complex exponent α. In other words, we have a convergent power series everywhere in α, hence a continuous function on α in the complex plane. The zero-dimensional generating functional is given by

$$z_3(\beta,\gamma;h) = \sqrt{\frac{2}{\pi}} \int_0^\infty d\varphi \, \exp(-\beta\varphi^p - \gamma\varphi^{-p}) \cosh h\varphi. \tag{82}$$

Also it is not possible to express $z_3(\beta,\gamma;h)$ in terms of known function. Let us express the zero-dimensional generating function $z_3(\beta,\gamma;h)$ in the absence of sources in a closed form and expand $\cosh(h\varphi)$ in power series, and in order to solve the resulting integrals we interchange the summation and the integration. Again $z^{(3)}(\beta,\gamma;h)$ that we obtained after use the power series expansion is the asymptotic expansion of $z_3(\beta,\gamma;h)$ and write that $z^{(3)}(\beta,\gamma;h) \sim z_3(\beta,\gamma;h)$. It is not difficult to find $z_3(\beta,\gamma;h)|_{h=0}$. Using the identity

$$\int_0^\infty dx\, x^{\nu-1} \exp(-\beta x^p - \gamma x^{-p}) = \frac{2}{p}(\frac{\gamma}{\beta})^{\frac{\nu}{2p}} K_{\frac{\nu}{p}}\left(2\sqrt{\beta\gamma}\right), \tag{83}$$

that is valid for $Re\,\beta > 0$ and $Re\,\gamma > 0$, we have that the zero-dimensional generating function $z_3(\beta,\gamma;h)$ in the absence of sources is given by

$$z_3(\beta,\gamma;h)|_{h=0} = \frac{1}{p}\sqrt{\frac{8}{\pi}}(\frac{\gamma}{\beta})^{\frac{1}{2p}} K_{\frac{1}{p}}(2\sqrt{\beta\gamma}), \tag{84}$$

and also using that $z^{(3)}(\beta,\gamma;h) \sim z_3(\beta,\gamma;h)$ we have

$$z_3(\gamma,\beta,h) = \sum_{k=0}^\infty h^{2k} c(p,k) (\frac{\gamma}{\beta})^{\frac{2k+1}{p}} K_{\frac{2k+1}{p}}(2\sqrt{\beta\gamma}), \tag{85}$$

where the coefficients $c(p,k)$ are given by

$$c(p,k) = \frac{1}{p}\sqrt{\frac{8}{\pi}}\frac{1}{2k!}. \tag{86}$$

After all, we have been analysed models where we obtained in the best situation asymptotic series representation for the ultra-local generating functions. The next step is to turn to the second question that we have raised. In which circumstances the ultra-local approximation is exact? To clarify the problem, let us present a interesting model where the ultra-local approximation is exact. In other words, the generating functional of all the Schwinger functions is given by the ultra-local generating functional $Z(h) \equiv Q_0(h)$. Let us discuss the Sinh-Gordon model given by Eq.(60). The following model that is non-renormalizable in the "weak" coupling expansion, where β and γ are the coupling constants and choosing $\sigma = 0$. It is clear that in the zero-dimensional generating fucntion $z_4(\beta,\gamma;h)$ in the absence of sources can be found in a closed form it is given by:

$$z_4(\beta,\gamma;h)|_{h=0} = \frac{2e^\beta}{\gamma} K_0(\beta). \tag{87}$$

It is interesting that for this kind of potential we can found a closed form for the zero-dimensional generating fuction $z_4(\beta,\gamma;h)$ in the presence of sources and is given by

$$z_4(\beta,\gamma;h) = \frac{2e^\beta}{\gamma} K_{\frac{h}{\gamma}}(\beta), \tag{88}$$

where $K_\nu(z)$ is the modified Bessel function of third kind. Substituting Eq.(87) and Eq.(88) in Eq.(65) we have for the Sinh-Gordon model that the ultra-local generating functional is given by

$$Q_0(h) = \mathcal{N} \exp\left(\frac{1}{a^d} \int d^d x\, ln(\frac{2e^\beta}{\gamma} K_{\frac{h}{\gamma}}(\beta))\right). \tag{89}$$

The normalization can be found using $Q_0(h)|_{h=0} = 1$. It is remarkable that in the Sinh-Gordon model it is possible to extract information from a finite number of terms of the infinite series representation of the generating functional in the strong coupling expansion. In the Sinh-Gordon model the ultra-local approximation is exact. Thus $Z(h) \equiv Q_0(h)$. Let us calculate the "free-energy" per unit volume $f(\beta,\gamma;h)$ for $\beta \neq 0$ and also $\gamma \neq 0$. It is clear that we have

$$f_{ren}(\beta,\gamma;h) = \ln\left(\frac{2e^\beta}{\gamma}\right) + \ln\left(K_{\frac{h}{\gamma}}(\beta)\right). \tag{90}$$

The first question that can be asked is related to the fact that $f_{ren}(\beta,\gamma;h)$ can diverges since the modified Bessel function of third kind $K_\nu(z)$ has zeros in the complex plane. From this discussion one needs information about the distribution of the zeros of the modified Bessel function of third kind in the complex plane. For real ν, $K_\nu(z)$ has no zeros in the region $|arg\,z| < \frac{\pi}{2}$ and in the complex plane with a cut along the segment $(-\infty, 0]$ the $K_\nu(z)$ has a finite number of zeros. Thus the free energy per unit volume in the the model is finite.

VI. BEYOND THE ULTRA-LOCAL APPROXIMATION

In this section we will sketch the formalism that can be used to obtain a regularized expression for $Z^1(h)$, going beyond the ultra-local approximation. We are using two different regularization procedures, and it is possible to identify the divergent contribution in each regularized expression and a renormalization procedure is implemented with an appropriate subtraction of the singular contribution. First we would like to discuss some related ideas. To solve the problem of the ultraviolet divergences in the weak-coupling expansion in the $\lambda\varphi^4$ theory, Bervillier et al [63] changed the kinematical term using a Gaussian regularized propagator, and the new "free" action is given by

$$S = \mu^2 \int d^d x \, \varphi(x) \exp(-\Delta\,\mu^2)\,\varphi(x), \tag{91}$$

where μ is a mass parameter that we have to introduce in the regularization procedure. The Gaussian propagator is given by $G(k) = \mu^{-2}\exp(-k^2\mu^2)$ and with the Gaussian regularized two-point function we can write the generating functional of all Schwinger functions. Since the theory with a Gaussian regularized propagator generates a perturbation series where all the Feynman diagrams are finite, we can study only the effect of the perturbative expansion that generates the vacuum persistence functional. Note that Bender et al [48] also discussed this kind of regularization scheme and also the θ function regularization scheme in the strong-coupling expansion. It is clear that it is possible to introduce a non-local regularized operator $D(\sigma)$ defined by

$$D(\sigma) = \mu^{-2\sigma+2}\,(-\Delta + m_0^2)^\sigma, \tag{92}$$

where σ is a regularization parameter or a "mixed" regularized operator. We would like to stress that such kind of study in another context was performed by Svaiter and Svaiter [64]. These authors developed a method to unify two unrelated regularization methods frequently employed to obtain the renormalized zero-point energy of quantum fields. Introducing a mixed cut-off function and studying the analytic properties of the regularized energy as a function of the two cut-off parameters, it was possible to not only relate the usual cut-off method and the analytic regularization method, but also unify both methods.

Let us use the ideas discussed above introducing a exponential cut-off and also an algebraic cut-off to regularize the kernel $K(m_0, x - y)$. For simplicity let us assume that $h(x) = h = cte$, and using Eq.(23) we have

$$Z^{(1)}(h) = -\frac{1}{2}\frac{\partial^2}{\partial h^2}\int d^d x \int d^d y \, K(m_0; x-y) Q_0(h). \tag{93}$$

The singularities of the above expression are coming from different terms. First the second derivative with respect to the source of the ultra-local generating functional is singular in the continuum limt. Since we have been discussed how to deal with the singularities of the ultra-local generating functional, let us study only the divergences coming from the kernel $K(m_0, x - y)$ integrated over the volume where we defined the scalar field. In the infinite volume limit, as we discussed a Fourier representation for the kernel $K(m_0; x - y)$ is given by

$$K(m_0; x-y) = \frac{1}{(2\pi)^d}\int d^d q\,(q^2 + m_0^2)\exp(iq(x-y)). \tag{94}$$

To evaluate the behavior of the kernel $K(m_0; x - y)$ for $|x - y|$ small and large, as we discussed let us introduce two different regulators. The divergent expression given by Eq.(94) can be regularized using for example a exponential cut-off function $f_1(m_0, \mu, \sigma; q^2)$ defined by

$$f_1(m_0,\mu,\sigma; q^2) = \exp(-\frac{\sigma}{\mu^2}(q^2+m_0^2)), \quad Re(\sigma) > 0, \tag{95}$$

or an algebraic cut-off function $f_2(m_0, \mu, \sigma; q^2)$ defined by

$$f_2(m_0,\mu,\sigma; q) = \mu^{-2\sigma}\,(q^2 + m_0)^\sigma, \quad Re(\sigma) < -\frac{d}{2} - 1. \tag{96}$$

Note that we introduced a mass parameter μ with dimension of inverse of length and the cut-off parameter σ is a dimensionless quantity.

Let us study first the exponential cut-off method. The regularized kernel $K(m_0, \mu, \sigma; x - y)$ is defined by

$$K(m_0,\mu,\sigma; x-y) = \frac{1}{(2\pi)^d}\int d^d q \, \exp(iq(x-y))\,(q^2+m_0^2)\exp(-\frac{\sigma}{\mu^2}(q^2+m_0^2)). \tag{97}$$

It is clear that we can write the regularized kernel as

$$K(m_0,\mu,\sigma; x-y) = -\frac{\mu^2}{(2\pi)^d}\frac{\partial}{\partial\sigma}\int d^d q \, \exp(iq(x-y)) \exp(-\frac{\sigma}{\mu^2}(q^2+m_0^2)). \tag{98}$$

Since the integral in Eq.(98) is Gaussian, can be performed and we obtain the following expression for the regularized kernel:

$$K(m_0,\mu,\sigma; x-y) = -\frac{\mu^{d+2}}{(2\sqrt{\pi})^d}\frac{\partial}{\partial\sigma}\left(\sigma^{-\frac{d}{2}}\exp(-\frac{\sigma}{\mu^2}m_0^2 - \frac{\mu^2}{4\sigma}(x-y)^2)\right). \tag{99}$$

Thus we have that the singular contribution in the limit where $\sigma \to 0$ is given by

$$K(m_0,\mu,\sigma; x-y) =$$
$$-\frac{1}{(2\sqrt{\pi})^d}\exp\left(-\frac{\sigma}{\mu^2}m_0^2 - \frac{\mu^2}{4\sigma}(x-y)^2\right)\left(-\frac{\mu^{d+4}}{4\sigma^{\frac{d}{2}+2}}(x-y)^2 + \frac{d}{2}\frac{\mu^{d+2}}{\sigma^{\frac{d}{2}+1}} + \frac{\mu^d}{\sigma^{\frac{d}{2}}}m_0^2\right). \tag{100}$$

We note that the negative powers portion of the Laurent series expansion of $K(m_0,\mu,\sigma; x-y)$ around $\sigma = 0$ has an infinite number of terms and the regularized expression has an essential singularity at $\sigma = 0$. But we know that if a function defined in the complex plane has an essential singularity at some z_0, then it takes on every value with one possible exception in every ϵ neighborhood of z_0 using the Picard theorem. Thus, let use an alternative method, i.e., the analytic regularization procedure, that we call an algebraic cut-off.

Using the algebric cut-off function $f_2(m_0,\mu,\sigma; q)$ the regularized kernel $K(m_0,\mu,\sigma; x-y)$ is defined by

$$K(m_0,\mu,\sigma; x-y) = \frac{\mu^{-2\sigma}}{(2\pi)^d}\int d^d q \, \exp(iq(x-y))(q^2+m_0^2)^{1+\sigma}. \tag{101}$$

The regularized kernel $K(m_0,\mu,\sigma; x-y)$ is convergent and analytic in the complex σ plane for $Re(\sigma) < -\frac{d}{2} - 1$. As in any cut-off method we have to take the limit $\sigma \to 0$, starting from $Re(\sigma) < -\frac{d}{2} - 1$. To perform the d-dimensional integration let us work in a d-dimensional polar coordinate system. Defining $|x-y| = r$ and $q = (q_1^2 + q_2^2 + .. + q_d^2)^{\frac{1}{2}}$ it is easy to show that the regularized kernel can be expressed in the following way:

$$K(m_0,\mu,\sigma; r) = \frac{\mu^{-2\sigma}}{(2\pi)^d}\frac{1}{r^{\frac{d}{2}-1}}\int_0^\infty dq\, q^{\frac{d}{2}}(q^2+m_0^2)^{1+\sigma}J_{\frac{d}{2}-1}(qr), \tag{102}$$

where $J_\nu(z)$ is the Bessel function of first kind of order ν. Let us analyse the cases $r \neq 0$ and the case where $r = 0$ separately to make our discussion more precise.

The case where $r = 0$ is trivial. We have for odd d that $K(m_0,\mu,\sigma; x \approx y)|_{\sigma=0} = 0$ and for even d it is easy to show that

$$K(m_0,\mu,\sigma; x \approx y) = \frac{\mu^{-2\sigma}}{(2\sqrt{\pi})^d}(m_0^2)^{\frac{d}{2}+\sigma+1}. \tag{103}$$

The more interesting case, where $r \neq 0$ can be solved evaluating the integral $I(\mu,\nu; a,b)$ defined by

$$I(\mu,\nu; a,b) = \int_0^\infty dx \frac{x^{\nu+1}}{(x^2+a^2)^{\mu+1}}J_\nu(bx), \quad for\, a > 0,\, b > 0. \tag{104}$$

To evaluate this integral first let us start from an integral representation for the Gamma function $\Gamma(z)$ using the identity

$$\frac{1}{(x^2+a^2)^{\mu+1}} = \frac{1}{\Gamma(\mu+1)}\int_0^\infty dt\, t^\mu \exp(-t(x^2+a^2)), \tag{105}$$

which is valid for $Re(\mu) > -1$ and to guaranty absolute convergence of the double integral that we are evaluating, we have to assume that the parameters μ and ν are defined in the region $-1 < Re(\nu) < 2\,Re(\mu) + \frac{1}{2}$. Then using an integral representation of the Macdonald's function and also the identity given by

$$\int_0^\infty dx\, x^{\nu+1}\exp(-a^2 x^2)\,J_\nu(bx) = \frac{b^\nu}{(2a^2)^{\nu+1}}\exp(-\frac{b^2}{4a^2}), \quad a > 0,\, b > 0, \tag{106}$$

it is possible to show that $I(\mu,\nu;a,b)$ is given by

$$I(\mu,\nu;a,b) = \frac{a^{\nu-\mu}b^{\mu}}{2^{\mu}\Gamma(\mu+1)}K_{\nu-\mu}(ab), \tag{107}$$

which is valid for $-1 < Re(\nu) < 2\,Re(\mu) + \frac{1}{2}$. Using the principle of analytic continuation we have that the regularized kernel is given by

$$K(m_0,\mu,\sigma;r) = \frac{\mu^{-2\sigma}}{(2\pi)^d}\left(\frac{m_0}{r}\right)^{\frac{d}{2}+\sigma+1}\frac{K_{\frac{d}{2}+\sigma+1}(m_0\,r)}{\Gamma(-1-\sigma)}. \tag{108}$$

We obtained that in the limit where $\sigma \to 0$ only the $r = 0$ case gives contribution to $Z^{(1)}(h)$, since the Gamma function $\Gamma(z)$ has simple poles at $z = 0, -1, -2...$ and $\frac{1}{\Gamma(z)}$ is an entire function of z. From our discussions we see that the divergences that appear in the strong coupling perturbative expansion are proportional to the volume of the domain where we defined the fields and consequently the ultra-local approximation is exact in the Sinh-Gordon model, since only the first term of the perturbative expansion, i.e the ultra-local term contributes to the generating functional.

VII. CONCLUSIONS

In this lecture we discussed the strong coupling expansion and the singularities of the perturbative expansion in the $(\lambda\varphi^4)_d$ model, quantum electrodynamics and also different scalar models on a d-dimensional Euclidean space. We have discussed the usual perturbative expansion and a non-standard perturbative approach that have been called in the literature the strong-coupling expansion. From the discussions we can see that the divergences which occur in any scalar model in quantum field theory fall into two distinct classes. The first one is related to the infinite volume-continuum hypotesis for the physical Euclidean space. The second is related to the functional form of the interaction action. We showed that in the $\lambda\varphi^4$ model and also quantum electrodynamics in the complex coupling constant plane there is a superposition between an essential singularity and a branch point.

As we discussed in the introduction, the ultraviolet limit of the regularized $\lambda\varphi^4$ in a lattice with a nearest neighborhood realization of the Laplacian must be Gaussian for $d > 4$. A different way to interpret this result is the following: in the formal representation of the Schwinger functional in the conventional perturbative expansion the Gaussian approximation must be exact. We obtained the folowing result. The ultraviolet limit of the Sinh-Gordon model in a lattice must be the ultra-local theory for any d. In other words, in the formal representation for the Schwinger functional given by $Z(h) = (1 + ..)\,Q_0(h)$ where $Q_0(h)$ is the ultra-local functional, in the Sinh-Gordon model the ultra-local approximation is exact, i.e. $Z(h) \equiv Q_0(h)$. It is interesting to extend such analysis for models with many coupling constants. For models with n independent coupling constants, the study of the analytic properties of the zero-dimensional generating function had to be investigated studying the partition function in a 2n dimensional complex space.

VIII. ACKNOWLEDGEMENTS

I would like to thank the hospitality of the Center of Theoretical Physics, Laboratory for Nuclear Science and Department of Physics, Massachusetts Institute of Technology, where part of this work was carried out. This paper was supported by Conselho Nacional de Desenvolvimento Cientifico e Tecnológico do Brazil (CNPq)

---

[1] R. J. Rivers, "Path Integral Methods in Quantum Field Theory", Cambridge University Press, Cambridge (1987).
[2] J. R. Klauder, "Beyond Conventional Quantization", Cambridge University Press, Cambridge (2000).
[3] G.'t Hooft and M.Veltman, Diagrammar, CERN report 73-9 (1973), reprinted in Nato Adv.Study Inst. serie B, vol 4b, 177.
[4] J. C. Colins, "Renormalization", Cambridge University Press, Cambridge (1983).
[5] W. Pauli and F. Villars, Rev. Mod. Phys. **21**, 434 (1949).
[6] L. F. Ashmore, Nuovo Cim. Lett. **9**, 289 (1972),
[7] C. G. Bollini and J. J. Giambiagi, Nuovo Cim. **B12**, 20 (1972).
[8] G. t'Hooft and M. Veltman, Nucl. Phys. **B44**, 189 (1972),

[9] G. Leibbrandt, Rev. Mod. Phys. **47**, 849 (1975).
[10] M. Gell-Mann and F. Low, Phys. Rev. **95**, 1300 (1954).
[11] C. G. Callan, Phys. Rev. **D2**, 1541, (1970).
[12] K. Symanzik, Comm. Math. Phys. **18**, 227 (1970).
[13] L. N. Cooper, Phys. Rev. **100**, 362 (1955).
[14] D. Gross and A. Neveu, Phys. Rev. **D10**, 3235 (1974).
[15] G. Parisi, Nuc. Phys. **B100**, 368 (1975).
[16] F. S. Dyson, Phys. Rev. **85**, 861 (1952).
[17] G. Parisi, Phys. Lett. **66B**, 167 (1977).
[18] E. Brezin, J. C. Le Guillou and J. Zinn-Justin, Phys. Rev. **D15**, 1544 (1977).
[19] I. Pomeranchuk, V. Sudakov and K. Ter Martirosjan, Phys. Rev. **103**, 784 (1954).
[20] L. D. Landau and I. Ya Pomeranchuk, Dokl. Akad. Nauk. **102**, 489 (1955).
[21] I. Ya Pomeranchuk, Nuovo Cimento **3**, 1186 (1956).
[22] H. D. Politzer, Phys. Rev. Lett. **30**, 1346 (1973).
[23] D. Gross and F. Wilczek, Phys. Rev. Lett. **30**, 1343 (1973).
[24] D. Gross and F. Wilczek, Phys. Rev. **D8**, 3633 (1973).
[25] K. Symanzik, Euclidean Quantum Field Theory in "Local Quantum Theory", Academic Press, N.Y. (1969) pp. 152-226.
[26] K. Wilson and J. Kogut, Phys. Rep. **12**, 75, (1974).
[27] K. Osterwalder and R. Schrader, Comm. Math. Phys. **31**, 83 (1973), **42**, 281 (1975).
[28] M. Aizeman, Phys. Rev. Lett. **47**, 1 (1981).
[29] J. Frohlich, Nucl. Phys. **B200**, 281 (1982).
[30] J. R. Klauder, hep-th/0209177.
[31] G. Gallavotti and V. Rivasseau, Ann. Inst. Henry Poincare, **40**, 185 (1984).
[32] A. D. Sokal, Ann. Inst. Henry Poincare, **37**, 317 (1982).
[33] S. Coleman, R. Jackiw and H. D. Politzer, Phys. Rev. **D10**, 2491 (1974).
[34] C. M. Bender and T. T. Wu, Phys. Rev. **D7**, 1620 (1973).
[35] J. Zinn-Justin, "Quantum Field Theory and Critical Phenomena" Oxford University Press, N.Y. (1996).
[36] J. M. Cornwall, R. Jackiw and E. Tomborilis, Phys. Rev. **D10**, 2428, (1974).
[37] P. K. Towsend, Phys. Rev. **D12**, 2269 (1975).
[38] G. N. J. Ananos, A. P. C. Malbouisson and N. F. Svaiter, Nucl. Phys **B547**, 271 (1999), G. N. J. Ananos and N. F. Svaiter, Mod. Phys. Lett. **A15**, 2235 (2000).
[39] J. Zinn-Justin, Phys. Rep. **70**, 111 (1981).
[40] A. Jaffe, Comm. Math. Phys. **1**, 127 (1965).
[41] C. de Calan and V. Rivasseau, Comm. Math. Phys. 82, 69 (1981), ibid **83**, 77 (1982).
[42] N. N. Kuri, Phys. Rev. **D12**, 2258 (1975), Phys. Rev. **D16**, 1754 (1977).
[43] G.'t Hooft, "Under the Speel of the Gauge Principle", World Scientific (1994).
[44] C. M. Bender, F. Cooper, G. S. Guralnik and D. H. Sharp, Phys. Rev. **D19**, 1865 (1979).
[45] N. Parga, D. Toussaint and J. R. Fulco, Phys. Rev. **D20**, 887 (1979).
[46] C. M. Bender, F. Cooper, G. S. Guralnik and D. H. Sharp, R. Roskies and M. L. Silverstein, Phys. Rev. **D20**, 1374 (1979).
[47] F. Cooper and R. Kenway, Phys. Rev. **D24**, 2706 (1981).
[48] C. Bender, F. Cooper, R. Kenway and L. M. Simmons, Phys. Rev. **D24**, 2693 (1981).
[49] J. R. Klauder, Phys. Rev. **D14**, 1952 (1976).
[50] J. R. Klauder, Ann. Phys. **117**, 19 (1979).
[51] E. R. Caianiello, G. Scarpetta, N. Cim. **22A**, 448 (1974), ibid. Lett. N. Cim. **11**, 283 (1974).
[52] R. Menikoff and D. H. Sharp, J. Math. Phys. **19**, 135 (1978).
[53] H. G. Dosch, Nucl. Phys. **96**, 525 (1975).
[54] R. J. Cant, R. J. Rivers, J. Phys. **A13**, 1623 (1980).
[55] C. Itzykson and J. M. Drouffe, "Statistical Field Theory", Cambridge University Press, (1989) (volume I, pp. 244-245).
[56] G. A. Baker, Jr and A. S. Wightman, in "Progress in Quantum Field Theory", Elsevier Science Publishers, B. V. (1986).
[57] I. S. Gradshteyn and I. M. Ryzhik, *Tables of Integrals, Series and Products*, (Academic Press Inc., New York 1980).
[58] A. P. C. Malbouisson, R. Portugal and N. F. Svaiter, Physica **A292**, 485 (2001).
[59] C. Itzykson, G. Parisi and J. B. Zuber, Phys. Rev. **D16**, 996 (1977).
[60] C. Itzykson in "Lectures Notes in Physics, Feynman Path Integrals", Proccedings in the International Colloquium, Marseille (1978), Springer-Verlag (1979).
[61] K. Knopp, "Theory and Application of Infinite Series", Dover Publications, Inc, N.Y. (1990).
[62] A. P. C. Malbouisson and N. F. Svaiter, J. Math. Phys. **37**, 4352 (1996).
[63] C. Bervillier, J. M. Drouffe, J. Zinn-Justin and C. Godreche, Phys. Rev. **D17**, 2144 (1978).
[64] N. F. Svaiter and B. F. Svaiter, J. Math. Phys. **32**, 175 (1991), N. F. Svaiter and B. F. Svaiter, Jour. Phys. **A25**, 979 (1992).

Space and Spacetime

M. Lachièze-Rey

Service d'Astrophysique, C.E. Saclay, F-91191 Gif sur Yvette Cedex, France

This lecture comprises two parts. The first one presents a short analysis of the evolution of ideas about space and spacetime. In particular, it suggests an analogy between the introduction of *homogeneous space* by Newton, and the introduction of *spacetime* in special relativity.

The second part provides, in the frame of general relativity, a prescription to define *space*, at a given moment, for an arbitrary observer in an arbitrary (sufficiently regular) curved spacetime. Based on synchronicity arguments, this prescription defines a foliation of spacetime, which provides a natural global reference frame (with space and time coordinates) for the observer, in spacetime, which remains Minkowskian along his world-line. This definition remains valid in curved spacetime, and/or for non inertial observers.

Application to Minkowski spacetime illustrates clearly the fact that different observers see different spaces. It allows, for instance, to define space everywhere without ambiguity, for the Langevin observer (involved in the Langevin pseudoparadox of twins). Applied to the Rindler observer (with uniform acceleration) it leads to the Rindler coordinates, whose choice is so justified with a physical basis. This allows to interpret the Unruh effect as due to the observer dependence of the space-time splitting. Finally an application if given for a rotating observer in circular motion.

We also apply this prescription in cosmology, to inertial observers in the Friedmann-Lemaître models: space constructed in this manner differs from the hypersurfaces of homogeneity, which do not obey the simultaneity requirement. I work out two examples: the Einstein – de Sitter model, in which space, for an inertial observer, is not flat nor homogeneous, and the de Sitter case.

I. FIRST PART: FROM SPACE TO SPACETIME

This part is concerned by the evolution of ideas about Space. To begin, the Cosmology of Aristotle is without a concept of *space*: he notion is replaced by that of a *place* (locus). The World is certainly not homogeneous : there is a center (coinciding with that of the Earth), there is a strong distinction between the *sublunar* part (below the Moon) and the *supralunar* one (the skies beyond the Moon); there is a hierarchy of concentric spheres, up to the last one, which is the *border* of the World.

Also, the geometry of the World is strongly anisotropic: the vertical dimension points towards the center of the World, and is fundamentally distinct from the horizontal ones. Terrestrial bodies fall along the vertical, when they tend to recover their "natural place", which coincides with the center of the Earth (because they contain the element "earth"). This privilegiates the vertical (light bodies also follow the vertical, in the upward direction). On the other hand, the celestial motions, and the surfaces of the celestial spheres may be seen as horizontal: the sky also is strongly anisotropic.

In the language of present physics, this geometry incorporates some aspects of the the (terrestrial) gravitation: rather than an identified physical effect, the latter appears as part of the geometry of the World. But this concerns only the *terrestrial* gravity since its universality will be discovered only two millenaries later.

A. The birth of modern physics

After Antiquity and Middle Age, the physics and astronomy of Aristotle are more and more criticized. After the works of Nicolas Copernic, Giordano Bruno, Tycho Brahe, Johannes Kepler, René Descartes,..., a new vision of the World is adopted. Its birth may be taken at the publication of the *Principia* by Isaac Newton. The apparitions of the new physics, of the new astronomy (now astrophysics) and of the new cosmology are strongly related to the notions of space, time, and Universe. The new born *Physical space* is assimilated to the mathematical *Euclidean* space, $\mathbb{R}^3$ for mathematicians. At the same moment, time is parametrized and geometrized as the real line $\mathbb{R}$.

Newton introduces the absolute space, with no physical property, other than being the passive frame for physical phenomena. The universal gravitation, his main discovery, is treated as a physical interaction - a *force* - not included in the geometry. Since gravitation becomes external to the geometry, this allows the *isotropisation* of space: all dimensions become equivalent.

The apparent specificity of the vertical dimension appears now as a *local* effect only, due to the peculiar direction of the gravitational acceleration here (see Vilain, 2003). It is non geometric in nature. The universality of the gravitation law, of the dynamics, and of all physical interactions leads to the concept of the *Universe*, at the basis of the modern physics. There is no more distinction between terrestrial and celestial physics.

Space and time

The relations between spatial and temporal dimensions are expressed by *kinematics*. In Newtonian physics, they remain independent: there is no *spacetime*, but separate space and time. The Newtonian kinematics is based on the Principle of inertia and on the law of composition of velocities by addition.

In fact, the Principle of inertia is strongly linked to the isotropy of space: in the absence of any privileged direction in space, a body cannot initiate a motion. This would imply the choice of such a direction, contrary to spatial isotropy. So, motion remains identical. Thus, the nature of the Principle of inertia, and of the resulting kinematics, appear geometrical.

Light and matter

The Newtonian time is universal, and geometrized, although without explicit introduction of the arrow of time. The Newtonian kinematics is based on the *Galilean* relativity of motion. But, during the XIXth century, it is realized that the addition law for velocities does not apply to light. In particular, the light velocity c remains constant in all conditions.

Are they two different kinematics, one for matter and an other one for light ? This would imply two different conceptions of space, of time, and of their links. This seems not acceptable. The crisis will be solved by special relativity.

B. Relativity

The theory of special relativity may be expressed by the replacement of *space and time* by [the Minkowski] *spacetime*. The temporal and spatial dimensions acquire a similar geometrical status, in a true 4 dimensional geometry, which is a *Lorentzian manifold*. In particular, there is a physical possibility to exchange spatial and temporal dimensions, by [hyperbolic] rotations called *boosts*. Spatial rotations and boosts form the *Lorentz group*, at the basis of the theory.

There is no absolute time, but every observer has a different *time line* (also called world line). There is no universal time line, common to every body, like in Newtonian physics. The composition of velocities is no more given by an addition, but it is the result of an hyperbolic rotation, called a *Lorentz transformation*. This is expressed by a new, *relativistic* kinematics. As it is well known, it leads to peculiar effects like time dilatation and length contraction. These relativistic effects (like the pseudoparadox of the Langevin twins) are of pure geometrical nature in spacetime.

Spacetime

The new kinematics is nothing else than the spacetime geometry. The Principle of Inertia acquires a new and simpler formulation: *Inertial motion is described by a straight line in spacetime*. This includes the prescription of constant velocity, that there is no need to add explicitly. The velocity of light c becomes an absolute quantity, now considered as a part of the structure of spacetime (this will remain true in general relativity). This new geometry corresponds to a gain of geometrical symmetry, as explained in the next section.

Like in Newtonian physics, gravitation is not incorporated in the geometry but remains excluded. This will be the task of general relativity to incorporate it again.

C. Symmetries of space and spacetime

The main point of this talk is to emphasize the analogy:

- The introduction of homogeneous and isotropic *space* by Newton may be seen as the replacement:
  2 horizontal dimensions + 1 vertical dimension
  → 3 equivalent dimensions of the homogeneous (3-d) space.

- The special relativity may be seen as the replacement:
  1 time dimension + 3 space dimensions
  → 4 (almost) equivalent dimensions
  of the homogeneous (4-d) Minkowski spacetime.

In the frame of Newtonian physics, no effect distinguishes a vertical or horizontal dimension, far from the Earth. A space traveller may define his "vertical direction" as that going from his feet to his head. But it will not coincide with the vertical of another space traveller.

In special relativity, no effect allows to distinguish a time direction among the directions of spacetime (although some of these directions are excluded as being purely spatial). A space traveller defines his "proper time", as the peculiar direction of its world line, in spacetime. But this direction will not coincide with the time direction of another space traveller, with a relative velocity.

In both cases, there is an increase of geometrical symmetry, expressed by the group theory:
Newton:
SO(2) [plane rotations] → SO(3) [space rotations]
Special relativity:
SO(3) [space rotations] → SO(1,3) = Lorentz group [spacetime rotations]

General relativity

Einstein incorporates the gravitation in the geometry of spacetime. The Minkowski spacetime is replaced by a more general spacetime, a Lorentzian (pseudo Riemannian) 4-dimensional manifold. The gravitation becomes part of its geometrical structure, expressed by the Riemann *curvature* [tensor]. In some sense, the latter plays the role of the elusive *gravitational ether*. Note that the pseudo Riemannian structure, usually expressed by the metric, may also be described by equivalent tools, like the [Levi-Civita] *connection*, the *tetrad coefficients*, which may appear more convenient.

This is not a return to the Aristotle's conception: the *universality* of gravitation is now taken into account. The gravitation is not necessarily that of the terrestrial field. It is not, in general, vertical and, on average (beside *local* irregularities), space remains isotropic. Locally (or rather infinitesimally, i.e., in the tangent space), the symmetries of Minkowski spacetime remain preserved.

II. SECOND PART: FROM SPACETIME TO SPACE

A. Do we need Space ?

The evolution of Physics has led us from space to spacetime. Special relativity learns us that we cannot define uniquely *space* in spacetime (see below), and independently of the choice of an observer. This remains the same in general relativity, with additional difficulties. This is the subject of this work.

On the other hand, we are used to perform and interpret most of our physical experiments and observations (in particular for quantum physics) in terms of space and time, rather than spacetime. Special relativity provides the possibility to define space an time without ambiguity *for an inertial observer*, in Minkowski spacetime. Although this is not exemplified, this possibility is related to synchronization procedures originally proposed by Einstein. This work considers the possibility to extend this procedure to non inertial observer, in Minkowski, or curved spacetimes.

General relativity and relativistic cosmology consider *spacetime* as the arena for physics, and it is an old question to define *space* and *time*. These notions are not covariant and all problems of general relativity and cosmology can be addressed without them, so that they may appear as rather academic. However, on the one hand, the literature refers often to *space*, for instance to affirm that it is flat (or not), or homogeneous (or not) in a given cosmological model. On the other hand, quantum physics, or its interpretation, requires most often a splitting of spacetime into space and time (for instance to define what is a frequency). This points out the necessity of a convenient definition of space in spacetime, or equivalently, the choice of a convenient *global reference frame*. The simple example of two inertial observers in Minkowski spacetime, with different velocities, shows that such a definition must be observer-dependent.

B. Global reference frame

An observer needs a frame to interpret the physics in his environment, in conformity with his physical intuition. In his immediate neighborhood, space is defined without ambiguity at any point of his world line (i.e., at any moment of his story) by orthogonality to his worldline, i.e., to his velocity u. But there are many different ways to extend this definition, i.e., to define space (and time) everywhere, in whole spacetime, or in the largest possible part of it: to define a global frame. Along the world line O, a global frame is constrained to be *Minkowskian* there. This is far from being sufficient to determine the choice. This paper provides a prescription which allows to associates to any observer (defined by his world line) an unique global reference frame (GRF) with special properties. A GRF may be seen as a foliation of spacetime into space plus time. The sheets are timelines (one of them is O). The transverse leaves (constrained to be orthogonal to the time-lines everywhere) are space-like 3-dimensional hypersurfaces, which are identified with the copies of space et the different moments. There is a large number of possibilities to define such foliations. A popular prescription (G) privilegiates the geodesic character of such an hypersurface. This corresponds to the *Fermi coordinates*: the spatial hypersurfaces are generated by the spacelike geodesics orthogonal to O. But, in many situations (even in very simple ones, like for the Langevin observer, see below), these spatial hypersurfaces intersect. This forbids a definition of time valid far from O: different values of time would be associated to the same event. Thus, in general, the validity of this prescription does not extend beyond a very local neighborhood of the observer (which may be sufficient for some applications). Moreover, there is no real motivation to impose a geodesic character, when O itself is not geodesic, i.e., for non inertial motion. Another possibility (H), very popular in cosmology, privilegiates *spatial homogeneity*: the spatial hypersurfaces of homogeneity orthogonal to the world line are selected. But such hypersurfaces do not exist in all spacetimes. Moreover, such a choice has no meaning when the observer himself breaks the spatial symmetries (by his acceleration or rotation for instance). Thus, this proposition has a low range of applications.

The prescription proposed here (S) does not suffer from these drawbacks. It is defined from a "simultaneity criterion", not verified in general by prescriptions G or H (or others), excepted in the immediate neighborhood of O. In the following I will underline space to refer to the result of this prescription: space at instant t is defined as the set of events that the observer $\overline{O}$ sees as *simultaneous* (see below for a precise definition) at his proper time t. An additional advantage of this prescription results from the fact that the synchronisation procedure depends only on the propagation of light-rays. Thus, only the *conformal* structure of the manifold (not the complete metric, excepted for the proper time of the observer) is used to construct the GRF. This prescription appears valid for a much broader class of spacetimes than those mentioned above (for instance, in the absence of spatial homogeneity). For any observer, inertial or not, in any spacetime (with some restrictive conditions, see below), this procedure allows a canonical global splitting of spacetime into space and time: space is defined uniquely everywhere as a "simultaneity space" Σ_τ, at a value τ of the proper time of the observer, which is so promoted as an universal time function. The Σ_τ never intersect, even in the situations where the Fermi hypersurfaces do, and they are defined even in the absence of spatial homogeneity. This extends the validity of the observer's proper time to the whole spacetime. In the Friedmann-Lemaître models, space does not coincide with some intuitive idea of what space could be (hypersurfaces of homogeneity).

Numerous attempts to define a quantization procedure in curved spacetime, and/or for non inertial observers (see, e.g., Birrel and Davis, 1982), involve, more or less explicitly, a space + time splitting of spacetime. This is especially important for giving a physical interpretation of quantum states in terms of frequencies or particles. For instance, I show below that the present procedure, when applied to the uniformely accelerated observer, leads to the widely used *Rindler coordinates* at the source of the *Unruh effect*. This definition of space provides a good justification otherwise absent, for the use of these coordinates (see also Dolby, 2000).

All the quantities appearing in this work are covariant. This includes all the observer dependent quantities like his velocity, acceleration, world-line and the special reference frame introduced here. In general relativity, an observable quantity (e.g., the energy) is a combination of a covariant quantity associated to the observer (e.g., its velocity u) with a covariant quantity associated to the observed system (e.g., its momentum-energy tensor). But this combination has a *local* character, i.e., it is simply a tensorial product (contraction). On the other hand, the quantum field theory involves *non local* observables, which may include integrals over [part of] space in Minkowski spacetime. Thus, an extension of quantum field theory to general relativity involves the definition of non local observables in curved spacetime, which requires most often a definition of space and time.

The goal of this work is to construct the reference frame associated to an observer in cosmological situation. Thus, it only concerns that part of spacetime which is causally related to him, in past and in future, what is called the "causal diamond". Throughout this paper, by an abuse of language, I call "spacetime" the causal diamond, i.e., the set $\mathcal{M}_0$ of events inside the particle horizon and the event horizon of the observer, if they exist. In the following, I will assume that the causal structure of spacetime admits only light cones without folding and conjugate points (no gravitational lensing; no multi-connected spacetime). These restrictions are appropriate for cosmology, and characterize a background spacetime convenient for quantization. They also appear with the other prescriptions, which all appear more restrictive that the one here. Moreover, the validity of the latter can be extended to many situations including conjugate points.

In Section 3, I implement the definition of space, and the related notions. I show how they allow to define a GRF convenient to the observer, and a congruence of canonically associated observers. Section 4 applies these results to observers in Minkowski spacetime. Section 5 considers inertial observers in the Friedmann-Lemaître cosmological models.

III. A GLOBAL REFERENCE FRAME FOR OBSERVERS

A. The accelerated Observer

In a spacetime $\mathcal{M}$, the most general observer is defined by his timelike world line $O(\tau)$, parametrized by proper time τ. The velocity $u(\tau) \equiv \partial_\tau$, defined everywhere on O, verifies $u \cdot u = 1$. Hereafter, we note the covariant derivative along the curve $\nabla_\tau X = u \cdot \nabla X = \dot{X}$ with a dot. Writing the acceleration

$$a(\tau) \equiv \dot{u} = A\, h_1, \text{ with } h_1 \cdot h_1 = -1,\ A \in \mathbb{R}^+, \qquad (1)$$

we may complete a moving tetrad along O with $h_0 \equiv u$, h_2 and h_3 defined by

$$\dot{a} \equiv \dot{A}\, h_1 + A\, \dot{h}_1, \quad \dot{h}_1 \equiv A\, u + R\, h_2, \tag{2}$$

$$(h_3)^\mu = \epsilon^{\mu\nu\rho\sigma}\, (h_0)_\nu\, (h_1)_\rho\, (h_2)_\sigma.$$

We have $h_\mu \cdot h_\nu = \eta_{\mu\nu}$. The unit vector h_2 characterizes the spatial rotation of the observer. For a non rotating (NR) observer (defined as having $R = 0$), h_2 may be chosen as an arbitrary vector orthogonal to h_0 and h_1. After calculations, the ON frame h naturally associated to the arbitrary observer, obeys the Frenet-Serret equations Synge (1967) Pauri & Vallisneri 2000

$$\dot{h}_0 = A\, h_1 \tag{3}$$
$$\dot{h}_1 = A\, h_0 + R\, h_2 \tag{4}$$
$$\dot{h}_2 = -R\, h_1 + C\, h_3 \tag{5}$$
$$\dot{h}_3 = -C\, h_2, \tag{6}$$

where A, R, C vary along O, in general. Transport along the world line corresponds to a Lorentz rotation, which may be seen as the combination of a boost, in the plane (u, a) and, for the rotating observer, a spatial rotation in a space like plane orthogonal to u and to the space-like vector $\omega \equiv C\, h_1 + R\, h_3$. The frame h is rotating with the observer, when the later does. We will associate to an observer an other non-rotating frame f.

1. The Fermi-derivative

For an arbitrary vector V field defined along O, we define the *Fermi-derivative* (along O) as

$$d_F V \equiv \dot{V} - [(u \cdot V)\, a - (a \cdot V)\, u], \tag{7}$$

$$(d_F V)^\mu = \dot{V}^\mu - a^{[\mu}\, u^{\nu]}\, V_\nu. \tag{8}$$

The vector is said to be *Fermi-transported* when $d_F V = 0$. It is easy to check that

$$d_F u = 0, \quad d_F h_1 = R\, h_2, \quad d_F h_2 = \dot{h}_2, \quad d_F h_3 = \dot{h}_3. \tag{9}$$

Thus, h_1 is Fermi transported only if the observer is NR. For the vectors h_μ, the transport (Lorentz rotation) along O combines a boost in the plane u, a with the spatial rotation $\mathcal{R}$ represented by the vectors u and ω:

$$V \mapsto \mathcal{R}(V): \quad [\mathcal{R}(V)]^\mu = u_\alpha\, \omega_\beta\, V_\gamma\, \epsilon^{\mu\alpha\beta\gamma}. \tag{10}$$

For the vectors above, we have $d_F h_\mu = \mathcal{R}(h_\mu)$, so that the Fermi derivative expresses their spatial rotation. Also, $\mathcal{R}(\omega) = \mathcal{R}(u) = 0$. Although the frame h defined above is rotating with the observer, it is possible to associate to O a non rotating frame $f = (f_\mu)$ (an "ideal gyroscope"), such that each f_μ is Fermi transported, i.e., $d_F f_\mu = 0$. We chose $f_0 = u$, but the spatial vectors do not coincide with the h_i when the observer is rotating. Then, the goal will be to extend the definition of f in the whole spacetime, or, at least, in an extended part of it.

B. Definition of space

A spacetime $\mathcal{M}$ admits many time-like foliations compatible with the world-line of a given observer O. I will show that synchronicity arguments allow to select an unique one, and provide therefore a global definition of space. In the whole paper, I denote $\tilde{v}$ the one-form metric-dual to a vector v, i.e., such that $\tilde{v}(v) = g(v,v) \equiv v \cdot v$.

As it is well known, it is impossible to define *absolute* simultaneity in special or general relativity. However, special relativity allows to define simultaneity in Minkowski spacetime, *from the point of view of an inertial observer*. In general relativity, this corresponds to a prescription of simultaneity or, better, *synchronicity*, which is *local* and *relative to an observer* (see, e.g., Landau and Lifshitz, 1966). This defines a *local* space + time splitting for this observer, with the only constraint that space and time are orthogonal where they meet on O. The construction S presented here extends this prescription beyond a local neighborhood, using perfectly operational arguments of synchronicity.

Given an observer O, I define Σ_τ, the *hypersurface of synchronicity* (HS) of O at proper time τ, as the set of events related by a null geodesics to both $O(\tau+\delta)$ and $O(\tau-\delta)$, where δ is an arbitrary interval of proper time for O:

$$\Sigma_\tau = \cup_\delta \; [I^{future}(\tau-\delta) \cap I^{past}(\tau+\delta)], \tag{11}$$

where $I^{past}(\tau)$ [resp. $I^{future}(\tau)$] denotes the null past [future] light-cone of the observer at proper time τ. The prescription S considers Σ_τ as the space for $O(\tau)$.

Given the restrictions above, the surfaces Σ_τ for different values of τ completely fill $\mathcal{M}_0$. This allows to extend the vector field u to the totality of $\mathcal{M}_0$ by requiring that it is everywhere unit ($u.u = 1$) and orthogonal to Σ_τ (Fig.1). The vector field u constitutes a foliation of $\mathcal{M}_0$, with the Σ_τ as transverse (orthogonal) surfaces. Each integral line of u can be labelled by its intersection $\mathbf{R}$ with Σ_1 (for instance), so that any point of $\mathcal{M}_0$ can be written $(\tau, \mathbf{R})$. The value $R(x) = \delta$ associated to any point x through (11) represents half the interval of observer's proper time between $\tau_1 = \tau + \delta$ and $\tau_2 = \tau - \delta$. These two events correspond to the emission of a flash light (or radar signal) which illuminates a cosmic object at x, and the observation of the resulting image by the observer, after mirror reflection. For any point x, $R(x) = \delta$ defines a natural radial space coordinate, that I call its " proper time interval " (PTI). The integral lines of u define an unique family of observers, that I will call *the canonical observers* associated to O. Care must be taken that they do not necessarily share the properties of O. For instance, they are not geodesic, even when O is, when there is expansion. The word lines generated by u are labelled by $\mathbf{R}$. Along them, the proper time t (with $t = \tau$ along O) corresponds to the *radar time* originally defined by Bondi, and reintroduced by Dolby (2000) and Dolby and Gull (2001). Since the congruence of "associated observers" is completely defined from the world line of the unique observer O, the foliation introduced here defines an unique space + time slicing, from the unique observer only.

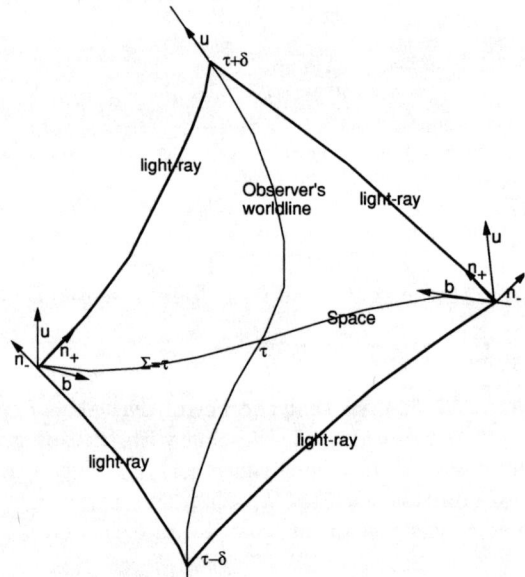

FIG. 1: At each point x, the light-rays from the observer (in the past and in the future) define the vectors n^+ and n^-. The velocity u of the observer is (non parallely) transported to give the vectors u and b at x.

C. Transport along the light-rays

The properties of space are defined from those of the world-line O, transported by the past and future light-rays. To explore them, we define the two null functions $\mathcal{N}_-(x)$ and $\mathcal{N}_+(x)$ such that the value of $\mathcal{N}_-(x)$ [resp. $\mathcal{N}_+(x)$] is the proper time $\tau - \delta$ of the observer O, when he emits a light-ray reaching x [resp. $\tau + \delta$, when he receives a light-ray emitted from x]. In other words the null hypersurface $\mathcal{N}_-(x) = \tau$ [resp. $\mathcal{N}_+(x) = \tau$] is the future [resp. past] light cone of the observer at proper time τ. We define their (null) (past and future) generators as $\tilde{n}_\pm = \nabla \mathcal{N}_\pm = d\mathcal{N}_\pm$. Both are future directed, and normalized so that the frequency emitted or received by the observer is unity (see below).

It is easy to show that Σ_τ is defined by the equation

$$T(x) \equiv [\mathcal{N}_-(x) + \mathcal{N}_+(x)]/2 = \tau. \tag{12}$$

For any point, T constitutes a natural time-coordinate. In addition we define the deformed cylindric hypersurface

$$R(x) \equiv [\mathcal{N}_+(x) - \mathcal{N}_-(x)]/2 = \delta \tag{13}$$

as the set of events at a constant PTI value δ from the observer, when he describes his world-line.

Given the normalization above, we have

$$dT = (\tilde{n}_- + \tilde{n}_+)/2 \text{ and } dR = (\tilde{n}_+ - \tilde{n}_-)/2. \tag{14}$$

It is easy to check that $dT \cdot dR = 0$, and

$$dT \cdot dT = -dR \cdot dR := N^{-2} = n_+ \cdot n_-/2, \tag{15}$$

which defines the *lapse function* N associated to this foliation. Since dT is orthogonal to Σ, we have $\tilde{u} = N\, dT$. From $u^2 = 1$, we have $N\, u \cdot dT = 1$. Since, along $\mathcal{O}$, $\tau = T$, this implies $N = 1$ on $\mathcal{O}$.

Everywhere (except on $\mathcal{O}$), we define

$$\tilde{b} \equiv N\, dR = \tilde{u} - N\,\tilde{n}_- = -\tilde{u} + N\,\tilde{n}_+. \tag{16}$$

We have $b^2 = -1$, $u \cdot b = 0$, $u + b = N\, n_+$ and $u - b = N\, n_-$. Thus, b is a unit space like vector, tangent to Σ_τ and orthogonal to the level surfaces of $R(x)$. In some sense, it points towards the observer O. In general, the vector b is not geodesic but it is *chorodesic*, due to the synchronicity property and the congruence of associated observers is *quasi-rigid* (see Bel, 1998).

1. Canonical observers

The vector field u is perfectly defined everywhere and characterizes the family of canonical observers. This family defines a "kinematics" in the sense of Smarr and York (1978). All the relevant formalism of projectors, lapse and shift functions, intrinsic curvature, etc. applies.

The vector fields u and b are not transported parallely along the light rays. In Lachièze-Rey (2001, hereafter MLR), I introduced the two vector fields U^+ and U^- which are, by definition, parallely transported along n^+ and n^- respectively, $n^\varepsilon . \nabla U^\varepsilon = 0$, and which both coincide with u along the world line of O (Fig.2). They allow to define a "future frame", and a "past frame" which, although not obeying synchronicity, appear convenient in some circumstances (Marzlin, 1994). "bisector frame", whose extension may provide a *local* surface of synchronicity for two different observers. This is for instance useful for the study of the quantum evolution of two interacting particles in spacetime (Ali et al., 1990).

2. Towards a global frame

This definition of space and time constitutes a first step towards that of a GRF but it is not the whole story. At each point of $\mathcal{M}_0$, we have defined a time like vector u and a 3-dimensional manifold that we consider as space. It remains to precise the spatial part of our frame. This can be done through the integral lines of the vector field b: through each point $x \in \mathcal{M}_0$, there is an unique line of this type. It crosses O at $O(\tau)$, with unit tangent vector $\hat{B}$. It will be possible to define angular coordinates from the 3 scalar products $\hat{B} \cdot f_i$. This will be developed in future work. In many cases of interest, the procedure becomes particularly simple: when spacetime has symmetries, like Minkowski spacetime or cosmological models; and when the observer has simple motion (inertial, non rotating, confined). We treat some specific cases below.

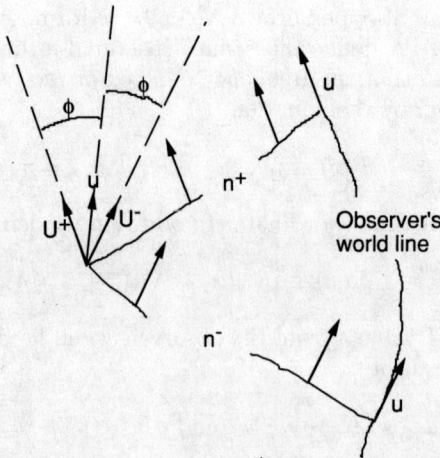

FIG. 2: The field u is not parallely transported. At each point x, U^+ is defined by the parallel transport of u from the future, i.e., by n^+, U^- is defined by the parallel transport of u from the past, i.e., by n^-.

3. Redshifts and Metric

Let us consider a congruence of (non necessarily canonical) objects with velocity $V(x)$ at the point (event) x in spacetime. Each of these objects, at x, is seen with a redshift $z^+ = (n^+(x) \cdot V(x))^{-1}$ by the observer (in the future) and sees the observer (in his past) with a redshift $z^- = n^-(x) \cdot V(x)$. For the congruence of *canonical* observers defined above, $V = u$, and $z^+ = N(x)$ and $z^- = N(x)^{-1}$. These observers are comoving with respect to the coordinate R, i.e., they keep a constant value of R. On the other hand, there is a unique congruence of objects for which $z^+ = 1$ [resp. $z^- = 1$], those with velocity U^+ [resp. U^-] (see MLR). Thus, N appears as the *lapse function* associated to the foliation, or to the congruence of associated observers. The usual ADM formalism allows to define time and space projectors, as well as the fundamental forms (metric and extrinsic curvature) on the surfaces Σ_τ (see, e.g., Smarr and York, 1978).

IV. OBSERVERS IN MINKOWSKI SPACETIME

A. Inertial observers

The inertial observer O (zero acceleration) has a velocity

$$u^0 = c,\ u^1 = s,\ u^2 = u^3 = 0,$$

where $c \equiv \cosh\psi$ and $s \equiv \sinh\psi$, the *rapidity* ψ being a constant. His world line is

$$x^0 = c\,\tau,\ x^1 = s\,\tau,\ x^2 = x^3 = 0. \tag{17}$$

Calculations of the light-ray trajectories (given in MLR) provide the surface Σ_τ as the plane of equation $c\,x^0 - s\,x^1 = \tau$, inclined by ψ with respect to the vertical, and thus orthogonal to O. Thus, space is different for all inertial observers.

B. The Langevin observer

The solution of the celebrated "Langevin's twin paradox" lies in geometry. I define a *Langevin observer* as an observer which is initially inertial, then (at $t = 0$) suffers an instantaneous acceleration, and then is inertial again (Fig.4). Such an observer is able to meet his twin, which remained always inertial, with a different lapse of proper time. Is is often quoted (see, e.g., Misner et al., 1973) that it is impossible to define space globally for such an observer. But the synchronicity prescription applies perfectly in this case, and provides an unambiguous definition of space for this observer. This has been firstly shown by Dolby and Gull (2001).

The trajectory is defined as

$$x^0 = \tau,\quad x^1 = x^2 = x^3 = 0,\qquad \text{for}\quad t < 0, \tag{18}$$
$$x^0 = c\,\tau,\quad x^1 = s\,\tau, x^2 = x^3 = 0,\qquad \text{for}\quad t > 0, \tag{19}$$

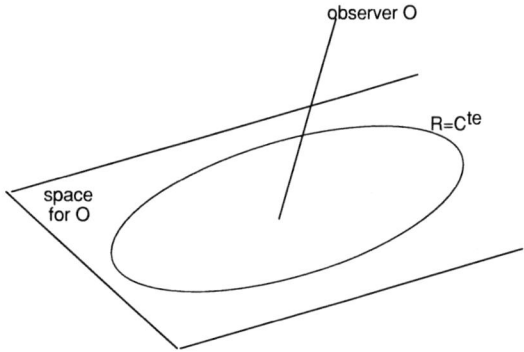

FIG. 3: For the inertial observer in Minkowski, with arbitrary velocity (rapidity ψ), space is the hyperplane with inclination ψ. We have drawn a curve $R = C^{te}$, in this plane.

with $c := \cosh \psi$ and $s := \sinh \psi$.

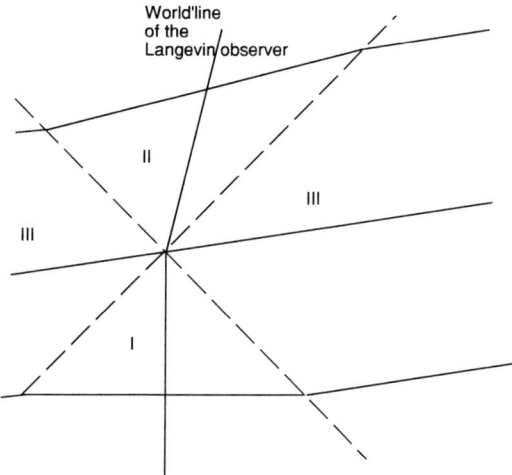

FIG. 4: World's line and (cuts of) space at various moments for the Langevin observer. Its light cone is indicated by the dashed lines.

The light cone of the observer at $t = 0$, $\mathcal{L}^0$, divides the spacetime into three sectors I, II and III (see Fig.4) corresponding to the past, future and spatially related regions to the acceleration point. Study of the light-rays and calculations (see MLR) lead to the values of $\mathcal{N}^\varepsilon(x)$, $T(x)$ and $R(x)$ in the three regions. The surfaces Σ_τ of equation $T(x) = \tau$ defines the spaces for the observer at proper time τ. Fig.4 gives their projections in the (x^0, x^1) plane: In Region I, they are straight horizontal lines, where $R(x) = x^1$. In Region II, they are lines inclined of ψ with respect to the vertical, and thus orthogonal to the world line of the observer in that region. In Region III, they are lines inclined of $\psi/2$ with respect to the vertical, i.e., at equal hyperbolic angle $\psi/2$ of the two previous lines (Fig.4).

For the observer at an arbitrary moment, space is made of a plane disk S^I [or S^{II}] up to the light cone $\mathcal{L}^0$ and is continued by a composite surface S^{III} beyond. Except at the single moment when the observer experiences the instantaneous acceleration, space is not flat, nor homogeneous.

This is the simplest example where our prescription S differs from F and H. As it is well known, it is impossible to extend the Fermi coordinates outside the conical regions, and no homogeneous hypersurfaces would be convenient. Thus, in this simple case, the prescription S is the only one providing a reference frame associated to the observer valid in the whole spacetime, to extend the validity of his proper time, and to consider unambiguous synchronicity procedures (a similar conclusion has been reached by Dolby and Gull, 2001).

C. The Rindler observer

The Rindler observer in Minkowski spacetime has constant acceleration a. His velocity is defined by

$$u^0 = \cosh(a\,\tau),\ u^1 = \sinh(a\,\tau) =,\ u^2 = u^3 = 0, \tag{20}$$

with acceleration $a^0 = a\,\sinh(a\,\tau)$, $a^1 = a\,\cosh(a\,\tau)$, $a^2 = a^3 = 0$ (with no loss of generality, I have taken the x^2 direction parallel to the acceleration). The world line has the equation $x^0 = a^{-1}\,\sinh(a\tau)$, $x^1 = a^{-1}\,\cosh(a\tau)$, $x^2 = x^3 = 0$, an hyperbola in spacetime. Since (anticipating) the solution requires $(x^1)^2 - (x^0)^2 > 0$, we can introduce the Rindler coordinates

$$x^0 := a^{-1}\,\exp(a\,\xi)\,\sinh(a\,\eta)$$

$$x^1 := a^{-1}\,\exp(a\,\xi)\,\cosh(a\,\eta). \tag{21}$$

The problem is usually treated in two dimensions, where it appears particularly simple and pedagogic: the hypersurface Σ_τ is the line of equation $\eta = \tau$, which implies $x^0 = \tanh\tau\, x^1$: a straight line through the origin. The level surfaces of R, $R(x) = \delta$ are the hyperbolae of equation $\xi = \delta$, or $(x^1)^2 - (x^0)^2 = a^{-2}\,\exp(2a\,\delta)$.

In four-dimensions, the hypersurface Σ_τ is the [flat] hyperplane through the origin, of equation $x^0 = \tanh\tau\, x^1$, which projects to the line seen in the previous section. The (spatial) metric on Σ_τ is given by

$$\begin{aligned}d\sigma^2 &= N^2\,dR^2 + (dx^2)^2 + (dx^3)^2 \\ &= a^{-2}\,[d\exp(a\,\xi)]^2 + (dx^2)^2 + (dx^3)^2 \\ &= (d\tfrac{x^1}{\cosh\tau})^2 + (dx^2)^2 + (dx^3)^2,\end{aligned} \tag{22}$$

the latter form showing that its hypersurface ($\tau = Ct$) is flat and homogeneous.

The surfaces of constant PTI δ are given by $\xi' = \delta$, or

$$2\,a^{-1}\,\cosh(a\,\delta)\,\sqrt{(x^1)^2 - (x^0)^2} - a^{-2} = (x^1)^2 - (x^0)^2 + (x^2)^2 + (x^3)^2. \tag{23}$$

This calculation shows that the widely used Rindler coordinates correspond in fact to the definition of space and time introduced here for the accelerated observer. This justifies their use, and sheds some light on the Unruh effect, which appears as a consequence of the different space-time splittings for the two observers (inertial and Rindler): they associate different frequencies to the same state, namely, the Minkowski inertial vacuum. This has led Pauri and Vallisneri (1999) classical (not quantum) origin for this effect, not discussed here.

D. Rotation in Minkowski spacetime

The previous observers are non rotating. Here we consider an observer which describes a circle in space (the radius of the circle is taken as an unit for all space and time coordinates). In spacetime, he describes the helix with coordinates

$$x^0 \equiv t = \gamma\,\tau,\ x^1 = \cos\Omega\tau,\ x^2 = \sin\Omega\tau,\ x^3 = 0,\ \gamma \equiv \sqrt{1 + \Omega^2}.$$

Its velocity is thus

$$u = (\gamma,\ -\Omega\,\sin\Omega\tau,\ \Omega\,\cos\Omega\tau,\ 0).$$

Derivation leads to

$$h_1 = (0, -\cos\Omega\tau, -\sin\Omega\tau, 0),\ A = \Omega^2,$$

$$R = \gamma\,\Omega,\ h_2 = (-\Omega, \gamma\,\sin\Omega\tau, -\gamma\,\cos\Omega\tau, 0),$$

$h_3 = (0, 0, 0, 1)$, $C = 0$, and the relations above are verified. It is easy to verify that u coincides, on O, with the Killing vector

$$\xi = w\,\partial_t + \Omega\,(x\,\partial_y - y\,\partial_x) = w\,\partial_t + \Omega\,\partial_\phi,$$

if we use the polar coordinates ρ, ϕ with $\tan\phi \equiv y/x$. Also, the acceleration a coincides (along O) with the Killing vector $\xi' = -\rho\,\Omega^2 \partial_\rho$. These Killing vectors Lie-transport each other, $\mathcal{L}_\xi \xi' = 0$.

1. The Fermi transported frame

Since the observer is NR, the frame h defined above is not Fermi-transported. To define a Fermi-transported frame f, we first put $f_0 = u$ and $f_3 = h_3$. We then define the 2 following as arbitrary combinations of h_1 and h_2, with orthonormality conditions and the requirement of Fermi-transport. This leads to

$$f_1 = \cos(R\,\tau + \psi)\,h_1 - \sin(R\,\tau + \psi)\,h_2, \quad f_2 = \sin(R\,\tau + \psi)\,h_1 + \cos(R\,\tau + \psi)\,h_2, \tag{24}$$

where ψ is an arbitrary phase. Note the apparition of the new frequency $R = \Omega\,\gamma$ linked with the Thomas precession (see for instance Misner et al., 1973, p. 175). Note also that $R\,\tau = \Omega\,x^0 = \Omega\,t$.

2. The synchronicity surface

There is no analytical expression of T as a function of the Cartesian coordinates. However, it is possible to coordonize the points of spacetime with τ, δ and two angular coordinates θ, ϕ. A 3-dimensional expression can be found in Pauri & Vallisneri (2000). Here we extend it in 4 dimensions:

$$t = \sin\Omega\delta\,\sin\theta + \gamma\,\tau, \tag{25}$$
$$x = \cos\Omega\tau\,[b(\delta)\,\cos\theta\,\cos\phi + \cos\Omega\delta] - \gamma\,\sin\Omega\tau\,\delta\,\sin\theta, \tag{26}$$
$$y = \sin\Omega\tau\,[b(\delta)\,\cos\theta\,\cos\phi + \cos\Omega\delta] + \gamma\,\cos\Omega\tau\,\delta\,\sin\theta, \tag{27}$$
$$z = \sin\phi\,\cos\theta\,b(\delta), \tag{28}$$

with $b(\delta) \equiv \sqrt{\gamma^2\,\delta^2 - \sin^2\Omega\delta}$. These equations also give the parametrization of the surface $R(x) = \delta$. Calculations of the transport by the light-rays give the extension of the velocity field

$$u^0 = \gamma\,\delta/b(\delta),\ u^1 = -\sin\Omega\tau\,\sin\Omega\delta/b(\delta),\ u^2 = \cos\Omega\tau\,\sin\Omega\delta/b(\delta),\ u^3 = 0, \tag{29}$$

the b field

$$b^0 = -\sin\Omega\delta\,\sin\theta/b(\delta) \tag{30}$$
$$b^1 = [-\cos\Omega\tau\,b(\delta)\,\cos\theta\,\cos\phi + \sin\Omega\tau\,\gamma\,\delta\,\sin\theta]/b(\delta) \tag{31}$$
$$b^2 = -[\sin\Omega\tau\,b(\delta)\,\cos\theta\,\cos\phi + \cos\Omega\tau\,\gamma\,\delta\,\sin\theta]/b(\delta) \tag{32}$$
$$b^3 = -\sin\phi\,\cos\theta, \tag{33}$$

and the lapse function

$$N = (\delta\gamma - \Omega\,\sin\Omega\delta\,b\,\cos\theta\,\cos\phi - \Omega\,\sin\Omega\delta\,\cos\Omega\delta)/b. \tag{34}$$

V. THE COSMOLOGICAL OBSERVER

Turning to cosmology, I consider the Friedmann-Lemaître models, i.e., spacetimes admitting spatial sections of maximal symmetry. There exists a special system of coordinates in which the metric takes the form

$$ds^2 = A(\eta)^2\,(d\eta^2 + [d\sigma^2 - S(\sigma)^2\,(d\alpha^2 + \sin^2\alpha\,d\beta^2)]), \tag{35}$$

where A is the usual scale factor, the expression between quotes is the metric of a spatial section with maximal symmetry (thus $\mathbb{R}^3$, S^3 or H^3) and η is the *conformal time*. Although different systems of coordinates would be as well convenient, I will perform calculations with the coordinates $(\eta, \sigma, \alpha, \beta)$. I consider here only a cosmological inertial observer (CIO) O_I which follows the line $\sigma = 0$, so that spherical symmetry is preserved. The proper time τ of the observer is defined by $d\tau = A\,d\eta$. The functions $\eta(\tau)$ and its inverse f such that $f[\eta(\tau)] := \tau$ will play an important role. Since $\eta > 0$, the CIO has a particle horizon and $\mathcal{M}_0$ is defined inside it, i.e., by $\sigma < \eta$.

For the CIO,

$$\mathcal{N}_\varepsilon(\eta, \sigma) = f[\eta + \varepsilon\,\sigma], \tag{36}$$

$$2T(\eta, \sigma) = f[\eta + \sigma] + f[\eta - \sigma], \tag{37}$$

$$2R(\eta,\sigma) = f[\eta+\sigma] - f[\eta-\sigma]. \tag{38}$$

Differentiation gives
$$n_\varepsilon = A^\varepsilon \, (d\eta + \varepsilon \, d\sigma), \tag{39}$$
where I have defined $A^\varepsilon(\eta,\sigma) := A(\eta + \varepsilon \, \sigma)$. Sum and difference lead to
$$dT = (A^+ + A^-)/2 \, d\eta + (A^+ - A^-)/2 \, d\sigma \tag{40}$$
and
$$dR = (A^+ - A^-)/2 \, d\eta + (A^+ + A^-)/2 \, d\sigma, \tag{41}$$
and thus
$$N^2(\eta,\sigma) = \frac{A(\eta)^2}{A^+ \, A^-}. \tag{42}$$

The parallel transport of the velocity of the CIO along the light rays leads to
$$\tilde{U}^\varepsilon = \frac{1}{A^\varepsilon} \, [((A^\varepsilon)^2 + A^2) \, d\eta + \varepsilon \, ((A^\varepsilon)^2 - A^2) \, d\sigma]. \tag{43}$$

3. Space for the inertial cosmological observer

Space, i.e., the surface Σ_τ, has the equation
$$f[\eta+\sigma] + f[\eta-\sigma] = 2\tau, \text{ with } \sigma < \eta. \tag{44}$$

Excepted in the case without expansion ($\eta = \tau$), this is not the surface $\eta = C^{te}$: space is not a spatial section with maximal symmetry, since the cosmic expansion breaks the spatial homogeneity (although not its isotropy when the observer is inertial). The spatial sections $\eta = C^{te}$, sometimes quoted as "space" do not verify the synchronicity condition (they verify a kind of synchronicity condition, but in the conformal time without physical relevance for the observer, rather than in its proper time).

Also, the cosmic expansion imprints a curvature onto space: even when spacetime admits spatial sections of constant curvature (like for instance flat in the Einstein – de Sitter case), this is not the case for the space. This appears clearly in the case of the Einstein – de Sitter model, for which detailed calculations are given in MLR. Space is limited by the horizon $\sigma = \eta$, or $T = R$. On the horizon, $A^- \to 0$, $N \to \infty$, and Space tends to become light-like.

4. Associated observers, time and distances

The *comoving* observers are defined by $\sigma = C^{te}$, and obey the equation $dT = \frac{A^+ + A^-}{A^+ - A^-} \, dR$. They differ from the associated (canonical) observers, which keep a constant PTI and obey the equation
$$dR = \frac{A^+ - A^-}{2} \, d\eta + \frac{A^+ + A^-}{2} \, d \mid \sigma \mid = 0. \tag{45}$$

An associated observer at the horizon is seen by the CIO with a redshift $z^+ \to \infty$.

All measurements made by an observer, local or not, refer to his proper time. Thus, when a CIO considers an event in spacetime, the relevant time to measure durations, or to date the event, is not t or η but T defined above (I recall that T and t coincide *on the world line of the CIO*).

On the other hand, the proper distance is intended to measure the interval between two objects considered *simultaneously*. This means, at a common value of time. But, again, no observer has access to the conformal time η. Thus, simultaneity (relative to the observer) must be defined not by η but by T as we have explained. This leads to use the *proper time distance* (I introduce this specific terminology to avoid confusion) between two objects, calculated by integration of the metric element, not along a spatial section $t = C^{te}$ (or $\eta = C^{te}$), but along Σ_T, i.e.,
$$d_{PT}(g) = \int_{\Sigma_T} ds = \int_{\Sigma_T} N \, dR. \tag{46}$$

The PT-distance is thus really the distance between two objects in space, at a given moment for the CIO. T and R appear as convenient coordinates for the CIO to measure space and time.

5. Inertial observer in de Sitter spacetime

The case of the de Sitter spacetime is particularly interesting since, because of its maximal symmetry, it has been widely considered as a frame for quantization. The metric is written as

$$ds^2 = dt^2 - \rho^2 (\cosh \rho^{-1} t)^2 [d\sigma^2 + \sin^2 \sigma \, d\Omega^2] \\ = A^2(\eta) [d\eta^2 - d\sigma^2 - \sin^2 \sigma \, d\Omega^2]. \tag{47}$$

The CIO is defined by $\sigma = 0$. The conformal time is $\eta = 2\tan^{-1}[e^{t/\rho}]$ and $A(\eta) = (\rho/2)[\tan(\eta/2) + 1/\tan(\eta/2)]$. The constant ρ characterizes the curvature of spacetime.

The proper time of the CIO is $t = \rho \ln[\tan(\eta/2)]$ so that $f(y) \equiv \rho \ln[\tan(y/2)]$. Calculation (see MLR) show that space, for the CIO at (proper) time τ, is given by

$$\tan \frac{\eta + \sigma}{2} \tan \frac{\eta - \sigma}{2} = e^{2\tau/\rho} \tag{48}$$

or

$$\sinh \frac{S - t}{\rho} = e^{2\tau/\rho} \sinh \frac{S + t}{\rho}, \tag{49}$$

where we defined $e^{S/\rho} := \tan(\sigma/2)$. Again, this is *not* the surface of constant (positive) curvature $t = C^{te}$. Thus we suggest to perform quantization with space and the orthogonal time (see Dolby and Gull 2001).

VI. DISCUSSION

The prescription S based on synchronicity defines space without ambiguity for any given observer, inertial or not, in arbitrary spacetime (without multi-crossing of null geodesics), including Minkowski and the Friedmann-Lemaître models. Space is relative to the observer, and well defined at each instant of its world line. This provides a foliation of spacetime, valid for this observer, which may be interpreted as a class of *canonically associated* observers, or a "kinematics" of spacetime (Smarr and York 1978). This provides also a natural reference frame, i.e., global space and time coordinates in the whole spacetime, which remains Minkowskian along the world line of the observer (thus, time coincides with its proper time there), thus convenient for physical measurements. In many cases (in particular for Rindler observers; see all references concerning the Rindler effect, and Sriramkumar and Padmanabhan, 1999), the coordinate system introduced here coincides with that used in the literature with no other justification than being "natural", and thus provides an *a posteriori* justification. Also, the prescription presented here applies to a range wider than other reference frames.

Application to Minkowski spacetime confirms that space and time differ for inertial observers with different velocities. It provides an unambiguous and global definition of space and time for the Langevin observer, for which the other prescriptions do not apply. For the Rindler observer (with uniform acceleration), space and time coordinates coincide with the usual Rindler coordinates. This provides a justification of their use. The corresponding interpretation of the Unruh effect involves the observer-dependent character of space and time.

In cosmology, this prescription provides, for the inertial observer in the general Friedmann-Lemaître model, an unambiguous definition of space, which *does not coincide with a spatial section of maximal symmetry*. Thus, in the Friedmann-Lemaître models, no inertial observers "sees" a homogeneous space. The lack of homogeneity of space is due to the curvature corresponding to the expansion law. In particular, space is not flat nor homogeneous (although the inertial character of the observer preserves its isotropy) in the Einstein – de Sitter model, sometimes called a "flat universe" ! I have also calculated space for the inertial observer in de Sitter spacetime, which, again, is not a hypersurface of maximal symmetry.

These results do not modify the cosmological formulae when they are expressed in a covariant form and do not involve a definition of space. However they change those interpretations of observational results, which involve a reference to space (like "*space* is homogeneous, flat" etc.). This modifies also the interpretation of the usual *proper distance*: it does not appear as the proper spatial interval between two events occurring at the same time, but rather as a mixed interval between two events which are not synchronous for the observer which performs the measurement (they would be synchronous if the observer's watch were indicating conformal time). The "proper time-distance", introduced here, represents a spatial interval between two events which are synchronous for the observer. It corresponds to the result of a practical measurement that the observer may perform with his watch indicating his proper time. Its value differs, in general, form the usual proper distance.

This prescription for space could have important implications for interpreting quantum effects in curved spacetime, and/or from the point of view of non inertial observers. Its application to the Rindler observer confirms the usual results of the Unruh effect, and provides a clearer comprehension. In other cases, the prescription adopted here differs from most attempts up to now, since the use of spatial sections with maximal symmetry (rather than space) does not obey the synchronicity requirements.

A new prescription for quantum field theory, based on the radar time and concepts very similar to those introduced here can be found in Gull's thesis (2000). This will also be the subject of a forthcoming paper. Also, subsequent work will explore in more detail the extension to arbitrary acceleration and rotation.

---

[1] Bel L. 1998, gr-qc/9812062
[2] Birrel N. D. and Davis P. C. W. 1982, *Quantum fields in curved space*, Cambridge University Press 1982
[3] Dolby C. 2000 A state-space based approach to quantum field theory in classical background fields, Thesis, available at **www.mrao.cam.ac.uk/~clifford/ publications/ abstracts/ carl_diss.html**
[4] Dolby C. E. and Gull S. F. 2001, Am. J. Phys. 69(2001) 1257-1261, gr-qc/ 0104077
[5] Ali S. T., Antoine J.-P. and Gazeau J.-P. 1990, Ann. Inst. Henri Poincaré, Vol. 52, n.1, 1990, p.83-111
[6] M. Lachièze-Rey 2001 (MLR), Space and Observers in Cosmology, Astronomy & Astrophysics, 376, 17-27 (arXiv: gr- qc/ 0107010)
[7] Landau L. and Lifshitz E. 1966, *Field Theory*, MIR (URSS), 1966
[8] Marzlin K.-P. 1994, gr- qc/ 9402010 v2
[9] Misner C. W., Thorne K. S. and Wheeler J. A. 1973, *Gravitation*, Freeman and co. 1973
[10] Pauri M. and Vallisneri M. 1999, gr-qc/ 9903052 v2
[11] Pauri M. and Vallisneri M. 2000, gr-qc/0006095
[12] Smarr L. and York J. W. Jr 1978 Phys. Rev. D, 17, 10, p. 2329 Synge J. L. 1967,
[13] Sriramkumar L. and Padmanabhan T. 1999, gr- qc/ 9903054 v2
[14] Vilain C. 2003 in *L'espace physique, entre mathématiques et philosophie*, proceedings of the 2001 Cargèse meeting, Lachièze-Rey editor, EDP Sciences, Paris 2003, in press.

P-branes, Extra Dimensions and their Observational Windows

VITALY N. MELNIKOV

*Center for Gravitation and Fundamental Metrology,
VNIIMS, 3-1 M. Ulyanovoy Str., Moscow, 117313, Russia;
Institute of Gravitation and Cosmology, Peoples' Friendship University of Russia
e-mail: rgs@com2com.ru, melnikov@rgs.phys.msu.su*

I. INTRODUCTION

The motivation for studying multidimensional models of gravitation and cosmology [1, 2] is quite apparent for several reasons. The main trend of modern physics is the unification of all known fundamental physical interactions: electromagnetic, weak, strong and gravitational. Beginning in the late 1960s there has been significant progress in unifying the weak and electromagnetic interactions, and more modest achievements in the unification of the electroweak and strong interactions via Grand Unified Theories (GUTs). There is also a rough outline for the unification of the particle physics interactions (electroweak and strong) with gravity in the form of superstring theories. Presently extensions and elaborations of the original superstring theories – theories of membranes, p-branes and the more vague M- and F-theories – are being created and studied. Since no self-consistent successful theory of this complete unification is currently available, it is desirable to study the common features of these theories, their applications to solving basic problems of gravity and cosmology, and their possible experimental consequences.

Multidimensional gravitational models, as well as scalar-tensor theories of gravity, are theoretical frameworks for describing possible temporal and scale variations of the fundamental physical constants [3, 4]. These ideas originated from the earlier papers of Milne (1935) and Dirac (1937) [5] on relations between the phenomena of micro- and macro-worlds. Additional interest to multidimensional models was caused by the ideas of large extra dimensions, brane world models etc. By applying multidimensional gravitational models to the basic problems of modern cosmology and black-hole physics, we hope to find answers to such problems as the nature of a possible cosmological constant, the acceleration of the Hubble expansion, isotropization and graceful exit problems, hierarchy and coincidence [98] problems, stability and the nature of the fundamental constants [4], the possible number of extra dimensions and their stable compactification [99, 100].

Multidimensional gravitational models are generalizations of 4D general relativity which is tested reliably for weak fields up to at least one part in 1000 and partially in strong fields via binary pulsars systems. Thus it is natural to inquire about the possible observational or experimental windows of these multidimensional extensions of general relativity. These windows include:

- Deviations from Newton's law, Coulomb's law, or new interactions [96].

- Variations of the effective gravitational constant with a time rate smaller than the Hubble one [94, 97].

- The existence of monopole modes in gravitational waves.

- Different behavior of strong field objects, such as multidimensional black holes, wormholes and p-branes [40].

- Cosmological tests for the number and nature of the extra dimensions such as variations in the microwave background from the standard cosmological models etc.

As no accepted unified model exists, the approach adopted will be to study simple, but general (from the point of view of number of dimensions) models based on multidimensional Einstein equations in vacuum or with different sources (cosmological constant, perfect and viscous fluids, scalar and electromagnetic fields, and fields of antisymmetric forms, which are related to p-branes). The main objective in our approach is to obtain exact self-consistent solutions (integrable models) for these models and then to analyze them in cosmological, spherically and axially symmetric cases. This general approach is a natural and reliable way to study highly nonlinear systems and strong field effects.

The history of the multidimensional approach begins with the well known papers of Kaluza and Klein on five-dimensional theories which initiated interest in investigations in multidimensional gravity. These ideas were continued by Jordan who suggested considering the more general case $g_{55} \neq \text{const}$, leading to a theory with an additional scalar field. These works were in some sense a source of inspiration for Brans and Dicke in their well known work on a scalar-tensor gravitational theory. After their work many investigations were performed in models with material or fundamental scalar fields, both conformal and non-conformal (see, for example, [3]).

A revival of the ideas of many dimensions started in the 1970s and continues now, mainly due to the development of unified theories. In the 1970s interest in multidimensional gravitational models was stimulated mainly by (i) the ideas of gauge theories leading to a non-Abelian generalization of the Kaluza-Klein approach and (ii) by supergravity theories. In the 1980s the supergravity theories were "replaced" by superstring models. Now it is driven by expectations connected with the overall M-theory. In all these theories, four-dimensional gravitational models with extra fields were obtained from some multidimensional model by a dimensional reduction based on the decomposition of the manifold

$$M = M^4 \times M_{\text{int}},$$

where M^4 is a four-dimensional manifold and M_{int} is some internal manifold (usually considered to be compact).

II. VACUUM AND FLUID MODELS IN DIVERSE DIMENSIONS

Much of the previous work dealt with multidimensional Einstein equations and with a block-diagonal cosmological or spherically symmetric metric, defined on the manifold $M = \mathbb{R} \times M_0 \times \ldots \times M_n$ of the form

$$g = -dt \otimes dt + \sum_{r=0}^{n} a_r^2(t) g^r$$

where (M_r, g^r) are Einstein spaces, $r = 0, \ldots, n$. In some of them a cosmological constant and simple scalar fields were also used [6]. Such models can be reduced to pseudo-Euclidean Toda-like systems with the Lagrangian [7]

$$L = \frac{1}{2} G_{ij} \dot{x}^i \dot{x}^j - \sum_{k=1}^{m} A_k e^{u_i^k x^i}$$

and the zero-energy constraint $E = 0$. In ref. [8] it was shown that there exists a special class of equations of state that gives rise to Euclidean Toda models. It should be noted that pseudo-Euclidean Toda-like systems are not well studied yet, and one of the goals is to carry out a more detailed investigation of such systems.

It is well known that cosmological solutions [1, 2] are closely related to the solutions exhibiting spherical symmetry, and relevant schemes to obtain these solutions are quite similar [1]. The first multidimensional generalization of such a type was considered by Kramer [9]. In the paper [10] the Schwarzschild solution was generalized to the case of n internal Ricci-flat spaces, showing that a black hole configuration takes place when the scale factors of internal spaces are constants. In [11] an analogous generalization was obtained for the Tangherlini solution, and an investigation of its singularities was performed in [12]. These solutions were also generalized to the electrovacuum case with and without a scalar field [13–15]. Here, it was also shown that BH's exist only when a scalar field is switched off. Deviations from Newton's and Coulomb's laws were obtained depending on the mass, charge and number of dimensions.

The stability of the various solutions was studied also. In ref. [15] an important theorem was obtained that shows that all non-black-hole configurations are unstable under even monopole perturbations. In [16] the extremely charged dilatonic black hole solution was generalized to a multicenter (Majumdar-Papapetrou) case when the cosmological constant is non-zero. For $D = 4$ the pioneering Majumdar-Papapetrou-type solutions with a conformal scalar field and an electromagnetic field were considered in refs. [17].

It is well known that part of any realistic multidimensional model should be a mechanism for extra dimension stabilization. This problem was a subject of numerous investigations. In the standard Kaluza-Klein approach cosmological models are taken in the form of warped product of Einstein spaces as internal spaces. Corresponding warp (scale) factors are assumed to be functions of external (our) space-time. If these scale factors are dynamical functions then it results in a variation of the fundamental physical constants. To be in agreement with observations, internal spaces should be static (or nearly static). The stability problem of these models with respect to conformal perturbations of the internal spaces was considered in detail in ref. [18]. It was shown that stability can be achieved with the help of an effective potential of a dimensionally reduced effective 4–D theory. Small conformal excitations of the internal spaces near minima of the effective potential have the form of massive minimal scalar fields developing in the external space-time. These particles were called gravitational excitons (gravexcitons).

A. p-brane model

Several classes of exact solutions for the multidimensional gravitational model governed by the Lagrangian

$$\mathcal{L} = R[g] - 2\Lambda - h_{\alpha\beta} g^{MN} \partial_M \varphi^\alpha \partial_N \varphi^\beta - \sum_a \frac{1}{n_a!} \exp(2\lambda_{a\alpha} \varphi^\alpha)(F^a)^2, \tag{1}$$

were considered as a next step. Here g is a metric, $F^a = dA^a$ are forms of ranks n_a, φ^α are scalar fields, and Λ is a cosmological constant.

The simplest D-dimensional theory with scalar field, 2-form and dilatonic coupling $\lambda^2 = (D-1)/(D-2)$ may be obtained by dimensionally reducing the $(D+1)$-dimensional Kaluza-Klein theory (in this case the scalar field φ is associated with the size of $(D+1)$ dimension).

For certain field contents with distinguished values of total dimension D, ranks n_a, dilatonic couplings λ_a and $\Lambda = 0$ such Lagrangians appear as "truncated" bosonic sectors (i.e. without Chern-Simons terms) of certain supergravity theories or the low-energy limit of superstring models [19]. It is now believed that all five string theories (I, IIA, IIB and the two heterotic ones with gauge groups $G = E_8 \times E_8$ and Spin(32)/Z_2) [19] as well as 11-dimensional supergravity [20] are limiting case of the conjectured M-theory [21, 22]. All these theories are conjectured to be related by a set of duality transformations: S, T and the more general U dualities [22].

It was proposed that IIB string may have its origin in a 12-dimensional theory, known as F-theory [23]. In [24] a low energy effective bosonic Lagrangian for F-theory was suggested. The field content of this 12-dimensional field model is the following: a metric, one scalar field with negative kinetic term, a 4-form and a 5-form. In [25] a chain of so-called B_D-models in dimensions $D = 11, 12, \ldots$ was suggested. B_D-models contain $l = D - 11$ scalar fields with negative kinetic terms (i.e. $h_{\alpha\beta}$ in (1) is negative definite) coupled to $(l+1)$ different forms of ranks $4, \ldots, 4 + l$. These models were constructed using p-brane intersection rules that will be discussed below. For $D = 11$ ($l = 0$) the B_D-model coincides with the truncated bosonic sector of $D = 11$ supergravity. For $D = 12$ ($l = 1$) it coincides with truncated $D = 12$ model from [24]. It was conjectured in [25] that these B_D-models for $D > 12$ may correspond to the low energy limit of some unknown F_D-theories (analogs of M and F-theories).

In [26] certain classes of p-brane solutions to field equations corresponding to the Lagrangian (1) were reviewed. These solutions have block-diagonal metrics defined on the D-dimensional product manifold as

$$g = e^{2\gamma} g^0 + \sum_{i=1}^{n} e^{2\phi^i} g^i, \qquad M_0 \times M_1 \times \ldots \times M_n, \tag{2}$$

where g^0 is a metric on M_0 and g^i are fixed Ricci-flat (or Einstein) metrics on M_i ($i > 0$). The moduli γ, ϕ^i and scalar fields φ^α are functions on M_0 and fields of forms are also governed by several scalar functions on M_0. Any

F^a is supposed to be a sum of monoms, corresponding to electric or magnetic p-branes (p-dimensional analogs of membranes), i.e. the so-called composite p-brane ansatz is considered. (In non-composite case there is no more than one monom for each F^a.) $p = 0$ corresponds to a particle, $p = 1$ to a string, $p = 2$ to a membrane. The p-brane worldvolume (worldline for $p = 0$, worldsurface for $p = 1$ etc.) is isomorphic to some product of internal manifolds: $M_I = M_{i_1} \times \ldots \times M_{i_k}$ where $1 \leq i_1 < \ldots < i_k \leq n$ and has dimension $p + 1 = d_{i_1} + \ldots + d_{i_k} = d(I)$, where $I = \{i_1, \ldots, i_k\}$ is a multiindex describing the location of the p-brane and $d_i = \dim M_i$. Any p-brane is described by the triplet, p-brane index $s = (a, v, I)$. Here a is the color index labeling the form F^a, $v = e(lectric), m(agnetic)$, and I is the multiindex defined above. For the electric and magnetic branes corresponding to form F^a the worldvolume dimensions are $d(I) = n_a - 1$ and $d(I) = D - n_a - 1$, respectively. The sum of these dimensions is $D - 2$. For $D = 11$ supergravity $d(I) = 3$ and $d(I) = 6$, corresponding to an electric $M2$-brane [27] and a magnetic $M5$-brane [28].

In [29] the model under consideration was reduced to a gravitating self-interacting σ-model with certain constraints imposed. The σ-model representation for the non-composite electric case was obtained earlier in [30, 31] (for the electric composite case see also [32]). Recently, a σ- model representation for non-block-diagonal metrics and two, intersecting branes was obtained [33].

The σ-model Lagrangian of ref. [29] has the form

$$\mathcal{L}_\sigma = R[g^0] - \hat{G}_{AB} g^{0\mu\nu} \partial_\mu \sigma^A \partial_\nu \sigma^B - \sum_s \varepsilon_s \exp(-2U^s) g^{0\mu\nu} \partial_\mu \Phi^s \partial_\nu \Phi^s - 2V, \tag{3}$$

where $(\sigma^A) = (\phi^i, \varphi^\alpha)$, V is a potential, $(\hat{G}_{AB})$ are components of (truncated) target space metric, and $\varepsilon_s = \pm 1$. Also

$$U^s = U^s_A \sigma^A = \sum_{i \in I_s} d_i \phi^i - \chi_s \lambda_{a_s \alpha} \varphi^\alpha \tag{4}$$

are linear functions, Φ^s are scalar functions on M_0 (corresponding to forms), and $s = (a_s, v_s, I_s)$. Here the parameter $\chi_s = +1$ for the electric brane ($v_s = e$) and $\chi_s = -1$ for the magnetic one ($v_s = m$).

A pure gravitational sector of the σ-model was considered earlier in [34–36] For p-brane applications g^0 is Euclidean, $(\hat{G}_{AB})$ is positive definite (for $d_0 > 2$) and $\varepsilon_s = -1$, for pseudo-Euclidean (electric and magnetic) p-branes in a pseudo-Euclidean space-time. The σ-model (3) may be also considered for the pseudo-Euclidean metric g^0 of signature $(-, +, \ldots, +)$ (e.g. in investigations of gravitational waves). In this case for a positive definite matrix $(\hat{G}_{AB})$ and $\varepsilon_s = 1$ there are non-negative kinetic energy terms.

The co-vectors U^s play a key role in studying the integrability of the field equations [29, 37, 38] and the possible existence of stochastic behavior near the singularity [39]. An important mathematical characteristic is the matrix of scalar products $(U^s, U^{s'}) = \hat{G}^{AB} U^s_A U^{s'}_B$, where $(\hat{G}^{AB}) = (\hat{G}_{AB})^{-1}$. The scalar products for co-vectors U^s were calculated in [29] as

$$(U^s, U^{s'}) = d(I_s \cap I_{s'}) + \frac{d(I_s) d(I_{s'})}{2 - D} + \chi_s \chi_{s'} \lambda_{a_s \alpha} \lambda_{a_{s'} \beta} h^{\alpha\beta},$$

where $(h^{\alpha\beta}) = (h_{\alpha\beta})^{-1}$; $s = (a_s, v_s, I_s)$, $s' = (a_{s'}, v_{s'}, I_{s'})$. They depend upon the brane intersections, the dimensions of brane worldvolumes and total dimension D, the scalar products of dilatonic coupling vectors, and the electromagnetic types of branes.

B. Cosmological and spherically symmetric solutions.

A family of general cosmological type p-brane solutions with n Ricci-flat internal spaces was considered in [38]. These solutions are defined up to solutions of Toda-type equations and may be obtained using the Lagrange dynamics following from the σ-model approach [25]. The solutions from [38] contain a subclass of spherically symmetric solutions (for $M_1 = S^{d_1}$). Special solutions with orthogonal and block-orthogonal sets of U-vectors were considered earlier in [25] and [40, 41], respectively. (For the non-composite case, see [42–44]) and references therein.)

C. Toda solutions.

In [25] the reduction of the p-brane cosmological type solutions of Toda-like systems was performed (see also [38]). General classes of p-brane solutions (cosmological and spherically symmetric) related to Euclidean Toda lattices associated with Lie algebras (mainly $\mathbf{A_m}$, $\mathbf{C_m}$) were obtained in [38, 45–49]. Special p-brane configurations were considered earlier in [50, 51].

D. Quantum cosmology.

When classical solutions exist, it is desirable to find the corresponding quantum solutions. In [25, 52] the Wheeler-DeWitt (WDW) equation for the quantum cosmology with composite electro-magnetic p-branes defined on a product of Einstein spaces was obtained (for non-composite electric case see also [43]). As in the pure gravitational case [53] this equation has a covariant and conformally covariant form. Moreover, in [25, 52] the WDW equation was integrated for intersecting p-branes with orthogonal U-vectors, when $n-1$ internal spaces are Ricci-flat and one is the Einstein space of a non-zero curvature (for the non-composite electric case see [43]). It should be mentioned also, that a slightly different approach with classical field of forms (and a rather special brane setup) was suggested in [51]. In [54] the solutions from [25] were used for constructing quantum analogs of black brane solutions.

E. Black brane solutions.

In [48, 49] a family of spherically-symmetric solutions from [38] was investigated and a subclass of black-hole configurations related to Toda-type equations with certain asymptotical conditions imposed was singled out. These black hole solutions are governed by functions $H_s(z) > 0$, defined on the interval $(0, (2\mu)^{-1})$ ($\mu > 0$ is the extremality parameter). The functions H_s obeyed the following set of differential equations (equivalent to Toda-type ones)

$$\frac{d}{dz}\left(\frac{(1-2\mu z)}{H_s}\frac{d}{dz}H_s\right) = \bar{B}_s \prod_{s'} H_{s'}^{-A_{ss'}},$$

Here $\bar{B}_s \neq 0$ and $(A_{ss'})$ is a quasi-Cartan matrix. It was shown, that for the positive definite scalar field metric $(h_{\alpha\beta})$ all p-branes in this solution should contain a time manifold [48, 49, 56]. In refs. [48, 49, 57] the following hypothesis was suggested: the functions H_s are polynomials when the intersection rules correspond to semisimple Lie algebras, i.e. when $(A_{ss'})$ is a Cartan matrix. This hypothesis was verified for Lie algebras: $\mathbf{A_m}$, $\mathbf{C_{m+1}}$, $m = 1, 2, \ldots$, in [48, 49, 103]. It was also confirmed by special black-hole "block orthogonal" solutions considered earlier in [40, 58, 59]. An analog of this conjecture for extremal black holes was considered earlier in [51]. In [48] explicit formulas for the solution corresponding to the algebra $\mathbf{A_2}$ were presented. These formulas were illustrated by two examples of $\mathbf{A_2}$-dyon solutions: a dyon in $D = 11$ supergravity (with $M2$ and $M5$ branes intersecting at a point) and a Kaluza-Klein dyon [102]. Special black brane solutions with orthogonal U-vectors were considered in [60, 61]. In [58, 62] some propositions related to (i) the interconnection between the Hawking temperature and the singularity behavior, and (ii) the multitemporal configurations were proved.

F. Variations of physical parameters and the PPN parameters.

In refs. [3, 4] the possible variation of physical constants as a result of extra dimensions and p-branes was considered. Quite recently some new estimations were made using modern data on Hubble parameter and the acceleration of the Universe in the two component dust plus (N - 1)-brane model [94, 97]. It led to possible G-dot over G ratio estimation on the level of 10^{-12} per year. Similar calculations for the time variation of G for these data within the general scalar-tensor theory led to the interval from $10^{-12} - 10^{-14}$ per year [95].

Also in refs. [90]-[93] the anholonomic frames method was applied to find anisotropic solutions to the multidimensional Einstein vacuum equations. These wormhole and blackhole solutions led in some cases to an effective variation of Newton's constant, G, and to anisotropic interactions. These could be used as signals for the presence and nature of the extra dimensions.

Another observational window on the extra dimensions and p-branes can be obtained by studying the parametrized post-Newtonian (PPN) or Eddington parameters β and γ. In [48, 57] the parameters β and γ for 4-dimensional section of the metric were calculated. It was shown that β does not depend upon the p-brane intersections, while γ does. These results agree with the earlier calculations for the block-orthogonal case [40, 59] (see also [63]). Recently, in [64] possible observational manifestations of static, spherically symmetric solutions for a class of multidimensional theories of gravity (which includes the low energy limits of supergravities and superstring theories as special cases) were considered. The choice of a physical conformal frame to be used for the description of observations was discussed. General expressions were given for (i) the Eddington parameters β and γ, characterizing the post-Newtonian gravitational field of a central body, (ii) p-brane black hole temperatures in different conformal frames and (iii) the modification of the Coulomb law by extra dimensions. It was concluded, in particular, that β and γ depend on the integration constants and can be therefore different for different central bodies. If, however, the Einstein frame is adopted for describing observations, $\gamma = 1$ is obtained. The modified Coulomb law was shown to be independent of the choice of a 4-dimensional conformal frame. It was also argued that the existence of specific multidimensional objects, T-holes, *etc.* could be potentially observable as bodies with mirror surfaces.

Extended situation with the three problems related to precise measurements of the gravitational constant and its possible time and range variations see in [105].

G. Stability of spherically symmetric solutions.

As a continuation of earlier stability studies of multidimensional solutions [15] it was shown in [65] that single-brane black hole solutions are stable under spherically symmetric perturbations, whereas similar solutions possessing naked singularities turn out to be catastrophically unstable (this conclusion may be also extended to some configurations with intersecting branes). Other possible instabilities in multidimensional models (*e.g.* caused by waves in extra dimensions) were given in [66].

H. Billiard representation near the singularity.

It is well-known, that the cosmological models with p-branes may have a "never ending" oscillating behavior near the cosmological singularity as takes place in the Bianchi-IX model. Remarkably, this oscillating behavior may be described using the so-called billiard representation near the singularity (for the multidimensional case see [68–70] and refs. therein). In [39, 71] the billiard representation for a cosmological model with a set of electro-magnetic composite p-branes in a theory with Lagrangian (1) and metric (2) was obtained. Some examples with billiards of a finite volume in multidimensional Lobachevsky space (*e.g.* triangle billiard imitating the Bianchi-IX model) and hence oscillating behavior near the singularity were considered. The U-vectors of (4) play a key role in the determination of possible oscillating behavior near the singularity. In this connection in ref. [72] conditions on the U-vectors were applied to $D = 10, 11$ supergravities and a never ending oscillatory behavior of the generic solution near the cosmological singularity was established.

I. Stability of multidimensional cosmological brane-world models.

As we wrote above, multidimensionality of our Universe follows naturally from theories unifying different fundamental interactions with gravity, e.g. M/string theory [73]. The idea has received a great deal of renewed attention over the last few years within the "brane-world" description of the Universe. In this approach the $SU(3) \times SU(2) \times U(1)$ standard model (SM) fields are localized on a 3−dimensional space-like hypersurface (brane) whereas the gravitational field propagates in the whole (bulk) space-time. The framework also implies that usual 4−dimensional physics is located on the brane (i.e. our Universe). Moreover, brane-world physics provides a possible solution of the hierarchy problem due to the well known connection between the Planck scale $M_{Pl(4)}$ and the fundamental scale $M_{*(4+D')}$ of the 4−dimensional and the $(4+D')$-dimensional space-time, respectively:

$$M_{Pl(4)}^2 \sim V_{D'} M_{*(4+D')}^{2+D'}. \qquad (5)$$

Here $V_{D'}$ denotes the volume of the compactified D' extra dimensions. It was realized in [74–76] that the localization of the SM fields on the brane allows to lower $M_{*(4+D')}$ down to the electroweak scale $M_{EW} \sim 1\text{TeV}$ without contradiction with present observations. Therefore, the compactification scale of the internal space can be of order

$$r \sim V_{D'}^{1/D'} \sim 10^{\frac{32}{D'}-17}\text{cm}. \qquad (6)$$

In this Arkani-Hamed–Dimopoulos–Dvali (ADD) model [74], physically acceptable values correspond to $D' \geq 3$ (see e.g. [77]), and for $D' = 3$ one arrives at a sub-millimeter compactification scale $r \sim 10^{-6}$cm of the internal space. Additionally, the geometry is assumed to be factorizable as in the standard Kaluza-Klein (KK) model. I.e., the topology is the direct product of a non-warped external space-time manifold and internal space manifolds with warp factors which depend on the external coordinates. Beside this, the M-theory inspired Randall–Sundrum (RS) scenario [78, 79] represents an interesting approach with non-factorizable geometry and $D' = 1$. Here, the 4−dimensional space-time is warped with a factor $\tilde{\Omega}$ which depends on the extra dimension and equation (5) is modified as follows: $M_{Pl(4)} \sim \tilde{\Omega}^{-1} M_{EW}$. In these framework, it is possible to reproduce 4–dimensional general relativity even if the extra–dimension is noncompact [79], due to the existence of a massless bound state of KK modes localized on the extra–dimension.

According to observations the internal space should be static or nearly static at least from the time of primordial nucleosynthesis, (otherwise the fundamental physical constants would vary). This means that at the present evolutionary stage of the Universe the compactification scale of the internal space should either be stabilized and trapped at the minimum of some effective potential, or it should be slowly varying (similar to the slowly varying cosmological constant in the quintessence scenario [80]). In both cases, small fluctuations over stabilized or slowly varying compactification scales (conformal scales/geometrical moduli) are possible.

Stabilization of extra dimensions (moduli stabilization) in models with large extra dimensions (ADD models) has been considered in a number of papers (an extended list of references on this topic can be found in [81]). In the corresponding approaches, as well as in the Kaluza-Klein scenario, a product topology of the $(4 + D')$−dimensional bulk space-time was constructed from Einstein spaces with scale (warp) factors depending only on the coordinates of the external 4−dimensional component. As a consequence, the conformal excitations have the form of massive scalar fields living in the external space-time. Within the framework of multidimensional cosmological models (MCM) such excitations were investigated in refs. [18], [81]-[86] where they were called gravitational excitons. Their physical meaning can be easily explained with the help of a simple 3–D model where 2–D spatial part has the cylindrical topology: $S^1 \times R^1$. Here, S^1 plays the role of the compact internal space and R^1 describes 1–D external space. Let us suppose that the size of S^1 is stabilized near some value by an effective potential. Then, conformal excitation of S^1 near its equilibrium position results in waves running along the cylinder (along R^1). Thus, any 1–D observer living on the cylinder (on R^1) will detect these oscillations as massive scalar fields. Obviously, this effect takes place for any multidimensional cosmological model with compact internal spaces. In general, it does not depend on presence or absence of branes in models. Masses of gravexcitons and equilibrium positions for the internal spaces depends on the form of the effective potential (on concrete topology and matter content of the model) (see [18], [81]-[86]).

Within the RS approach, stability of brane-world models with respect to conformal excitations was investigated in paper [87]. It was shown that models with the Poincaré and the de Sitter branes are unstable because they have negative mass squared of gravexcitons whereas models with the Anti de Sitter branes have positive gravexciton mass squared and are stable. It was also shown that 4–D effective cosmological and gravitational constants on branes as well as gravexciton masses undergo hierarchy: they have different values on different branes.

J. The fundamental constants variations in brane-world models.

Obviously, the standard matter (SM) particles may escape from the brane into a bulk resulting in the violation of the energy-momentum and charge conservation laws in the brane [88]. Such effect can take place if SM particles interact with bulk fields. For example, non-minimal coupling between bulk dilatonic scalar field φ and SM fields on the brane with the following action of interaction:

$$S_{int} = \int_{M_4} d^4x \sqrt{|h|} \, f(\varphi)\{-T + L_m[\varphi, h]\} \tag{7}$$

(where h, T and L_m are induced metric, tension and SM Lagrangian on the brane, respectively) leads to an effective conservation equation for the matter on the brane of the form

$$(f(\varphi)T_\mu^\nu[h])_{;\nu} = \varphi_{,\mu}(df/d\varphi)L_m[h]. \tag{8}$$

This equation shows that the matter is conserved on the brane if the dilaton field is either minimally coupled to the SM ($f \equiv \text{const}$) or stabilized on the brane ($\varphi|_{brane} \to \text{const}$).

¿From other hand, it is well known that interaction of the form $f(\varphi)L_m \equiv f(\varphi)F^2$ with 4-dimensional electromagnetic field F results in variation of the fine structure constant α:

$$\frac{\dot{\alpha}}{\alpha} = \frac{\dot{f}}{f}, \tag{9}$$

where the dot denotes differentiation with respect to time.

The most of the dilatonic models are motivated by string theories which, at a low-energy limit, usually have the Liouville-type potentials: $f(\varphi) = \exp(b\varphi)$ with $b \sim \mathcal{O}(1)$. However, the experimental bounds on $|\dot{\alpha}/\alpha|$ leads to the following limits on the parameter b [89]: $|b| \leq 10^{-3}$. Thus, the dilatonic models with non-minimal coupling to the SM fields on the brane are ruled out by this estimate for theories with $b \sim \mathcal{O}(1)$ [89]. In order to avoid the problem of the fundamental constant variation in the non-minimal dilatonic brane-world models, it is natural to suppose that the dilaton is stabilized on the brane (before primordial nucleosynthesis), i.e. $\varphi \to \varphi_0 \equiv \text{const}$ where φ_0 corresponds to a stable solution of the equation of motion on the brane.

Some relation of our p-brane approach to thick brane models and the treatment of nonnewtonian interactions see in [104].

III. CONCLUSIONS

One of the aims of these studies, besides the general theoretical framework, was and is to give possible experimentally detectable signatures for the presence and nature of the extra dimensions, especially possible time and range (Yukawa-type) variations of the gravitational constant or changes in values of the PPN parameters with respect to the GR ones. Such theoretical studies can help to focus the various experimental searches (both "table top" experiments, space (like STEP and SEE [105]) and collider experiments) currently being conduct and those planned for the near future. They can also suggest new experiments for determining if extra dimensions exist. Although the demonstration of the existence of extra dimensions may not have immediate practical applications, it would certainly be one of the major discoveries in fundamental science.

IV. ACKNOWLEDGMENTS

The author would like to express his gratitude to Prof. Mario Novello for his hospitality during the stay in Rio de Janeiro, CBPF, in April-June and Portobello, in August of 2002.

Partial support by the DFG grant 436 RUS 113/678/3-1 and the hospitality of Prof. Dr. H.Dehnen at the University of Konstanz was highly appreciated.

---

[1] V.N. Melnikov, "Multidimensional Classical and Quantum Cosmology and Gravitation. Exact Solutions and Variations of Constants". CBPF-NF-051/93, Rio de Janeiro, 1993; V.N. Melnikov, in: "Cosmology and Gravitation", ed. M. Novello, Editions Frontieres, Singapore, 1994, p. 147.

[2] V.N. Melnikov, "Multidimensional Cosmology and Gravitation", CBPF-MO-002/95, Rio de Janeiro, 1995, 210 p.; V.N. Melnikov. In: *Cosmology and Gravitation. II*, ed. M. Novello, Editions Frontieres, Singapore, 1996, p. 465.

[3] K.P. Staniukovich and V.N. Melnikov, "Hydrodynamics, Fields and Constants in the Theory of Gravitation", Energoatomizdat, Moscow, 1983, 256 pp. (in Russian).

[4] V.N. Melnikov, *Int. J. Theor. Phys.* **33**, 1569 (1994).

[5] P.A.M. Dirac, *Nature* **139**, 323 (1937).

[6] U. Bleyer, V.D. Ivashchuk, V.N. Melnikov and A.I. Zhuk, "Multidimensional classical and quantum Wormholes in models with cosmological constant", gr-qc/9405020; *Nucl. Phys.* **B 429**, 117 (1994).

[7] V.D. Ivashchuk and V.N. Melnikov, *Int. J. Mod. Phys.* **D 3** (1994), 795; gr-qc/9403063.

[8] V.R. Gavrilov, V.D. Ivashchuk and V.N. Melnikov, *J. Math. Phys.* **36**, 5829 (1995).

[9] D. Kramer, *Acta Physica Polonica* **2**, F. 6, 807 (1969).

[10] K.A. Bronnikov and V.D. Ivashchuk. In: Abstr. 8th Sov. Grav. Conf., Erevan, EGU, 1988, p.156.

[11] S.B. Fadeev, V.D. Ivashchuk and V.N. Melnikov, *Phys. Lett.* **A 161**, 98 (1991).

[12] V.D. Ivashchuk and V.N. Melnikov, "On singular solutions in multidimensional gravity", hep-th/9612089; *Grav. and Cosmol.* **1**, 204 (1996).

[13] S.B. Fadeev, V.D. Ivashchuk and V.N. Melnikov, *Chinese Phys. Lett.* **8**, 439 (1991).

[14] V.D. Ivashchuk and V.N. Melnikov, *Class. Quant. Grav.*, **11**, 1793 (1994).

[15] K.A. Bronnikov and V.N. Melnikov, *Annals of Physics (N.Y.)* **239**, 40 (1995).

[16] V.D. Ivashchuk and V.N. Melnikov, "Extremal dilatonic black holes in string-like model with cosmological term", *Phys. Lett.* **B 384**, 58 (1996).

[17] N.M. Bocharova, K.A. Bronnikov and V.N. Melnikov, *Vestnik MGU (Moscow Univ.)*, **6**, 706 (1970)(in Russian, English transl.: Moscow Univ. Phys. Bull., **25**, 6, 80 (1970)) — the first MP-type solution with a conformal scalar field; K.A. Bronnikov, *Acta Phys. Polonica* , **B4**, 251 (1973); K.A. Bronnikov and V.N. Melnikov, in *Problems of Theory of Gravitation and Elementary Particles*, **5**, 80 (1974) (in Russian) — the first MP-type solution with conformal scalar and electromagnetic fields.

[18] U. Günther and A. Zhuk, *Phys. Rev.* **D56**, 6391 (1997).

[19] M.B. Green, J.H. Schwarz and E. Witten, "Superstring Theory" (Cambridge University Press., Cambridge, 1987).

[20] E. Cremmer, B. Julia, and J. Scherk, *Phys. Lett.* **B76**, 409 (1978).

[21] E. Witten, "String theory dynamics in various dimensions", *Nucl. Phys.* **B 443**, 85 (1995); hep-th/9503124.

[22] C. Hull and P. Townsend, "Unity of superstring dualities", *Nucl. Phys.* **B 438**, 109 (1995); hep-th/9410167; P. Horava and E. Witten, "Heterotic and type I string dynamics from eleven dimensions", *Nucl. Phys.* **B 460**, 506 (1996); hep-th/9510209.

[23] C. Vafa, "Evidence for F-theory", hep-th/9602022; *Nucl. Phys.* **B 469**, 403 (1996).

[24] N. Khviengia, Z. Khviengia, H. Lü, C.N. Pope, "Towards a field theory of F-theory", hep-th/9703012; *Class. Quant. Grav.*, **15**, 759-773 (1998).

[25] V.D. Ivashchuk and V.N. Melnikov, Multidimensional classical and quantum cosmology with intersecting p-branes, hep-th/9708157; *J. Math. Phys.*, **39**, 2866-2889 (1998).

[26] V.D. Ivashchuk and V.N. Melnikov, "Exact solutions in multidimensional gravity with antisymmetric forms", topical review, *Class. Quant. Grav.*, **18**, R1-R66 (2001); hepth/0110274.

[27] M.J. Duff and K.S. Stelle, *Phys. Lett.* **B 253**, 113 (1991).

[28] R. Güven, *Phys. Lett.* **B 276**, 49 (1992); *Phys. Lett.* **B 212**, 277 (1988).

[29] V.D. Ivashchuk and V.N. Melnikov, "Sigma-model for the Generalized Composite p-branes", hep-th/9705036; *Class. Quantum Grav.* **14**, 3001-3029 (1997); Corrigenda *ibid.* **15** (12), 3941 (1998).

[30] V.D. Ivashchuk and V.N. Melnikov, "Intersecting p-Brane Solutions in Multidimensional Gravity and M-Theory", hep-th/9612089; *Grav. and Cosmol.* **2**, No 4, 297-305 (1996).

[31] V.D. Ivashchuk and V.N. Melnikov, *Phys. Lett. B* **403**, 23-30 (1997).

[32] V.D. Ivashchuk, V.N. Melnikov and M. Rainer, "Multidimensional Sigma-Models with Composite Electric p-branes", gr-qc/9705005; *Gravit. Cosm.* **4**, No 1(13), 73-82 (1998).

[33] D.V. Gal'tsov and O.A. Rytchkov, "Generating branes via sigma models", *Phys. Rev.* **D 58**, 122001 (1998); hep-th/9801180.

[34] V.A. Berezin, G. Domenech, M.L. Levinas, C.O. Lousto and N.D. Umerez, *Gen. Rel. Grav.*, **21**, 1177 (1989).

[35] M. Rainer and A. Zhuk, *Phys. Rev.*, **D 54**, 6186-6192 (1996).

[36] V.D. Ivashchuk and V.N. Melnikov, "Multidimensional Gravity with Einstein Internal Spaces", hep-th/9612054; *Grav. and Cosmol.*, **2**, No 3 (7), 211-220 (1996).

[37] V.D. Ivashchuk and V.N. Melnikov, "Madjumdar-Papapetrou Type Solutions in Sigma-model and Intersecting p-branes", *Class. Quant. Grav.* **16**, 849 (1999); hep-th/9802121.

[38] V.D.Ivashchuk and S.-W. Kim. "Solutions with intersecting p-branes related to Toda chains", *J. Math. Phys.*, **41** 444 (2000); hep-th/9907019

[39] V.D. Ivashchuk and V.N. Melnikov, "Billiard representation for multidimensional cosmology with intersecting p-branes near the singularity". *J. Math. Phys.*, **41**, No 8, 6341-6363 (2000); hep-th/9904077.

[40] V.D.Ivashchuk and V.N.Melnikov. Multidimensional cosmological and spherically symmetric solutions with intersecting p-branes. In Lecture Notes in Physics, Vol. 537, "Mathematical and Quantum Aspects of Relativity and Cosmology Proceedings of the Second Samos Meeting on Cosmology, Geometry and Relativity held at Pythagoreon, Samos, Greece, 1998, eds: S. Cotsakis, G.W. Gibbons., Berlin, Springer, 2000; gr-qc/9901001.

[41] V.D.Ivashchuk and V.N.Melnikov, "Cosmological and Spherically Symmetric Solutions with Intersecting p-branes". *J. Math. Phys.*, 1999, **40** (12), 6558-6576.

[42] K.A. Bronnikov, M.A. Grebeniuk, V.D. Ivashchuk and V.N. Melnikov, "Integrable Multidimensional Cosmology for Intersecting p-branes", *Grav. and Cosmol.*, **3**, No 2(10), 105-112 (1997).

[43] M.A. Grebeniuk, V.D. Ivashchuk and V.N. Melnikov, "Integrable Multidimensional Quantum Cosmology for Intersecting p-Branes", *Grav. and Cosmol.*, **3**, No 3 (11), 243-249 (1997), gr-qc/9708031.

[44] K.A. Bronnikov, U. Kasper and M. Rainer, "Intersecting Electric and Magnetic p-Branes: Spherically Symmetric Solutions", *Gen. Rel. Grav.*, **31** 1681 (1999); gr-qc/9708058.

[45] V.R. Gavrilov and V.N. Melnikov, Toda Chains with Type A_m Lie Algebra for Multidimensional Classical Cosmology with Intersecting p-branes, In : Proceedings of the International seminar "Curent topics in mathematical cosmology", (Potsdam, Germany , 30 March - 4 April 1998), Eds. M. Rainer and H.-J. Schmidt, World Scientific, 1998, p. 310; hep-th/9807004.

[46] V.R. Gavrilov and V.N. Melnikov, "Toda Chains Associated with Lie Algebras A_m in Multidimensional Gravitation and Cosmology with Intersecting p-branes", *Theor. Math. Phys.* **123**, No 3, 374-394 (2000) (in Russian).

[47] S. Cotsakis, V.R. Gavrilov and V.N. Melnikov, "Spherically Symmetric Solutions for p-Brane Models Associated with Lie Algebras", *Grav. and Cosmol.* **6**, No 1 (21), 66-75 (2000).

[48] V.D.Ivashchuk and V.N.Melnikov. "Black hole p-brane solutions for general intersection rules". *Grav. and Cosmol.*, **6**, No 1 (21), 27-40 (2000); hep-th/9910041.

[49] V.D.Ivashchuk and V.N.Melnikov. Toda p-brane black holes and polynomials related to Lie algebras. *Class. Quant. Grav.*, **17** 2073-2092 (2000); math-ph/0002048.

[50] H. Lü and C.N. Pope, "$SL(N+1,R)$ Toda solitons in supergravities", hep-th/9607027; *Int. J. Mod. Phys.* **A 12**, 2061-2074 (1997).

[51] H. Lü, J. Maharana, S. Mukherji and C.N. Pope, "Cosmological Solutions, p-branes and the Wheeler De Witt Equation", hep-th/9707182; *Phys. Rev.* **D 57** 2219-2229 (1997).

[52] V.D. Ivashchuk and V.N. Melnikov, "Multidimensional Quantum Cosmology with Intersecting p-branes", *Hadronic J.* **21**, 319-335 (1998).

[53] V.D. Ivashchuk, V.N. Melnikov and A.I. Zhuk, *Nuovo Cimento*, **B 104**, 575 (1989).

[54] V.D. Ivashchuk, M. Kenmoku and V.N. Melnikov, "On quantum analogues of p-brane black hole", *Grav. Cosmol.* **6**, No 3 (23), 225-232 (2000); gr-qc/0101043.

[55] V. Dzhunushaliev and D. Singleton, "Non-differentiable degrees of freedom: fluctuating metric signature", *Class. Quant. Grav.*, **18**, 1787 (2001); Entropy, 4, 3 (2002)

[56] K.A. Bronnikov, "Gravitating Brane Systems: Some General Theorems", gr-qc/9806102; *J. Math. Phys.* **40**, 924 (1999).

[57] V.D.Ivashchuk and V.N.Melnikov. "P-brane black Holes for General Intersections". *Grav. and Cosmol.* **5**, No 4 (20), 313-318 (1999); gr-qc/0002085.

[58] K.A. Bronnikov, "Block-orthogonal Brane systems, Black Holes and Wormholes", hep-th/9710207; *Grav. and Cosmol.* **4**, No 1 (13), 49 (1998).

[59] S. Cotsakis, V.D. Ivashchuk and V.N. Melnikov, "P-branes Black Holes and Post-Newtonian Approximation", *Grav. and Cosmol.* **5**, No 1 (17), 52-57 (1999); gr-qc/9902148..

[60] I.Ya. Aref'eva, M.G. Ivanov and I.V. Volovich, "Non-extremal intersecting p-branes in various dimensions", hep-th/9702079; *Phys. Lett.* **B 406**, 44-48 (1997).

[61] N. Ohta, "Intersection rules for non-extreme p-branes", hep-th/9702164; *Phys. Lett.* **B 403**, 218-224 (1997).

[62] K.A. Bronnikov, V.D. Ivashchuk and V.N. Melnikov, "The Reissner-Nordström Problem for Intersecting Electric and Magnetic p-Branes", gr-qc/9710054; *Grav. and Cosmol.*, **3**, No 3 (11), 203-212 (1997).

[63] V.D. Ivashchuk, V.S. Manko and V.N. Melnikov, "Post-Newtonian parameters for general black hole and spherically symmetric p-brane solutions", *Grav. Cosmol.* **6**, No 3 (23), 219-224 (2000); gr-qc/0101044.

[64] K.A. Bronnikov, V.N. Melnikov, "On observational predictions from multidimensional gravity", gr-qc/0103079; to appear

in **GRG**.

[65] K.A. Bronnikov and V.N. Melnikov, "p-Brane Black Holes as Stability Islands", *Nucl. Phys.* **B 584**, 436-458 (2000).
[66] R. Gregory and R. Laflamme, *Phys. Rev. Lett.* **70**, 2387 (1993); hep-th/9301052.
[67] K.S. Stelle, "Lectures on supergravity p-branes", hep-th/9701088.
[68] V.D. Ivashchuk and V.N. Melnikov, "Billiard representation for multidimensional cosmology with multicomponent perfect fluid near the singularity", *Class. Quantum Grav.* **12**, 809 (1995).
[69] V.D. Ivashchuk, A.A. Kirillov and V.N. Melnikov, *Izv. Vuzov (Fizika)*, No 11, 107-111 (1994) (in Russian).
[70] V.D. Ivashchuk, A.A. Kirillov and V.N. Melnikov, *Pis'ma ZhETF* **60**, No 4, 225-229 (1994) (in Russian).
[71] V.D. Ivashchuk and V.N. Melnikov, "Billiard representation for multidimensional cosmology with p-branes near the singularity". In Advanced Series in Astrophysica and Cosmology-Vol. 10. "The Chaotic Universe", Proc. of the Second ICRA Network Workshop, Eds. V.G. Gurzadyan and R. Ruffini, 1999, World Scientific, Singapore, p. 509-524.
[72] T. Damour and M. Henneaux, *Phys. Rev. Lett.* **85**, 920 (2000).
[73] M.B. Green, J.H. Schwarz and E. Witten, Superstring theory, Cambridge: Cambridge University Press, 1987; J. Polchinski, String theory, Cambridge: Cambridge University Press, 1998.
[74] N. Arkani-Hamed, S. Dimopoulos and G. Dvali, *Phys. Lett.* **B429**, 263 (1998), hep-ph/9803315.
[75] I. Antoniadis, N. Arkani-Hamed, S. Dimopoulos and G. Dvali, *Phys. Lett.* **B436**, 257 (1998), hep-ph/9804398.
[76] N. Arkani-Hamed, S. Dimopoulos and G.J. March-Russell, *Phys. Rev.* **D63**, 064020 (2001), hep-th/9809124.
[77] C.D. Hoyle et al, *Phys. Rev. Let.* **86**, 1418 (2001), hep-ph/0011014; G. Dvali, G. Gabadadze, X. Hou and E. Sefusatti, *See-saw modification of gravity*, hep-th/0111266.
[78] L. Randall and R. Sundrum, *Phys. Rev. Let.* **83**, 3370 (1999), hep-ph/9905221.
[79] L. Randall and R. Sundrum, *Phys. Rev. Lett.* **83**, 4690 (1999), hep-th/9906064.
[80] L. Wang, R.R. Caldwell, J.P. Ostriker and P.J. Steinhardt, *Astrophys. J.* **530**, 17 (2000), astro-ph/9901388.
[81] U. Günther, P. Moniz and A. Zhuk, *Phys. Rev.* **D66**, 044014 (2002), hep-th/0205148.
[82] U. Günther and A. Zhuk, *Stable compactification and gravitational excitons from extra dimensions*, (Proc. Workshop "Modern Modified Theories of Gravitation and Cosmology", Beer Sheva, Israel, June 29 - 30, 1997), *Hadronic Journal*, **21**, 279 (1998), gr-qc/9710086;
[83] U. Günther, S. Kriskiv and A. Zhuk, *Grav. and Cosmology*, **4**, 1 (1998), gr-qc/9801013;
[84] U. Günther and A. Zhuk, *Class. Quant. Grav.* **15**, 2025 (1998), gr-qc/9804018.
[85] U. Günther and A. Zhuk, *Phys. Rev.* **D61**, 124001 (2000), hep-ph/0002009.
[86] U. Günther and A. Zhuk, *Class. Quant. Grav.* **18**, 1441 (2001), hep-ph/0006283.
[87] M. Bouhmadi-López and A. Zhuk, *Phys. Rev.* **D65**, 044009 (2002), hep-th/0107227.
[88] S.L. Dubovsky, V.A. Rubakov and P.G. Tinyakov, *Phys. Rev.* **D62**, 105011 (2000), hep-th/0006046; *JHEP*, **0008** 041 (2000), hep-ph/0007179.
[89] A. Zhuk, "Restrictions on dilatonic brane-world models", hep-ph/0204195; to appear in *Int. Journ. Mod. Phys.* **D** (2002).
[90] S. Vacaru, D. Singleton, V. Botan, and D. Dotenco, "Locally anisotropic wormholes and flux tubes in 5D gravity", *Phys. Lett.* **B519**, 249 (2001)
[91] S. Vacaru and D. Singleton, "Ellipsoidal cylindrical, bipolar and toroidal wormholes in 5D gravity", *J. Math. Phys.*, **43**, 2486 (2002)
[92] S. Vacaru and D. Singleton, "Warped, anisotropic wormhole/soliton configurations in vacuum 5D gravity", *Class. Quant. Grav.*, **19**, 2793 (2002)
[93] S. Vacaru and D. Singleton, "Warped solitonic deformations and propagation of black holes in 5D vacuum gravity", *Class. Quant. Grav.*, **19**, 3583 (2002)
[94] V. N. Melnikov and V. D. Ivashchuk, in *Proceed. JGRG-11* (Tokyo, 2002).
[95] K.A. Bronnikov, V.N. Melnikov and M. Novello, *Gravit. and Cosm.*, **8**, N1-2, Suppl, 18-21 (2002)
[96] V. de Sabbata, V.N. Melnikov and P.I. Pronin, *Progr. Theor. Phys*, **88**,623 1992
[97] V.N. Melnikov, *Int. J. Mod. Phys. D* (2002)
[98] V.R. Gavrilov and V.N. Melnikov, *Gravit. and Cosm.*, **7**, N4(28), 301 (2001)
[99] A.A.Kirillov and V.N.Melnikov, *Phys. Rev. D*, **52**, 723 (1995)
[100] A.A.Kirillov and V.N.Melnikov, *Phys.Lett.B*, **389**, 221 (1996)
[101] V.D.Ivashchuk, V.N.Melnikov and A.B. Selivanov, *Gravit. and Cosm.* **7**, N 4(28), 308 (2001)
[102] V.N.Melnikov et al. *Gravit. and Cosm.* **7**, N 4(28), 343 (2001)
[103] M.A. Grebeniuk, V.D.Ivashchuk and V.N.Melnikov, *Phys.Lett.B* (2002)
[104] V.D.Ivashchuk and V.N.Melnikov , *Gravit. and Cosm.* **7**, N 3(27), 241 (2001)
[105] V.N.Melnikov, "Gravity as a Key Problem of the Millenium". Proc.2000 NASA/JPL Conference on Fundamental Physics in Microgravity,CD-version, NASA Document D-21522, 2001, p.4.1-4.17, Solvang, CA, USA; gr-qc/0007067.

EXPERIMENTAL STATUS OF CORRECTIONS TO NEWTONIAN GRAVITY INSPIRED BY THE EXTRA DIMENSIONAL PHYSICS

V. M. MOSTEPANENKO[48]

*Departamento de Física, Universidade Federal da Paraíba,
C.P. 5008, CEP 58059-970, João Pessoa, Pb-Brazil*

I. INTRODUCTION

It is well known that the gravitational interaction is described on different basis than all the other physical interactions. Up to the present there is no any unified description of gravitation and gauge interactions of the Standard Model which would be satisfactory both physically and mathematically. Gravitational interaction persistently avoids unification with the other interactions. There is an evident lack of experimental data in gravitational physics. Newtonian law of gravitation, which is also valid with high precision in the framework of the Einstein General Relativity Theory, is not verified at the separations less than 1 mm. Surprisingly, at the separations less than 1 μm corrections to the Newtonian gravitation are not excluded experimentally that are ten orders of magnitude greater than the Newtonian force itself. What this means is the general belief, that the Newtonian law of gravitation is obeyed up to Planckean separation distances, is nothing more than a large scale extrapolation. It is meaningful also that the Newtonian gravitational constant G is determined with much less accuracy than the other fundamental physical constants. In spite of all attempts the results of recent experiments on the precision measurement of G are in contradiction [1].

Prediction of non-Newtonian corrections to the law of gravitation comes from the extra dimensional unification schemes of High Energy Physics. According to this schemes, which go back to Kaluza[2] and Klein [3], the true dimensionality of physical space is larger than 3 with the extra dimensions being spontaneously compactified at the Planckean length-scale. At the separation distances several times larger than a compactification scale, the Yukawa-type corrections to the Newtonian gravitational potential must arise. This prediction would be of only academic interest if to take account of the extreme smallness of the Planckean length $l_{Pl} = \sqrt{G} \sim 10^{-33}$ cm (we use units with $\hbar = c = 1$) and the excessively high value of the Planckean energy $M_{Pl} = 1/\sqrt{G} = 10^{19}$ GeV. Recently, however, the low energy (high compactification length) unification schemes were proposed [4, 5]. In the framework of these schemes the "true", multidimensional, Planckean energy takes a moderate value $M_* = 10^3$ GeV=1 TeV and the value of a compactification scale belongs to a submillimeter range. It is amply clear that in the same range the Yukawa-type corrections to the Newtonian gravitation are expected[6, 7] and this prediction can be verified experimentally.

Much public attention given to non-Newtonian gravitation is generated not only by the extra dimensional physics. The new long-range forces which can be considered as corrections to the Newtonian law of gravitation are produced also by the exchange of light and massless hypothetical elementary particles between the atoms of closely spaced macrobodies. Such particles (like axion, scalar axion, dilaton, graviphoton, moduli, arion etc.) are predicted by many extensions to the Standard Model and practically inavoidable in the modern theory of elementary particles and their interactions [8]. The long-range forces produced due to the exchange of hypothetical particles can be considered as some corrections to the Newtonian gravitation leading to the same phenomenological consequences as in the case of extra spatial dimensions.

In the present report we summarize the best constraints on the corrections to Newtonian gravitation obtained from the recent laboratory experiments (we do not discuss the astrophysical constraints or satellite experiments in preparation). The main attention is paid to the gravitational experiments of the Eötvos- and Cavendish-types (the new constraints following from the Casimir and van der Waals force measurements are more detailly discussed in Ref. [9]). The paper is organized as follows. In Sec. 2 the types of potentials describing the corrections to Newtonian gravitation are briefly outlined. In Sec. 3 the best constraints on the parameters of these potentials following from the gravitational experiments of Eötvos- and Cavendish-type are summarized. In Sec. 4 the constraints following from the Casimir and van der Waals force measurements are collected. In Sec. 5 several conclusions are formulated. The laboratory experiments are demonstrated to have the potential for obtaining more strong constraints on the corrections to Newtonian gravitation in near future.

II. DESCRIPTION OF THE CORRECTIONS TO NEWTONIAN GRAVITATION IN TERMS OF POTENTIALS

The usual Newton's law of gravitation is only valid in a 4-dimensional space-time. If the extra dimensions exist, it will be modified by some corrections. In models with large but compact extra dimensions (like those proposed in Ref. [4]) the gravitational potential between two point particles with masses m_1 and m_2 separated by a distance

$r \gg R_*$, where R_* is a compactification scale, is given by[6, 7]

$$V(r) = -\frac{Gm_1m_2}{r}\left(1 + \alpha_G e^{-r/\lambda}\right). \tag{1}$$

The first term in the right-hand side of Eq. (1) is the Newtonian contribution, whereas the second term represents the Yukawa-type correction. Here G is the Newtonian gravitational constant, α_G is a dimensionless constant depending on the nature of extra dimensions and λ is the interaction range of a correction.

At small separation distances $r \ll R_*$ the usual Newtonian law of gravitation should be generalized to

$$V(r) = -\frac{G_{4+n}m_1m_2}{r^{n+1}} \tag{2}$$

in order to preserve the continuity of the force lines in a $(4+n)$-dimensional space-time. Here G_{4+n} is the underlying multidimensional gravitational constant connected with the usual one by the relation $G_{4+n} \sim GR_*^n$.

In fact the characteristic energy scale in multidimensional space-time is given by the multidimensional Planckean mass $M_* = 1/G_{4+n}^{1/(2+n)}$, and the compactification scale is given by[4]

$$R_* = \frac{1}{M_*}\left(\frac{M_{Pl}}{M_*}\right)^{2/n} \sim 10^{\frac{32}{n}-17}\,\text{cm}, \tag{3}$$

where $M_{Pl} = 1/\sqrt{G}$ is the usual Planckean mass, and $M_* = 10^3$ GeV as was told in Introduction. Then, for $n = 1$ (one extra dimension) one finds from Eq. (3) $R_* \sim 10^{15}$ cm. If to take into account that, as was shown in Refs. [6, 7], $\alpha_G \sim 10$ and $\lambda \sim R_*$, this possibility must be rejected on the basis of solar system tests of Newtonian gravity [10]. If, however, $n = 2$ one obtains from Eq. (3) $R_* \sim 1$ mm, and for $n = 3$ $R_* \sim 5$ nm. For these scales the corrections of form (1) to Newtonian gravity are not excluded experimentally.

The other type of multidimensional models considers noncompact but warped extra dimensions. In these models the leading contribution to the gravitational potential is given by[5, 11]

$$U(r) = -\frac{Gm_1m_2}{r}\left(1 + \frac{2}{3k^2r^2}\right), \tag{4}$$

where $r \gg 1/k$ and $1/k$ is the so-called warping scale. Here the correction to the Newtonian gravitation depends on the separation distance inverse proportionally to the second power of separation.

As was told in Introduction, many extensions to the Standard Model predict the hypothetical long-range forces, distinct from gravitation and electromagnetism, caused by the exchange of light and massless elementary particles between the atoms of macrobodies. Under appropriate parametrization of the interaction constant these forces also can be considered as some corrections to the Newtonian gravitation. The velocity independent part of the effective potential due to the exchange of hypothetical particles between two atoms can be calculated by means of Feynman rules. For the case of massive particles with mass $\mu = 1/\lambda$ (λ is their Compton wavelength) the effective potential takes the Yukawa form

$$V_{Yu}(r) = -\alpha N_1 N_2 \frac{1}{r} e^{-r/\lambda}, \tag{5}$$

where $N_{1,2}$ are the numbers of nucleons in the atomic nuclei, α is a dimensionless interaction constant. If to introduce a new constant $\alpha_G = \alpha/(Gm_p^2) \approx 1.7 \times 10^{38}\alpha$ (m_p being a nucleon mass) and consider the sum of potential (5) and Newton's gravitational potential one returns back to the potential (1).

For the case of exchange of one massless particle the effective potential is just the usual Coulomb potential which is inverse proportional to separation. The effective potentials inverse proportional to higher powers of a separation distance appear if the exchange of even number of pseudoscalar particles is considered. The power-type potentials with higher powers of a separation are obtained also in the exchange of two neutrinos, two goldstinos or other massless fermions [12]. The resulting interaction potential acting between two atoms can be represented in the form[13]

$$U(r) = -\Lambda_l N_1 N_2 \frac{1}{r}\left(\frac{r_0}{r}\right)^{l-1}, \tag{6}$$

where $r_0 = 1\,\text{F} = 10^{-15}$ m is introduced for the proper dimensionality of potentials with different l, and Λ_l with $l = 1, 2, 3, \ldots$ are the dimensionless constants.

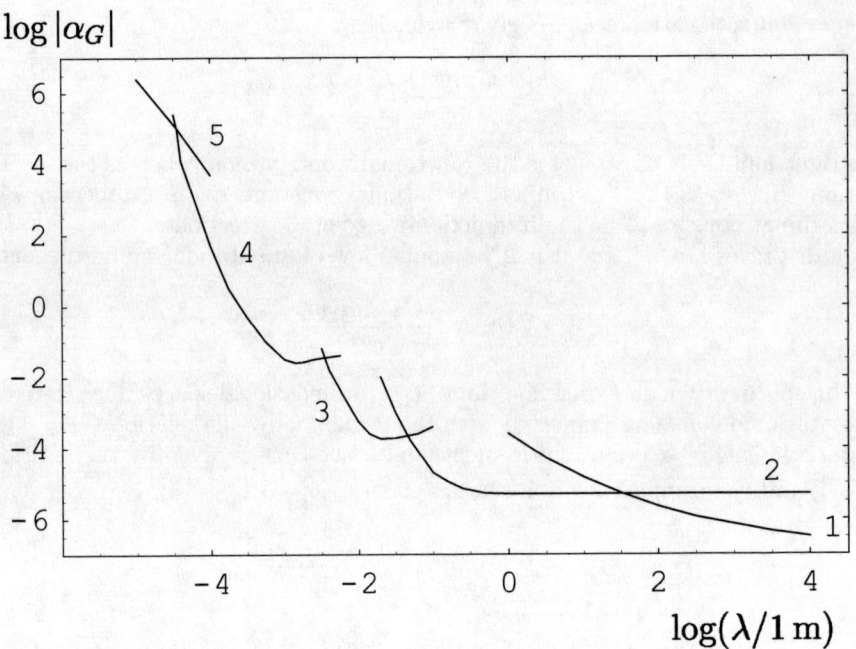

FIG. 1: Constraints on the Yukawa-type corrections to Newtonian gravitation. Curves 1,2 follow from the Eötvos-type experiments, and curves 3,4 follow from the Cavendish-type experiments. The beginning of curve 5 shows constraints from the measurements of the Casimir force. Permitted regions on (λ, α_G)-plane lie beneath the curves.

If to introduce a new set of constants $\Lambda_l^G = \Lambda_l/(Gm_p^2)$ and consider the sum of (6) and Newton's gravitational potential one obtains

$$U_l(r) = -\frac{Gm_1 m_2}{r}\left[1 + \Lambda_l^G \left(\frac{r_0}{r}\right)^{l-1}\right]. \tag{7}$$

This equation represents the power-type hypothetical interaction as a correction to the Newtonian gravitation. The potential (4) following from the extra dimensional physics is obtained from Eq. (7) with $l = 3$. Note that the case $l = 3$ corresponds also to two arions exchange between electrons [12].

III. CONSTRAINTS FROM GRAVITATIONAL EXPERIMENTS

Constraints on the corrections to Newtonian gravitation can be obtained from the experiments of Eötvos- and Cavendish-type. In the Eötvos-type experiments the difference between inertial and gravitational masses of a body is measured, i.e. the equivalence principle is verified. The existence of an additional hypothetical force which is not proportional to the masses of interacting bodies can lead to the appearance of the effective difference between inertial and gravitational masses. Therefore some constraints on hypothetical interactions emerge from the experiments of Eötvos type.

The typical result of the Eötvos-type experiments is that the relative difference between the accelerations imparted by the Earth, Sun or some laboratory attractor to various substances of the same mass is less than some small number. Many Eötvos-type experiments were performed (see, e.g., Refs. [14–17]). By way of example, in Ref. [16] the above relative difference of accelerations was to be less than 10^{-11}.

The results of the most precise Eötvos-type experiments can be found in Refs. [18, 19]. They permit to obtain the best constraints on the constants of hypothetical long-range interactions inspired by extra dimensions or by the exchange of light and massless elementary particles (see Fig. 1).

The constraints under consideration can be obtained also from the Cavendish-type experiments. In these experiments the deviations of the gravitational force F from Newtonian law are measured (see, e.g., Refs. [20–25]). The characteristic value of deviations in the case of two point-like bodies a distance r apart can be described by the parameter

$$\varepsilon = \frac{1}{rF}\frac{d}{dr}\left(r^2 F\right), \tag{8}$$

which is equal exactly to zero in the case of pure Newtonian force. For example, in Refs. [22, 23] $|\varepsilon| \leq 10^{-4}$ at the separation distances $r \sim 10^{-2} - 1$ m. This can be used to constrain the size of corrections to the Newtonian gravitation. The results of the most recent Cavendish-type experiment can be found in Ref. [26].

Let us now outline the strongest constraints on the corrections to Newtonian gravitation obtained up to date from the gravitational experiments. The constraints on the parameters of Yukawa-type correction, given by Eq. (1), are presented in Fig. 1. In this figure, the regions of (λ, α_G)-plane above the curves are prohibited by the results of the experiment under consideration, and the regions below the curves are permitted. By the curves 1 and 2 the results of the best Eötvos-type experiments are shown (Refs. [19] and [18], respectively). Curve 4 represents constraints obtained from the Cavendish-type experiment of Ref. [26]. At the intersection of curves 2 and 4 the better constraints are given by curve 3 following from the results of older Cavendish-type experiment of Ref. [25]. As is seen from Fig. 1, rather strong constraints on the Yukawa-type corrections to Newtonian gravitation ($\alpha_G < 10^{-5}$) are obtained only within the interaction range $\lambda > 0.1$ m. With decreasing λ the strength of constraints falls off, so that at $\lambda = 0.1$ mm $\alpha_G < 100$. By the beginning of curve 5 the constraints are shown following from the Casimir force measurements (see Sec. 4).

Now we consider constraints on the power-type corrections to Newtonian gravitation given by Eq. (7). The best of them follow from the Eötvos- and Cavendish-type experiments. They are collected in Table 1.

TABLE I: Constraints on the constants of power-type potentials.

l	$|\Lambda_l|_{\max}$	$|\Lambda_l^G|_{\max}$	Source
1	6×10^{-48}	1×10^{-9}	Ref. [27]
2	2.4×10^{-30}	4×10^8	Ref. [18]
3	7×10^{-17}	1.2×10^{22}	Refs. [25, 28, 29]
4	7.5×10^{-4}	1.3×10^{35}	Refs. [23, 29]
5	1.2×19^9	2×10^{47}	Refs. [23, 29]

For $l = 1, 2$ the constraints presented in Table 1 are obtained from the Eötvos-type experiments, and for $l = 3, 4, 5$ from the Cavendish-type ones. It is seen that the strength of constraints falls greatly with the increase of l.

IV. CONSTRAINTS FROM CASIMIR AND VAN DER WAALS FORCE MEASUREMENTS

As is seen from Sec. 3, for larger interaction range the best constraints on the corrections to Newtonian gravitation follow from the Eötvos-type experiments and for lesser interaction range from the Cavendish-type ones. With the further decrease of the interaction range the strength of constraints following from the gravitational experiments greatly reduces. Within a micrometer separations, the Casimir and van der Waals force [30–32] becomes the dominant force between two macrobodies. As was shown first in Ref. [33] for the case of Yukawa-type interactions in the interaction range $\lambda < 10^{-4}$ m and in Ref. [34] for the power-type ones, the measurements of the van der Waals and Casimir forces lead to the strongest constraints on non-Newtonian gravity (see the discussion about the Casimir effect as a test for non-Newtonian gravitation in Ref. [35]).

Currently a lot of precision experiments on the measurement of the Casimir and van der Waals force has been performed (see Ref. [36] for a review). Also the extensive theoretical study of different corrections to the Casimir force due to surface roughness, finite conductivity of a boundary metal and nonzero temperature gave the possibility to compute the theoretical value of this force with high precision. At the moment the agreement between theory and experiment at a level of 1% is achieved for the smallest experimental separation distances [36]. This permitted to obtain stronger constraints on the corrections to Newtonian gravitation from the results of the Casimir force measurements [37–43]. Here we briefly present only the strongest constraints of this type (see Ref. [9] for more detailed discussion of different experiments and prospects for future).

In Fig. 2 the strongest constraints on the Yukawa-type correction to Newtonian gravitation in the interaction range $\lambda < 10^{-4}$ m are presented. This figure is complementary to Fig. 1 and coveres the interaction ranges with smaller λ. The numeration of curves continues Fig. 1. By this means curve 4 is the end of the curve 4 in Fig. 1, and curves 5a,b obtained for $\alpha_G > 0$, respectively, $\alpha_G < 0$ are the continuations of curve 5 in Fig. 1. Curve 6 which bridges a gap between modern experiments follows from old Casimir force measurements between dielectrics[36]. Curve 7 was obtained[42] by the use of the most precision new experiment of Ref. [45]. Curve 9 follows[41] from the experiment of Ref. [46], and curve 9 presents the constraints obtained from old van der Waals force measurements between dielectrics[36]. By curve 10 a typical prediction of extra dimensional theories is shown. Remind that for three extra dimensions Eq. (3) gives $R_* \sim 5$ nm, and the interaction range λ is of the same order.

As is seen from Fig. 2, the present strength of constraints is not sufficient to confirm or to reject the predictions of extra dimensional physics with the compactification scale $R_* < 0.1$ mm. However, Fig. 2 gives the possibility to

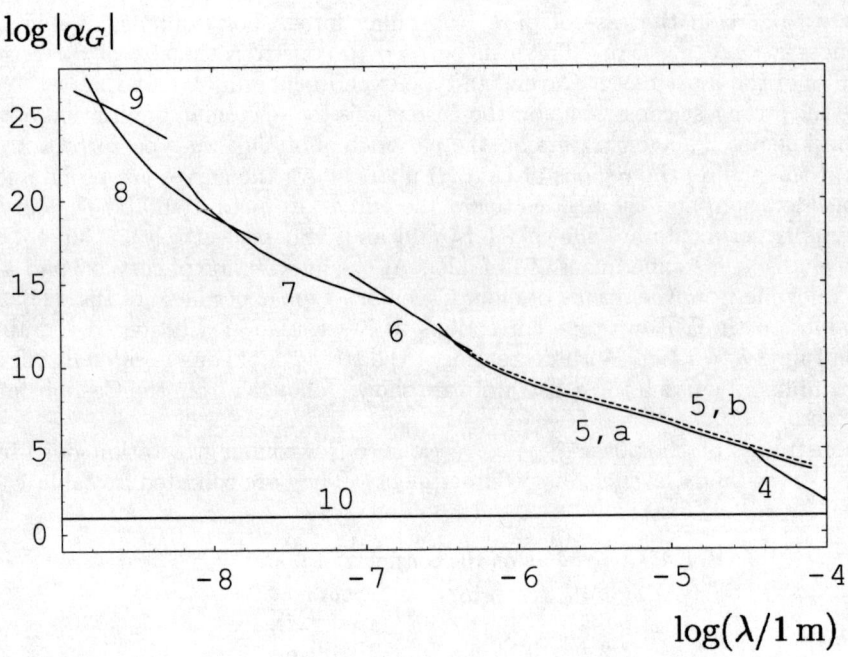

FIG. 2: Constraints on the Yukawa-type corrections to Newtonian gravitation. Curves 5–8 follow from the Casimir, and curve 9 from the van der Waals force measurements. The typical prediction of extra dimensional physics is shown by curve 10.

set constraints on the parameters of light hypothetical particles, moduli, for instance. Such particles are predicted in superstring theories and are characterized by the interaction range from one micrometer to one centimeter [47].

V. CONCLUSIONS

The above discussion permits to conclude that the laboratory experiments of the Eötvos- and Cavendish-type, and also on measurement of the Casimir and van der Waals force give the possibility to constrain corrections to Newtonian gravitation. Until recent times rather strong constraints were obtained within the interaction range $\lambda > 1$ mm. For smaller λ large work should be done in order to obtain stronger constraints. In this respect the experiments on the Casimir and van der Waals force measurements deserve more attention. So far these experiments were not especially designed to obtain stronger constraints on the corrections to Newtonian gravitation. The obtained strengthening up to 4500 times[42] is only a by-product of the recent Casimir force measurements.

A great deal needs to be done before more strong constraints could be gained from the Casimir force measurements. The most evident suggestion is to use the test bodies of larger size, made of heavier metals at increased separation distance. Also a new dynamical experiment was proposed[35, 42] designed specifically to search for the new forces rather than to test the Casimir force. There is evidently a great potential in the possibility to obtain stronger constraints on the corrections to Newtonian gravitation from the laboratory experiments of different kinds.

Acknowledgements

The author is indebted to M. Bordag, E. Fischbach, B. Geyer, G. L. Klimchitskaya, D. E. Krause and M. Novello for helpful discussions and collaboration. He is grateful to CNPq for financial support.

---

[1] G. T. Gillies, *Rep. Prog. Phys.* **60**, 151 (1997).
[2] Th. Kaluza, *Sitzungsber. Preuss. Akad. Wiss. Berlin Math. Phys.* **K1**, 966 (1921).
[3] O. Klein, *Z. Phys.* **37**, 895 (1926).
[4] N. Arkani-Hamed, S. Dimopoulos and G. Dvali, *Phys. Rev.* **D59**, 086004 (1999).

[5] L. Randall and R. Sundrum, *Phys. Rev. Lett.* **83**, 3370 (1999).
[6] E. G. Floratos and G. K. Leontaris, *Phys. Lett.* **B465**, 95 (1999).
[7] A. Kehagias and K. Sfetsos, *Phys. Lett.* **B472**, 39 (2000).
[8] J. Kim, *Phys. Rep.* **150**, 1 (1987).
[9] G. L. Klimchitskaya and U. Mohideen, *Int. J. Mod. Phys.* **A17**, contribution to this volume (2002).
[10] E. Fischbach and C. L. Talmadge, *The Search for Non-Newtonian Gravity* (Springer-Verlag, New York, 1999).
[11] L. Randall and R. Sundrum, *Phys. Rev. Lett.* **83**, 4690 (1999).
[12] V. M. Mostepanenko and I. Yu. Sokolov, *Phys. Rev.* **D47**, 2882 (1993).
[13] G. Feinberg and J. Sucher, *Phys. Rev.* **D20**, 1717 (1979).
[14] C. W. Stubbs, E. G. Adelberger, F. J. Raab, J. H. Gundlach, B. R. Heckel, K. D. McMurry, H. E. Swanson and R. Watanabe, *Phys. Rev. Lett.* **58**, 1070 (1987).
[15] C. W. Stubbs, E. G. Adelberger, B. R. Heckel, W. F. Rogers, H. E. Swanson, R. Watanabe, J. H. Gundlach and F. J. Raab, *Phys. Rev. Lett.* **62**, 609 (1989).
[16] B. R. Heckel, E. G. Adelberger, C. W. Stubbs, Y. Su, H. E. Swanson and G. Smith, *Phys. Rev. Lett.* **63**, 2705 (1989).
[17] V. B. Braginskii and V. I. Panov, *Sov. Phys. JETP* **34**, 463 (1972).
[18] G. L. Smith, C. D. Hoyle, J. H. Gundlach, E. G. Adelberger, B. R. Heckel and H. E. Swanson, *Phys. Rev.* **D61**, 022001 (2000).
[19] Y. Su, B. R. Heckel, E. G. Adelberger, J. H. Gundlach, M. Harris, G. L. Smith and H. E. Swanson, *Phys. Rev.* **D50**, 3614 (1994).
[20] S. C. Holding, F. D. Stacey and G. J. Tuck, *Phys. Rev.* **D33**, 3487 (1986).
[21] F. D. Stacey, G. J. Tuck, G. I. Moore, S. C. Holding, B. D. Goodwin and R. Zhou, *Rev. Mod. Phys.* **59**, 157 (1987).
[22] Y. T. Chen, A. H. Cook and A. J. F. Metherell, *Proc. R. Soc. London* **A394**, 47 (1984).
[23] V. P. Mitrofanov and O. I. Ponomareva, *Sov. Phys. JETP* **67**, 1963 (1988).
[24] G. Müller, W. Zurn, K. Linder and N. Rosch, *Phys. Rev. Lett.* **63**, 2621 (1989).
[25] J. K. Hoskins, R. D. Newman, R. Spero and J. Schultz, *Phys. Rev.* **D32**, 3084 (1985).
[26] C. D. Hoyle, U. Schmidt, B. R. Heckel, E. G. Adelberger, J. H. Gundlach, D. J. Kapner and H. E. Swanson, *Phys. Rev. Lett.* **86**, 1418 (2001).
[27] J. H. Gundlach, G. L. Smith, E. G. Adelberger, B. R. Heckel and H. E. Swanson, *Phys. Rev. Lett.* **78**, 2523 (1997).
[28] V. M. Mostepanenko and I. Yu. Sokolov, *Phys. Lett.* **A146**, 373 (1990).
[29] E. Fischbach and D. E. Krause, *Phys. Rev. Lett.* **83**, 3593 (1999).
[30] P. W. Milonni, *The Quantum Vacuum* (Academic Press, San Diego, 1994).
[31] V. M. Mostepanenko and N. N. Trunov, *The Casimir Effect and Its Applications* (Clarendon Press, Oxford, 1997).
[32] K. A. Milton, *The Casimir Effect* (World Scientific, Singapore, 2001).
[33] V. A. Kuz'min, I. I. Tkachev and M. E. Shaposhnikov, *JETP Lett. (USA)* **36**, 59 (1982).
[34] V. M. Mostepanenko and I. Yu. Sokolov, *Phys. Lett.* **A125**, 405 (1987).
[35] D. E. Krause and E. Fischbach, in *Gyros, Clocks, and Interferometers: Testing Relativistic Gravity in Space*, ed. C. Lämmerzahl *et al.* (Springer-Verlag, Berlin, 2001), pp. 292–309.
[36] M. Bordag, U. Mohideen and V. M. Mostepanenko, *Phys. Rep.* **353**, 1 (2001).
[37] M. Bordag, B. Geyer, G. L. Klimchitskaya and V. M. Mostepanenko, *Phys. Rev.* **D58**, 075003 (1998).
[38] M. Bordag, B. Geyer, G. L. Klimchitskaya and V. M. Mostepanenko, *Phys. Rev.* **D60**, 055004 (1999).
[39] M. Bordag, B. Geyer, G. L. Klimchitskaya and V. M. Mostepanenko, *Phys. Rev.* **D62**, 011701(R) (2000).
[40] J. C. Long, H. W. Chan and J. C. Price, *Nucl. Phys.* **B539**, 23 (1999).
[41] V. M. Mostepanenko and M. Novello, *Phys. Rev.* **D63**, 115003 (2001).
[42] E. Fischbach, D. E. Krause, V. M. Mostepanenko and M. Novello, *Phys. Rev.* **D64**, 075010 (2001).
[43] V. M. Mostepanenko, *Int. J. Mod. Phys.* **A17**, 722 (2002).
[44] S. K. Lamoreaux, *Phys. Rev. Lett.* **78**, 5 (1997); **81**, 5475(E) (1998).
[45] B. W. Harris, F. Chen and U. Mohideen, *Phys. Rev.* **A62**, 052109 (2000).
[46] T. Ederth, *Phys. Rev.* **A62**, 062104 (2000).
[47] S. Dimopoulos and G. F. Guidice, *Phys. Lett.* **B379**, 105 (1996).
[48] On leave from A. Friedmann Laboratory for Theoretical Physics, St.Petersburg, Russia. E-mail: mostep@fisica.ufpb.br

Variable Cosmological Term

Irina Dymnikova
*Department of Mathematics and Computer Science,
University of Warmia and Mazury, Zolnierska 14, 10-561 Olsztyn, Poland*[*]

In the spherically symmetric case the dominant energy condition, together with the requirement of regularity of a density and finiteness of the mass, defines the family of asymptotically flat globally regular solutions to the Einstein minimally coupled equations which includes the class of metrics asymptotically de Sitter as $r \to 0$ and asymptotically Schwarzschild as $r \to \infty$. A source term connects smoothly de Sitter vacuum in the origin with the Minkowski vacuum at infinity and corresponds to anisotropic vacuum defined macroscopically by the algebraic structure of its stress-energy tensor invariant under boosts in the radial direction. Dependently on parameters, geometry describes vacuum nonsingular black and white holes, and self-gravitating particle-like structures. ADM mass for this class is related to both de Sitter vacuum trapped inside an object and to breaking of space-time symmetry. This class of metrics is easily extended to the case of nonzero cosmological constant at infinity. The source term connects then smoothly two de Sitter vacua and corresponds to extension of the Einstein cosmological term $\Lambda g_{\mu\nu}$ to an r-dependent cosmological term $\Lambda_{\mu\nu}$. In this approach a constant scalar Λ associated with a vacuum density $\Lambda = 8\pi G \rho_{vac}$, becomes a tensor component Λ^t_t associated explicitly with a density component of a perfect fluid tensor whose vacuum properties follow from its symmetry and whose variability follows from the Bianchi identities. In this review we outline and discuss $\Lambda_{\mu\nu}$ geometry and its applications.

I. INTRODUCTION

Einstein's biggest fault - The basic idea of the Einstein General Relativity was that a matter is responsible for a geometry so that a free motion in a gravitational field is the motion along geodesics of the geometry generated by distribution of matter. He called it G-field (gravity as geometry) and described by the equations

$$G_{\mu\nu} = -\kappa T_{\mu\nu} \tag{1}$$

Here $\kappa = 8\pi G c^{-4}$ is the Einstein gravitational constant. One of Einstein's primary motivations [1] was the Mach principle according to which inertial motion is due to global distribution of matter - some matter has the property of inertia only because there exists also some other matter in the Universe [2].

Following this idea Einstein expected that in a consequent theory, matter would determine any geometry, in particular that reasonable regular solution to his equations (1) would describe inertial motion as dictated by global matter distribution. When he found that the Minkowski geometry is the regular solution to (1) perfectly describing inertial motion in absence of any matter (test particles follow geodesics which are straight lines of the Newton Law of inertia), he modified his equations (1) by adding a cosmological term $\Lambda g_{\mu\nu}$ in the hope that modified equations

$$G_{\mu\nu} + \Lambda g_{\mu\nu} = -\kappa T_{\mu\nu} \tag{2}$$

would have reasonable regular solutions describing inertial motion properly only when matter is present. His reasoning was that if matter is the source of inertia, then in case of its absence there should not be any inertia [3]. The primary task of Λ was thus actually to eliminate inertia in the case when matter is absent by eliminating regular G-field solutions in case when $T_{\mu\nu} = 0$.

Soon after introducing $\Lambda g_{\mu\nu}$, de Sitter found quite reasonable regular solution [4]

$$ds^2 = \left(1 - \frac{\Lambda}{3}r^2\right)dt^2 - \frac{dr^2}{\left(1 - \frac{\Lambda}{3}r^2\right)} - r^2 d\Omega^2 \tag{3}$$

for the case when a matter is absent

$$G_{\mu\nu} + \Lambda g_{\mu\nu} = 0 \tag{4}$$

---

[*]Electronic address: irina@matman.uwm.edu.pl

which made evident that a matter is not necessary to produce the property of inertia [5].

During the next decade Einstein used $\Lambda g_{\mu\nu}$ as a universal repulsion whose task was to make a universe static, until successes of FRW cosmology confirmed by Hubble's discovery of the Universe expansion, enforced him to conclude that Λ is quite useless and he abandoned it as "the biggest fault in his carreer" [5].

Approximately at the same time Eddington, discussing in his book geodesics of de Sitter geometry (3), asked - *May be this is why the Universe is expanding?* [6].

The cosmological aspect of Λ story somehow left in shadow the primary sense of the Einstein idea of introducing Λ as some thing which has something in common with inertia.

De Sitter vacuum - The Einstein equations (4) can be written in the form

$$G_\mu^\nu = -\Lambda \delta_\mu^\nu \qquad (5)$$

which associates the Λ-term $\Lambda g_{\mu\nu}$ with a stress-energy tensor (SET)

$$\Lambda \delta_\mu^\nu = \kappa T_\mu^\nu = \kappa \rho_{vac} \delta_\mu^\nu \qquad (6)$$

identified as a vacuum SET [7] by its algebraic structure.

In the Petrov classification of SET's by their algebraic structure [8], the stress-energy tensor (6) is denoted as [IIII], all eigenvalues equal. Due to such a structure, any reference frame is comoving with a medium characterized by a SET of type (6). A comoving reference frame for an observer moving through (6) is comoving also with (6) (as any other), as a result an observer cannot measure his velocity with respect to vacuum (6) [7].

In de Sitter geometry Λ must be constant by virtue of the Bianchi identities $G^{\mu\nu}_{;\nu} = 0$. De Sitter vacuum described by the stress-energy tensor (6) has therefore constant energy density

$$\rho_{vac} = \kappa^{-1} \Lambda \qquad (7)$$

Observational case for Λ - Properties of de Sitter geometry ultimately advanced Λ to provide the generic reason for the Universe accelerated expansion [9] producing a huge growth of the scale factor sufficient to explain various puzzles of the standard big bang cosmology (for review see [10, 11]). At the inflationary stage the vacuum density corresponding to Λ is estimated for the GUT scale vacuum, $E_{GUT} \sim 10^{15}$GeV, as $\rho_{vac} \sim 10^{77}$g cm^{-3}. On the other hand, astronomical observations convincingly testify for the today value of cosmological constant corresponding to $\rho_{vac} \sim 0.7 \rho_{total} \sim 10^{-30}$g cm^{-3} [12].

In this talk we outline the possible way to make Λ variable. We address this question in the frame of the Einstein non-minimally coupled equations (1) and the Petrov classification for stress-energy tensors [8]. Actually the question sounds - What kind of behavior follows from equations (1) when certain reasonable requirements imposed on a source term [13]. Model-independent analysis [13, 14] leads to a vacuum geometry with the regular center where de Sitter vacuum replaces a Schwarzschild singularity while a source term (identified as vacuum SET by its algebraic structure) connects smoothly two de Sitter vacua with different values of cosmological constant. $\Lambda_{\mu\nu}$-geometry [15] (for review [16–18]) is able to describe vacuum cosmologies dominated by evolving Λ [19, 20] as well as vacuum objects due to clustering Λ: vacuum non-singular black hole [21] and self-gravitating particlelike structure [22], called G-lump whose zero-point vacuum energy coincides up to a coefficient with the Hawking temperature from de Sitter horizon [13]. The connection between Λ and inertia appears in this geometry as a by-product [13].

This talk is organized as follows. In Section 2 we show how a vacuum spherically-symmetric space-time with the regular de Sitter center appears among solutions to (1). Section 3 presents the cosmological term $\Lambda_{\mu\nu}$ in the spherically symmetric case. Section 4 summarizes applications of $\Lambda_{\mu\nu}$ geometry of clustering type - vacuum nonsingular black and white hole and G-lump. Section 5 is devoted to connection between cosmological term and mass. In Section 6 we outline $\Lambda_{\mu\nu}$ dominated cosmologies. Section 7 contains summary and discussion.

II. SPHERICALLY SYMMETRIC SPACE-TIME WITH THE REGULAR DE SITTER CENTER

Basic equations -

A static spherically symmetric line element can be written in the form [23]

$$ds^2 = e^{\mu(r)}dt^2 - e^{\nu(r)}dr^2 - r^2 d\Omega^2 \qquad (8)$$

where $d\Omega^2$ is the metric of a unit 2-sphere. The metric coefficients satisfy the Einstein equations

$$\kappa T_t^t = \kappa \rho(r) = e^{-\nu}\left(\frac{\nu'}{r} - \frac{1}{r^2}\right) + \frac{1}{r^2} \qquad (9)$$

$$\kappa T_r^r = -\kappa p_r(r) = -e^{-\nu}\left(\frac{\mu'}{r} + \frac{1}{r^2}\right) + \frac{1}{r^2} \qquad (10)$$

$$\kappa T_\theta^\theta = \kappa T_\phi^\phi = -\kappa p_\perp(r) =$$

$$-e^{-\nu}\left(\frac{\mu''}{2} + \frac{\mu'^2}{4} + \frac{(\mu'-\nu')}{2r} - \frac{\mu'\nu'}{4}\right) \qquad (11)$$

Here $\rho(r) = T_t^t$ is the energy density (we adopted $c=1$ for simplicity), $p_r(r) = -T_r^r$ is the radial pressure, and $p_\perp(r) = -T_\theta^\theta = -T_\phi^\phi$ is the tangential pressure for anisotropic perfect fluid [23]. The prime denotes differentiation with respect to r. Integration of Eq.(9) gives

$$e^{-\nu(r)} = g(r) = 1 - \frac{2GM(r)}{r}; \quad M(r) = 4\pi\int_0^r \rho(x)x^2 dx \qquad (12)$$

whose asymptotic for large r is $e^{-\nu} = 1 - 2Gm/r$, with the mass parameter m given by

$$m = 4\pi\int_0^\infty \rho(r)r^2 dr \qquad (13)$$

Equations (9)-(10) give the Oppenheimer equation [24]

$$\kappa(T_t^t - T_r^r) = \kappa(p_r + \rho) = \frac{e^{-\nu}}{r}(\nu' + \mu') \qquad (14)$$

Using equations (9)-(11) we derive hydrodynamic equation which generalizes the Tolman-Oppenheimer-Volkoff equation [25] to the case of different principal pressures. It reads [13, 26]

$$p_\perp = p_r + \frac{r}{2}p_r' + (\rho + p_r)\frac{GM(r) + 4\pi Gr^3 p_r}{2(r - 2GM(r))} \qquad (15)$$

The geometry is described by two metric functions, $\mu(r)$ and $\nu(r)$. The Oppenheimer equation (14) is of special interest since it connects the derivative $\mu'(r) + \nu'(r)$ with the quantity $p_r + \rho$ which allows us to investigate behavior of metric functions by applying the energy conditions.

To investigate the system with the different principal pressures we impose the following requirements.

Requirements:
1 Regularity of a density $\rho(r)$
2 Finiteness of the mass parameter m
3 Dominant Energy Condition (DEC) on $T_{\mu\nu}$

The dominant energy condition, $T^{00} \geq |T^{ab}|$ for each $a,b = 1,2,3$, holds if and only if [27]

$$\rho \geq 0; \quad -\rho \leq p_k \leq \rho; \quad k = 1,2,3 \qquad (16)$$

and implies that the energy density as measured by any local observer moving along a time-like curve, is non-negative, and each principal pressure does not exceed the energy density.

The Weak Energy Condition (WEC) contained in the DEC, reads $T_{\mu\nu}\xi^\mu\xi^\nu \geq 0$ for any time-like vector ξ^μ and holds if and only if

$$\rho \geq 0, \quad \rho + p_k \geq 0, \quad k = 1,2,3 \qquad (17)$$

The requirements 1-3 imposed on a system (9)-(11), enforce the following behavior [13, 14]
Behavior -
Finiteness of a mass (13) leads to $\nu(r) = 0$ as $r \to \infty$, and requires a density profile $\rho(r)$ to vanish at infinity quicker than r^{-3}. DEC requires p_k to vanish equally fast or faster as $r \to \infty$. Then, by Oppenheimer equation (14), $\mu' = 0$ and $\mu =$const at infinity. By rescaling the time coordinate we get the standard boundary condition $\mu \to 0$ as $r \to \infty$ [25] which ensures asymptotic flatness needed to identify (13) as the ADM mass [25].

Regularity of density $\rho(r=0) < \infty$, requires the mass function $M(r)$ to vanish as r^3 when $r \to 0$, as a result $\nu(r) \to 0$ as $r \to 0$. It leads also, by DEC, to regularity of pressures, then $p_r + \rho < \infty$ leads to $\nu' + \mu' = 0$ and $\nu + \mu = \mu(0)$ at $r = 0$ with $\mu(0)$ playing the role of the family parameter.

The weak energy condition defines, by the Oppenheimer equation (14), the sign of the sum $\mu'+\nu'$. In the case when $e^{\nu(r)} > 0$ everywhere, it demands $\mu'+\nu' \geq 0$ everywhere. In case when $e^{\nu(r)}$ changes sign, the function $T_t^t - T_r^r$ is zero at the horizons. In the regions inside the horizons, the radial coordinate r is time-like and T_t^t represents a tension, $p_r = -T_t^t$, along the axes of the space-like 3-cylinders of constant time r=const [28], then $T_t^t - T_r^r = -(p_r + \rho)$, and the WEC still demands $\nu' + \mu' \geq 0$ there. As a result the function $\mu + \nu$ is growing from $\mu = \mu(0)$ at $r = 0$ to $\mu = 0$ at $r \to \infty$, which gives $\mu(0) \leq 0$ [13].

The example of solution from this family is boson stars [29] (for review [30]) which are regular configurations without horizons generated by a self-gravitating massive scalar field whose SET is essentially anisotropic, $p_r \neq p_\perp$.

The range of family parameter $\mu(0)$ includes $\mu(0) = 0$. In this case the function $\nu(r) + \mu(r)$ is zero at $r = 0$ and at $r \to \infty$, its derivative is non-negative, it follows that $\nu(r) = -\mu(r)$ everywhere.

For this class of metrics behavior at $r \to 0$ is dictated by the WEC. It is easily to prove [13] that the function $\mu(r) + \nu(r)$, which is equal zero everywhere for $0 \leq r < \infty$, cannot have extremum at $r = 0$, therefore $\mu'' + \nu'' = 0$ at $r = 0$. It leads to $p_r + \rho = 0$ at $r = 0$, and in the limit $r \to 0$ Eq.(15) gives $p_\perp = -\rho - \frac{r}{2}\rho'$. The DEC and regularity of ρ requires $p_k + \rho < \infty$ and thus $|\rho'| < \infty$. The equation of state near the center becomes $p = -\rho$, which gives de Sitter asymptotic as $r \to 0$ [13]

$$ds^2 = \left(1 - \frac{r^2}{r_0^2}\right)dt^2 - \frac{dr^2}{\left(1 - \frac{r^2}{r_0^2}\right)} - r^2 d\Omega^2 \tag{18a}$$

$$T_{\mu\nu} = \rho_0 g_{\mu\nu}; \qquad \rho_0 = \kappa^{-1}\Lambda; \qquad r_0^2 = \frac{3}{\Lambda} \tag{18b}$$

where $\rho_0 = \rho(r=0)$ and Λ is the cosmological constant which appeared at the origin although was not present in the basic equations.

The WEC, $p_\perp + \rho \geq 0$, demands monotonic decreasing of a density profile, $\rho' \leq 0$. The simple analysis shows that a metric from this class cannot have more than two horizons [13].

Summarizing we see that the requirements 1-3 lead to the existence of the class of metrics

$$ds^2 = g(r)dt^2 - \frac{dr^2}{g(r)} - r^2 d\Omega^2 \tag{19a}$$

$$g(r) = 1 - \frac{R_g(r)}{r}; \qquad R_g(r) = 2GM(r) \tag{19b}$$

with $M(r)$ given by Eq.(12).

Any metric from this class has the form shown in Fig.1.

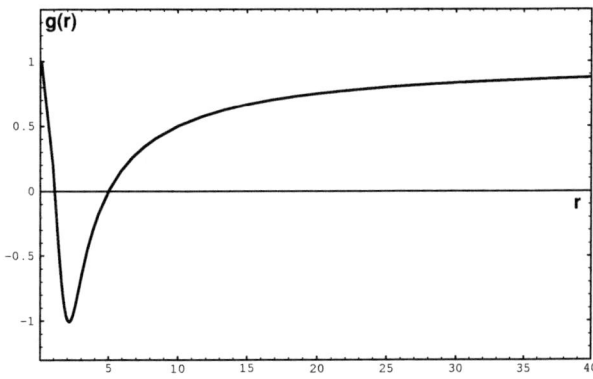

FIG. 1: The metric function $g(r)$ for the class (19) [22].

It is asymptotically de Sitter as $r \to 0$, asymptotically Schwarzschild at infinity

$$ds^2 = \left(1 - \frac{r_g}{r}\right) - \frac{dr^2}{\left(1 - \frac{r_g}{r}\right)} - r^2 d\Omega^2; \qquad r_g = 2Gm \tag{20}$$

and has not more than two horizons: a black hole event horizon r_+ and an internal Cauchy horizon r_- [28].

This class of metrics is easily extended to the case of nonzero background cosmological constant λ by adding $\lambda g_{\mu\nu}$ to the field equations which leads to the metric function $g(r)$ of the form [31]

$$g(r) = 1 - \frac{2GM(r)}{r} - \frac{\lambda r^2}{3} \qquad (21)$$

plotted in the Fig.2. Its asymptotics are de Sitter behavior at both origin and infinity, with λ as $r \to \infty$ and with

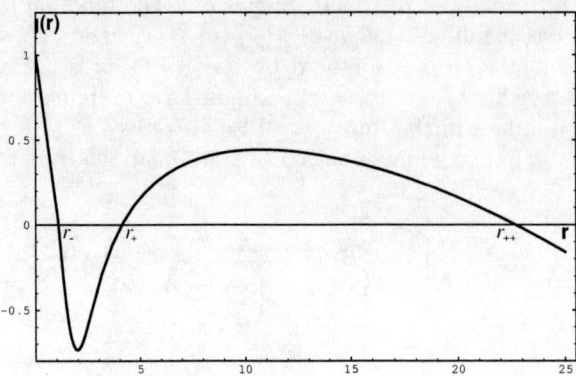

FIG. 2: The metric function (21) [31].

$\Lambda + \lambda$ as $r \to 0$.

III. COSMOLOGICAL TERM IN THE SPHERICALLY SYMMETRIC CASE

For the class of metrics (19) a source term has the algebraic structure

$$T_t^t = T_r^r; \quad T_\theta^\theta = T_\phi^\phi \qquad (22)$$

and the equation of state

$$p_r = -\rho; \quad p_\perp = -\rho - \frac{r}{2}\rho' \qquad (23)$$

In the Petrov classification scheme such a stress-energy tensor is denoted by [(II)(II)]. It is invariant under rotations in the (r,t) plane. Therefore an observer moving through a medium (22) cannot in principle measure the radial component of his velocity. The stress-energy tensor with the algebraic structure (22) has an infinite set of comoving reference frames and is identified as describing a spherically symmetric vacuum $T_{\mu\nu}^{vac}$, invariant under boosts in the radial direction [21].

For considered class of metrics it must be asymptotically de Sitter as $r \to 0$. It connects de Sitter vacuum $T_{\mu\nu} = \rho_0 g_{\mu\nu}$ in the origin with the Minkowski vacuum $T_{\mu\nu} = 0$ at infinity for the case (19) or two de Sitter vacua with different values of cosmological constant in the case (21):

$$\kappa^{-1}\Lambda g_{\mu\nu} \quad \leftarrow \quad T_{\mu\nu}^{vac} \quad \to \quad \kappa^{-1}\lambda g_{\mu\nu} \qquad (24)$$

There was discussion in the literature (see, e.g., [32]) where to put the Einstein cosmological term $\Lambda g_{\mu\nu}$. Keeping it on the left-hand side of the Einstein equations (2), one treats Λ as a geometrical entity. Shifting it to the right-hand side of (2), one treat $\Lambda g_{\mu\nu}$ as de Sitter vacuum with constant energy density $\rho_{vac} = \kappa^{-1}\Lambda$.

Here we started from the Einstein equation (1) and found the case when we have on the right-hand side a stress-energy tensor describing a spherically symmetric anisotropic vacuum with variable density and pressures. Nothing would prevent from shifting it to the left-hand side and treating as a variable cosmological term [15]

$$\Lambda_{\mu\nu} = \kappa T_{\mu\nu}^{vac} \qquad (25)$$

evolving from $\Lambda_{\mu\nu} = \Lambda g_{\mu\nu}$ at $r = 0$ to $\Lambda_{\mu\nu} = \lambda g_{\mu\nu}$ as $r \to \infty$, and satisfying the equation of state (23) with

$$\kappa \rho^\Lambda(r) = \Lambda_t^t; \quad \kappa p_r^\Lambda(r) = -\Lambda_r^r; \kappa p_\perp^\Lambda(r) = -\Lambda_\theta^\theta = -\Lambda_\phi^\phi \qquad (26)$$

Einstein equation (1) in this case can be written as

$$G^\nu_\mu + \Lambda^\nu_\mu = 0 \qquad (27)$$

and the Bianchi identities determine the evolution of $\Lambda_{\mu\nu}$

$$\Lambda^\mu_{\nu;\mu} = 0 \qquad (28)$$

A cosmological tensor $\Lambda_{\mu\nu}$ represents the extension of the algebraic structure of the Einstein cosmological term $\Lambda g_{\mu\nu}$. The cosmological term $\Lambda_{\mu\nu}$ includes $\Lambda g_{\mu\nu}$ as the particular case ρ_{vac}=const when the full symmetry is restores, and remains $\Lambda g_{\mu\nu}$ as proper asymptotics of $\Lambda_{\mu\nu}$ at both regular center and infinity. In such an extension the cosmological constant - scalar associated with the vacuum density (7) - becomes a tensor component Λ^t_t associated *explicite* with the density component of vacuum tensor whose variability is dictated by the Bianchi identities.

De Sitter-Schwarzschild geometry - In the case when geometry is asymptotically Schwarzschild at $r \to \infty$, type of configuration depends on the mass m of an object. The key point of de Sitter-Schwarzschild geometry is the existence of two horizons. A critical value of a mass parameter exists, m_{crit}, at which the horizons come together and which puts a lower limit on a black hole mass [22]. De Sitter-schwarzschild configurations are plotted in Fig.3 for the case of the density profile [21]

$$\rho(r) = \rho_0 e^{-r^3/r_0^2 r_g} \qquad (29)$$

which can be interpreted [22] as due to vacuum polarization in the spherically symmetric gravitational field described semiclassically by the Schwinger formula $w \sim exp(-F_{crit}/F)$ (see, e.g., [33]) with tidal forces due to Riemann curvature, $F \sim r_g/r^3$ and $F_{crit} \sim r_0^{-2}$. The idea goes back to the Zeldovich interpretation of Λ as due to gravitational interaction of virtual particles in vacuum [34], and to the Poisson and Israel elegant picture of the Schwarzschild-de Sitter transition in which *geometry can be self-regulatory* and describable semiclassically down a few Planckian radii by the Einstein equations with a source term representing vacuum polarization effects [28]. The metric function $g(r)$

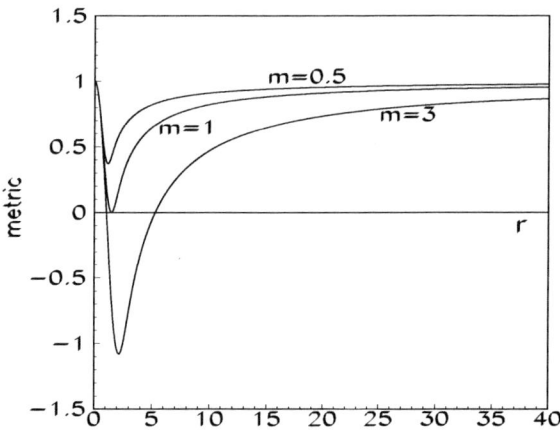

FIG. 3: De Sitter-Schwarzschild configurations. Mass parameter is normalized to m_{crit}.

for this case is given by (19b) with the r−dependent gravitational radius

$$R_g(r) = r_g\left(1 - e^{-r^3/r_0^2 r_g}\right) \qquad (30)$$

and the critical value of the mass parameter is

$$m_{crit} \simeq 0.3 m_{Pl}\sqrt{\rho_{Pl}/\rho_0} \qquad (31)$$

For masses $M \geq m_{crit}$ de Sitter-Schwarzschild geometry describes (for any particular density profile satisfying requirements 1-3) a black hole without a singularity.

Vacuum nonsingular black hole - The global structure of space-time, shown in Fig.4 [22], contains an infinite sequence of black and white holes whose future and past singularities (see Fig.5) are replaced with regular cores $\mathcal{RC}$ asymptotically de Sitter as $r \to 0$, and asymptotically flat universes $\mathcal{U}$. The Penrose-Carter diagram is plotted in coordinates related to the photon radial geodesics. The surfaces $\mathcal{J}^-$ and $\mathcal{J}^+$ represent their past and future infinities. The event horizon r_+ and the Cauchy horizon r_-, are formed by the outgoing and ingoing photon geodesics $r_\pm =$const. Interpretation of the internal horizon r_- as the Cauchy horizon is due to existence of geodesics inextendible to the past which implies the absence of a global Cauchy surface in the geometry. Vacuum nonsingular black and white

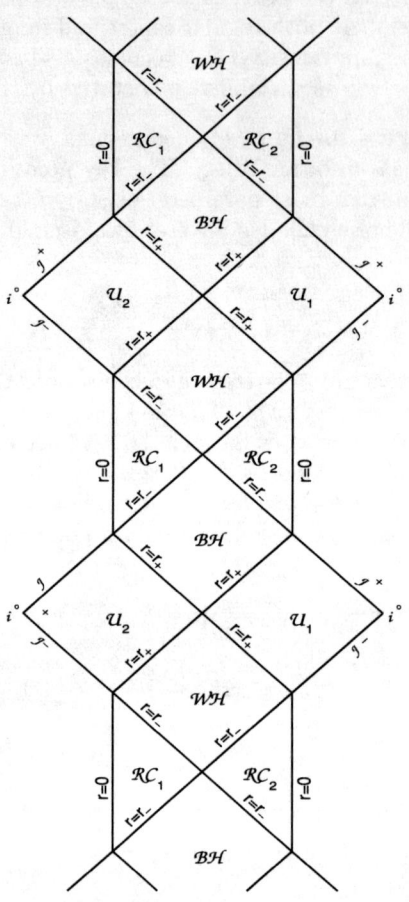

FIG. 4: Penrose-Carter diagram for Λ black hole.

holes can be called ΛBH and ΛWH, since in place of Schwarzschild singularities (Fig.5) an infinite ladder of structures appears asymptotically de Sitter ($\Lambda g_{\mu\nu}$) at approaching each regular surface $r = 0$.

Vacuum nonsingular white hole - Replacing a Schwarzschild singularity with the regular core $\mathcal{RC}$ transforms the space-like singular surfaces $r = 0$ both in the future of Schwarzschild BH and in the past of Schwarzschild WH (see Fig.5), into the time-like regular surfaces $r = 0$ in the future of a ΛBH and in the past of a ΛWH. In a sense this rehabilitates a white hole whose existence in a singular version has been forbidden by the cosmic censorship since a singularity open into the future of a universe U breaks the predictability in U [35].

The regular core $\mathcal{RC}$ in the past of a ΛWH models an early evolution of an expanding universe [19]. In the coordinates related to radial geodesics of non-relativistic test particles, the static metric (19) transforms into the Lemaitre-type metric

$$ds^2 = d\tau^2 - (1 - g(r(R,\tau)))dR^2 - r^2(R,\tau)d\Omega^2 \tag{32}$$

Near the regular surface $r = 0$ corresponding to $R + \tau = -\infty$ (see Fig.6), the metric (32) transforms into the FRW form with the de Sitter scale factor $a(\tau) \sim \cosh(H_0\tau)$ for $f(R) < 0$, $a(\tau) \sim \exp(H_0\tau)$ for $f(R) = 0$, $a(\tau) \sim \sinh(H_0\tau)$

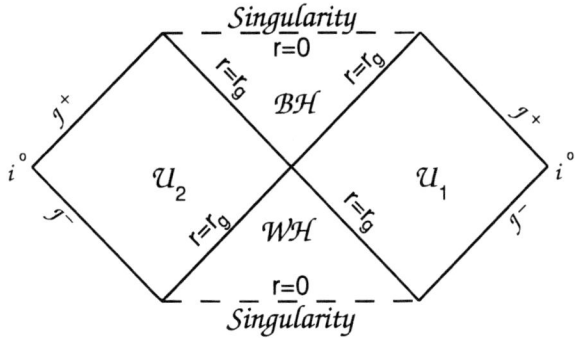

FIG. 5: Penrose-Carter diagram for Schwarzschild BH.

for $f(R) > 0$, where H_0 is the Hubble parameter corresponding to the initial value of Λ. The evolution starts with the nonsingular nonsimultaneous de Sitter bang [19]. The inflationary start is followed by a Kasner-type stage of

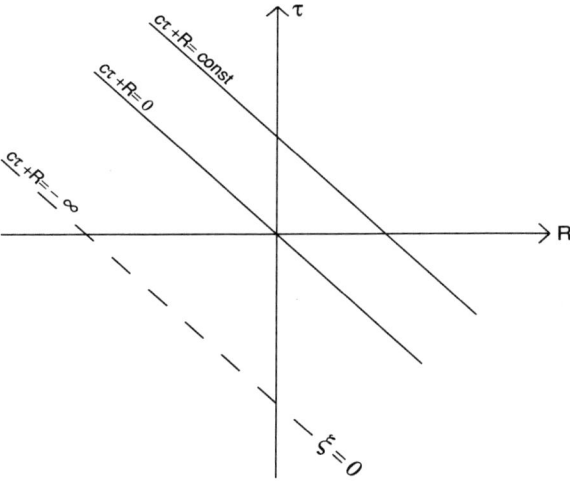

FIG. 6: The Lemaitre metric for a nonsingular white hole. Surfaces $r = const$ are plotted for the dimensionless radius ξ. The surface $\xi = 0$ is the big bang surface.

anisotropic expansion at which most of a universe mass is produced [19].

Baby universes inside a ΛBH - The idea of a universe inside a black hole has been suggested by Farhi and Guth (FG) in 1987 [36] as the idea of creation of a universe in the laboratory starting from a false vacuum bubble in the Minkowski space. They studied an expanding spherical de Sitter bubble separated by a thin wall from the outside region of the Schwarzschild geometry. The global structure of spacetime in this case (Fig.7) implies that the expanding bubble must be associated with an initial spacelike singularity which represents a singular initial value configuration. Therefore Farhi and Guth concluded that the initial singularity would be an unavoidable obstacle to creation of a universe in the laboratory [36]. A year later Poisson and Israel found that the Cauchy horizon must exist in this geometry (H_C at the Fig.1) [28]. In !989 arising a new universe inside a black hole has been considered by Frolov, Markov and Mukhanov [37] in the context of the hypothesis that the curvature is limited by the Planckian scale and that at this scale the equation of state becomes $p = -\rho$. The difference of this approach from the approach of Farhi and Guth was that an assumption of existence of a global Cauchy surface can be violated due to existence of the Cauchy horizon [37].

The case of direct de Sitter-Schwarzschild matching (Fig.7) clearly corresponds to arising of a closed or semiclosed world inside a black hole [37].

In general case of a distributed density profile (the global structure of space-time as shown in Fig.4), the physical situation near the surface $r = 0$ is similar to that considered by Farhi and Guth. This region, which is the part of the regular core $\mathcal{RC}$, differs from the Farhi-Guth bubble only by that a density profile is r-dependent. On the other hand, the global structure Fig.4 differs essentially from the global structure Fig.7 of direct matching. In the case of

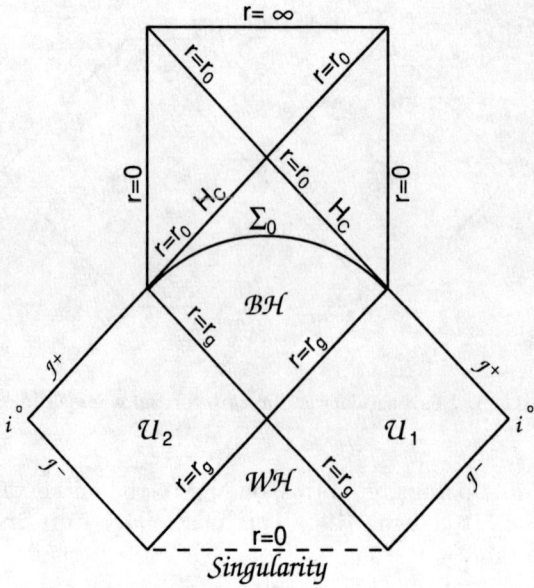

FIG. 7: Penrose-Carter diagram for the case of the direct de Sitter-Schwarzschild matching.

de Sitter-Schwarzschild geometry, closed or semiclosed world can arise in any of the $\mathcal{RC}$ structures in the future of a ΛBH. For the GUT scale $E_{GUT} \sim 10^{15}$GeV the probability of a single tunnelling event is $D \sim \exp\left(-\frac{2}{3}10^{16}\right)$ [19, 36], however in the case of an infinite number of regions $\mathcal{RC}$ inside a ΛBH, probability of quantum birth of a baby universe in one of them is much bigger than probability of a single tunnelling event. Instability of a Λ white hole leads to possibilities other than considered by Farhi and Guth. Instability of Schwarzschild white hole is related to physical processes (particle creation) near a singularity (see, e.g., [33] and references therein). In the case of a Λ white hole its quantum instability is related to instability of the de Sitter vacuum near the surface $r = 0$. Instability of the de Sitter vacuum is well studied, both with respect to particle creation and with respect to the quantum birth of a universe (for references see [19]).

The possibility of a multiple birth of causally disconnected universes from the de Sitter background was noticed in 1975 in the Ref. [9]. In 1982 such a possibility has been investigated by Gott III who considered creation of a universe as a quantum barrier penetration leading to an open FRW cosmology [38]. The case of arising of open universes from de Sitter vacuum is illustrated by Fig.9 [38]. The events E and E' are creation of causally disconnected universes. The curved lines are world lines of comoving observers. At the spacelike surface AB the phase transition occurs from the inflationary to the radiation dominated stage [9, 38]. In the case of a ΛBH the region ECB in Fig.9 corresponds to the region $\mathcal{RC}_1$ in Fig.4, and the region BFD corresponds to a part of the region $\mathcal{RC}_2$. The regions $\mathcal{RC}_1$ and $\mathcal{RC}_2$ in the de Sitter-Schwarzschild spacetime are entirely disjoint from each other for the same reason as the regions $\mathcal{U}_1$ and $\mathcal{U}_2$. Birth of open (or flat) baby universes inside a ΛBH looks very similar to the picture shown in Fig.9 and can occur in any of an infinite number of the regions $\mathcal{RC}$ inside a ΛBH.

Quantum birth of an open or flat universe is possible when a primordial quantum fluctuation contains an admixture of strings or some other quintessence with the equation of state $p = -\rho/3$. In case when there is also some small admixture of radiation, quantum tunnelling occurs from a discrete energy level with a nonzero quantized temperature. The probability of tunnelling in this case is much larger, $D_{from\ a\ level} \sim \exp\left(-\frac{2}{3}10^7\right)$ [40].

The probability of a single tunnelling event is small, but an infinite number of white holes inside a ΛBH enhances essentially the probability of quantum birth of a baby universe inside it as a result of quantum instability of de Sitter vacuum [19].

Let us note also that an obstacle related to the initial singularity does not arise this case, since both future and past singularities are replaced with the regular surfaces $r = 0$. On the other hand, possibility of influence on a created universe is restricted by existence of the Cauchy horizon in de Sitter-Schwarzschild geometry.

Thermodynamics of vacuum nonsingular black hole - A ΛBH emits Hawking radiation from both horizons with the Gibbons-Hawking temperature $T = \hbar\Gamma(2\pi kc)^{-1}$ [41] where k is the Boltzmann constant and Γ is the surface gravity which satisfies the equation for the Killihg vector K_α on the horizons, $K_{\alpha;\beta}K^\beta = \Gamma K_\alpha$. For de

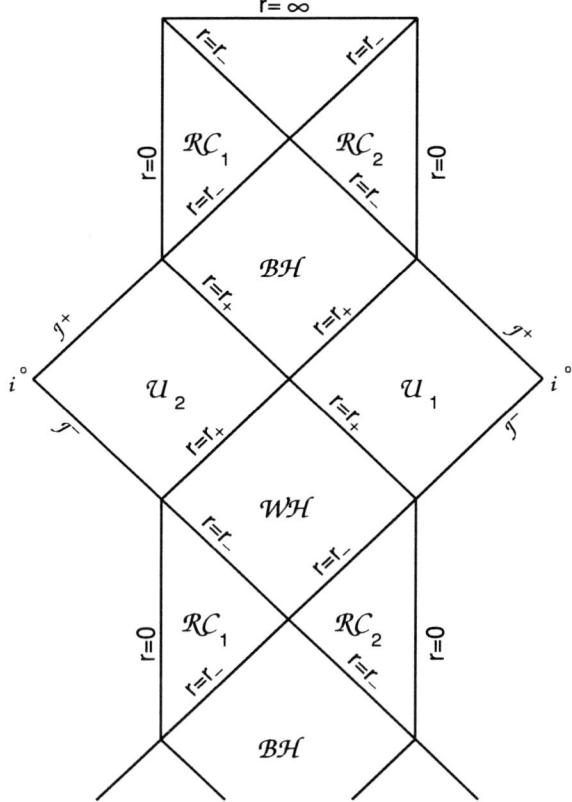

FIG. 8: The global structure of space-time for the case of a birth of a closed or semiclosed world inside a Λ black hole.

Sitter-Schwarzschild black hole with two horizons the surface gravity is given by [22]

$$\Gamma = \frac{c^2}{2}\left[\frac{R_g(r_h)}{r_h^2} - \frac{R'_g(r_h)}{r_h}\right]; \quad r_h = r_+, r_- \tag{33}$$

The temperature on the internal horizon is negative since the surface gravity characterizing a force that must be exerted at infinity to hold a test particle in place at the horizon [25], is negative on the internal horizon r_- due to repulsive character of gravity near r_-.

Temperature on the BH horizon, T_+ is positive. Temperature-mass diagram for the BH horizon is shown in Fig.10 [22]. Its form is generic for de Sitter-Schwarzschild geometry. The temperature on the BH horizon drops to zero at $m = m_{crit}$, while the Schwarzschild asymptotic requires $T_+ \to 0$ as $m \to \infty$. As a result the temperature-mass curve has a maximum between m_{crit} and $m \to \infty$. In a maximum the specific heat is broken and changes its sign testifying to a second-order phase transition in the course of Hawking evaporation and suggesting symmetry restoration to the de Sitter group in the origin [42]. For particular form of the density profile (29) the temperature is given by [22]

$$T_h = \frac{\hbar c}{4\pi k r_0}\left[\frac{r_0}{r_h} - \frac{3r_h}{r_0}\left(1 - \frac{r_h}{r_g}\right)\right] \tag{34}$$

The value of the mass parameter at the maximum and the temperature of the phase transition are

$$m_{tr} \simeq 0.38 m_{Pl}\sqrt{\rho_{Pl}/\rho_0}; \quad T_{tr} \simeq 0.2 m_{Pl}\sqrt{\rho_{Pl}/\rho_0}$$

G-lump - In the course of Hawking evaporation a ΛBH loses its mass and configuration evolves towards a self-gravitating particle-like vacuum structure without horizons (see Fig.3), globally regular and globally neutral, called G-lump [13]. It resembles Coleman's lumps - non-singular, non-dissipative solutions of finite energy, holding themselves together by their own self-interaction [43]. The idea actually goes back to the Einstein profound proposal to describe an elementary particle by a regular solution of nonlinear field equations as a "bunched field" located in the confined region where field tension and energy are particularly high [44].

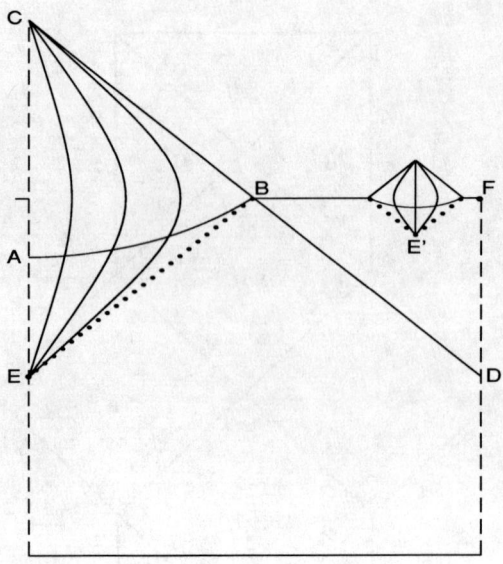

FIG. 9: Penrose-Carter diagram corresponding to the case of a quantum birth of baby universes inside a ΛBH (reproduced from Ref.[38] with permission from Nature).

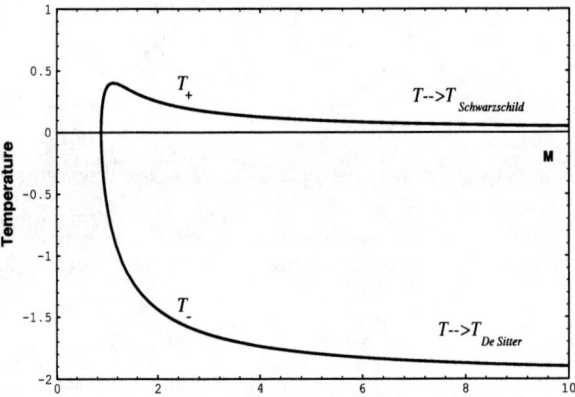

FIG. 10: Temperature-mass diagram for ΛBH.

G- lump is the regular solution to the Einstein equations, perfectly localized (see Fig.11) in a region where field tension and energy are particularly high (this is the region of the former singularity). It holds itself together by gravity due to balance between gravitational attraction outside and gravitational repulsion inside of zero-gravity surface $r = r_c$ beyond which the strong energy condition of singularities theorems is violated. The surface of zero gravity is defined by $p_\perp(r_c) = 0$ [22]. It is depicted in Fig.12 together with horizons and the surface $r = r_s$ of zero scalar curvature $R(r_s) = 0$ which represents the characteristic curvature size in the de Sitter-Schwarzschild geometry. In the case of the density profile (29) the characteristic size r_s is given by

$$r_s = \left(\frac{4}{3}r_0^2 r_g\right)^{1/3} = \left(\frac{m}{\pi\rho_0}\right)^{1/3} \tag{35}$$

and confines about 3/4 of the total mass m [22].

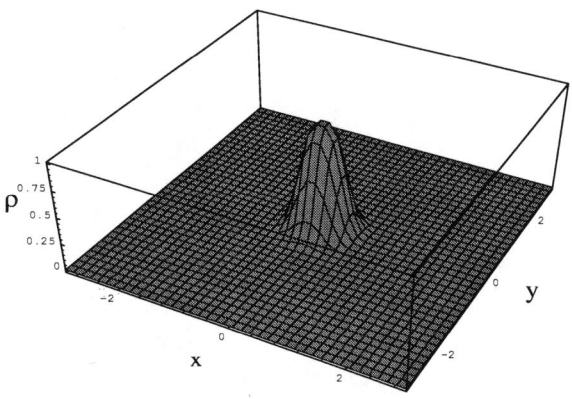

FIG. 11: G-lump in the case $r_g = 0.1 r_0$ ($m \simeq 0.06 m_{crit}$).

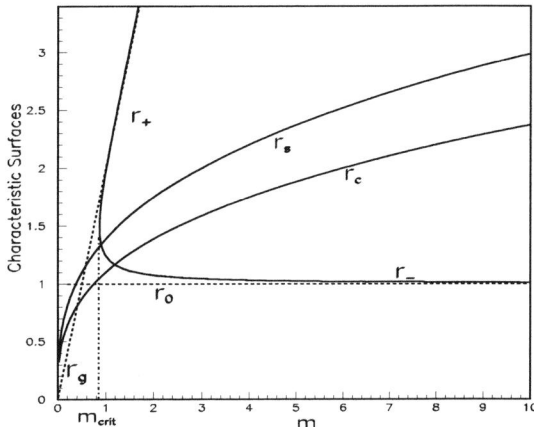

FIG. 12: Horizons of ΛBH, surfaces $r = r_s$ and $r = r_c$.

IV. COSMOLOGICAL TERM AS A SOURCE OF MASS

The ADM mass of both G-lump and ΛBH is directly connected to cosmological term $\Lambda_{\mu\nu}$ by the ADM formula (13) which in this case reads [13]

$$m = (2G)^{-1} \int_0^\infty \Lambda_t^t(r) r^2 dr \qquad (36)$$

In de Sitter-Schwarzschild geometry the ADM mass (13) is identified as gravitational mass by the Schwarzschild asymptotic which is asymptotically flat at infinity. By the equivalence principle, gravitational mass is equal to inertial mass [2]. As a result, the inertial mass which is the measure of inertia, is related for this class of metrics, to both de Sitter vacuum trapped inside an object and reducing of space-time symmetry from the de Sitter group at its center to the Lorentz group at infinity through radial boosts in between.

This fact does not depend on proposed extension of a cosmological term from $\Lambda g_{\mu\nu}$ to $\Lambda_{\mu\nu}$. Whichever would be a matter source (22) for this class of metrics and its interpretation (as associated to $\Lambda_{\mu\nu}$ or not), mass is related to de Sitter vacuum inside of an object, and symmetry of source term $T_{\mu\nu}$ responsible for metrics from this class, is broken due to its symmetry properties (22). More precisely, symmetry of $T_{\mu\nu}$ is reduced from the full Lorentz group to the Lorentz boosts in the radial direction only. Together with asymptotic flatness this allows us to introduce a distinguished point as the center of an object whose ADM mass is defined by (36) [13].

This picture agrees with the basic idea of the Higgs mechanism for generation of mass via spontaneous breaking of symmetry of a scalar field vacuum from a false vacuum state $T_{\mu\nu} = V(0) g_{\mu\nu}$ to a true vacuum state $T_{\mu\nu} = 0$. In both cases de Sitter vacuum is involved and vacuum symmetry is broken. The gravitational potential $g(r)$ resembles a

Higgs potential (see Fig.13). The gravitational potential $g(r)$ is generic, and the de Sitter vacuum supplies a particle

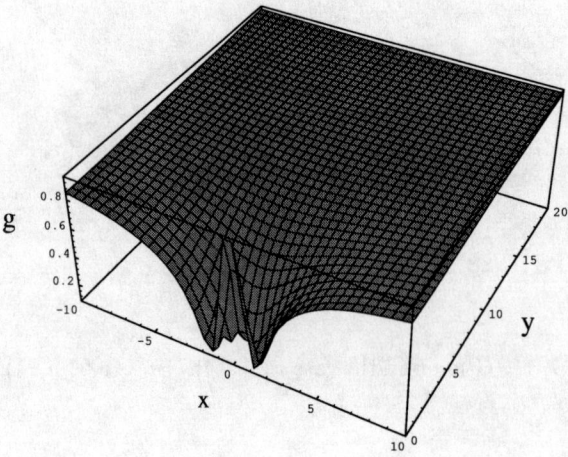

FIG. 13: The gravitational potential $g(r)$ for the case of G-lump with the mass a little bit less than m_{crit}.

with mass via smooth breaking of space-time symmetry from the de Sitter group in its center to the Lorentz group at its infinity.

This picture leads to the natural assumption [45] that whatever would be particular mechanism involving de Sitter vacuum in mass generation, a fundamental particle (a particle which does not display substructure, like a lepton or quark) thought of as an extended object, may have an internal vacuum core (at the scale where it gets mass) with de Sitter vacuum trapped inside. Its geometrical size is defined by gravity. Since geometry of such an object is de Sitter at the origin and Minkowski at infinity, its geometrical size can be estimated by de Sitter-Schwarzschild geometry.

The characteristic curvature size in this geometry given by (35) depends on vacuum density at $r = 0$ and represents modification of the Schwarzschild radius r_g depending only on the mass parameter m, to the case when a singularity is replaced with the de Sitter vacuum. While application of the Schwarzschild radius to elementary particle scale is highly speculative (obtained estimates are many orders of magnitude less than l_{Pl}, e.g., for electron $r_g \sim 10^{-57}$cm), the characteristic curvature size r_s of de Sitter-Schwarzschild geometry gives reasonable numbers close to experimental limits. In Fig.14 geometrical sizes for leptons estimated by de Sitter-Schwarzschild curvature radius r_s with vacuum of the electroweak scale, are compared with electromagnetic (EM) and electroweak (EW) experimental upper limits. The black triangles on the middle curve are for theoretical estimates of lower limits by geometrical size r_s [45]. Calculating characteristic geometrical size (35) for leptons whose masses are related to de Sitter vacuum as a false vacuum of the Higgs mechanism at the electroweak scale $E_{EW} \simeq 246$GeV, we get the estimates on the lower limits of sizes for electron, μ and τ leptons [45]

$$r_e \simeq 1.5 \times 10^{-18} cm; r_\mu \simeq 0.9 \times 10^{-17} cm;$$

$$r_\tau \simeq 2.3 \times 10^{-17} cm \tag{37}$$

Zero-point vacuum energy for G-lump - The question always discussed in connection with cosmological term is zero-point vacuum energy. For G-lump, which clearly represents an elementary spherically symmetric excitation of a vacuum defined macroscopically by its symmetry (22), a zero-point energy is estimated by its lowest quantum energy level. Since the de Sitter vacuum is trapped within a G-lump, we can model it by a spherical bubble whose density decreases with distance. The coordinates r, t are transformed to coordinates R, τ by [14]

$$\frac{\partial r}{\partial \tau} = \sqrt{\frac{R_g(r)}{r} + f(R)}; \frac{\partial t}{\partial \tau} = \frac{\sqrt{1 + f(R)}}{1 - \frac{R_g(r)}{r}};$$

$$\frac{\partial r}{\partial R} = \sqrt{\frac{R_g(r)}{r}(1 + f(R))}; \frac{\partial \tau}{\partial t} = \sqrt{1 + f(R)};$$

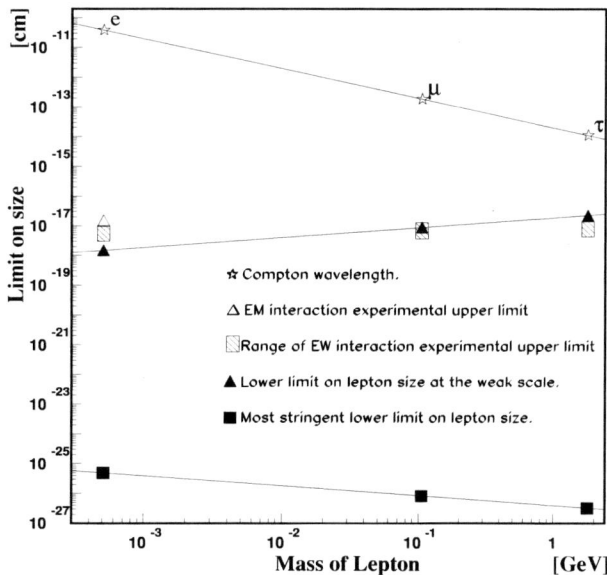

FIG. 14: Characteristic sizes for leptons [45].

$$\frac{\partial t}{\partial R} = \frac{\sqrt{\frac{R_g(r)}{r}(\frac{R_g(r)}{r} + f(R))}}{1 - \frac{R_g(r)}{r}}; \quad \frac{\partial \tau}{\partial r} = -\frac{\sqrt{\frac{R_g(r)}{r} + f(R)}}{1 - \frac{R_g(r)}{r}};$$

$$\frac{\partial R}{\partial r} = \sqrt{\frac{r}{R_g(r)}} \frac{\sqrt{1 + f(R)}}{1 - \frac{R_g(r)}{r}}; \quad \frac{\partial R}{\partial t} = -\sqrt{\frac{r}{R_g(r)} f(R) + 1}$$

where $f(R)$ is an arbitrary function satisfying $1 + f(R) > 0$. Geometry of a bubble is then described by the metric

$$ds^2 = d\tau^2 - \frac{2GM(r(R,\tau))}{r(R,\tau)} - r^2(R,\tau)d\Omega^2 \tag{38}$$

The equation of motion derived from the Einstein equations for a Lemaitre anisotropic model [23] reads [19]

$$\dot{r}^2 + 2r\ddot{r} - \kappa\rho(r)r^2 = f(R) \tag{39}$$

It has the first integral [13]

$$\dot{r}^2 - \frac{2GM(r)}{r} = f(R) \tag{40}$$

which resembles the equation of a particle in the potential

$$V(r) = -\frac{GM(r)}{r} \tag{41}$$

with the constant of integration $f(R)$ playing the role of the total energy $f = 2E$. The Lagrangian in this case is $\mathcal{L} = \dot{r}^2/2 - V(r)$, the Hamiltonian is $\mathcal{H} = \dot{r}^2/2 + V(r)$, and the conjugate momentum is $p = \partial\mathcal{L}/\partial\dot{r} = \dot{r}$.

A spherical bubble can be described by the minisuperspace model with a single degree of freedom, the bubble radius r [46]. By the standard procedure of quantization the momentum operator is introduced as $\hat{p} = -il_{Pl}^2 d/dr$ and the equation (40) transforms into the Wheeler-DeWitt equation in the minisuperspace which reduces to the one-dimensional Schrödinger equation

$$\frac{\hbar^2}{2m_{Pl}}\frac{d^2\psi}{dr^2} - (V(r) - E)\psi = 0 \tag{42}$$

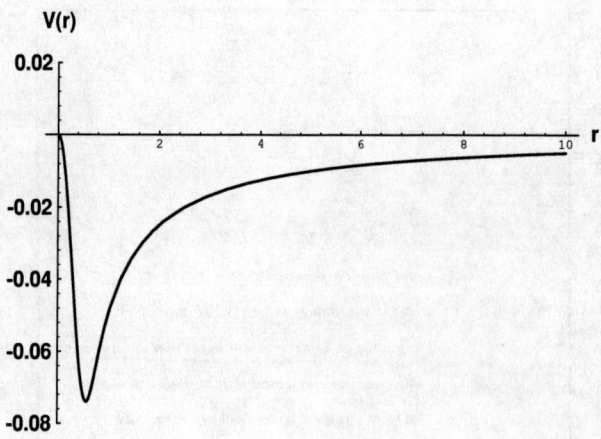

FIG. 15: The plot of the potential for G-lump with $r_g = 0.1 r_0$ ($m \simeq 0.07 m_{crit}$).

with the potential (41) shown in Fig.15.

Near the minimum $r = r_m$ the potential $V(r)$ is approximated by $V(r) = V(r_m) + 4\pi G p_\perp(r_m)(r - r_m)^2$ [13], and Eq.(42) reduces to the equation for a harmonic oscillator

$$\frac{d^2\psi}{dx^2} - \frac{m_{Pl}^2}{\hbar^2}\omega^2 x^2 \psi + \frac{2m_{Pl}\tilde{E}}{\hbar^2}\psi = 0 \qquad (43)$$

where $x = r - r_m$, $\tilde{E} = E - V(r_m)$, $\omega^2 = \Lambda c^2 \tilde{p}_\perp(r_m)$, and $\tilde{p}_\perp$ is the ransversal pressure normalized to ρ_0; for the density profile (29) $\tilde{p}_\perp(r_m) \simeq 0.2$. The energy spectrum

$$E_n = \hbar\omega\left(n + \frac{1}{2}\right) - \frac{GM(r_m)}{r_m} E_{Pl} \qquad (44)$$

is shifted down by the minimum of the potential $V(r_m)$ which represents the binding energy. The energy of zero-point vacuum mode [13]

$$\tilde{E}_0 = \frac{\sqrt{3\tilde{p}_\perp}}{2}\frac{\hbar c}{r_0} \qquad (45)$$

never exceeds the binding energy. The formula (45) remarkably resembles the Hawking temperature from the de Sitter horizon $kT_H = \frac{1}{2\pi}\frac{\hbar c}{r_0}$ [41].

V. $\Lambda_{\mu\nu}$ GEOMETRY

In the case of non-zero value of cosmological constant, the metric Eq.(21) shown in Fig.2, can have not more than three horizons [20]. These are the internal Cauchy horizon r_-, the black hole event horizon r_+, and the cosmological horizon r_{++}. The number of horizons depends on the mass parameter m and on the parameter $q = \sqrt{\Lambda/\lambda}$. In Fig.16 the horizon-mass diagram is plotted for the case of density profile (29). In the range of masses $M_{cr1} < m < M_{cr2}$, geometry has three horizons and describes a vacuum cosmological nonsingular black hole. The global structure of space-time is shown in Fig.17. The global structure of space-time contains an infinite sequence of ΛBH and ΛWH, their future and past regular cores asymptotically de Sitter with $\Lambda g_{\mu\nu}$ as $r \to 0$, and asymptotically de Sitter universes with $\lambda g_{\mu\nu}$ as $r \to \infty$ in the regions CC between cosmological horizons and space-like infinities. Rectangular regions confined by the surfaces $r = 0$ and $r = \infty$ do not belong to the diagram.

Three-horizon configuration represents the nonsingular modification of the Kottler-Trefftz solution [47] known as the Schwarzschild-de Sitter geometry [41].

The case $m = M_{cr2}$ ($r_+ = r_{++}$) is the nonsingular modification of the Nariai geometry of an extreme black hole [48]. The global structure of space-time is shown in Fig.18 [20]. The case $m = M_{cr1}$ ($r_+ = r_-$) is another extreme black hole state which appears due to replacing a former singularity with a regular core $\mathcal{RC}$. The critical value of mass M_{cr1} at which $r_- = r_+$, puts the lower limit for a black hole mass. It practically does not depend on the parameter

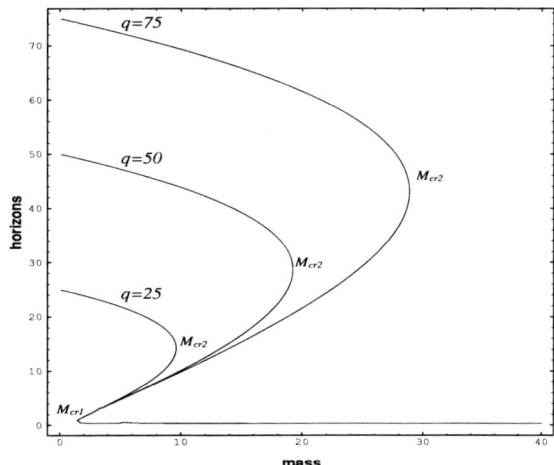

FIG. 16: Horizon-mass diagram of a $\Lambda_{\mu\nu}$ geometry.

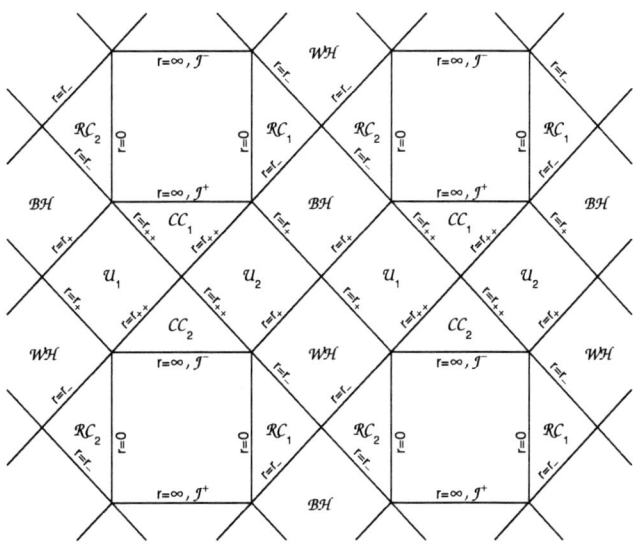

FIG. 17: Penrose-Carter diagram for $\Lambda_{\mu\nu}$ geometry with three horizons.

$q = \sqrt{\Lambda/\lambda}$ and is given by (31). Global structure of space-time is shown in Fig.19 [20]. Five types of configurations described by $\Lambda_{\mu\nu}$ geometry are shown in Fig.20 for the case $q \equiv \sqrt{\Lambda/\lambda} = 10$. The case $M = 2$ can be seen as a particle-like structure at the de Sitter background $\lambda g_{\mu\nu}$. On the other hand, the global structure in this case is the same as for de Sitter geometry, so that in cosmological coordinates it represents cosmological model of de Sitter type on global structure but with cosmological density Λ^t_t evolving from Λ at the beginning to λ at late times. Similar case is for $M = 9$. Those two cases differ only on dynamics (nontrivial behavior in the R region in the first case and in the T region in the second).

VI. $\Lambda_{\mu\nu}$ DOMINATED COSMOLOGIES

All cosmological models dominated by variable cosmological term $\Lambda_{\mu\nu}$ belong to the Lemaitre class of cosmologies with anisotropic perfect fluid. They are described by the line element

$$ds^2 = c^2 d\tau^2 - e^{\mu(R,\tau)} dR^2 - r^2(R,\tau) d\Omega^2 \qquad (46)$$

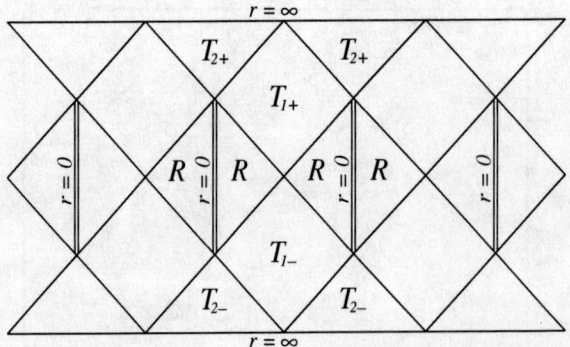

FIG. 18: Global structure of $\Lambda_{\mu\nu}$ geometry with the double horizon $r_+ = r_{++}$.

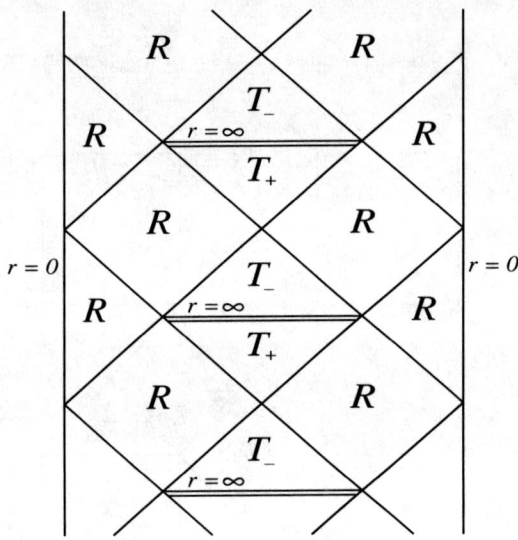

FIG. 19: Global structure of $\Lambda_{\mu\nu}$ geometry with the double horizon $r_- = r_+$.

and governed by equations [19]

$$\kappa p_r^\Lambda r^2 = e^{-\mu} r'^2 - 2r\ddot{r} - \dot{r}^2 - 1 \tag{47}$$

$$2\kappa p_\perp^\Lambda r = 2e^{-\mu} r'' - e^{-\mu} r' \mu' - \dot{\mu}\dot{r} - \ddot{\mu} r \tag{48}$$

$$\kappa \rho^\Lambda r^2 = -e^{-\mu}\left(2rr'' + r'^2 - rr'\mu'\right) + \left(r\dot{r}\dot{\mu} + \dot{r}^2 + 1\right) \tag{49}$$

$$e^{\mu(R,\tau)} = \frac{r'^2}{1 + f(R)} \tag{50}$$

The dot denotes differentiation with respect to τ and the prime with respect to R. Principal pressures satisfying the equation of state (23), are plotted in Fig.21 [19]. With using Eq.(50) the equation (47) leads to the equation of motion describing the evolution [19]

$$\dot{r}^2 + 2r\ddot{r} + \kappa p_r^\Lambda r^2 = f(R) \tag{51}$$

It has the first integral

$$\dot{r}^2 = A + R_g(r(R,\tau)) + f(R)r \tag{52}$$

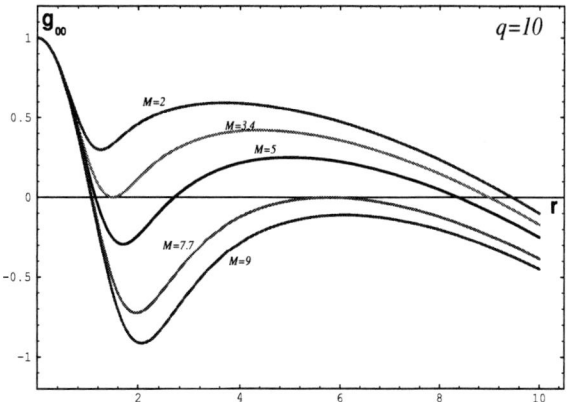

FIG. 20: $\Lambda_{\mu\nu}$ configurations for the case $q = 10$. The parameter M is a mass normalized to $(3/G^2\Lambda)^{1/2}$.

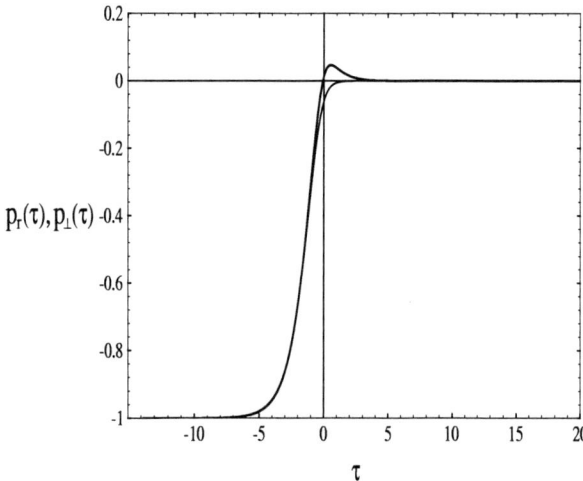

FIG. 21: Radial and tangential pressures, $p_r^\Lambda < p_\perp^\Lambda$.

and the second integral

$$\tau - \tau_0(R) = \int_{r_0}^{r} \sqrt{\frac{x}{A + R_g(x) + f(R)}} dx \tag{53}$$

Here $\tau_0(R)$ is an arbitrary function (constant of integration over τ parametrized by R) which is called the "bang-time function" (see references in [19]). For comparison, in the case of the Tolman-Bondi dust-filled model the evolution is described by $r(R, \tau) = (9GM(R)/2)^{1/3}(\tau - \tau_0(R))^{2/3}$, where $\tau_0(R)$ represents the big bang singularity surface for which $r(R, \tau) = 0$.

The basic feature of $\Lambda_{\mu\nu}$ dominated cosmologies is that near the bang surface $r = 0$ corresponding to $R + \tau = -\infty$ (see Fig.6), the metric (46) transforms to the FRW form for any $f(R)$ and reads [19, 20]

$$ds^2 = c^2 d\tau^2 - a^2(\tau)(d\chi^2 + \sin^2\chi d\Omega^2) \tag{54}$$

with the de Sitter scale factor $a(\tau) \sim \cosh(H_0\tau)$ for $f(R) < 0$, $a(\tau) \sim \sinh(H_0\tau)$ for $f(R) > 0$, $a(\tau) \sim \exp(H_0\tau)$ for $f(R) = 0$, where H_0 is the value of the Hubble parameter corresponding to the initial value of Λ. In the case of $\Lambda_{\mu\nu}$ geometry the regular surface $r(R, \tau) = 0$ is time-like (see Figs.17-19). As a result evolution starts from the nonsingular non-simultaneous de Sitter bang [19, 20].

The inflationary stage (nonsimultaneous de Sitter bang) is followed by anisotropic Kasner type stage with contraction in the radial direction and expansion in the tangential direction [19, 20]

$$ds^2 = d\tau^2 - F(R)(\tau + R)^{-2/3}dR^2 - a(\tau + R)^{4/3}d\Omega^2 \tag{55}$$

where $F(R)$ is a smooth regular function and A is the constant expressed in the model parameters (Λ, λ, M). This is the stage when acceleration of the "scale factor" r changes quickly and drastically. In Fig.22 it is shown for the case of a spatially flat model with $f(R) = 0$ [19]. In the spatially flat case the Schwarzschild (ADM) mass M coincides

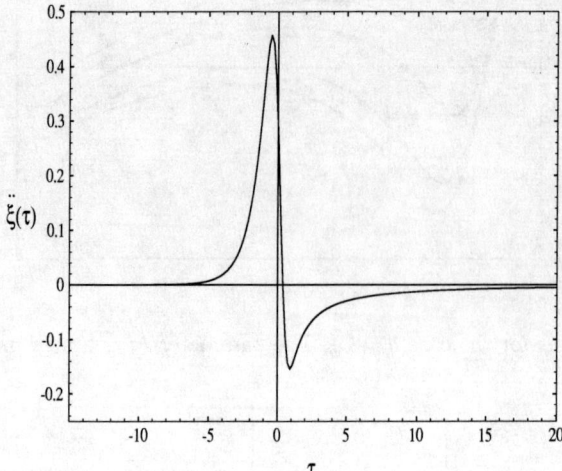

FIG. 22: The acceleration of the "scale factor" $r(\tau - \tau_0)$ normalized to $(GM/\Lambda)^{1/3}$.

with the total proper mass which is the sum of masses of all particles with radial coordinates less than R which is ultimately given by the Bondi formula [49]

$$\mathcal{M} = 4\pi \int_0^r \rho(x) x^2 dx$$

The behavior of a mass normalized to the Schwarzschild mass M is shown in the Fig.23 [19]. During inflationary

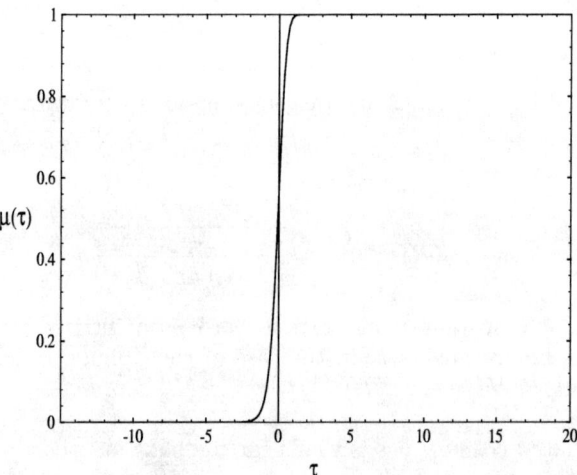

FIG. 23: Plot of the mass function $\mu = \mathcal{M}/M$.

stage the mass increases as r^3. At the next anisotropic stage it is growing abruptly towards M. The growth in a mass is connected with the fall of $\rho^\Lambda = \kappa^{-1} \Lambda_t^t$, i.e., with decay of the initial vacuum energy (the growth of a universe mass by many orders of magnitude in the course of decay of the de Sitter vacuum was first noticed in the Ref [9]).

For a certain class of observers $\Lambda_{\mu\nu}$ dominated models can be specified as Kantowski-Sachs models with regular R regions [20]. For the Kantowski-Sachs observers evolution starts from horizons with a highly anisotropic "null bang" where the volume of the spatial section vanishes. A null bung surface seems singular to a comoving observer although it is perfectly regular in the $\Lambda_{\mu\nu}$ geometry. These results are extended to the case of the planar and pseudospherical spatial symmetries. Nonsingular $\Lambda_{\mu\nu}$ dominated cosmologies are Bianchi type I in the planar case and hyperbolic analogs of the Kantowski-Sachs models in the pseudospherical case. At late times all $\Lambda_{\mu\nu}$ dominated models approach de Sitter asymptotic with $\lambda < \Lambda$ [20].

VII. SUMMARY AND DISCUSSION

The idea of replacing a Schwarzschild singularity with de Sitter vacuum goes back to the 1966 papers of Sakharov [50] who considered $p = -\rho$ as the equation of state for superhigh densities, and of Gliner who suggested that a vacuum associated with the Einstein cosmological term (μ-vacuum in his terms) could be a final state in a gravitational collapse [7].

Realizing this idea by direct matching of Schwarzschild metric outside to de Sitter metric inside a short transitional space-like layer of the Planckian depth [37, 51–53] results in metrics which typically have a jump at the junction surface.

The main point outlined here is the existence of the class of globally regular analytic solutions to the minimally coupled GR equations with a source term of the algebraic structure (22) interpreted as spherically symmetric anisotropic vacuum with variable density and pressures $T_{\mu\nu}^{vac}$ associated with a time-dependent and spatially inhomogeneous cosmological term $\Lambda_{\mu\nu} = \kappa T_{\mu\nu}^{vac}$, whose asymptotic in the origin, dictated by the weak energy condition, is the Einstein cosmological term $\Lambda g_{\mu\nu}$.

$\Lambda_{\mu\nu}$ geometry describes generic properties of any configuration satisfying (22) and requirements 1-3, obligatory for any particular model in the same sense as de Sitter geometry is obligatory for any matter source satisfying (6).

Where to put cosmological term - Overduin and Cooperstock distinguished two approaches to $\Lambda g_{\mu\nu}$ existing in the literature [32]. In the first approach $\Lambda g_{\mu\nu}$ is shifted onto the right-hand side of the Einstein field equation (2) and treated as a dynamical part of the matter content. This approach (characterized as connected to dialectic materialism of the Soviet physics school), goes back to Gliner who interpreted $\Lambda g_{\mu\nu}$ as a vacuum stress tensor [7], to Zel'dovich who connected Λ with gravitational interaction of virtual particles [34], and to Linde who suggested that in principle Λ can vary [54]. In contrast, idealistic approach prefers to keep Λ on the left-hand side of Eq. (2) as geometrical entity and treat is as a constant of nature.

This classification suggests that any variable Λ would have to be identified with a matter. The first candidate for a matter source of $\Lambda_{\mu\nu}$ would be gravitational vacuum polarization in the spirit of Zel'dovich idea, and Poisson and Israel self-regulatory picture. On the other hand, nothing prevents from shifting $T_{\mu\nu}^{vac}$ from the right-hand side of Einstein equations (1) where we found it, to the left-hand side and treat $\Lambda_{\mu\nu} = 8\pi G T_{\mu\nu}^{vac}$ as evolving geometrical entity in the spirit of Wheeler's geometrodynamics [55]. [65]

The question of where to put Λ looks like kind of philosophical question of what is primary. If we remind that dialectic materialism is nothing but application of Hegel's dialectic idealism to matter, then one is tempted to approach Λ by the Hegel laws of a new triad, as a new quality ether appearing in the new turn of cognitive helix, a Lorentz-invariant ether with respect to which one cannot measure velocity in principle [13].

Actually at the moment there is more information about which matter sources do not couple with a variable Λ. They are specified by "no-go" theorems.

$\Lambda_{\mu\nu}$ and "no-go" theorems -

The question whether a regular black hole can be obtained as a false vacuum configuration described by the action

$$S = \int d^4x \sqrt{-g}\left[R + (\partial\phi)^2 - 2V(\phi)\right] \tag{56}$$

where R is the scalar curvature, $(\partial\phi)^2 = g^{\mu\nu}\partial_\mu\phi\partial_\nu\phi$, with various forms for a scalar field potential $V(\phi)$, is regulated by the "no-go" theorems: Asymptotically flat regular black hole solutions are absent in the theory (56) with any non-negative potential $V(\phi)$ [58]. This result has been extended to the case of any $V(\phi)$ and any asymptotic and then generalized to the case of a theory with the action $S = \int d^4x \sqrt{-g}\left[R + F[(\partial\phi)^2, \phi]\right]$, where F is an arbitrary function [59], to the multi-scalar theories of sigma-model type, and to scalar-tensor and curvature-nonlinear gravity theories [60]. The only possible regular solutions are either de Sitter-like with a single cosmological horizon or those without horizons, including asymptotically flat ones; the latter do not exist for $V(\phi) \geq 0$, so that the set of causal false vacuum structures is the same as known for $\phi = const$ case, namely Minkowski (or anti-de Sitter), Schwarzschild, de Sitter, and Schwarzschild-de Sitter [59, 60], and thus does not include de Sitter-Schwarzschild configurations.

In the case of *complex* massive scalar field the regular structures can be obtained in the minimally coupled theory with positive $V(\phi)$ [61]. These are boson stars ([30] and references therein), but in this case algebraic structure of the stress-energy tensor [30] does not satisfy Eq.(22), and asymptotic at $r = 0$ is not de Sitter.

Possible sources of $\Lambda_{\mu\nu}$ -
However situation is not completely hopeless.

In 1991 Strominger demonstrated the possibility of natural, not *ad hoc*, arising of de Sitter core inside a black hole in the model of two-dimensional dilaton gravity conformally coupled to N scalar fields [62].

Morgan has considered a black hole in a simple model for quantum gravity in which quantum effects are represented by an upper cutoff on the curvature, and obtained de Sitter-like past and future cores replacing singularities, although in this model (similar to the case of direct de Sitter-Schwarzschild matching in the model with the limiting curvature [37]) it appeared impossible to splice them together exactly [63].

The examples of structures whose stress-energy tensor is of type (22) which can be associated with $\Lambda_{\mu\nu}$, are found in nonlinear electrodynamics [64]. In this case the Lagrangian $L(F) = L(F_{\mu\nu}F^{\mu\nu})$, with a correct weak field limit, leads to nontrivial spherically symmetric regular solutions. They exist if and only if the electric charge is zero and $L(F)$ tends to a finite limit as $F \to \infty$ [64]. Properties and examples of such configurations which include magnetic black hole and solitonlike objects (monopoles), are discussed in Ref [64].

$\Lambda_{\mu\nu}$ versus quintessence -
The key difference of $\Lambda_{\mu\nu}$ from the quintessence which is introduced as a time-varying spatially inhomogeneous component of matter content with negative pressure is in the algebraic structure of stress tensors. Quintessence is defined by the equation of state $p = -\alpha\rho$ with $\alpha < 1$ [39]. This corresponds to such a stress-energy tensor $T_{\mu\nu}$ for which a comoving reference frame is defined uniquely. The quintessence represents thus a non-vacuum negative-pressure isotropic alternative to a cosmological constant Λ while the cosmological tensor $\Lambda_{\mu\nu}$ represents the extension of the algebraic structure of the Einstein cosmological term $\Lambda g_{\mu\nu}$ which makes it variable and anisotropic.

Acknowledgement

This work was supported by the Polish Committee for Scientific Research through the Grant 5P03D.007.20, and through the grant for UWM.

---

[1] A.Einstein, Sitzungs. Ber. Berl. Akad. Wiss. **142** (1917)

[2] D.W.Sciama, *The Physical Foundations of General Relativity*, Doubleday and Co., NY, Ch.2 (1969)

[3] H.Bondi, *Assumption and Myth in Physical Theory*, Cambridge Univ. Press, Ch. 4 (1967)

[4] W. De Sitter, Mon. Not. R. Astr. Soc. **78** 3 (1917)

[5] S. Weinberg, Rev. Mod. Phys. **61** 1 (1989)

[6] A.S.Eddington, *The Mathematical Theory of Relativity*, Cambridge Univ.Press (1924)

[7] E.B.Gliner, Sov. Phys. JETP **22** 378 (1966)

[8] A.Z.Petrov, *Einstein Spaces*, Pergamon Press (1969)

[9] E.B.Gliner, I.G.Dymnikova, Sov. Astron. Lett. **1** 93 (1975); I.G.Dymnikova, Sov. Phys. JETP **63** 1111 (1986)

[10] K.A.Olive, Phys. Rep. **190**, 307 (1990)

[11] E.W.Kolb and M.S.Turner, *The Early Universe*, Addison-Wesley (1990)

[12] L. Krauss and M. Turner, astro-ph/9504003 (1995); N. A. Bahcall, J. P. Ostriker, S. Perlmutter, P. J. Steinhardt, Science **284**, 1481 (1999)

[13] I.Dymnikova, Class. Quant. Grav. **19** 1 (2002); gr-qc/0112052

[14] I.Dymnikova, "Spherically symmetric space-time with the regular de Sitter center" (2002), to be published

[15] I.G.Dymnikova, Phys. Lett. **B472** 33 (2000); gr-qc/9912116, and references therein

[16] I.Dymnikova, in *Woprosy Matematicheskoj Fiziki i Prikladnoj Matematiki*, Ed.E.Tropp, St.Petersburg (2000), p.29; gr-qc/0010016

[17] I.Dymnikova, Gravitation and Cosmlogy **8**, suppl 131 (2002); gr-qc/0201058

[18] I.Dymnikova, in *General Relativity, Cosmology and Gravitational Lensing*, Eds. G.marmo, C.Rubano, P.Scudellaro, Bibliopolis, Napoli (2002)

[19] I.Dymnikova, A.Dobosz, M.Fil'chenkov, A.Gromov, Phys. Lett. **B506** 351 (2001); gr-qc/0102032

[20] K.Bronnikov, A.Dobosz, I.Dymnikova, "Nonsingular vacuum cosmologies with a variable cosmological term", to be published (2002)

[21] I.Dymnikova, Gen. Rel. Grav. **24** 235 (1992); CAMK preprint 216 (1990)

[22] I.G.Dymnikova, Int. J. Mod. Phys. **D5** 529 (1996)

[23] R.C.Tolman, *Relativity, Thermodynamics and Cosmology*, Clarendon Press, Oxford (1969)

[24] J.R.Oppenheimer, H.Snyder, Phys. Rev. **56** 455 (1939)

[25] R.M.Wald, *General relativity*, Univ.Chicago (1984)

[26] R.L.Bowers, E.P.T.Liang, Astrophys.J. **188** 657 (1974)

[27] S.W.Hawking and G.F.R.Ellis, *The large scale structure of space-time*, Cambridge Univ. Press (1973)

[28] E.Poisson, W.Israel, Class. Quant. Grav.**5** L201 (1988)

[29] D.J.Kaup, Phys. Rev.**172** 1331 (1968); R.Ruffini, S.Bonazzola, Phys. Rev.**187** 1767 (1969)
[30] E.W.Mielke and F.E.Schunk, gr-qc/9801063 (1998); Nucl. Phys. **B 564** 185 (2000)
[31] I.Dymnikova, B.Soltysek, Gen. Rel. Grav. **30** 1775 (1998);I. Dymnikova, B. Soltysek, in "Particles, Fields and Gravitation", Ed. J. Rembielinsky, 460 (1998)
[32] J.M.Overduin and F.I.Cooperstock, Phys. Rev. **D58** 043506 (1998)
[33] I.D.Novikov, V.P.Frolov, *Physics of Black Holes*, Kluwer Acad. (1986)
[34] Ya.B.Zel'dovich, Sov. Phys. Lett. **6** 883 (1968)
[35] R.Penrose, in: *General Relativity: An Einstein Centenary Survey*, Eds. S.W.Hawking, W.Israel, Cambridge Univ. Press (1979), p.581
[36] E.Farhi, A.Guth, Phys. Lett. **B183** 149 (1987)
[37] V.P.Frolov, M.A.Markov, and V.F.Mukhanov, Phys. Rev. **D41** 3831 (1990).
[38] J.R.Gott III, Nature **295** 304 (1982)
[39] R.R.Caldwell, R.Dave, P.J.Steinhardt, Phys. Rev. Lett. **80** 1582 (1998)
[40] I.Dymnikova, M.Fil'chenkov, Phys.Lett. **B 545, 3-4**, 214 (2002); gr-qc/0209065; Gravitation and Cosmology **8**, suppl 19 (2002); gr-qc/0009025
[41] G.W.Gibbons, S.W.Hawking, Phys. Rev.**D15** 2738 (1977)
[42] I.Dymnikova, in *Internal Structure of Black Holes and Spacetime Singularities*, Eds. M. Burko and A. Ori, Inst. Phys. Publ., Bristol and Philadelphia, and The Israel Physical Society, p. 422 (1997)
[43] S.Coleman, in *New Phenomena in Subnuclear Physics*, Ed. A.Zichichi, Plenum Press, p. 297 (1977)
[44] A.Einstein, Sci. Amer. **182** 13 (1950)
[45] I.Dymnikova, J.Ulbricht, J.Zhao, in *Quantum Electrodynamics and Physics of the Vacuum*, Ed. G.Cantatore, AIP, 255 (2001) and references therein; I.Dymnikova, A.Hasan, J.Ulbricht, J.Zhao, Gravitation and Cosmology **7**, 122 (2001)
[46] A.Vilenkin, Phys. Rev. **D30** 509 (1984); **D50**, 2581 (1994)
[47] F. Kottler, Encykl. Math. Wiss. **22a**, 231 (1922); E. Trefftz, Math. Ann. **86**, 317 (1922)
[48] H.Nariai, Sci.Rep.Tohoku Univ. **35**, 62 (1951)
[49] H.Bondi, Mon. Not. R. Astron. Soc. **107**,410 (1947.
[50] A.D.Sakharov, Sov. Phys. JETP **22** 241 (1966)
[51] M.A.Markov, JETP Lett. **36** 265 (1982); Ann. Phys. **155** 333 (1984)
[52] M.R.Bernstein, Bull. Amer. Phys. Soc. **16** 1016 (1984)
[53] W.Shen, S.Zhu, Phys. Lett. **A126** 229 (1988)
[54] A.D.Linde, Sov. Phys. Lett. **19** 183 (1974)
[55] J.A.Wheeler, Ann. Phys. **2** 604 (1957)
[56] E.B.Gliner, I.G.Dymnikova, Phys. Rev. **D28** 1278 (1983) and references therein
[57] A.D.Linde, astro-ph/9601004 (1996)
[58] D.V.Gal'tsov and J.P.S.Lemos, Class. Quant. Grav. **18** 1715 (2001)
[59] K.A.Bronnikov, Phys. Rev. **D 64** 064013 (2001)
[60] K.A.Bronnikov and G.N.Shikin, gr-qc/0109027 (2001)
[61] F.E.Schunck, astro-ph/9802258
[62] A.Strominger, Phys. Rev. **D46** 4396 (1992)
[63] D.Morgan, Phys. Rev. **D43** 3144 (1991)
[64] K.A.Bronnikov, Phys. Rev. **D 63** 044005 (2001)
[65] Let us note that GR equation (1) can be written in the four-indices form $G_{\alpha\beta\gamma\delta} = -8\pi T_{\alpha\beta\gamma\delta}$ as the equivalence relations that put the matter and geometry in direct algebraic correspondence [56].

Cosmology from Topological Defects

Alejandro Gangui

Instituto de Astronomía y Física del Espacio, Ciudad Universitaria, 1428 Buenos Aires, Argentina, and Dept. de Física, Universidad de Buenos Aires, Ciudad Universitaria - Pab. 1, 1428 Buenos Aires, Argentina.

The potential role of cosmic topological defects has raised interest in the astrophysical community for many years now. In this set of notes, we give an introduction to the subject of cosmic topological defects and some of their possible observable signatures. We begin with a review of the basics of general defect formation and evolution, we briefly comment on some general features of conducting cosmic strings and vorton formation, as well as on the possible role of defects as dark energy, to end up with cosmic structure formation from defects and some specific imprints in the cosmic microwave background radiation from simulated cosmic strings. A detailed, pedagogical explanation of the mechanism underlying the recently discovered tiny level of polarization in the cosmic microwave background by the DASI collaboration is also given, and a first comparison with some predictions from defects is provided.

I. INTRODUCTION

On a cold day, ice forms quickly on the surface of a pond. But it does not grow as a smooth, featureless covering. Instead, the water begins to freeze in many places independently, and the growing plates of ice join up in random fashion, leaving zig–zag boundaries between them. These irregular margins are an example of what physicists call "topological defects" – *defects* because they are places where the crystal structure of the ice is disrupted, and *topological* because an accurate description of them involves ideas of symmetry embodied in topology, the branch of mathematics that focuses on the study of continuous surfaces.

Current theories of particle physics likewise predict that a variety of topological defects would almost certainly have formed during the early evolution of the universe. Just as water turns to ice (a phase transition) when the temperature drops, so the interactions between elementary particles run through distinct phases as the typical energy of those particles falls with the expansion of the universe. When conditions favor the appearance of a new phase, it generally crops up in many places at the same time, and when separate regions of the new phase run into each other, topological defects are the result. The detection of such structures in the modern universe would provide precious information on events in the earliest instants after the Big Bang. Their absence, on the other hand, would force a major revision of current physical theories.

The aim of this set of Lectures is to introduce the reader to the subject of cosmology from topological defects. We begin with a review of the basics of defect formation and evolution, to get a grasp of the overall picture. We will see that defects are generically predicted to exist in most interesting models of high energy physics trying to describe the early universe. The basic elements of the standard cosmology, with its successes, shortcomings, and new developments, are covered elsewhere in this volume. See for example the lecture notes by Rocky Kolb on Astroparticle Physics, Ed Copeland's material on String / M-Theory Cosmology, and Jim Bartlett's Observational Cosmology. So we will not devote much space to these topics here. Rather, we will focus on some specific subjects. We will first briefly comment on conducting cosmic strings and one of their most important predictions for cosmology, namely, the existence of equilibrium configurations of string loops, dubbed vortons. We will then pass on to study some key signatures that a network of defects would produce on the cosmic microwave background (CMB) radiation, *e.g.*, the CMB bispectrum of the temperature anisotropies from a simulated model of cosmic strings. Miscellaneous topics also reviewed below are, for example, the way in which these cosmic entities lead to large–scale structure formation and some astrophysical footprints left by the various defects, and we will discuss the possibility of isolating their effects by astrophysical observations. Also, we include a short, detailed discussion of CMB polarization and some brief comparison with the predictions from cosmic defects.

Many areas of modern research directly related to cosmic defects are unfortunately not covered in these notes. The subject is now so vast -and beyond the possibilities of a single review- that we suggest the reader to consult some of the excellent recent literature already available. So, have a look, for example, to the report by Achúcarro & Vachaspati [2000] for a treatment of semilocal and electroweak strings [1], and to [Vachaspati, 2001] for a review of certain topological defects, like monopoles, domain walls and, again, electroweak strings, virtually not covered here. For conducting defects, cosmic strings in particular, see for example [Gangui & Peter, 1998] for a brief overview of many different astrophysical and cosmological phenomena, [Gangui, 2001b] for an updated treatment of conducting cosmic strings and one of their most important predictions for cosmology, namely, the existence of equilibrium configurations of string loops, dubbed vortons. Finally, refer to the comprehensive colorful lecture notes by Carter [1997] on the dynamics of branes with applications to conducting cosmic strings and vortons. If your are in cosmological structure formation, Durrer [2000] presents a good review of modern developments on global topological defects and their relation to CMB anisotropies, while Magueijo & Brandenberger [2000] give a set of imaginative lectures with an update on local string models of large-scale structure formation and also baryogenesis with cosmic defects. Finally, Durrer, Kunz & Melchiorri [2002] give a complete update of cosmic structure formation with global defects, including detailed analyses of correlators, mixed models, and the resulting matter and CMB power spectra.

The interdisciplinary subject of topological defects in the cosmos and the lab is nicely covered in the proceedings of the school held *aux* Houches on topological defects and non-equilibrium dynamics, edited by Bunkov & Godfrin [2000]; the ensemble of lectures in this volume, together with the recent review by Kibble [2002], give an exhaustive illustration of this fast developing area of research, which includes various fields of physics, like low–temperature condensed–matter, liquid crystals, astrophysics and high–energy physics. Finally, all of the above can also be found in the concise review by Hindmarsh & Kibble [1995], particularly concerned with the physics and cosmology of cosmic strings, and in the monograph by Vilenkin & Shellard [2000] on cosmic strings and other topological defects.

A. How defects form

A central concept of particle physics theories attempting to unify all the fundamental interactions is the concept of symmetry breaking. As the universe expanded and cooled, first the gravitational interaction, and subsequently all other known forces would have begun adopting their own identities. In the context of the standard hot Big Bang theory the spontaneous breaking of fundamental symmetries is realized as a phase transition in the early universe. Such phase transitions have several exciting cosmological consequences and thus provide an important link between particle physics and cosmology.

There are several symmetries which are expected to break down in the course of time. In each of these transitions the space–time gets 'oriented' by the presence of a hypothetical force field called the 'Higgs field', named for Peter Higgs, pervading all the space. This field orientation signals the transition from a state of higher symmetry to a final state where the system under consideration obeys a smaller group of symmetry rules. As an every–day analogy we may consider the transition from liquid water to ice; the formation of the crystal structure ice (where water molecules are arranged in a well defined lattice), breaks the symmetry possessed when the system was in the higher temperature liquid phase, when every direction in the system was equivalent. In the same way, it is precisely the orientation in the Higgs field which breaks the highly symmetric state between particles and forces.

Having built a model of elementary particles and forces, particle physicists and cosmologists are today embarked on a difficult search for a theory that unifies all the fundamental interactions. As we mentioned, an essential ingredient in all major candidate theories is the concept of symmetry breaking. Experiments have determined that there are four physical forces in nature; in addition to gravity these are called the strong, weak and electromagnetic forces. Close to the singularity of the hot Big Bang, when energies were at their highest, it is believed that these forces were unified in a single, all–encompassing interaction. As the universe expanded and cooled, first the gravitational interaction, then the strong interaction, and lastly the weak and the electromagnetic forces would have broken out of the unified scheme and adopted their present distinct identities in a series of symmetry breakings.

Theoretical physicists are still struggling to understand how gravity can be united with the other interactions, but for the unification of the strong, weak and electromagnetic forces plausible theories exist. Indeed, force–carrying particles whose existence demonstrated the fundamental unification of the weak and electromagnetic forces into a primordial "electroweak" force – the W and Z bosons – were discovered at CERN, the European accelerator laboratory, in 1983. In the context of the standard Big Bang theory, cosmological phase transitions are produced by the spontaneous breaking of a fundamental symmetry, such as the electroweak force, as the universe cools. For example, the electroweak

---

[1] Animations of semilocal and electroweak string formation and evolution can be found at http://www.nersc.gov/~borrill/

interaction broke into the separate weak and electromagnetic forces when the observable universe was 10^{-12} seconds old, had a temperature of 10^{15} degrees Kelvin, and was only one part in 10^{15} of its present size. There are also other phase transitions besides those associated with the emergence of the distinct forces. The quark-hadron confinement transition, for example, took place when the universe was about a microsecond old. Before this transition, quarks – the particles that would become the constituents of the atomic nucleus – moved as free particles; afterward, they became forever bound up in protons, neutrons, mesons and other composite particles.

As we said, the standard mechanism for breaking a symmetry involves the hypothetical Higgs field that pervades all space. As the universe cools, the Higgs field can adopt different ground states, also referred to as different vacuum states of the theory. In a symmetric ground state, the Higgs field is zero everywhere. Symmetry breaks when the Higgs field takes on a finite value (see Figure 1).

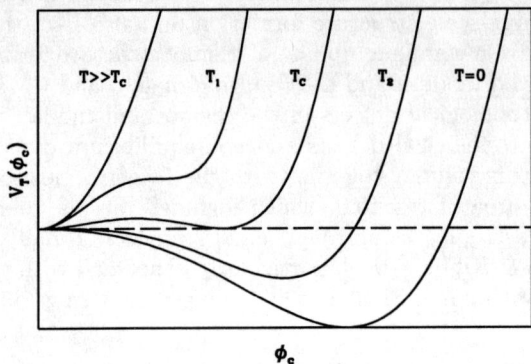

FIG. 1: *Temperature–dependent effective potential for a first–order phase transition for the Higgs field. For very high temperatures, well above the critical one T_c, the potential possesses just one minimum for the vanishing value of the Higgs field. Then, when the temperature decreases, a whole set of minima develops (it may be two or more, discrete or continuous, depending of the type of symmetry under consideration). Below T_c, the value $\phi = 0$ stops being the global minimum and the system will spontaneously choose a new (lower) one, say $\phi = \eta \exp(i\theta)$ (for complex ϕ) for some angle θ and nonvanishing η, amongst the available ones. This choice signals the breakdown of the symmetry in a cosmic phase transition and the generation of random regions of conflicting field orientations θ. In a cosmological setting, the merging of these domains gives rise to cosmic defects.*

Kibble [1976] first saw the possibility of defect formation when he realized that in a cooling universe phase transitions proceed by the formation of uncorrelated domains that subsequently coalesce, leaving behind relics in the form of defects. In the expanding universe, widely separated regions in space have not had enough time to 'communicate' amongst themselves and are therefore not correlated, due to a lack of causal contact. It is therefore natural to suppose that different regions ended up having arbitrary orientations of the Higgs field and that, when they merged together, it was hard for domains with very different preferred directions to adjust themselves and fit smoothly. In the interfaces of these domains, defects form. Such relic 'flaws' are unique examples of incredible amounts of energy and this feature attracted the minds of many cosmologists.

B. Phase transitions and finite temperature field theory

Phase transitions are known to occur in the early universe. Examples we mentioned are the quark to hadron (confinement) transition, which QCD predicts at an energy around 1 GeV, and the electroweak phase transition at about 250 GeV. Within grand unified theories (GUT), aiming to describe the physics beyond the standard model, other phase transitions are predicted to occur at energies of order 10^{15} GeV; during these, the Higgs field tends to fall towards the minima of its potential while the overall temperature of the universe decreases as a consequence of the expansion.

A familiar theory to make a bit more quantitative the above considerations is the $\lambda |\phi|^4$ theory,

$$\mathcal{L} = \frac{1}{2}|\partial_\mu \phi|^2 + \frac{1}{2}m_0^2|\phi|^2 - \frac{\lambda}{4!}|\phi|^4 , \qquad (1)$$

with $m_0^2 > 0$. The second and third terms on the right hand side yield the usual 'Mexican hat' potential for the complex scalar field. For energies much larger than the critical temperature, T_c, the fields are in the so-called 'false' vacuum: a highly symmetric state characterized by a vacuum expectation value $\langle |\phi| \rangle = 0$. But when energies decrease the symmetry is spontaneously broken: a new 'true' vacuum develops and the scalar field rolls down the potential and sits onto one of the degenerate new minima. In this situation the vacuum expectation value becomes $\langle|\phi|\rangle^2 = 6m_0^2/\lambda$.

Research done in the 1970's in finite–temperature field theory [Weinberg, 1974; Dolan & Jackiw, 1974; Kirzhnits & Linde, 1974] has led to the result that the temperature–dependent effective potential can be written down as

$$V_T(|\phi|) = -\frac{1}{2}m^2(T)|\phi|^2 + \frac{\lambda}{4!}|\phi|^4 \qquad (2)$$

with $T_c^2 = 24m_0^2/\lambda$, $m^2(T) = m_0^2(1 - T^2/T_c^2)$, and $\langle|\phi|\rangle^2 = 6m^2(T)/\lambda$. We easily see that when T approaches T_c from below the symmetry is restored, and again we have $\langle|\phi|\rangle = 0$. In condensed–matter jargon, the transition described above is second–order [Mermin, 1979].[2]

C. The Kibble mechanism

The model described in the last subsection is an example in which the transition may be second–order. As we saw, for temperatures much larger than the critical one the vacuum expectation value of the scalar field vanishes at all points of space, whereas for $T < T_c$ it evolves smoothly in time towards a non vanishing $\langle|\phi|\rangle$. Both thermal and quantum fluctuations influence the new value taken by $\langle|\phi|\rangle$ and therefore it has no reasons to be uniform in space. This leads to the existence of domains wherein the $\langle|\phi(\vec{x})|\rangle$ is coherent and regions where it is not. The consequences of this fact are the subject of this subsection.

Phase transitions can also be first–order proceeding via bubble nucleation. At very high energies the symmetry breaking potential has $\langle|\phi|\rangle = 0$ as the only vacuum state. When the temperature goes down to T_c a set of vacua, degenerate to the previous one, develops. However this time the transition is not smooth as before, for a potential barrier separates the old (false) and the new (true) vacua (see, e.g. Figure 1). Provided the barrier at this small temperature is high enough, compared to the thermal energy present in the system, the field ϕ will remain trapped in the false vacuum state even for small ($< T_c$) temperatures. Classically, this is the complete picture. However, quantum tunneling effects can liberate the field from the old vacuum state, at least in some regions of space: there is a probability per unit time and volume in space that at a point $\vec{x}$ a bubble of true vacuum will nucleate. The result is thus the formation of bubbles of true vacuum with the value of the field in each bubble being independent of the value of the field in all other bubbles. This leads again to the formation of domains where the fields are correlated, whereas no correlation exits between fields belonging to different domains. Then, after creation the bubble will expand at the speed of light surrounded by a 'sea' of false vacuum domains. As opposed to second–order phase transitions, here the nucleation process is extremely inhomogeneous and $\langle|\phi(\vec{x})|\rangle$ is not a continuous function of time.

Let us turn now to the study of correlation lengths and their rôle in the formation of topological defects. One important feature in determining the size of the domains where $\langle|\phi(\vec{x})|\rangle$ is coherent is given by the spatial correlation of the field ϕ. Simple field theoretic considerations [see, e.g., Copeland, 1993] for long wavelength fluctuations of ϕ lead to different functional behaviors for the correlation function $G(r) \equiv \langle\phi(r_1)\phi(r_2)\rangle$, where we noted $r = |r_1 - r_2|$. What is found depends radically on whether the wanted correlation is computed between points in space separated by a distance r much smaller or much larger than a characteristic length $\xi^{-1} = m(T) \simeq \sqrt{\lambda}\,|\langle\phi\rangle|$, known as the *correlation length*. Then, we have $G(r) \simeq \frac{T_c}{4\pi r}\exp(-\frac{r}{\xi})$ for $r >> \xi$, while $G(r) \simeq \frac{T^2}{2\pi^2}$ for $r << \xi$.

This tells us that domains of size $\xi \sim m^{-1}$ arise where the field ϕ is correlated. On the other hand, well beyond ξ no correlations exist and thus points separated apart by $r >> \xi$ will belong to domains with in principle arbitrarily different orientations of the Higgs field. This in turn leads, after the merging of these domains in a cosmological setting, to the existence of defects, where field configurations fail to match smoothly.

However, when $T \to T_c$ we have $m \to 0$ and so $\xi \to \infty$, suggesting perhaps that for all points of space the field ϕ becomes correlated. This fact clearly violates causality. The existence of particle horizons in cosmological models (proportional to the inverse of the Hubble parameter H^{-1}) constrains microphysical interactions over distances beyond this causal domain. Therefore we get an upper bound to the correlation length as $\xi < H^{-1} \sim t$.

The general feature of the existence of uncorrelated domains has become known as the Kibble mechanism [Kibble, 1976] and it seems to be generic to most types of phase transitions.

---

[2] In a first–order phase transition the order parameter (e.g., $< |\phi| >$ in our case) is not continuous. It may proceed by bubble nucleation [Callan & Coleman, 1977; Linde, 1983b] or by spinoidal decomposition [Langer, 1992]. Phase transitions can also be continuous second–order processes. The 'order' depends sensitively on the ratio of the coupling constants appearing in the Lagrangian.

D. A survey of topological defects

Different models for the Higgs field lead to the formation of a whole variety of topological defects, with very different characteristics and dimensions. Some of the proposed theories have symmetry breaking patterns leading to the formation of 'domain walls' (mirror reflection discrete symmetry): incredibly thin planar surfaces trapping enormous concentrations of mass–energy which separate domains of conflicting field orientations, similar to two–dimensional sheet–like structures found in ferromagnets. Within other theories, cosmological fields get distributed in such a way that the old (symmetric) phase gets confined into a finite region of space surrounded completely by the new (non–symmetric) phase. This situation leads to the generation of defects with linear geometry called 'cosmic strings'. Theoretical reasons suggest these strings (vortex lines) do not have any loose ends in order that the two phases not get mixed up. This leaves infinite strings and closed loops as the only possible alternatives for these defects to manifest themselves in the early universe[3].

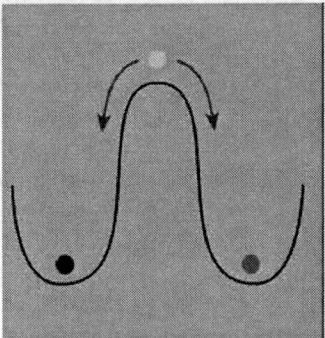 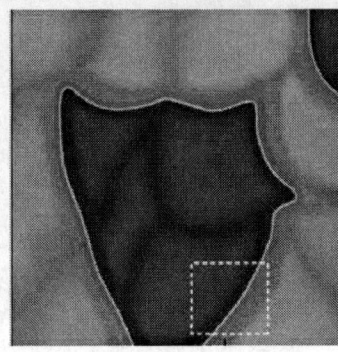 

FIG. 2: *In a simple model of symmetry breaking, the initial symmetric ground state of the Higgs field (central dot, middle panel) can fall into the left- or right-hand valley of a double-well energy potential (light and dark dots). In a cosmic phase transition, regions of the new phase appear randomly and begin to grow and eventually merge as the transition proceeds toward completion (middle). Regions in which the symmetry has broken the same way can coalesce, but where regions that have made opposite choices encounter each other, a topological defect known as a domain wall forms (right). Across the wall, the Higgs field has to go from one of the valleys to the other (in the left panel), and must therefore traverse the energy peak. This creates a narrow planar region of very high energy, in which the symmetry is locally unbroken* [127].

With a bit more abstraction scientists have even conceived other (semi) topological defects, called 'textures'. These are conceptually simple objects, yet, it is not so easy to imagine them for they are just global field configurations living on a three–sphere vacuum manifold (the minima of the effective potential energy), whose non linear evolution perturbs spacetime. Turok [1989] was the first to realize that many unified theories predicted the existence of peculiar Higgs field configurations known as (texture) knots, and that these could be of potential interest for cosmology. Several features make these defects interesting. In contrast to domain walls and cosmic strings, textures have no core and thus the energy is more evenly distributed over space. Secondly, they are unstable to collapse and it is precisely this last feature which makes these objects cosmologically relevant, for this instability makes texture knots shrink to a microscopic size, unwind and radiate away all their energy. In so doing, they generate a gravitational field that perturbs the surrounding matter in a way which can seed structure formation.

E. Conditions for their existence: topological criteria

Let us now explore the conditions for the existence of topological defects. It is widely accepted that the final goal of particle physics is to provide a unified gauge theory comprising strong, weak and electromagnetic interactions (and some day also gravitation). This unified theory is to describe the physics at very high temperatures, when the age of the universe was slightly bigger than the Planck time. At this stage, the universe was in a state with the

---

[3] 'Monopole' is another possible topological defect; we defer its discussion to the next subsection. Cosmic strings bounded by monopoles is yet another possibility in GUT phase transitions of the kind, e.g., $\mathbf{G} \to \mathbf{K} \times U(1) \to \mathbf{K}$. The first transition yields monopoles carrying a magnetic charge of the $U(1)$ gauge field, while in the second transition the magnetic field in squeezed into flux tubes connecting monopoles and antimonopoles [Langacker & Pi, 1980].

$\pi_0(\mathcal{M}) \neq 1$	$\mathcal{M}$ *disconnected*	DOMAIN WALLS
$\pi_1(\mathcal{M}) \neq 1$	*non contractible loops* in $\mathcal{M}$	COSMIC STRINGS
$\pi_2(\mathcal{M}) \neq 1$	*non contractible 2-spheres* in $\mathcal{M}$	MONOPOLES
$\pi_3(\mathcal{M}) \neq 1$	*non contractible 3-spheres* in $\mathcal{M}$	TEXTURES

TABLE I: The topology of $\mathcal{M}$ determines the type of defect that will arise.

highest possible symmetry, described by a symmetry group **G**, and the Lagrangian modeling the system of all possible particles and interactions present should be invariant under the action of the elements of **G**.

As we explained before, the form of the finite temperature effective potential of the system is subject to variations during the cooling down evolution of the universe. This leads to a chain of phase transitions whereby some of the symmetries present in the beginning are not present anymore at lower temperatures. The first of these transitions may be described as **G**→**H**, where now **H** stands for the new (smaller) unbroken symmetry group ruling the system. This chain of symmetry breakdowns eventually ends up with SU(3)×SU(2)×U(1), the symmetry group underlying the 'standard model' of particle physics.

A broken symmetry system (with a Mexican-hat potential for the Higgs field) may have many different minima (with the same energy), all related by the underlying symmetry. Passing from one minimum to another is included as one of the symmetries of the original group **G**, and the system will not change due to one such transformation. If a certain field configuration yields the lowest energy state of the system, transformations of this configuration by the elements of the symmetry group will also give the lowest energy state. For example, if a spherically symmetric system has a certain lowest energy value, this value will not change if the system is rotated.

The system will try to minimize its energy and will spontaneously choose one amongst the available minima. Once this is done and the phase transition achieved, the system is no longer ruled by **G** but by the symmetries of the smaller group **H**. So, if **G**→**H** and the system is in one of the lowest energy states (call it S_1), transformations of S_1 to S_2 by elements of **G** will leave the energy unchanged. However, transformations of S_1 by elements of **H** will leave S_1 *itself* (and not just the energy) unchanged. The many distinct ground states of the system $S_1, S_2, \ldots$ are given by all transformations of **G** that are *not* related by elements in **H**. This space of distinct ground states is called the *vacuum manifold* and denoted $\mathcal{M}$. So, $\mathcal{M}$ is the space of all elements of **G** in which elements related by transformations in **H** have been identified. Mathematicians call it the *coset space* and denote it **G/H**. We then have $\mathcal{M} = $ **G/H**.

The importance of the study of the vacuum manifold lies in the fact that it is precisely the *topology* of $\mathcal{M}$ what determines the type of defect that will arise. Homotopy theory tells us how to map $\mathcal{M}$ into physical space in a non–trivial way, and what ensuing defect will be produced. For instance, the existence of non contractible loops in $\mathcal{M}$ is the requisite for the formation of cosmic strings. In formal language this comes about whenever we have the first homotopy group $\pi_1(\mathcal{M}) \neq 1$, where **1** corresponds to the trivial group. If the vacuum manifold is disconnected we then have $\pi_0(\mathcal{M}) \neq 1$, and domain walls are predicted to form in the boundary of these regions where the field ϕ is away from the minimum of the potential. Analogously, if $\pi_2(\mathcal{M}) \neq 1$ it follows that the vacuum manifold contains non contractible two–spheres, and the ensuing defect is a monopole. Textures arise when $\mathcal{M}$ contains non contractible three–spheres and in this case it is the third homotopy group, $\pi_3(\mathcal{M})$, the one that is non trivial. We summarize this in Table I .

II. DEFECTS IN THE UNIVERSE

Generically topological defects will be produced if the conditions for their existence are met. Then for example if the unbroken group **H** contains a disconnected part, like an explicit U(1) factor (something that is quite common in many phase transition schemes discussed in the literature), monopoles will be left as relics of the transition. This is due to the fundamental theorem on the second homotopy group of coset spaces [Mermin, 1979], which states that for a simply–connected covering group **G** we have[4]

$$\pi_2(\mathbf{G}/\mathbf{H}) \cong \pi_1(\mathbf{H}_0) \ , \qquad (3)$$

---

[4] The isomorfism between two groups is noted as $\cong$. Note that by using the theorem we therefore can reduce the computation of π_2 for a coset space to the computation of π_1 for a group. A word of warning: the focus here is on the physics and the mathematically-oriented reader should bear this in mind, especially when we will become a bit sloppy with the notation. In case this happens, consult the book [Steenrod, 1951] for a clear exposition of these matters.

with $\mathbf{H}_0$ being the component of the unbroken group connected to the identity. Then we see that since monopoles are associated with unshrinkable surfaces in $\mathbf{G}/\mathbf{H}$, the previous equation implies their existence if $\mathbf{H}$ is multiply-connected. The reader may guess what the consequences are for GUT phase transitions: in grand unified theories a semi-simple gauge group $\mathbf{G}$ is broken in several stages down to $\mathbf{H} = \mathrm{SU}(3) \times \mathrm{U}(1)$. Since in this case $\pi_1(\mathbf{H}) \cong \mathcal{Z}$, the integers, we have $\pi_2(\mathbf{G}/\mathbf{H}) \neq \mathbf{1}$ and therefore gauge monopole solutions exist [Preskill, 1979].

A. Local and global monopoles and domain walls

Monopoles are yet another example of stable topological defects. Their formation stems from the fact that the vacuum expectation value of the symmetry breaking Higgs field has random orientations ($\langle \phi^a \rangle$ pointing in different directions in group space) on scales greater than the horizon. One expects therefore to have a probability of order unity that a monopole configuration will result after the phase transition (cf. the Kibble mechanism). Thus, about one monopole per Hubble volume should arise and we have for the number density $n_{monop} \sim 1/H^{-3} \sim T_c^6/m_P^3$, where T_c is the critical temperature and m_P is Planck mass, when the transition occurs. We also know the entropy density at this temperature, $s \sim T_c^3$, and so the monopole to entropy ratio is $n_{monop}/s \simeq 100(T_c/m_P)^3$. In the absence of non-adiabatic processes after monopole creation this constant ratio determines their present abundance. For the typical value $T_c \sim 10^{14}$ GeV we have $n_{monop}/s \sim 10^{-13}$. This estimate leads to a present $\Omega_{monop} h^2 \simeq 10^{11}$, for the superheavy monopoles $m_{monop} \simeq 10^{16}$ GeV that are created[5]. This value contradicts standard cosmology and the presently most attractive way out seems to be to allow for an early period of inflation: the massive entropy production will hence lead to an exponential decrease of the initial n_{monop}/s ratio, yielding Ω_{monop} consistent with observations.[6] In summary, the broad-brush picture one has in mind is that of a mechanism that could solve the monopole problem by 'weeping' these unwanted relics out of our sight, to scales much bigger than the one that will eventually become our present horizon today.

Note that these arguments do not apply for global monopoles as these (in the absence of gauge fields) possess long-range forces that lead to a decrease of their number in comoving coordinates. The large attractive force between global monopoles and antimonopoles leads to a high annihilation probability and hence monopole over-production does not take place. Simulations performed by Bennett & Rhie [1990] showed that global monopole evolution rapidly settles into a scale invariant regime with only a few monopoles per horizon volume at all times.

Given that global monopoles do not represent a danger for cosmology one may proceed in studying their observable consequences. The gravitational fields of global monopoles may lead to matter clustering and CMB anisotropies. Given an average number of monopoles per horizon of ~ 4, Bennett & Rhie [1990] estimate a scale invariant spectrum of fluctuations $(\delta\rho/\rho)_H \sim 30 G\eta^2$ at horizon crossing[7]. In a subsequent paper they simulate the large-scale CMB anisotropies and, upon normalization with COBE–DMR, they get roughly $G\eta^2 \sim 6 \times 10^{-7}$ in agreement with a GUT energy scale η [Bennett & Rhie, 1993]. However, as we will see in the CMB sections below, current estimates for the angular power spectrum of global defects do not match the most recent observations, their main problem being the lack of power on the degree angular scale once the spectrum is normalized to COBE on large scales [Durrer et al., 1996; Durrer et al., 2002].

Let us concentrate now on domain walls, and briefly try to show why they are not welcome in any cosmological context [at least in the simple version we here consider – there is always room for more complicated (and contrived) models]. If the symmetry breaking pattern is appropriate at least one domain wall per horizon volume will be formed. The mass per unit surface of these two-dimensional objects is given by $\sim \lambda^{1/2} \eta^3$, where λ as usual is the coupling constant in the symmetry breaking potential for the Higgs field. Domain walls are generally horizon-sized and therefore their mass is given by $\sim \lambda^{1/2} \eta^3 H^{-2}$. This implies a mass energy density roughly given by $\rho_{DW} \sim \eta^3 t^{-1}$ and we may readily see now how the problem arises: the critical density goes as $\rho_{crit} \sim t^{-2}$ which implies $\Omega_{DW}(t) \sim (\eta/m_P)^2 \eta t$. Taking a typical GUT value for η we get $\Omega_{DW}(t \sim 10^{-35} \mathrm{sec}) \sim 1$ *already* at the time of the phase transition. It is not hard to imagine that today this will be at variance with observations; in fact we get $\Omega_{DW}(t \sim 10^{18} \mathrm{sec}) \sim 10^{52}$.

---

[5] These are the actual figures for a gauge SU(5) GUT second-order phase transition. Preskill [1979] has shown that in this case monopole antimonopole annihilation is not effective to reduce their abundance. Guth & Weinberg [1983] did the case for a first-order phase transition and drew qualitatively similar conclusions regarding the excess of monopoles.

[6] The inflationary expansion reaches an end in the so-called reheating process, when the enormous vacuum energy driving inflation is transferred to coherent oscillations of the inflaton field. These oscillations will in turn be damped by the creation of light particles (e.g., via preheating) whose final fate is to thermalise and reheat the universe.

[7] The spectrum of density fluctuations on smaller scales has also been computed. They normalize the spectrum at $8 h^{-1}$ Mpc and agreement with observations lead them to assume that galaxies are clustered more strongly than the overall mass density, this implying a 'biasing' of a few [see Bennett, Rhie & Weinberg, 1993 for details].

This indicates that models where domain walls are produced are tightly constrained, and the general feeling is that it is best to avoid them altogether [see Kolb & Turner, 1990 for further details; see also Dvali et al., 1998, Pogosian & Vachaspati, 2000 [8] and Alexander et al., 1999 for an alternative solution].

B. Are defects inflated away?

It is important to realize the relevance that the Kibble's mechanism has for cosmology; nearly every sensible grand unified theory (with its own symmetry breaking pattern) predicts the existence of defects. We know that an early era of inflation helps in getting rid of the unwanted relics. One could well wonder if the very same Higgs field responsible for breaking the symmetry would not be the same one responsible for driving an era of inflation, thereby diluting the density of the relic defects. This would get rid not only of (the unwanted) monopoles and domain walls but also of any other (cosmologically appealing) defect. Let us follow [Brandenberger, 1993] and sketch why this actually does not occur. Take first the symmetry breaking potential of Eq. (2) at zero temperature and add to it a harmless ϕ–independent term $3m^4/(2\lambda)$. This will not affect the dynamics at all. Then we are led to

$$V(\phi) = \frac{\lambda}{4!} \left(\phi^2 - \eta^2\right)^2 , \qquad (4)$$

with $\eta = (6m^2/\lambda)^{1/2}$ the symmetry breaking energy scale, and where for the present heuristic digression we just took a real Higgs field. Consider now the equation of motion for ϕ,

$$\ddot{\phi} \simeq -\frac{\partial V}{\partial \phi} = -\frac{\lambda}{3!}\phi^3 + m^2\phi \approx m^2\phi , \qquad (5)$$

for $\phi << \eta$ very near the false vacuum of the effective Mexican hat potential and where, for simplicity, the expansion of the universe and possible interactions of ϕ with other fields were neglected. The typical time scale of the solution is $\tau \simeq m^{-1}$. For an inflationary epoch to be effective we need $\tau >> H^{-1}$, i.e., a sufficiently large number of e–folds of slow–rolling solution. Note, however, that after some e–folds of exponential expansion the curvature term in the Friedmann equation becomes subdominant and we have $H^2 \simeq 8\pi G\, V(0)/3 \simeq (2\pi m^2/3)(\eta/m_P)^2$. So, unless $\eta > m_P$, which seems unlikely for a GUT phase transition, we are led to $\tau << H^{-1}$ and therefore the amount of inflation is not enough for getting rid of the defects generated during the transition by hiding them well beyond our present horizon.

Recently, there has been a large amount of work in getting defects, particularly cosmic strings, after post-inflationary preheating. Reaching the latest stages of the inflationary phase, the inflaton field oscillates about the minimum of its potential. In doing so, parametric resonance may transfer a huge amount of energy to other fields leading to cosmologically interesting nonthermal phase transitions. Just like thermal fluctuations can restore broken symmetries, here also, these large fluctuations may lead to the whole process of defect formation again. Numerical simulations employing potentials similar to that of Eq. (4) have shown that strings indeed arise for values $\eta \sim 10^{16}$ GeV [Tkachev et al., 1998, Kasuya & Kawasaki, 1998]. Hence, preheating after inflation helps in generating cosmic defects.

C. Cosmic strings

Cosmic strings are without any doubt the topological defect most thoroughly studied, both in cosmology and solid–state physics (vortices). The canonical example, also describing flux tubes in superconductors, is given by the Lagrangian

$$\mathcal{L} = -\frac{1}{4}F_{\mu\nu}F^{\mu\nu} + \frac{1}{2}|D_\mu\phi|^2 - \frac{\lambda}{4!}\left(|\phi|^2 - \eta^2\right)^2 , \qquad (6)$$

with $F_{\mu\nu} = \partial_{[\mu}A_{\nu]}$, where A_ν is the gauge field and the covariant derivative is $D_\mu = \partial_\mu + ieA_\mu$, with e the gauge coupling constant. This Lagrangian is invariant under the action of the Abelian group $\mathbf{G} = \mathrm{U}(1)$, and the spontaneous breakdown of the symmetry leads to a vacuum manifold $\mathcal{M}$ that is a circle, S^1, i.e., the potential is minimized for $\phi = \eta \exp(i\theta)$, with arbitrary $0 \leq \theta \leq 2\pi$. Each possible value of θ corresponds to a particular 'direction' in the field space.

---

[8] Animations of monopoles colliding with domain walls can be found in 'LEP' page at http://theory.ic.ac.uk/~LEP/figures.html

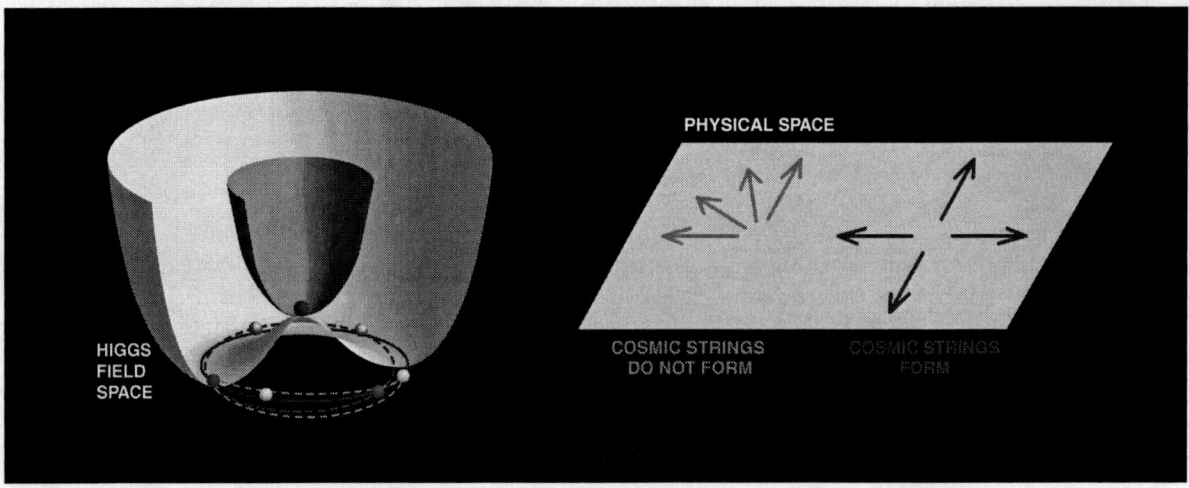

FIG. 3: *The complex scalar Higgs field evolves in a temperature-dependent potential $V(\phi)$. At high temperatures (narrow surface in the left panel) the vacuum expectation value of the field lies at the bottom of V. For lower temperatures, the potential adopts the "Mexican hat" form (wider surface) and the field spontaneously chooses one amongst the new available (degenerate) lowest energy states (a circle along the valley of the hat). This isolates a single value/direction for the phase of the field, spontaneously breaking the symmetry possessed by the system at high energies. Different regions of the universe, with no causal connection, will end up having arbitrarily different directions for the field (arrows on the right panel). As separate regions of broken symmetry merge, it is not always possible for the field orientations to match. It may happen that a closed loop in physical space intersects regions where the Higgs phase varies from 0 to 2π (dark arrows, corresponding to the dark dashed-line on the left panel). In that situation, a cosmic string will pass somewhere inside the loop. On the contrary, light-grey arrows (and light-grey dashed-line on the left panel) show a situation where no string is formed after the phase transition [127].*

Now, as we have seen earlier, due to the overall cooling down of the universe, there will be regions where the scalar field rolls down to different vacuum states. The choice of the vacuum is totally independent for regions separated apart by one correlation length or more, thus leading to the formation of domains of size $\xi \sim \eta^{-1}$. When these domains coalesce they give rise to edges in the interface. If we now draw a imaginary circle around one of these edges and the angle θ varies by 2π then by contracting this loop we reach a point where we cannot go any further without leaving the manifold $\mathcal{M}$. This is a small region where the variable θ is not defined and, by continuity, the field should be $\phi = 0$. In order to minimize the spatial gradient energy these small regions line up and form a line-like defect called cosmic string.

The width of the string is roughly $m_\phi^{-1} \sim (\sqrt{\lambda}\eta)^{-1}$, m_ϕ being the Higgs mass. The string mass per unit length, or tension, is $\mu \sim \eta^2$. This means that for GUT cosmic strings, where $\eta \sim 10^{16}$ GeV, we have $G\mu \sim 10^{-6}$. We will see below that the dimensionless combination $G\mu$, present in all signatures due to strings, is of the right order of magnitude for rendering these defects cosmologically interesting.

There is an important difference between global and gauge (or local) cosmic strings: local strings have their energy confined mainly in a thin core, due to the presence of gauge fields A_μ that cancel the gradients of the field outside of it. Also these gauge fields make it possible for the string to have a quantized magnetic flux along the core. On the other hand, if the string was generated from the breakdown of a *global* symmetry there are no gauge fields, just Goldstone bosons, which, being massless, give rise to long-range forces. No gauge fields can compensate the gradients of ϕ this time and therefore there is an infinite string mass per unit length.

Just to get a rough idea of the kind of models studied in the literature, consider the case $\mathbf{G} = SO(10)$ that is broken to $\mathbf{H} = SU(5) \times \mathcal{Z}_2$. For this pattern we have $\pi_1(\mathcal{M}) = \mathcal{Z}_2$, which is clearly non trivial and therefore cosmic strings are formed [Kibble et al., 1982].[9]

---

[9] In the analysis one uses the fundamental theorem stating that, for a simply-connected Lie group $\mathbf{G}$ breaking down to $\mathbf{H}$, we have $\pi_1(\mathbf{G}/\mathbf{H}) \cong \pi_0(\mathbf{H})$; see [Hilton, 1953].

D. String loops and scaling

We saw before the reasons why gauge monopoles and domain walls were a bit of a problem for cosmology. Essentially, the problem was that their energy density decreases more slowly than the critical density with the expansion of the universe. This fact resulted in their contribution to Ω_{def} (the density in defects normalized by the critical density) being largely in excess compared to 1, hence in blatant conflict with modern observations. The question now arises as to whether the same might happened with cosmic strings. Are strings dominating the energy density of the universe? Fortunately, the answer to this question is *no*; strings evolve in such a way to make their density $\rho_{\text{strings}} \propto \eta^2 t^{-2}$. Hence, one gets the same temporal behavior as for the critical density. The result is that $\Omega_{\text{strings}} \sim G\mu \sim (\eta/m_P)^2 \sim 10^{-6}$ for GUT strings, *i.e.*, we get an interestingly small enough, constant fraction of the critical density of the universe and strings never upset standard observational cosmology.

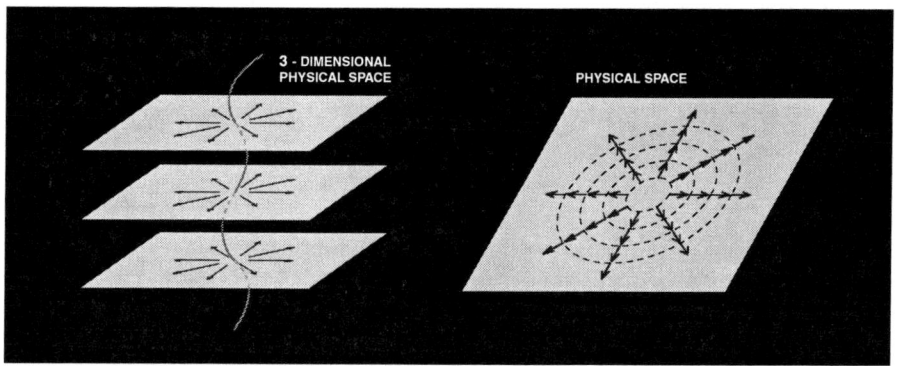

FIG. 4: *We can now extend the mechanism shown in the previous figure to the full three-dimensional space. Regions of the various planes that were traversed by strings can be superposed to show the actual location of the cosmic string (left panel). The figure on the right panel shows why we are sure a string crosses the plane inside the loop in physical space (the case with dark arrows in the previous figure). Continuity of the field imposes that if we gradually contract this loop the direction of the field will be forced to wind "faster". In the limit in which the loop reduces to a point, the phase is no longer defined and the vacuum expectation value of the Higgs field has to vanish. This corresponds to the central tip of the Mexican hat potential in the previous figure and is precisely the locus of the false vacuum. Cosmic strings are just that, narrow, extremely massive line-like regions in physical space where the Higgs field adopts its high-energy false vacuum state [127].*

Now, why this is so? The answer is simply the efficient way in which a network of strings looses energy. The evolution of the string network is highly nontrivial and loops are continuously chopped off from the main infinite strings as the result of (self) intersections within the infinite–string network. Once they are produced, loops oscillate due to their huge tension and slowly decay by emitting gravitational radiation. Thus, energy is transferred from the cosmic string network to radiation.[10]

It turns out from simulations that most of the energy in the string network (roughly a 80%) is in the form of infinite strings. Soon after formation one would expect long strings to have the form of random-walk with characteristic step given by the correlation length ξ. Also, the typical distance between long string segments should also be of order ξ. Monte Carlo simulations show that these strings are Brownian on sufficiently large scales, which means that the length ℓ of a string is related to the end-to-end distance d of two given points along the string (with d $\gg \xi$) in the form

$$\ell = \text{d}^2/\xi. \qquad (7)$$

What remains of the energy is given in the form of closed loops with no preferred length scale (a scale invariant distribution) which implies that the number density of loops having sizes between R and $R + dR$ follows just from dimensional analysis

$$dn_{\text{loops}} \propto \frac{dR}{R^4} \qquad (8)$$

---

[10] High–resolution cosmic string simulations can be found in the Cambridge cosmology page at http://www.damtp.cam.ac.uk/user/gr/public/cs_evol.html

which is just another way of saying that $n_{\text{loops}} \propto 1/R^3$, loops behave like normal nonrelativistic matter. The actual coefficient, as usual, comes from string simulations.

There are both analytical and numerical indications in favor of the existence of a stable "scaling solution" for the cosmic string network. After generation, the network quickly evolves in a self similar manner with just a few infinite string segments per Hubble volume and Hubble time. A heuristic argument for the scaling solution due to Vilenkin [1985] is as follows.

If we take $\nu(t)$ to be the mean number of infinite string segments per Hubble volume, then the energy density in infinite strings $\rho_{\text{strings}} = \rho_{\text{s}}$ is

$$\rho_{\text{s}}(t) = \nu(t)\eta^2 t^{-2} = \nu(t)\mu t^{-2}. \tag{9}$$

Now, ν strings will typically have ν intersections, and so the number of loops $n_{\text{loops}}(t) = n_{\text{l}}(t)$ produced per unit volume will be proportional to ν^2. We find

$$dn_{\text{l}} \sim \nu^2 R^{-4} dR. \tag{10}$$

Hence, recalling now that the loop sizes grow with the expansion like $R \propto t$ we have

$$\frac{dn_{\text{l}}(t)}{dt} \sim p\nu^2 t^{-4} \tag{11}$$

where p is the probability of loop formation per intersection, a quantity related to the intercommuting probability, both roughly of order 1. We are now in a position to write an energy conservation equation for strings plus loops in the expanding universe. Here it is

$$\frac{d\rho_{\text{s}}}{dt} + \frac{3}{2t}\rho_{\text{s}} \sim -m_{\text{l}}\frac{dn_{\text{l}}}{dt} \sim -\mu t \frac{dn_{\text{l}}}{dt} \tag{12}$$

where $m_{\text{l}} = \mu t$ is just the loop mass and where the second on the left hand side is the dilution term $3H\rho_{\text{s}}$ for an expanding radiation–dominated universe. The term on the right hand side amounts to the loss of energy from the long string network by the generation of small closed loops. Plugging Eqs. (9) and (11) into (12) Vilenkin finds the following kinetic equation for $\nu(t)$

$$\frac{d\nu}{dt} - \frac{\nu}{2t} \sim -p\frac{\nu^2}{t} \tag{13}$$

with $p \sim 1$. Thus if $\nu \gg 1$ then $d\nu/dt < 0$ and ν tends to decrease in time, while if $\nu \ll 1$ then $d\nu/dt > 0$ and ν increases. Hence, there will be a stable solution with $\nu \sim$ a few.

E. Global textures

Whenever a global non–Abelian symmetry is spontaneously and completely broken (*e.g.* at a grand unification scale), global defects called textures are generated. Theories where this global symmetry is only partially broken do not lead to global textures, but instead to global monopoles and non–topological textures. As we already mentioned global monopoles do not suffer the same constraints as their gauge counterparts: essentially, having no associated gauge fields, the long–range forces between pairs of monopoles lead to the annihilation of their eventual excess and as a result monopoles scale with the expansion. On the other hand, non–topological textures are a generalization that allows the broken subgroup **H** to contain non–Abelian factors. It is then possible to have π_3 trivial as in, *e.g.*, SO(5)→SO(4) broken by a vector, for which case we have $\mathcal{M} = S^4$, the four–sphere [Turok, 1989]. Having explained this, let us concentrate in global topological textures from now on.

Textures, unlike monopoles or cosmic strings, are not well localized in space. This is due to the fact that the field remains in the vacuum everywhere, in contrast to what happens for other defects, where the field leaves the vacuum manifold precisely where the defect core is. Since textures do not possess a core, all the energy of the field configuration is in the form of field gradients. This fact is what makes them interesting objects *only* when coming from global theories: the presence of gauge fields A_μ could (by a suitable reorientation) compensate the gradients of ϕ and yield $D_\mu \phi = 0$, hence canceling out (gauging away) the energy of the configuration[11].

---

[11] This does not imply, however, that the classical dynamics of a gauge texture is trivial. The evolution of the ϕ–A_μ system will be

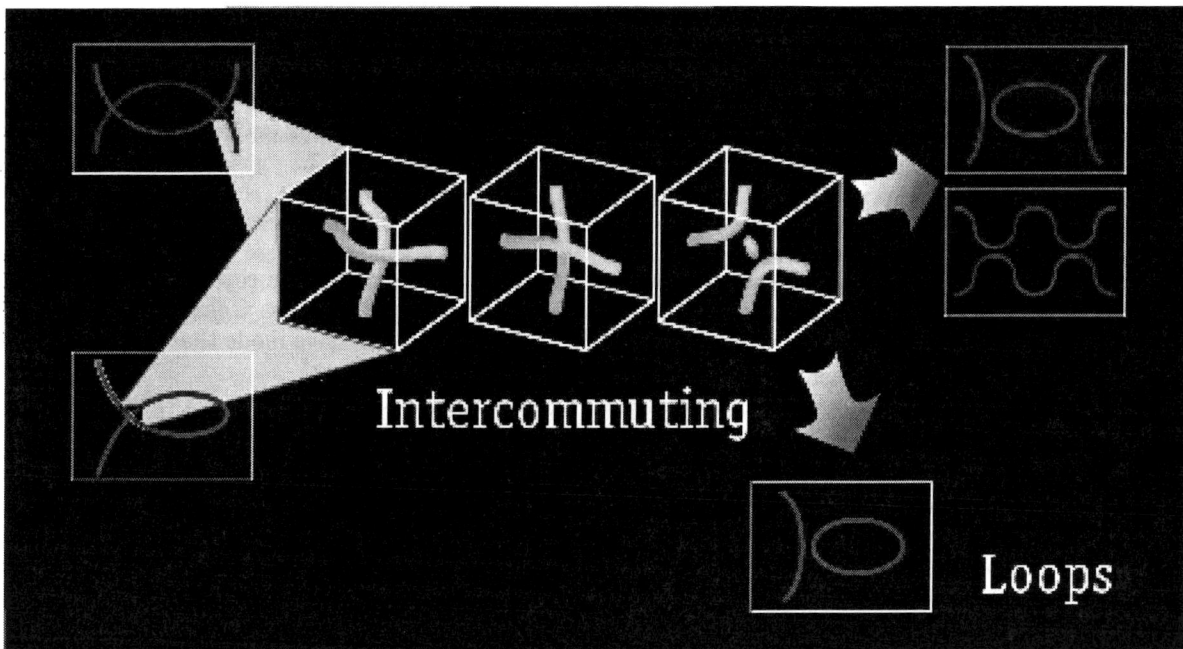

FIG. 5: *Global string interactions leading to loop formation. Whenever two string segments intersect, they reconnect or intercommute (upper part of the figure). Analogously, if a string intersects itself, it can break off a closed loop (bottom part of the figure). In both cases, the interacting string segments first suffer a slight deformation (due to the long–range forces present for global strings), they subsequently fuse and finally exchange partners. A ephemeral unstable amount of energy in the form of a small loop remains in the middle where the energy is high enough to place the Higgs field in the false vacuum. It then quickly collapses, radiating away its energy. The situation is roughly the same for local strings, as simulations have shown [127].*

One feature endowed by textures that really makes these defects peculiar is their being unstable to collapse. The initial field configuration is set at the phase transition, when ϕ develops a nonzero vacuum expectation value. ϕ lives in the vacuum manifold $\mathcal{M}$ and winds around $\mathcal{M}$ in a non–trivial way on scales greater than the correlation length, $\xi \lesssim t$. The evolution is determined by the nonlinear dynamics of ϕ. When the typical size of the defect becomes of the order of the horizon, it collapses on itself. The collapse continues until eventually the size of the defect becomes of the order of η^{-1}, and at that point the energy in gradients is large enough to raise the field from its vacuum state. This makes the defect unwind, leaving behind a trivial field configuration. As a result ξ grows to about the horizon scale, and then keeps growing with it. As still larger scales come across the horizon, knots are constantly formed, since the field ϕ points in different directions on $\mathcal{M}$ in different Hubble volumes. This is the scaling regime for textures, and when it holds simulations show that one should expect to find of order 0.04 unwinding collapses per horizon volume per Hubble time [Turok, 1989]. However, unwinding events are not the most frequent feature [Borrill et al., 1994], and when one considers random field configurations without an unwinding event the number raises to about 1 collapse per horizon volume per Hubble time.

F. Evolution of global textures

We mentioned earlier that the breakdown of any non–Abelian global symmetry led to the formation of textures. The simplest possible example involves the breakdown of a global SU(2) by a complex doublet ϕ^a, where the latter may be expressed as a four–component scalar field, i.e., $a = 1\ldots 4$. We may write the Lagrangian of the theory much in the same way as it was done in Eq. (6), but now we drop the gauge fields (thus the covariant derivatives become partial derivatives). Let us take the symmetry breaking potential as follows, $V(\phi) = \frac{\lambda}{4}\left(|\phi|^2 - \eta^2\right)^2$. The situation in

---

determined by the competing tendencies of the global field to unwind and of the gauge field to compensate the ϕ gradients. The result depends on the characteristic size L of the texture: in the range $m_\phi^{-1} \ll L \ll m_A^{-1} \sim (e\eta)^{-1}$ the behavior of the gauge texture resembles that of the global texture, as it should, since in the limit m_A very small ($e \to 0$) the gauge texture turns into a global one [Turok & Zadrozny, 1990].

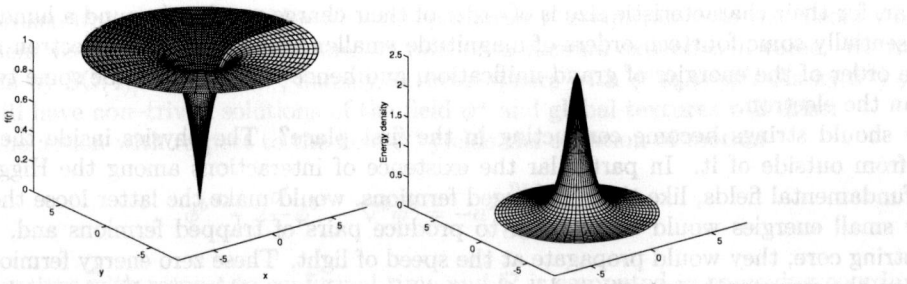

FIG. 6: *Higgs field and energy profiles for Goto–Nambu cosmic strings. The left panel shows the amplitude of the Higgs field around the string. The field vanishes at the origin (the false vacuum) and attains its asymptotic value (normalized to unity in the figure) far away from the origin. The phase of the scalar field (changing from 0 to 2π) is shown by the shading of the surface. In the right panel we show the energy density of the configuration. The maximum value is reached at the origin, exactly where the Higgs is placed in the false vacuum. [Hindmarsh & Kibble, 1995].*

Finally, we can mention a few condensed–matter 'cousins' of Goto–Nambu strings: flux tubes in superconductors [Abrikosov, 1957] for the nonrelativistic version of gauge strings (Φ corresponds to the Cooper pair wave function). Also, vortices in superfluids, for the nonrelativistic version of global strings (Φ corresponds to the Bose condensate wave function). Moreover, the only two relevant scales of the problem we mentioned above are the Higgs mass m_s and the gauge vector mass m_v. Their inverse give an idea of the characteristic scales on which the fields acquire their asymptotic solutions far away from the string 'location'. In fact, the relevant core widths of the string are given by m_s^{-1} and m_v^{-1}. It is the comparison of these scales that draws the dividing line between two qualitatively different types of solutions. If we define the parameter $\beta = (m_s/m_v)^2$, superconductivity theory says that $\beta < 1$ corresponds to Type I behavior while $\beta > 1$ corresponds to Type II. For us, $\beta < 1$ implies that the characteristic scale for the vector field is smaller than that for the Higgs field and so magnetic field B flux lines are well confined in the core; eventually, an n–vortex string with high winding number n stays stable. On the contrary, $\beta > 1$ says that the characteristic scale for the vector field exceeds that for the scalar field and thus B flux lines are not confined; the n–vortex string will eventually split into n vortices of flux $2\pi/q$. In summary:

$$\beta = (\frac{m_s}{m_v})^2 \begin{cases} < 1 \; n\text{--vortex stable } (B \text{ flux lines confined in core}) \; - \text{Type I} \\ > 1 \; \text{Unstable}: \text{ splitting into } n \text{ vortices of flux } 2\pi/q \; - \text{Type II} \end{cases} \quad (18)$$

IV. STRUCTURE FORMATION FROM DEFECTS

A. Cosmic strings

In this section we will provide just a quick description of the remarkable cosmological features of cosmic strings. Many of the proposed observational tests for the existence of cosmic strings are based on their gravitational interactions. In fact, the gravitational field around a straight static string is very unusual [Vilenkin, 1981]. As is well known, the Newtonian limit of Einstein field equations with source term given by $T_\nu^\mu = \text{diag}(\rho, -p_1, -p_2, -p_3)$ in terms of the Newtonian potential Φ is given by $\nabla^2 \Phi = 4\pi G(\rho + p_1 + p_2 + p_3)$, just a statement of the well known fact that pressure terms also contribute to the 'gravitational mass'. For an infinite string in the z–direction one has $p_3 = -\rho$, i.e., strings possess a large relativistic tension (negative pressure). Moreover, averaging on the string core results in vanishing pressures for the x and y directions yielding $\nabla^2 \Phi = 0$ for the Poisson equation. This indicates that space is flat outside of an infinite straight cosmic string and therefore test particles in its vicinity should not feel any gravitational attraction.

In fact, a full general relativistic analysis confirms this and test particles in the space around the string feel no Newtonian attraction; however there exists something unusual, a sort of wedge missing from the space surrounding the string and called the 'deficit angle', usually noted Δ, that makes the topology of space around the string that of a cone. To see this, consider the metric of a source with energy–momentum tensor [Vilenkin 1981, Gott 1985]

$$T_\mu^\nu = \delta(x)\delta(y)\text{diag}(\mu, 0, 0, T) \, . \quad (19)$$

In the case with $T = \mu$ (a rather simple equation of state) this is the effective energy–momentum tensor of an unperturbed string with string tension μ as seen from distances much larger than the thickness of the string (a

Goto–Nambu string). However, real strings develop small–scale structure and are therefore not well described by the Goto–Nambu action. When perturbations are taken into account T and μ are no longer equal and can only be interpreted as effective quantities for an observer who cannot resolve the perturbations along its length. And in this case we are left without an effective equation of state. Carter [1990] has proposed that these 'noisy' strings should be such that both its speeds of propagation of perturbations coincide. Namely, the transverse (wiggle) speed $c_{\rm T} = (T/\mu)^{1/2}$ for extrinsic perturbations should be equal to the longitudinal (woggle) speed $c_{\rm L} = (-dT/d\mu)^{1/2}$ for sound–type perturbations. This requirement yields the new equation of state

$$\mu T = \mu_0^2 \tag{20}$$

and, when this is satisfied, it describes the energy-momentum tensor of a wiggly string as seen by an observer who cannot resolve the wiggles or other irregularities along the string [Carter 1990, Vilenkin 1990].

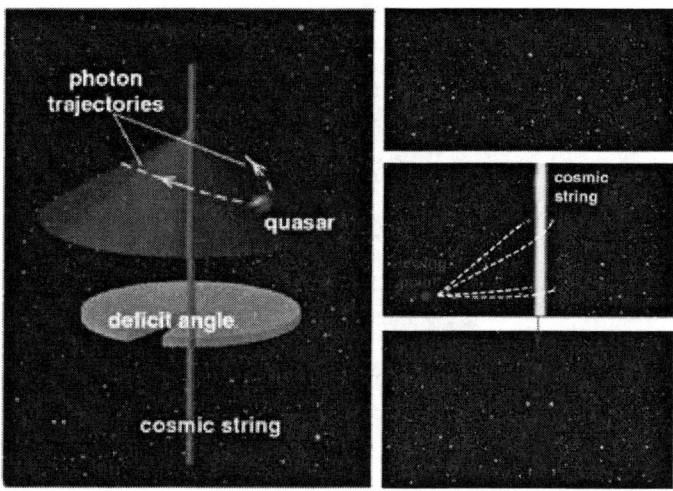

FIG. 7: *Cosmic strings affect surrounding spacetime by removing a small angular wedge, creating a conelike geometry (left). Space remains flat everywhere, but a circular path around the string encompasses slightly less than 360 degrees. The deficit angle is tiny, about 10^{-5} radian. To an observer, the presence of a cosmic string would be betrayed by its effect on the trajectory of passing light rays, which are deflected by an amount equal to the deficit angle. The resultant gravitational lensing reveals itself in the doubling of images of objects behind the string (right panel).*

The gravitational field around the cosmic string [neglecting terms of order $(G\mu)^2$] is found by solving the linearized Einstein equations with the above T_μ^ν. One gets

$$h_{00} = h_{33} = 4G(\mu - T)\ln(r/r_0), \tag{21}$$

$$h_{11} = h_{22} = 4G(\mu + T)\ln(r/r_0), \tag{22}$$

where $h_{\mu\nu} = g_{\mu\nu} - \eta_{\mu\nu}$ is the metric perturbation, the radial distance from the string is $r = (x^2 + y^2)^{1/2}$, and r_0 is a constant of integration.

For an ideal, straight, unperturbed string, the tension and mass per unit length are $T = \mu = \mu_0$ and one gets

$$h_{00} = h_{33} = 0, \quad h_{11} = h_{22} = 8G\mu_0 \ln(r/r_0). \tag{23}$$

By a coordinate transformation one can bring this metric to a locally flat form

$$ds^2 = dt^2 - dz^2 - dr^2 - (1 - 8G\mu_0)r^2 d\phi^2, \tag{24}$$

which describes a conical and flat (Euclidean) space with a wedge of angular size $\Delta = 8\pi G\mu_0$ (the deficit angle) removed from the plane and with the two faces of the wedge identified.

1. Wakes and gravitational lensing

We saw above that test particles[14] at rest in the spacetime of the straight string experience no gravitational force, but if the string moves the situation radically changes. Two particles initially at rest while the string is far away, will suddenly begin moving towards each other after the string has passed between them. Their head–on velocities will be proportional to Δ or, more precisely, the particles will get a boost $v = 4\pi G\mu_0 v_s \gamma$ in the direction of the surface swept out by the string. Here, $\gamma = (1 - v_s^2)^{-1/2}$ is the Lorentz factor and v_s the velocity of the moving string. Hence, the moving string will built up a *wake* of particles behind it that may eventually form the 'seed' for accreting more matter into sheet–like structures [Silk & Vilenkin 1984].

FIG. 8: *By deflecting the trajectory of ordinary matter, strings offer an interesting means of forming large-scale structure. A string sweeping through a distribution of interstellar dust will draw particles together in its wake, giving them lateral velocities of a few kilometers per second. The trail of the moving string will become a planar region of high-density matter, which, after gravitational collapse, could turn into thin, sheetlike distributions of galaxies [Image courtesy of Pedro Avelino and Paul Shellard].*

Also, the peculiar topology around the string makes it act as a cylindric gravitational lens that may produce double images of distant light sources, *e.g.*, quasars. The angle between the two images produced by a typical GUT string would be $\propto G\mu$ and of order of a few seconds of arc, independent of the impact parameter and with no relative magnification between the images [see Cowie & Hu, 1987, for a recent observational attempt].

The situation gets even more interesting when we allow the string to have small–scale structure, which we called wiggles above, as in fact simulations indicate. Wiggles not only modify the string's effective mass per unit length, μ, but also built up a Newtonian attractive term in the velocity boost inflicted on nearby test particles. To see this, let us consider the formation of a wake behind a moving wiggly string. Assuming the string moves along the x-axis, we can describe the situation in the rest frame of the string. In this frame, it is the particles that move, and these flow past the string with a velocity v_s in the opposite direction. Using conformally Minkowskian coordinates we can express the relevant components of the metric as

$$ds^2 = (1 + h_{00})[dt^2 - (dx^2 + dy^2)], \tag{25}$$

---

[14] If one takes into account the own gravitational field of the particle living in the spacetime around a cosmic string, then the situation changes. In fact, the presence of the conical 'singularity' introduced by the string distorts the particle's own gravitational field and results in the existence of a weak attractive force proportional to $G^2\mu m^2/r^2$, where m is the particle's mass [Linet, 1986].

where the missing wedge is reproduced by identifying the half-lines $y = \pm 4\pi G\mu x$, $x \geq 0$. The linearized geodesic equations in this metric can be written as

$$2\ddot{x} = -(1 - \dot{x}^2 - \dot{y}^2)\partial_x h_{00}, \tag{26}$$

$$2\ddot{y} = -(1 - \dot{x}^2 - \dot{y}^2)\partial_y h_{00}, \tag{27}$$

where over-dots denote derivatives with respect to t. Working to first order in $G\mu$, the second of these equations can be integrated over the unperturbed trajectory $x = v_s t$, $y = y_0$. Transforming back to the frame in which the string has a velocity v_s yields the result for the velocity impulse in the y-direction after the string has passed [Vachaspati & Vilenkin, 1991; Vollick, 1992]

$$v = -\frac{2\pi G(\mu - T)}{v_s \gamma} - 4\pi G\mu v_s \gamma \tag{28}$$

The second term is the velocity impulse due to the conical deficit angle we saw above. This term will dominate for large string velocities, case in which big planar wakes are predicted. In this case, the string wiggles will produce inhomogeneities in the wake and may easy the fragmentation of the structure. The 'top-down' scenario of structure formation thus follows naturally in a universe with fast-moving strings. On the contrary, for small velocities, it is the first term that dominates over the deflection of particles. The origin of this term can be easily understood [Vilenkin & Shellard, 2000]. From Eqn. (21), the gravitational force on a non-relativistic particle of mass m is $F \sim mG(\mu - T)/r$. A particle with an impact parameter r is exposed to this force for a time $\Delta t \sim r/v_s$ and the resulting velocity is $v \sim (F/m)\Delta t \sim G(\mu - T)/v_s$.

B. Textures

During the radiation era, and when the correlation length is already growing with the Hubble radius, the texture field has energy density $\rho_{texture} \sim (\nabla\phi)^2 \sim \eta^2/H^{-2}$, and remains a fixed fraction of the total density $\rho_c \sim t^{-2}$ yielding $\Omega_{texture} \sim G\eta^2$. This is the scaling behavior for textures and thus we do not need to worry about textures dominating the universe.

But as we already mentioned, textures are unstable to collapse, and this collapse generates perturbations in the metric of spacetime that eventually lead to large scale structure formation. These perturbations in turn will affect the photon geodesics leading to CMB anisotropies, the clearest possible signature to probe the existence of these exotic objects being the appearance of hot and cold *spots* in the microwave maps. Due to their scaling behavior, the density fluctuations induced by textures on any scale at horizon crossing are given by $(\delta\rho/\rho)_H \sim G\eta^2$. CMB temperature anisotropies will be of the same amplitude. Numerically-simulated maps, with patterns smoothed over $10°$ angular scales, by Bennett & Rhie [1993] yield, upon normalization to the *COBE-DMR* data, a dimensionless value $G\eta^2 \sim 10^{-6}$, in good agreement with a GUT phase transition energy scale. It is fair to say, however, that the texture scenario is having problems in matching current data on smaller scales [see, e.g., Durrer, 2000].

C. Defects as dark energy

There is recent mounting evidence that our current universe is being dominated by a unexpectedly large amount of dark energy [e.g., Riess et al., 1998; Perlmutter et al., 1999]. Recent observations with type Ia supernovae, together with other astrophysical tests, suggest that more than 65 percent of the critical energy density is made up by some yet unknown energy component.

Cosmic defects can also be seen as a novel form of dark energy. For example, a tangled web of cosmic strings with fixed mass per unit length, which self-intersects without having reconnection. Non intercommuting strings means no production of loops, and therefore the main channel for loosing energy is not active. The model proposed in [Vilenkin, 1984] has the mean mass density in strings scaling as $\rho_{strings} \propto (ta(t))^{-1}$ instead of $\rho_{strings} \propto \eta^2 t^{-2}$ as we saw above. From this, one has $\rho_{strings} t^2 \propto t^{1/2}$ in the radiation-dominated era and $\rho_{strings} t^2 \propto t^{1/3}$ during matter domination, which means that the energy in strings $\Omega_{strings}$ grows with time and, after a certain t_s, strings would dominate the universe. With t_s falling in the matter-domination era, we have $\rho_{strings}/\rho(t_s) \sim (t_{eq}/t_*)^{1/2}(t_s/t_{eq})^{1/3}G\mu \sim 1$, with the background $\rho \propto 1/Gt^2$. In the case $0 < z \lesssim 3$, roughly $t_s \sim 10^{17}$ sec., with $t_{eq} \sim 10^{11}$ sec. and $t_* \sim m_P/\eta^2$, we get $G\mu \sim 10^{-20}$ for the characteristic energy scale of these non-intercommuting strings.

After t_s the Friedmann's equation can be cast as $(\dot{a}/a)^2 \propto G\rho_{\text{strings}} \propto G/ta(t)$, which implies that the scale factor goes as $a(t) \propto t$ and then $\rho_{\text{strings}} \propto 1/t^2$. Now, recalling the local energy conservation law $\dot{\rho} = -3(\rho+p)\dot{a}/a$, and applying it for a dark "x" component, $w_x = p_x/\rho_x$, we get $\rho_x \propto a^{-3(1+w_x)}$. If this dark component is made up by strings, one then deduces that it should be $w_x = -1/3$. Of course, this gives $\ddot{a} = 0$ for the scale factor, so it cannot explain the recent acceleration phase. It nevertheless goes in the right direction.

Similar arguments have been studied for other defects, like textures [Davis, 1987] and can also be devised for domain walls [Zel'dovich et al., 1974; Battye et al., 1999], in this latter case yielding $w_x = -2/3$ which points closer to the observational "equation of state" currently selected by the analysis of the different astrophysical surveys. For these and other reasons, with the words of the recent authoritative review by Peebles & Ratra [2002], the class of cosmic defect models is worth bearing in mind.

V. CMB SIGNATURES FROM DEFECTS

If cosmic defects have really formed in the early universe and some of them are still within our present horizon today, the anisotropies in the CMB they produce would have a characteristic signature. Strings, for example, would imprint the background radiation in a very particular way due to the Doppler shift that the background radiation suffers when a string intersects the line of sight. The conical topology of space around the string will produce a differential redshift of photons passing on different sides of it, resulting in step–like discontinuities in the effective CMB temperature, given by $\frac{\Delta T}{T} \approx 8\pi G\mu v_s \gamma$ with, as before, $\gamma = (1-v_s^2)^{-1/2}$ the Lorentz factor and v_s the velocity of the moving string. This 'stringy' signature was first studied by Kaiser & Stebbins [1984] and Gott [1985] (see Figure 9).

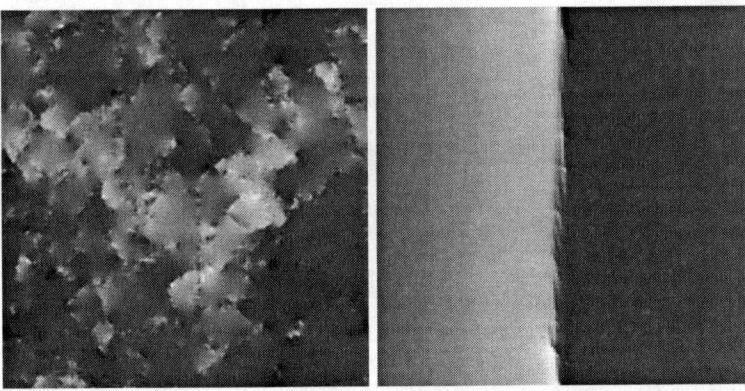

FIG. 9: *The Kaiser-Stebbins effect for cosmic strings. A string network evolves into a self-similar scaling regime, perturbing matter and radiation during its evolution. The effect on the CMB after recombination leads to distinct steplike discontinuities on small angular scales that were first studied by Kaiser & Stebbins [1984]. The left panel shows a simulated patch of the sky that fits in one of the pixels of the COBE experiment. Hence, higher resolution observatories are needed in order to detect strings. The right panel shows a patch on the CMB sky of order 20' across. However, recent studies indicate that this clean tell-tale signal gets obscured at subdegree angular scales due to the temperature fluctuations generated before recombination. [Magueijo & Ferreira 1997].*

Anisotropies of the CMB are directly related to the origin of structure in the universe. Galaxies and clusters of galaxies eventually formed by gravitational instability from primordial density fluctuations, and these same fluctuations left their imprint on the CMB. Recent balloon [de Bernardis, et al., 2000; Hanany, et al., 2000] and ground-based interferometer [Halverson, et al., 2001] experiments have produced reliable estimates of the power spectrum of the CMB temperature anisotropies. While they helped eliminate certain candidate theories for the primary source of cosmic perturbations, the power spectrum data is still compatible with the theoretical estimates of a relatively large variety of models, such as ΛCDM, quintessence models or some hybrid models including cosmic defects.

There are two main classes of models of structure formation –*passive* and *active* models. In passive models, density inhomogeneities are set as initial conditions at some early time, and while they subsequently evolve as described by Einstein–Boltzmann equations, no additional perturbations are seeded. On the other hand, in active models the sources of density perturbations are time–dependent.

All specific realizations of passive models are based on the idea of inflation. In simplest inflationary models it is assumed that there exists a weakly coupled scalar field ϕ, called the inflaton, which "drives" the (quasi) exponential expansion of the universe. The quantum fluctuations of ϕ are stretched by the expansion to scales beyond the horizon, thus "freezing" their amplitude. Inflation is followed by a period of thermalization, during which standard forms of matter and energy are formed. Because of the spatial variations of ϕ introduced by quantum fluctuations, thermalization occurs at slightly different times in different parts of the universe. Such fluctuations in the thermalization time give rise to density fluctuations. Because of their quantum nature and because of the fact that initial perturbations are assumed to be in the vacuum state and hence well described by a Gaussian distribution, perturbations produced during inflation are expected to follow Gaussian statistics to a high degree [Gangui, Lucchin, Matarrese & Mollerach, 1994], or either be products of Gaussian random variables. This is a fairly general prediction that will be tested shortly with MAP and more thoroughly in the future with Planck.[15]

Active models of structure formation are motivated by cosmic topological defects with the most promising candidates being cosmic strings. As we saw in previous sections, it is widely believed that the universe underwent a series of phase transitions as it cooled down due to the expansion. If our ideas about grand unification are correct, then some cosmic defects should have formed during phase transitions in the early universe. Once formed, cosmic strings could survive long enough to seed density perturbations. Defect models possess the attractive feature that they have no parameter freedom, as all the necessary information is in principle contained in the underlying particle physics model. Generically, perturbations produced by active models are not expected to be Gaussian distributed [Gangui, Pogosian & Winitzki, 2001a].

A. CMB power spectrum from strings

The narrow main peak and the presence of the second and the third peaks in the CMB angular power spectrum, as measured by BOOMERANG, MAXIMA and DASI [de Bernardis, et al., 2000; Hanany, et al., 2000; Halverson, et al., 2001], is an evidence for coherent oscillations of the photon–baryon fluid at the beginning of the decoupling epoch [see, e.g., Gangui, 2001]. While such coherence is a property of all passive model, realistic cosmic string models produce highly incoherent perturbations that result in a much broader main peak. This excludes cosmic strings as the primary source of density fluctuations unless new physics is postulated, e.g. models with a varying speed of light [Avelino & Martins, 2000]. In addition to purely active or passive models, it has been recently suggested that perturbations could be seeded by some combination of the two mechanisms. For example, cosmic strings could have formed just before the end of inflation and partially contributed to seeding density fluctuations. It has been shown [Contaldi, et al., 1999; Battye & Weller, 2000; Bouchet, et al., 2001] that such hybrid models can be rather successful in fitting the CMB power spectrum data.

Calculating CMB anisotropies sourced by topological defects is a rather difficult task. In inflationary scenario the entire information about the seeds is contained in the initial conditions for the perturbations in the metric. In the case of cosmic defects, perturbations are continuously seeded over the period of time from the phase transition that had produced them until today. The exact determination of the resulting anisotropy requires, in principle, the knowledge of the energy–momentum tensor [or, if only two point functions are being calculated, the unequal time correlators, Pen, Seljak, & Turok, 1997] of the defect network and the products of its decay at all times. This information is simply not available! Instead, a number of clever simplifications, based on the expected properties of the defect networks (e.g. scaling), are used to calculate the source. The latest data from BOOMERANG and MAXIMA experiments clearly disagree with the predictions of these simple models of defects [Durrer, Gangui & Sakellariadou, 1996].

The shape of the CMB angular power spectrum is determined by three main factors: the geometry of the universe, coherence and causality. The curvature of the universe directly affects the paths of light rays coming to us from the surface of last scattering. In a closed universe, because of the lensing effect induced by the positive curvature, the same physical distances between points on the sky would correspond to larger angular scales. As a result, the peak structure in the CMB angular power spectrum would shift to larger angular scales or, equivalently, to smaller values of the multipoles ℓ's.

The prediction of the cosmic string model of [Pogosian & Vachaspati, 1999] for $\Omega_{\text{total}} = 1.3$ is shown in Figure 10. As can be seen, the main peak in the angular power spectrum can be matched by choosing a reasonable value for Ω_{total}. However, even with the main peak in the right place the agreement with the data is far from satisfactory. The peak is significantly wider than that in the data and there is no sign of a rise in power at $l \approx 600$ as the actual data seems to suggest [Hanany, et al., 2000]. The sharpness and the height of the main peak in the angular spectrum can be enhanced by including the effects of gravitational radiation [Contaldi, Hindmarsh & Magueijo, 1999] and wiggles [Pogosian & Vachaspati, 1999]. More precise high–resolution numerical simulations of string networks in realistic cosmologies with a large contribution from Ω_Λ are needed to determine the exact amount of small–scale structure on

---

[15] Useful CMB resources can be found at http://www.mpa-garching.mpg.de/~banday/CMB.html

the strings and the nature of the products of their decay [Landriau & Shellard, 2002]. It is, however, unlikely that including these effects alone would result in a sufficiently narrow main peak and some presence of a second peak. This

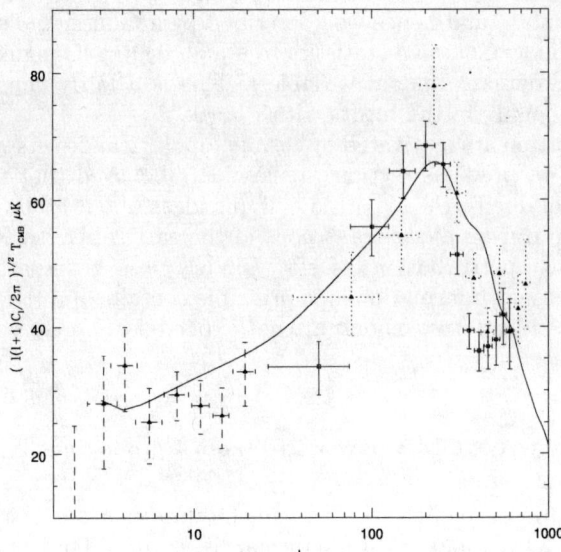

FIG. 10: *The CMB power spectrum produced by the wiggly string model of [Pogosian & Vachaspati, 1999] in a closed universe with $\Omega_{\rm total} = 1.3$, $\Omega_{\rm baryon} = 0.05$, $\Omega_{\rm CDM} = 0.35$, $\Omega_\Lambda = 0.9$, and $H_0 = 65$ km s^{-1}Mpc^{-1} [Pogosian, 2001].*

brings us to the issues of causality and coherence and how the random nature of the string networks comes into the calculation of the anisotropy spectrum. Both experimental and theoretical results for the CMB power spectra involve calculations of averages. When estimating the correlations of the observed temperature anisotropies, it is usual to compute the average over all available patches on the sky. When calculating the predictions of their models, theorists find the average over the *ensemble* of possible outcomes provided by the model.

In inflationary models, as in all passive models, only the initial conditions for the perturbations are random. The subsequent evolution is the same for all members of the ensemble. For wavelengths higher than the Hubble radius, the linear evolution equations for the Fourier components of such perturbations have a growing and a decaying solution. The modes corresponding to smaller wavelengths have only oscillating solutions. As a consequence, prior to entering the horizon, each mode undergoes a period of phase "squeezing" which leaves it in a highly coherent state by the time it starts to oscillate. Coherence here means that all members of the ensemble, corresponding to the same Fourier mode, have the same temporal phase. So even though there is randomness involved, as one has to draw random amplitudes for the oscillations of a given mode, the time behavior of different members of the ensemble is highly correlated. The total spectrum is the ensemble–averaged superposition of all Fourier modes, and the predicted coherence results in an interference pattern seen in the angular power spectrum as the well-known acoustic peaks.

In contrast, the evolution of the string network is highly non-linear. Cosmic strings are expected to move at relativistic speeds, self–intersect and reconnect in a chaotic fashion. The consequence of this behavior is that the unequal time correlators of the string energy–momentum vanish for time differences larger than a certain coherence time (τ_c in Figure 11). Members of the ensemble corresponding to a given mode of perturbations will have random temporal phases with the "dice" thrown on average once in each coherence time. The coherence time of a realistic string network is rather short. As a result, the interference pattern in the angular power spectrum is completely washed out.

Causality manifests itself, first of all, through the initial conditions for the string sources, the perturbations in the metric and the densities of different particle species. If one assumes that the defects are formed by a causal mechanism in an otherwise smooth universe then the correct initial condition are obtained by setting the components of the stress–energy pseudo–tensor $\tau_{\mu\nu}$ to zero [Veeraraghavan & Stebbins, 1990; Pen, Spergel & Turok, 1994]. These are the same as the isocurvature initial conditions [Hu, Spergel & White, 1997]. A generic prediction of isocurvature models (assuming perfect coherence) is that the first acoustic peak is almost completely hidden. The main peak is then the second acoustic peak and in flat geometries it appears at $\ell \approx 300 - 400$. This is due to the fact that after entering the horizon a given Fourier mode of the source perturbation requires time to induce perturbations in the photon density. Causality also implies that no superhorizon correlations in the string energy density are allowed. The correlation length of a "realistic" string network is normally between 0.1 and 0.4 of the horizon size.

An interesting study was performed by Magueijo, Albrecht, Ferreira & Coulson [1996], where they constructed a toy model of defects with two parameters: the coherence length and the coherence time. The coherence length was taken to be the scale at which the energy density power spectrum of the strings turns from a power law decay for large values of k into a white noise at low k. This is essentially the scale corresponding to the correlation length of the string network. The coherence time was defined in the sense described in the beginning of this section, in particular, as the time difference needed for the unequal time correlators to vanish. Their study showed (see Figure 11) that by

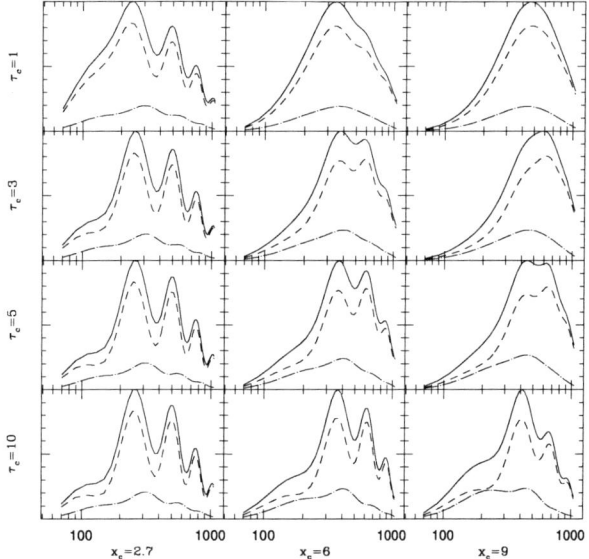

FIG. 11: *The predictions of the toy model of Magueijo, et al. [1996] for different values of parameters x_c, the coherence length, and τ_c, the coherence time. $x_c \propto \eta/\lambda_c(\eta)$, where η is the conformal time and $\lambda_c(\eta)$ is the correlation length of the network at time η. One can obtain oscillations in the CMB power spectrum by fixing either one of the parameters and varying the other.*

accepting any value for one of the parameters and varying the other (within the constraints imposed by causality) one could reproduce the oscillations in the CMB power spectrum. Unfortunately for cosmic strings, at least as we know them today, they fall into the parameter range corresponding to the upper right corner in Figure 11.

In order to get a better fit to present–day observations, cosmic strings must either be more coherent or they have to be stretched over larger distances, which is another way of making them more coherent. To understand this imagine that there was just one long straight string stretching across the universe and moving with some given velocity. The evolution of this string would be linear and the induced perturbations in the photon density would be coherent. By increasing the correlation length of the string network we would move closer to this limiting case of just one long straight string and so the coherence would be enhanced.

The question of whether or not defects can produce a pattern of the CMB power spectrum similar to, and including the acoustic peaks of, that produced by the adiabatic inflationary models was repeatedly addressed in the literature [Contaldi, Hindmarsh & Magueijo 1999; Magueijo, et al. 1996; Liddle, 1995; Turok, 1996; Avelino & Martins, 2000]. In particular, it was shown [Magueijo, et al. 1996; Turok, 1996] that one can construct a causal model of active seeds which for certain values of parameters can reproduce the oscillations in the CMB spectrum. The main problem today is that current realistic models of cosmic strings fall out of the parameter range that is needed to fit the observations. At the moment, only the (non-minimal) models with either a varying speed of light or hybrid contribution of strings+inflation are the only ones involving topological defects that to some extent can match the observations. One possible way to distinguish their predictions from those of inflationary models would be by computing key non–Gaussian statistical quantities, such as the CMB bispectrum.

B. CMB bispectrum from active models

Different cosmological models differ in their predictions for the statistical distribution of the anisotropies beyond the power spectrum. Future MAP and Planck satellite missions will provide high-precision data allowing definite estimates of non-Gaussian signals in the CMB. It is therefore important to know precisely which are the predictions of all candidate models for the statistical quantities that will be extracted from the new data and identify their specific signatures.

Of the available non-Gaussian statistics, the CMB bispectrum, or the three-point function of Fourier components of the temperature anisotropy, has been perhaps the one best studied in the literature [Gangui & Martin, 2000a]. There are a few cases where the bispectrum may be deduced analytically from the underlying model. The bispectrum can be estimated from simulated CMB sky maps; however, computing a large number of full-sky maps resulting from defects is a much more demanding task. Recently, a precise numerical code to compute it, not using CMB maps and similar to the CMBFAST code[16] for the power spectrum, was developed in [Gangui, Pogosian & Winitzki 2001b]. What follows below is an account of this work.

In a few words, given a suitable model, one can generate a statistical *ensemble* of realizations of defect matter perturbations. We used a modified Boltzmann code based on CMBFAST to compute the effect of these perturbations on the CMB and found the bispectrum estimator for a given realization of sources. We then performed statistical averaging over the ensemble of realizations to compute the expected CMB bispectrum. (The CMB power spectrum was also obtained as a byproduct.) As a first application, we then computed the expected CMB bispectrum from a model of simulated string networks first introduced by Albrecht *et al.* [1997] and further developed in [Pogosian & Vachaspati, 1999] and in [Gangui, Pogosian & Winitzki 2001].

We assume that, given a model of active perturbations, such as a string simulation, we can calculate the energy-momentum tensor $T_{\mu\nu}(\mathbf{x}, \tau)$ for a particular realization of the sources in a finite spatial volume V_0. Here, $\mathbf{x}$ is a 3-dimensional coordinate and τ is the cosmic time. Many simulations are run to obtain an ensemble of random realizations of sources with statistical properties appropriate for the given model. The spatial Fourier decomposition of $T_{\mu\nu}$ can be written as

$$T_{\mu\nu}(\mathbf{x}, \tau) = \sum_{\mathbf{k}} \Theta_{\mu\nu}(\mathbf{k}, \tau) e^{i\mathbf{k}\mathbf{x}} , \tag{29}$$

where $\mathbf{k}$ are discrete. If V_0 is sufficiently large we can approximate the summation by the integral

$$\sum_{\mathbf{k}} \Theta_{\mu\nu}(\mathbf{k}, \tau) e^{i\mathbf{k}\mathbf{x}} \approx \frac{V_0}{(2\pi)^3} \int d^3\mathbf{k}\, \Theta_{\mu\nu}(\mathbf{k}, \tau) e^{i\mathbf{k}\mathbf{x}} , \tag{30}$$

and the corresponding inverse Fourier transform will be

$$\Theta_{\mu\nu}(\mathbf{k}, \tau) = \frac{1}{V_0} \int_{V_0} d^3\mathbf{x}\, T_{\mu\nu}(\mathbf{x}, \tau) e^{-i\mathbf{k}\mathbf{x}} . \tag{31}$$

Of course, the final results, such as the CMB power spectrum or bispectrum, do not depend on the choice of V_0. To ensure this independence, we shall keep V_0 in all expressions where it appears below.

It is conventional to expand the temperature fluctuations over the basis of spherical harmonics,

$$\Delta T/T(\hat{\mathbf{n}}) = \sum_{lm} a_{lm} Y_{lm}(\hat{\mathbf{n}}), \tag{32}$$

where $\hat{\mathbf{n}}$ is a unit vector. The coefficients a_{lm} can be decomposed into Fourier modes,

$$a_{lm} = \frac{V_0}{(2\pi)^3} (-i)^l 4\pi \int d^3\mathbf{k}\, \Delta_l(\mathbf{k}) Y_{lm}^*(\hat{\mathbf{k}}). \tag{33}$$

Given the sources $\Theta_{\mu\nu}(\mathbf{k}, \tau)$, the quantities $\Delta_l(\mathbf{k})$ are found by solving linearized Einstein-Boltzmann equations and integrating along the line of sight, using a code similar to CMBFAST [Seljak & Zaldarriaga, 1996]. This standard procedure can be written symbolically as the action of a linear operator $\hat{B}_l^{\mu\nu}(k)$ on the source energy-momentum tensor, $\Delta_l(\mathbf{k}) = \hat{B}_l^{\mu\nu}(k)\Theta_{\mu\nu}(\mathbf{k}, \tau)$, so the third moment of $\Delta_l(\mathbf{k})$ is linearly related to the three-point correlator of $\Theta_{\mu\nu}(\mathbf{k}, \tau)$. Below we consider the quantities $\Delta_l(\mathbf{k})$, corresponding to a set of realizations of active sources, as given. The numerical procedure for computing $\Delta_l(\mathbf{k})$ was developed in [Albrecht *et al.* 1997] and in [Pogosian & Vachaspati, 1999].

---

[16] http://physics.nyu.edu/matiasz/CMBFAST/cmbfast.html

The third moment of a_{lm}, namely $\langle a_{l_1 m_1} a_{l_2 m_2} a_{l_3 m_3} \rangle$, can be expressed as

$$(-i)^{l_1+l_2+l_3} (4\pi)^3 \frac{V_0^3}{(2\pi)^9} \int d^3\mathbf{k}_1 d^3\mathbf{k}_2 d^3\mathbf{k}_3 Y^*_{l_1 m_1}(\hat{\mathbf{k}}_1) Y^*_{l_2 m_2}(\hat{\mathbf{k}}_2) Y^*_{l_3 m_3}(\hat{\mathbf{k}}_3) \langle \Delta_{l_1}(\mathbf{k}_1) \Delta_{l_2}(\mathbf{k}_2) \Delta_{l_3}(\mathbf{k}_3) \rangle. \tag{34}$$

A straightforward numerical evaluation of Eq. (34) from given sources $\Delta_l(\mathbf{k})$ is prohibitively difficult, because it involves too many integrations of oscillating functions. However, we shall be able to reduce the computation to integrations over scalars [a similar method was employed in Komatsu & Spergel, 2001 and in Wang & Kamionkowski, 2000]. Due to homogeneity, the 3-point function vanishes unless the triangle constraint is satisfied,

$$\mathbf{k}_1 + \mathbf{k}_2 + \mathbf{k}_3 = 0. \tag{35}$$

We may write

$$\langle \Delta_{l_1}(\mathbf{k}_1) \Delta_{l_2}(\mathbf{k}_2) \Delta_{l_3}(\mathbf{k}_3) \rangle = \delta^{(3)}(\mathbf{k}_1 + \mathbf{k}_2 + \mathbf{k}_3) P_{l_1 l_2 l_3}(\mathbf{k}_1, \mathbf{k}_2, \mathbf{k}_3), \tag{36}$$

where the three-point function $P_{l_1 l_2 l_3}(\mathbf{k}_1, \mathbf{k}_2, \mathbf{k}_3)$ is defined only for values of $\mathbf{k}_i$ that satisfy Eq. (35). Given the scalar values k_1, k_2, k_3, there is a unique (up to an overall rotation) triplet of directions $\hat{\mathbf{k}}_i$ for which the RHS of Eq. (36) does not vanish. The quantity $P_{l_1 l_2 l_3}(\mathbf{k}_1, \mathbf{k}_2, \mathbf{k}_3)$ is invariant under an overall rotation of all three vectors $\mathbf{k}_i$ and therefore may be equivalently represented by a function of *scalar* values k_1, k_2, k_3, while preserving all angular information. Hence, we can rewrite Eq. (36) as

$$\langle \Delta_{l_1}(\mathbf{k}_1) \Delta_{l_2}(\mathbf{k}_2) \Delta_{l_3}(\mathbf{k}_3) \rangle = \delta^{(3)}(\mathbf{k}_1 + \mathbf{k}_2 + \mathbf{k}_3) P_{l_1 l_2 l_3}(k_1, k_2, k_3). \tag{37}$$

Then, using the simulation volume V_0 explicitly, we have

$$P_{l_1 l_2 l_3}(k_1, k_2, k_3) = \frac{(2\pi)^3}{V_0} \langle \Delta_{l_1}(\mathbf{k}_1) \Delta_{l_2}(\mathbf{k}_2) \Delta_{l_3}(\mathbf{k}_3) \rangle. \tag{38}$$

Given an arbitrary direction $\hat{\mathbf{k}}_1$ and the magnitudes k_1, k_2 and k_3, the directions $\hat{\mathbf{k}}_2$ and $\hat{\mathbf{k}}_3$ are specified up to overall rotations by the triangle constraint. Therefore, both sides of Eq. (38) are functions of scalar k_i only. The expression on the RHS of Eq. (38) is evaluated numerically by averaging over different realizations of the sources *and* over permissible directions $\hat{\mathbf{k}}_i$; below we shall give more details of the procedure.

Substituting Eqs. (37) and (38) into (34), Fourier transforming the Dirac delta and using the Rayleigh identity, we can perform all angular integrations analytically and obtain a compact form for the third moment,

$$\langle a_{l_1 m_1} a_{l_2 m_2} a_{l_3 m_3} \rangle = \mathcal{H}^{m_1 m_2 m_3}_{l_1 l_2 l_3} \int r^2 dr\, b_{l_1 l_2 l_3}(r), \tag{39}$$

where, denoting the Wigner $3j$-symbol by $\begin{pmatrix} l_1 & l_2 & l_3 \\ m_1 & m_2 & m_3 \end{pmatrix}$, we have

$$\mathcal{H}^{m_1 m_2 m_3}_{l_1 l_2 l_3} \equiv \sqrt{\frac{(2l_1+1)(2l_2+1)(2l_3+1)}{4\pi}} \begin{pmatrix} l_1 & l_2 & l_3 \\ 0 & 0 & 0 \end{pmatrix} \begin{pmatrix} l_1 & l_2 & l_3 \\ m_1 & m_2 & m_3 \end{pmatrix}, \tag{40}$$

and where we have defined the auxiliary quantities $b_{l_1 l_2 l_3}$ using spherical Bessel functions j_l,

$$\begin{aligned} b_{l_1 l_2 l_3}(r) &\equiv \frac{8}{\pi^3} \frac{V_0^3}{(2\pi)^3} \int k_1^2 dk_1\, k_2^2 dk_2\, k_3^2 dk_3 \\ &\quad \times j_{l_1}(k_1 r) j_{l_2}(k_2 r) j_{l_3}(k_3 r) P_{l_1 l_2 l_3}(k_1, k_2, k_3). \end{aligned} \tag{41}$$

The volume factor V_0^3 contained in this expression is correct: as shown in the next section, each term Δ_l includes a factor $V_0^{-2/3}$, while the average quantity $P_{l_1 l_2 l_3}(k_1, k_2, k_3) \propto V_0^{-3}$ [cf. Eq. (38)], so that the arbitrary volume V_0 of the simulation cancels.

Our proposed numerical procedure therefore consists of computing the RHS of Eq. (39) by evaluating the necessary integrals. For fixed $\{l_1 l_2 l_3\}$, computation of the quantities $b_{l_1 l_2 l_3}(r)$ is a triple integral over scalar k_i defined by Eq. (41); it is followed by a fourth scalar integral over r [Eq. (39)]. We also need to average over many realizations of sources to obtain $P_{l_1 l_2 l_3}(k_1, k_2, k_3)$. It was not feasible for us to precompute the values $P_{l_1 l_2 l_3}(k_1, k_2, k_3)$ on a grid before integration because of the large volume of data: for each set $\{l_1 l_2 l_3\}$ the grid must contain $\sim 10^3$ points for each k_i. Instead, we precompute $\Delta_l(\mathbf{k})$ from one realization of sources and evaluate the RHS of Eq. (38) on that data as an *estimator* of $P_{l_1l_2l_3}(k_1,k_2,k_3)$, averaging over allowed directions of $\hat{\mathbf{k}}_i$. The result is used for integration in Eq. (41).

Because of isotropy and since the allowed sets of directions $\hat{\mathbf{k}}_i$ are planar, it is enough to restrict the numerical calculation to directions $\hat{\mathbf{k}}_i$ within a fixed two-dimensional plane. This significantly reduces the amount of computations and data storage, since $\Delta_l(\mathbf{k})$ only needs to be stored on a two-dimensional grid of $\mathbf{k}$.

In estimating $P_{l_1l_2l_3}(k_1,k_2,k_3)$ from Eq. (38), averaging over directions of $\hat{\mathbf{k}}_i$ plays a similar role to ensemble averaging over source realizations. Therefore if the number of directions is large enough (we used 720 for cosmic strings), only a moderate number of different source realizations is needed. The main numerical difficulty is the highly oscillating nature of the function $b_{l_1l_2l_3}(r)$. The calculation of the bispectrum for cosmic strings presented in the next Section requires about 20 days of a single-CPU workstation time per realization.

We note that this method is specific for the bispectrum and cannot be applied to compute higher-order correlations. The reason is that higher-order correlations involve configurations of vectors $\mathbf{k}_i$ that are not described by scalar values k_i and not restricted to a plane. For instance, a computation of a 4-point function would involve integration of highly oscillating functions over four vectors $\mathbf{k}_i$ which is computationally infeasible.

From Eq. (39) we derive the CMB angular bispectrum $C_{l_1l_2l_3}$, defined as [Gangui & Martin, 2000b]

$$\langle a_{l_1m_1}a_{l_2m_2}a_{l_3m_3}\rangle = \begin{pmatrix} l_1 & l_2 & l_3 \\ m_1 & m_2 & m_3 \end{pmatrix} C_{l_1l_2l_3}. \tag{42}$$

The presence of the 3j-symbol guarantees that the third moment vanishes unless $m_1+m_2+m_3=0$ and the l_i indices satisfy the triangle rule $|l_i-l_j| \leq l_k \leq l_i+l_j$. Invariance under spatial inversions of the three-point correlation function implies the additional 'selection rule' $l_1+l_2+l_3=$ even, in order for the third moment not to vanish. Finally, from this last relation and using standard properties of the 3j-symbols, it follows that the angular bispectrum $C_{l_1l_2l_3}$ is left unchanged under any arbitrary permutation of the indices l_i.

In what follows we will restrict our calculations to the angular bispectrum $C_{l_1l_2l_3}$ in the 'diagonal' case, i.e. $l_1=l_2=l_3=l$. This is a representative case and, in fact, the one most frequently considered in the literature. Plots of the power spectrum are usually done in terms of $l(l+1)C_l$ which, apart from constant factors, is the contribution to the mean squared anisotropy of temperature fluctuations per unit logarithmic interval of l. In full analogy with this, the relevant quantity to work with in the case of the bispectrum is

$$G_{lll} = l(2l+1)^{3/2} \begin{pmatrix} l & l & l \\ 0 & 0 & 0 \end{pmatrix} C_{lll}. \tag{43}$$

For large values of the multipole index l, $G_{lll} \propto l^{3/2} C_{lll}$. Note also what happens with the 3j-symbols appearing in the definition of the coefficients $\mathcal{H}^{m_1m_2m_3}_{l_1l_2l_3}$: the symbol $\begin{pmatrix} l_1 & l_2 & l_3 \\ m_1 & m_2 & m_3 \end{pmatrix}$ is absent from the definition of $C_{l_1l_2l_3}$, while in Eq. (43) the symbol $\begin{pmatrix} l & l & l \\ 0 & 0 & 0 \end{pmatrix}$ is squared. Hence, there are no remnant oscillations due to the alternating sign of $\begin{pmatrix} l & l & l \\ 0 & 0 & 0 \end{pmatrix}$.

However, even more important than the value of C_{lll} itself is the relation between the bispectrum and the cosmic variance associated with it. In fact, it is their comparison that tells us about the observability 'in principle' of the non-Gaussian signal. The cosmic variance constitutes a theoretical uncertainty for all observable quantities and comes about due to the fact of having just one realization of the stochastic process, in our case, the CMB sky [Scaramella & Vittorio, 1991].

The way to proceed is to employ an estimator $\hat{C}_{l_1l_2l_3}$ for the bispectrum and compute the variance from it. By choosing an unbiased estimator we ensure it satisfies $C_{l_1l_2l_3} = \langle \hat{C}_{l_1l_2l_3}\rangle$. However, this condition does not isolate a unique estimator. The proper way to select the *best unbiased* estimator is to compute the variances of all candidates and choose the one with the smallest value. The estimator with this property was computed in [Gangui & Martin, 2000b] and is

$$\hat{C}_{l_1l_2l_3} = \sum_{m_1,m_2,m_3} \begin{pmatrix} l_1 & l_2 & l_3 \\ m_1 & m_2 & m_3 \end{pmatrix} a_{l_1m_1}a_{l_2m_2}a_{l_3m_3}. \tag{44}$$

The variance of this estimator, assuming a mildly non-Gaussian distribution, can be expressed in terms of the angular power spectrum C_l as follows

$$\sigma^2_{\hat{C}_{l_1l_2l_3}} = C_{l_1}C_{l_2}C_{l_3}(1+\delta_{l_1l_2}+\delta_{l_2l_3}+\delta_{l_3l_1}+2\delta_{l_1l_2}\delta_{l_2l_3}). \tag{45}$$

The theoretical signal-to-noise ratio for the bispectrum is then given by

$$(S/N)_{l_1l_2l_3} = |C_{l_1l_2l_3}/\sigma_{\hat{C}_{l_1l_2l_3}}|. \tag{46}$$

In turn, for the diagonal case $l_1 = l_2 = l_3 = l$ we have

$$(S/N)_l = |C_{lll}/\sigma_{\hat{C}_{lll}}|. \tag{47}$$

Incorporating all the specifics of the particular experiment, such as sky coverage, angular resolution, etc., will allow us to give an estimate of the particular non-Gaussian signature associated with a given active source and, if observable, indicate the appropriate range of multipole l's where it is best to look for it.

C. CMB bispectrum from strings

To calculate the sources of perturbations we have used an updated version of the cosmic string model first introduced by Albrecht et al. [1997] and further developed in [Pogosian & Vachaspati, 1999], where the wiggly nature of strings was taken into account. In these previous works the model was tailored to the computation of the two-point statistics (matter and CMB power spectra). When dealing with higher-order statistics, such as the bispectrum, a different strategy needs to be employed.

In the model, the string network is represented by a collection of uncorrelated straight string segments produced at some early epoch and moving with random uncorrelated velocities. At every subsequent epoch, a certain fraction of the number of segments decays in a way that maintains network scaling. The length of each segment at any time is taken to be equal to the correlation length of the network. This and the root mean square velocity of segments are computed from the velocity-dependent one-scale model of Martins & Shellard [1996]. The positions of segments are drawn from a uniform distribution in space, and their orientations are chosen from a uniform distribution on a two-sphere.

The total energy of the string network in a volume V at any time is $E = N\mu L$, where N is the total number of string segments at that time, μ is the mass per unit length, and L is the length of one segment. If L is the correlation length of the string network then, according to the one-scale model, the energy density is $\rho = E/V = \mu/L^2$, where $V = V_0 a^3$, the expansion factor a is normalized so that $a = 1$ today, and V_0 is a constant simulation volume. It follows that $N = V/L^3 = V_0/\ell^3$, where $\ell = L/a$ is the comoving correlation length. In the scaling regime ℓ is approximately proportional to the conformal time τ and so the number of strings $N(\tau)$ within the simulation volume V_0 falls as τ^{-3}.

To calculate the CMB anisotropy one needs to evolve the string network over at least four orders of magnitude in cosmic expansion. Hence, one would have to start with $N \gtrsim 10^{12}$ string segments in order to have one segment left at the present time. Keeping track of such a huge number of segments is numerically infeasible. A way around this difficulty was suggested in Ref.[3], where the idea was to consolidate all string segments that decay at the same epoch. The number of segments that decay by the (discretized) conformal time τ_i is

$$N_d(\tau_i) = V_0 \left(n(\tau_{i-1}) - n(\tau_i)\right), \tag{48}$$

where $n(\tau) = [\ell(\tau)]^{-3}$ is the number density of strings at time τ. The energy-momentum tensor in Fourier space, $\Theta^i_{\mu\nu}$, of these $N_d(\tau_i)$ segments is a sum

$$\Theta^i_{\mu\nu} = \sum_{m=1}^{N_d(\tau_i)} \Theta^{im}_{\mu\nu}, \tag{49}$$

where $\Theta^{im}_{\mu\nu}$ is the Fourier transform of the energy-momentum of the m-th segment. If segments are uncorrelated, then

$$\langle \Theta^{im}_{\mu\nu} \Theta^{im'}_{\sigma\rho} \rangle = \delta_{mm'} \langle \Theta^{im}_{\mu\nu} \Theta^{im}_{\sigma\rho} \rangle \tag{50}$$

and

$$\langle \Theta^{im}_{\mu\nu} \Theta^{im'}_{\sigma\rho} \Theta^{im''}_{\gamma\delta} \rangle = \delta_{mm'} \delta_{mm''} \langle \Theta^{im}_{\mu\nu} \Theta^{im}_{\sigma\rho} \Theta^{im}_{\gamma\delta} \rangle. \tag{51}$$

Here the angular brackets $\langle \ldots \rangle$ denote the ensemble average, which in our case means averaging over many realizations of the string network. If we are calculating power spectra, then the relevant quantities are the two-point functions of $\Theta^i_{\mu\nu}$, namely

$$\langle \Theta^i_{\mu\nu} \Theta^i_{\sigma\rho} \rangle = \langle \sum_{m=1}^{N_d(\tau_i)} \sum_{m'=1}^{N_d(\tau_i)} \Theta^{im}_{\mu\nu} \Theta^{im'}_{\sigma\rho} \rangle. \tag{52}$$

Eq. (50) allows us to write

$$\langle \Theta^i_{\mu\nu} \Theta^i_{\sigma\rho} \rangle = \sum_{m=1}^{N_d(\tau_i)} \langle \Theta^{im}_{\mu\nu} \Theta^{im}_{\sigma\rho} \rangle = N_d(\tau_i) \langle \Theta^{i1}_{\mu\nu} \Theta^{i1}_{\sigma\rho} \rangle, \tag{53}$$

where $\Theta^{i1}_{\mu\nu}$ is of the energy-momentum of one of the segments that decay by the time τ_i. The last step in Eq. (53) is possible because the segments are statistically equivalent. Thus, if we only want to reproduce the correct power spectra in the limit of a large number of realizations, we can replace the sum in Eq. (49) by

$$\Theta^i_{\mu\nu} = \sqrt{N_d(\tau_i)} \Theta^{i1}_{\mu\nu}. \tag{54}$$

The total energy-momentum tensor of the network in Fourier space is a sum over the consolidated segments:

$$\Theta_{\mu\nu} = \sum_{i=1}^{K} \Theta^i_{\mu\nu} = \sum_{i=1}^{K} \sqrt{N_d(\tau_i)} \Theta^{i1}_{\mu\nu}. \tag{55}$$

So, instead of summing over $\sum_{i=1}^{K} N_d(\tau_i) \gtrsim 10^{12}$ segments we now sum over only K segments, making K a parameter. For the three-point functions we extend the above procedure. Instead of Eqs. (52) and (53) we now write

$$\langle \Theta^i_{\mu\nu} \Theta^i_{\sigma\rho} \Theta^i_{\gamma\delta} \rangle = \langle \sum_{m=1}^{N_d(\tau_i)} \sum_{m'=1}^{N_d(\tau_i)} \sum_{m''=1}^{N_d(\tau_i)} \Theta^{im}_{\mu\nu} \Theta^{im'}_{\sigma\rho} \Theta^{im''}_{\gamma\delta} \rangle = \sum_{m=1}^{N_d(\tau_i)} \langle \Theta^{im}_{\mu\nu} \Theta^{im}_{\sigma\rho} \Theta^{im}_{\gamma\delta} \rangle = N_d(\tau_i) \langle \Theta^{i1}_{\mu\nu} \Theta^{i1}_{\sigma\rho} \Theta^{i1}_{\gamma\delta} \rangle \tag{56}$$

Therefore, for the purpose of calculation of three-point functions, the sum in Eq. (49) should now be replaced by

$$\Theta^i_{\mu\nu} = [N_d(\tau_i)]^{1/3} \Theta^{i1}_{\mu\nu}. \tag{57}$$

Both expressions in Eqs. (54) and (57), depend on the simulation volume, V_0, contained in the definition of $N_d(\tau_i)$ given in Eq. (48). This is to be expected and is consistent with our calculations, since this volume cancels in expressions for observable quantities.

Note also that the simulation model in its present form does not allow computation of CMB sky maps. This is because the method of finding the two- and three-point functions as we described involves "consolidated" quantities $\Theta^i_{\mu\nu}$ which do not correspond to the energy-momentum tensor of a real string network. These quantities are auxiliary and specially prepared to give the correct two- or three-point functions after ensemble averaging.

In Fig. 12 we show the results for $G^{1/3}_{lll}$ [cf. Eq. (43)]. It was calculated using the string model with 800 consolidated segments in a flat universe with cold dark matter and a cosmological constant. Only the scalar contribution to the anisotropy has been included. Vector and tensor contributions are known to be relatively insignificant for local cosmic strings and can safely be ignored in this model [3, 92][17]. The plots are produced using a single realization of the string network by averaging over 720 directions of $\mathbf{k}_i$. The comparison of $G^{1/3}_{lll}$ (or equivalently $C^{1/3}_{lll}$) with its cosmic variance [cf. Eq. (45)] clearly shows that the bispectrum (as computed from the present cosmic string model) lies hidden in the theoretical noise and is therefore undetectable for any given value of l.

Let us note, however, that in its present stage the string code employed in these computations describes Brownian, wiggly long strings in spite of the fact that long strings are very likely not Brownian on the smallest scales, as recent field–theory simulations indicate. In addition, the presence of small string loops [Wu, et al., 1998] and gravitational radiation into which they decay were not yet included in this model. These are important effects that could, in principle, change the above predictions for the string-generated CMB bispectrum on very small angular scales.

The imprint of cosmic strings on the CMB is a combination of different effects. Prior to the time of recombination strings induce density and velocity fluctuations on the surrounding matter. During the period of last scattering these fluctuations are imprinted on the CMB through the Sachs-Wolfe effect, namely, temperature fluctuations arise because relic photons encounter a gravitational potential with spatially dependent depth. In addition to the Sachs-Wolfe effect, moving long strings drag the surrounding plasma and produce velocity fields that cause temperature anisotropies due to Doppler shifts. While a string segment by itself is a highly non-Gaussian object, fluctuations induced by string

---

[17] The contribution of vector and tensor modes is large in the case of global strings [Turok, Pen & Seljak, 1998; Durrer, Gangui & Sakellariadou, 1996].

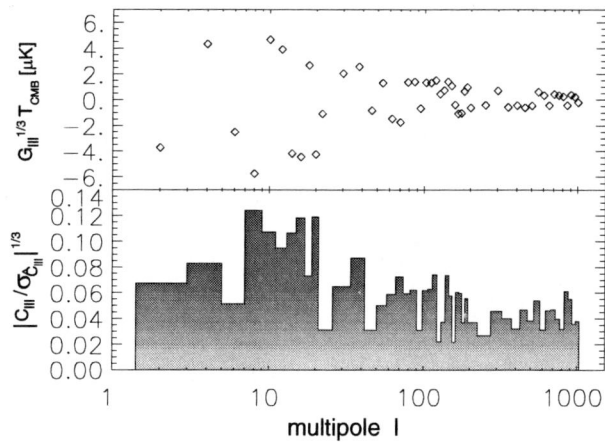

FIG. 12: *The CMB angular bispectrum in the 'diagonal' case ($G_{lll}^{1/3}$) from wiggly cosmic strings in a spatially flat model with cosmological parameters $\Omega_{\rm CDM} = 0.3$, $\Omega_{\rm baryon} = 0.05$, $\Omega_\Lambda = 0.65$, and Hubble constant $H = 0.65 km s^{-1} Mpc^{-1}$ [upper panel]. In the lower panel we show the ratio of the signal to theoretical noise $|C_{lll}/\sigma_{\hat{C}_{lll}}|^{1/3}$ for different multipole indices. Normalization follows from fitting the power spectrum to the BOOMERANG and MAXIMA data.*

segments before recombination are a superposition of effects of many random strings stirring the primordial plasma. These fluctuations are thus expected to be Gaussian as a result of the central limit theorem.

As the universe becomes transparent, strings continue to leave their imprint on the CMB mainly due to the Kaiser & Stebbins [1984] effect. As we mentioned in previous sections, this effect results in line discontinuities in the temperature field of photons passing on opposite sides of a moving long string.[18] However, this effect can result in non-Gaussian perturbations only on sufficiently small scales. This is because on scales larger than the characteristic inter-string separation at the time of the radiation-matter equality, the CMB temperature perturbations result from superposition of effects of many strings and are likely to be Gaussian. Avelino *et al.* [1998] applied several non-Gaussian tests to the perturbations seeded by cosmic strings. They found the density field distribution to be close to Gaussian on scales larger than $1.5(\Omega_M h^2)^{-1}$ Mpc, where Ω_M is the fraction of cosmological matter density in baryons and CDM combined. Scales this small correspond to the multipole index of order $l \sim 10^4$.

D. CMB polarization

The possibility that the CMB be polarized was first discussed by Martin Rees in 1968, in the context of anisotropic Universe models. In spite of his optimism, and after many attempts during more than thirty years, including some important upper limits [e.g., Keating, et al. 2001; Hedman, et al. 2001, 2002], there has been no positive detection of the polarization field until the DASI detection, a couple of months ago [Leitch et al. 2002; Kovac et al. 2002].

Unlike previous experiments, DASI reached the required sensitivity to make a sounding discovery. On the same line, MAP will have the capability to detect the polarization field and this to a level of better than 10 μK in its low frequency channels. Polarization is an important probe both for cosmological models and for the more recent history of our nearby Universe. It arises from the interactions of CMB photons with free electrons; hence, polarization can *only* be produced at the last scattering surface (its amplitude depends on the duration of the decoupling process) and, unlike temperature fluctuations, it is largely unaffected by variations of the gravitational potential after last scattering. Future measurements of polarization will thus provide a clean view of the inhomogeneities of the Universe at about 400,000 years after the Bang.

For understanding polarization, a couple of things should be clear. First, the energy of the photons is small compared to the mass of the electrons. Then, the CMB frequency does not change, since the electron recoil is negligible. Second, the change in the CMB polarization (i.e., the orientation of the oscillating electric field $\vec{E}$ of the radiation) occurs due to a certain transition, called *Thomson scattering*. The transition probability per unit time is proportional to a

---

[18] The extension of the Kaiser-Stebbins effect to polarization will be treated below. In fact, Benabed and Bernardeau [2000] have recently considered the generation of a B-type polarization field out of E-type polarization, through gravitational lensing on a cosmic string.

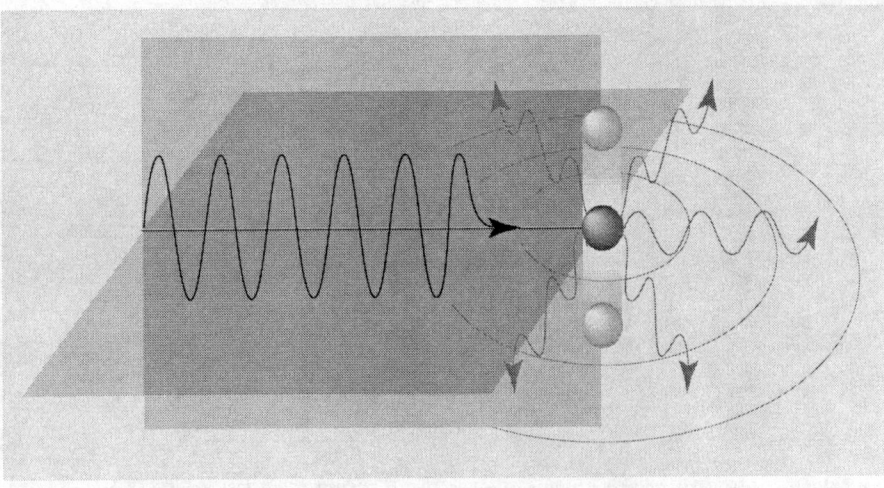

FIG. 13: *An electromagnetic linearly polarized wave (dark wave inciding from the left) oscillates in a given (vertical) plane. Reaching an electron (represented by a ball) the wave induces the electron to also oscillate, making it emit radiation. This resulting electromagnetic wave is concentrated essentially in the plane orthogonal to the movement of the electron and it is polarized like the incident wave [127].*

combination of the old ($\hat{\epsilon}_\alpha^{\text{in}}$) and new ($\hat{\epsilon}_\alpha^{\text{out}}$) directions of polarization in the form $|\hat{\epsilon}_\alpha^{\text{in}} \cdot \hat{\epsilon}_\alpha^{\text{out}}|^2$. In other words, the initial direction of polarization will be favored. Third, an oscillating $\vec{E}$ will push the electron to also oscillate; the latter can then be seen as a dipole (not to be confused with the CMB dipole), and dipole radiation emits preferentially perpendicularly to the direction of oscillation. These 'rules' will help us understand why the CMB should be linearly polarized.

Previous to the recombination epoch, the radiation field is unpolarized. In unpolarized light the electric field can be decomposed into the two orthogonal directions (along, say, $\hat{x}$ and $\hat{z}$) perpendicular to the line of propagation ($\hat{y}$). The electric field along $\hat{\epsilon}_{\hat{z}}^{\text{in}}$ (suppose $\hat{z}$ is vertical) will make the electron oscillate also vertically. Hence, the dipolar radiation will be maximal over the horizontal xy-plane. Analogously, dipole radiation due to the electric field along $\hat{x}$ will be on the yz-plane. If we now look from the side (e.g., from $\hat{x}$, on the horizontal plane and perpendicularly to the incident direction $\hat{y}$) we will see a special kind of scattered radiation. From our position we cannot perceive the radiation that the electron oscillating along the $\hat{x}$ direction would emit, just because this radiation goes to the yz-plane, orthogonal to us. Then, it is *as if* only the vertical component ($\hat{\epsilon}_{\hat{z}}^{\text{in}}$) of the incoming electric field would cause the radiation we perceive. From the above rules we know that the highest probability for the polarization of the outgoing radiation $\hat{\epsilon}_\alpha^{\text{out}}$ will be to be aligned with the incoming one $\hat{\epsilon}_{\hat{z}}^{\text{in}}$, and therefore it follows that the outgoing radiation will be *linearly* polarized. Now, as both the chosen incoming direction and our position as observers were arbitrary, the result will not be modified if we change them. Thomson scattering will convert unpolarized radiation into linearly polarized one.

This however is not the end of the story. To get the total effect we need to consider all possible directions from which photons will come to interact with the target electron, and sum them up. We see easily that for an initial isotropic radiation distribution the individual contributions will cancel out: just from symmetry arguments, in a spherically symmetric configuration no direction is privileged, unlike the case of a net linear polarization which would select one particular direction.

Fortunately, we know the CMB is *not exactly* isotropic; to the millikelvin precision the dominant mode is dipolar. So, what about a CMB dipolar distribution ? Although spatial symmetry does not help us now, a dipole will not generate polarization either. Take, for example, the radiation incident onto the electron from the left to be more intense than the radiation incident from the right, with average intensities above and below (that's a dipole); it then suffices to sum up all contributions to see that no net polarization survives. However, if the CMB has a *quadrupolar* variation in temperature (that *it has*, first discovered by COBE, to tens of μK precision) then there will be an excess of vertical polarization from left- and right-incident photons (assumed hotter than the mean) with respect to the horizontal one from top and bottom light (cooler). From any point of view, orthogonal contributions to the final polarization will be different, leaving a net linear polarization in the scattered radiation.

There is one more point to emphasize. Before recombination, ionized matter, electrons and radiation formed a single fluid. In it, the inertia was provided by massive nucleons whilst the pressure was that of radiation. And this fluid supported sound waves. In fact, the gravitational clumping tendency of the effective mass in the perturbations was resisted by the restoring radiation pressure, and therefore gravity-driven acoustic oscillations in both the fluid

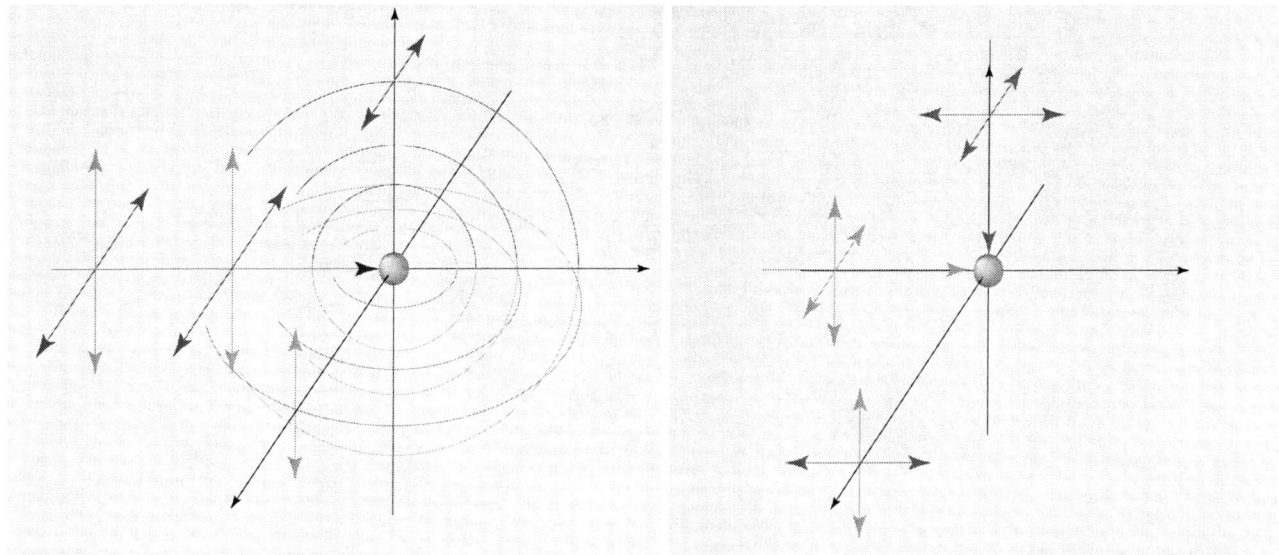

FIG. 14: *Left panel: non-polarized electromagnetic wave can be decomposed into the sum of two linearly-polarized waves, one along the line of sight (in dark-grey), the other along a perpendicular direction (light-grey). Scattered radiation due to the first wave is contained in the plane orthogonal to the line of sight and cannot be detected. Only the second component (in light-grey) reaches the observer and it is polarized as the incident wave. Right panel: when the charged particle receives non polarized waves from different directions, it will re-emit the radiation, polarized also along different directions, to the observer. If the original radiation is not isotropic (say, the dark-grey arrows from above are bigger than the light-grey ones from the left), then one of the resulting waves (in dark-grey) will be slightly more intense than the other, and the observer will perceive a net excess of linear polarization [127].*

density and local velocity appeared.

Whereas the acoustic peaks in the temperature anisotropies correspond to the compression and rarefaction maxima of the oscillating plasma, the polarization field responds to the local quadrupole moment during the decoupling process. But this local quadrupole is mainly due to the Doppler shifts induced by the velocity field of the plasma [Zaldarriaga & Harari, 1995]. That is why we know with certainty that polarization shows the uncontaminated dynamics of the primordial seeds at recombination.

Within standard recombination models the predicted level of linear polarization on large scales is tiny (see Figure 15): the quadrupole generated in the radiation distribution as the photons travel between successive scatterings is too small. Multiple scatterings make the plasma very homogeneous and only wavelengths that are small enough (big ℓ's) to produce anisotropies over the (rather short) mean free path of the photons will lead to a significant quadrupole, and thus also to polarization. Indeed, if the CMB photons last scattered at $z \sim 1100$, the SCDM model with $h = 1$ predicts no more than 0.05 μK on scales greater than a few degrees. Hence, measuring polarization at these scales represents an experimental challenge.

However, CMB polarization increases remarkably around the degree-scale in standard models. In fact, for $\theta < 1°$ a bump with superimposed acoustic oscillations reaching $\sim 5\mu$K is generically forecasted. On these scales, like for the temperature anisotropies, the polarization field shows acoustic oscillations. However, polarization spectra are sharper: temperature fluctuations receive contributions from both density (dominant) and velocity perturbations and these, being out of phase in their oscillation, partially cancel each other. On the other hand, polarization is mainly produced by velocity gradients in the baryon-photon fluid before last scattering, which also explains why temperature and polarization peaks are located differently. Moreover, acoustic oscillations depend on the *nature* of the underlying perturbation; hence, we do not expect scalar acoustic sound-waves in the baryon-photon plasma, propagating with characteristic adiabatic sound speed $c_S \sim c/\sqrt{3}$, close to that of an ideal radiative fluid, to produce the same peak-frequency as that produced by gravitational waves, which propagate with the speed of light c (see Fig.15).

The main technical complication with polarization (characterized by a tensor field) is that it is not invariant under rotations around a given direction on the sky, unlike the temperature fluctuation that is described by a scalar quantity and invariant under such rotations. The level of linear polarization is conveniently expressed in terms of the *Stokes parameters* Q and U. It turns out that there is a clever combination of these parameters that results in scalar quantities (in contrast to the above noninvariant tensor description) but with different transformation properties under spatial inversions (*parity* transformations). Then, inspired by classical electromagnetism, any polarization pattern on the

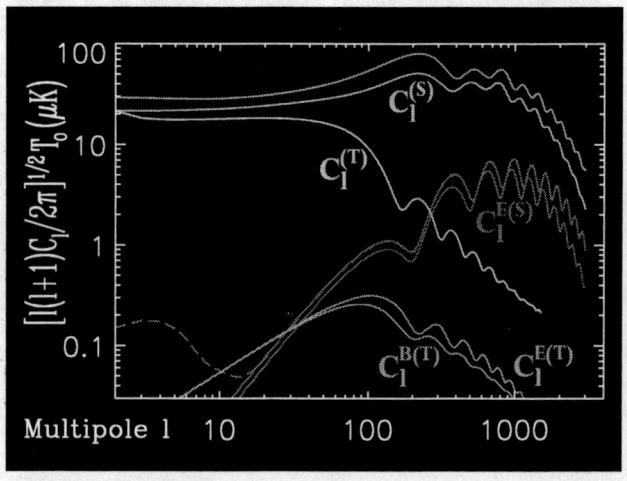 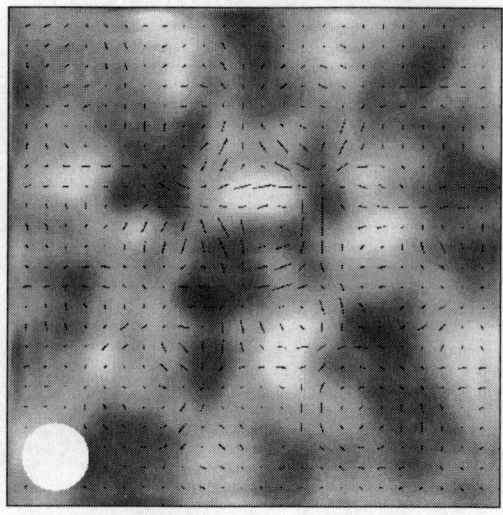

FIG. 15: Left panel: CMB Polarization for two different models. The first two (unlabeled) curves are the angular spectra derived for a ΛCHDM model, both with (dashed line) and without reionization. The temperature anisotropy spectrum from scalar perturbations (proportional to $[C_\ell]^{1/2}$) is virtually unchanged for both ionization histories. The polarization spectrum ($\propto [C_\ell^{E(S)}]^{1/2}$), although indistinguishable for $\ell \gtrsim 20$, dramatically changes for small ℓ's; in this model the Universe is reionized suddenly at low redshift with optical depth $\tau_c = 0.05$ [note that recent first-year data from WMAP indicates that $\tau_c = 0.17 \pm 0.04$]. The other curves represent a SCDM model but with a high tensor-mode amplitude, T/S=1 at the quadrupole ($\ell = 2$) level, with scale-invariant spectral indices $n_S = 1$ and $n_T = 0$. Separate scalar (noted $C_\ell^{(S)}$) and tensor ($C_\ell^{(T)}$) contributions to temperature anisotropies are shown (top curves). Scalar modes only generate E-type polarization ($C_\ell^{E(S)}$), which is smaller than the corresponding curve of the ΛCHDM model both due to differences in the models (notably $\Lambda \neq 0$ for the latter curves) and due to the influence of tensors on the normalization at small ℓ. E- and B-type polarization from tensor modes are also shown, respectively $C_\ell^{E(T)}$ and $C_\ell^{B(T)}$. Model spectra were computed with CMBFAST and are normalized to $\delta T_{\ell=10} = 27.9 \mu K$. Right panel: image of the intensity and polarization of the CMB made with the DASI telescope. The small temperature variations of the CMB are shown in false color, with light-grey hot and dark-grey cold. The polarization at each spot in the image is shown by a black line. The length of the line shows the strength of the polarization and the orientation of the line indicates the direction in which the radiation is polarized. The size of the white spot in the lower left corner approximates the angular resolution of the observations [127].

sky can be separated into 'electric' (scalar, unchanged under parity transformation) and 'magnetic' (pseudo-scalar, changes sign under parity) components (E- and B-type polarization, respectively).

1. CMB polarization from global defects

One then expands these different components in terms of spherical harmonics, very much like we did for temperature anisotropies, getting coefficients a_ℓ^m for E and B polarizations and, from these, the multipoles $C_\ell^{E,B}$. The interesting thing is that (for symmetry reasons) scalar-density perturbations will *not* produce any B polarization (a pseudo-scalar), that is $C_\ell^{B(S)} = 0$. We see then that an unambiguous detection of some level of B-type fluctuations will be a signature of the existence (and of the amplitude) of a background of gravitational waves ! [Seljak & Zaldarriaga, 1997] (and, if present, also of rotational modes, like in models with topological defects).

Linear polarization is a symmetric and traceless 2x2 tensor that requires 2 parameters to fully describe it: Q, U Stokes parameters. These depend on the orientation of the coordinate system on the sky. It is convenient to use $Q + iU$ and $Q - iU$ as the two independent combinations, which transform under right-handed rotation by an angle ϕ as $(Q + iU)' = e^{-2i\phi}(Q + iU)$ and $(Q - iU)' = e^{2i\phi}(Q - iU)$. These two quantities have spin-weights 2 and -2 respectively and can be decomposed into spin ± 2 spherical harmonics $\pm 2 Y_{lm}$

$$(Q + iU)(\hat{n}) = \sum_{lm} a_{2,lm} \, _2Y_{lm}(\hat{n}) \tag{58}$$

$$(Q - iU)(\hat{n}) = \sum_{lm} a_{-2,lm} \, _{-2}Y_{lm}(\hat{n}). \tag{59}$$

Spin s spherical harmonics form a complete orthonormal system for each value of s. Important property of spin-weighted basis: there exists spin raising and lowering operators $\eth$ and $\bar{\eth}$. By acting twice with a spin lowering and raising operator on $(Q + iU)$ and $(Q - iU)$ respectively one obtains quantities of spin 0, which are *rotationally invariant*. These quantities can be treated like the temperature and no ambiguities connected with the orientation of coordinate system on the sky will arise. Conversely, by acting with spin lowering and raising operators on usual harmonics spin s harmonics can be written explicitly in terms of derivatives of the usual spherical harmonics. Their action on $_{\pm 2}Y_{lm}$ leads to

$$\bar{\eth}^2(Q+iU)(\hat{n}) = \sum_{lm}\left(\frac{[l+2]!}{[l-2]!}\right)^{1/2} a_{2,lm}Y_{lm}(\hat{n}) \tag{60}$$

$$\eth^2(Q-iU)(\hat{n}) = \sum_{lm}\left(\frac{[l+2]!}{[l-2]!}\right)^{1/2} a_{-2,lm}Y_{lm}(\hat{n}). \tag{61}$$

With these definitions the expressions for the expansion coefficients of the two polarization variables become [Seljak & Zaldarriaga, 1997]

$$a_{2,lm} = \left(\frac{[l-2]!}{[l+2]!}\right)^{1/2}\int d\Omega\, Y^*_{lm}(\hat{n})\bar{\eth}^2(Q+iU)(\hat{n}) \tag{62}$$

$$a_{-2,lm} = \left(\frac{[l-2]!}{[l+2]!}\right)^{1/2}\int d\Omega\, Y^*_{lm}(\hat{n})\eth^2(Q-iU)(\hat{n}). \tag{63}$$

Instead of $a_{2,lm}$, $a_{-2,lm}$ it is convenient to introduce their linear *electric* and *magnetic* combinations

$$a_{E,lm} = -\frac{1}{2}(a_{2,lm} + a_{-2,lm}) \qquad a_{B,lm} = \frac{i}{2}(a_{2,lm} - a_{-2,lm}). \tag{64}$$

These two behave differently under *parity* transformation: while E remains unchanged B changes the sign, in analogy with electric and magnetic fields.

To characterize the statistics of the CMB perturbations only four power spectra are needed, those for $X = T, E, B$ and the cross correlation between T and E. The cross correlation between B and E or B and T vanishes because B has the opposite parity of T and E. As usual, the spectra are defined as the rotationally invariant quantities

$$C_{Xl} = \frac{1}{2l+1}\sum_m \langle a^*_{X,lm}a_{X,lm}\rangle \qquad C_{Cl} = \frac{1}{2l+1}\sum_m \langle a^*_{T,lm}a_{E,lm}\rangle \tag{65}$$

in terms of which on has

$$\langle a^*_{X,l'm'}a_{X,lm}\rangle = C_{Xl}\,\delta_{l'l}\delta_{m'm} \tag{66}$$

$$\langle a^*_{T,l'm'}a_{E,lm}\rangle = C_{Cl}\,\delta_{l'l}\delta_{m'm} \tag{67}$$

$$\langle a^*_{B,l'm'}a_{E,lm}\rangle = \langle a^*_{B,l'm'}a_{T,lm}\rangle = 0. \tag{68}$$

According to what was said above, one expects some amount of polarization to be present in all possible cosmological models. However, symmetry breaking models giving rise to topological defects differ from inflationary models in several important aspects, two of which are the relative contributions from scalar, vector and tensor modes and the coherence of the seeds sourcing the perturbation equations. In the local cosmic string case one finds that in general scalar modes are dominant, if one compares to vector and tensor modes in the usual decomposition of perturbations. The situation with global topological defects is radically different and this leads to a very distinctive signature in the polarization field.

Temperature and polarization spectra for various symmetry breaking models were calculated by Seljak, Pen & Turok [1997] and are shown in figure 16. Both electric and magnetic components of polarization are shown for a variety of global defects. They also plot for comparison the corresponding spectra in a typical inflationary model, namely, the standard CDM model ($h = 0.5$, $\Omega = 1$, $\Omega_{\text{baryon}} = 0.05$) but with equal amount of scalars and tensors perturbations (noted $T/S = 1$) which maximizes the amount of B component from inflationary models. In all the models they assumed a standard reionization history. The most interesting feature they found is the large magnetic mode polarization, with a typical amplitude of $\sim 1\mu K$ on degree scales [exactly those scales probed by Hedman, et al., 2001]. For multipoles below $\ell \sim 100$ the contributions from E and B are roughly equal. This differs strongly from the inflationary model predictions, where B is much smaller than E on these scales even for the extreme case of $T/S \sim 1$.

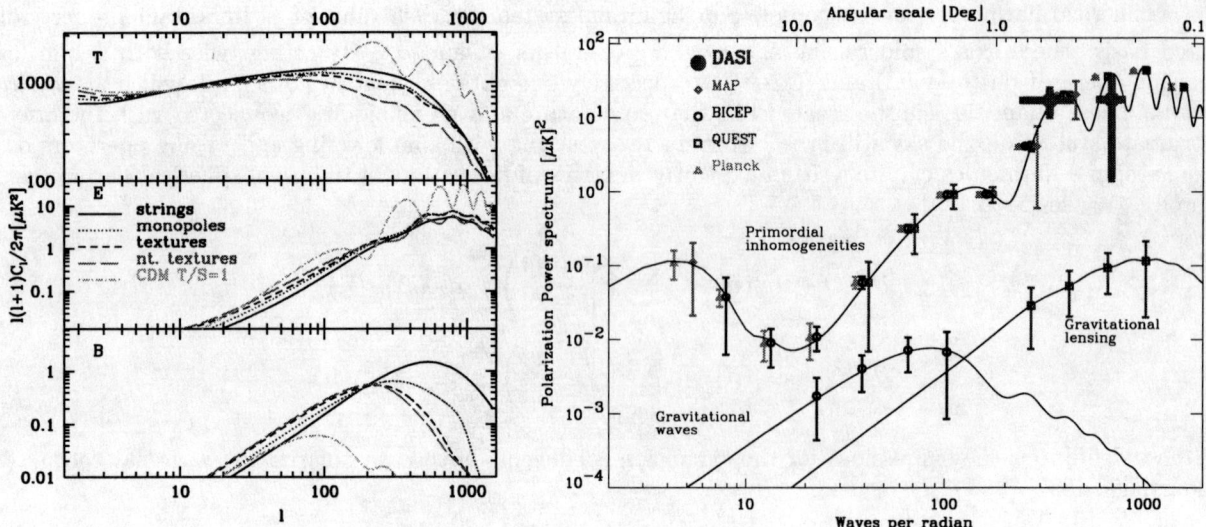

FIG. 16: Left panel: Power spectra of temperature (T), electric type polarization (E) and magnetic type polarization (B) for global strings, monopoles, textures and nontopological textures [taken from Seljak. et al., 1997]. The corresponding spectra for a standard CDM model with $T/S = 1$ is also shown for comparison. B polarization turns out to be notably larger for all global defects considered if compared to the corresponding predictions of inflationary models on small angular scales. Right panel: current and future polarization data by Hivon & Kamionkowski [2002]. Top curve shows the prediction for the polarization from primordial inhomogeneities produced by inflation. The large-angle bump in this curve is the enhancement from early star formation (reionization). The lower curves are for inflationary gravitational-wave and gravitational-lensing signals. Recently detected DASI data points are shown in red while the rest are expected data points for future experiments with more sensitivity.

Inflationary models only generate scalar and tensor modes, while global defects also have a significant contribution from vector modes. As we mentioned above, scalar modes only generate E, vector modes predominantly generate B, while for tensor modes E and B are comparable with B being somewhat smaller. Together this implies that B can be significantly larger in symmetry breaking models than in inflationary models. In figure 16 we also show the recent discovery of a tiny level of polarization by the DASI collaboration together with predictions for future experiments, assuming an inflationary origin for the temperature perturbation and polarization signals [19].

2. String lensing and CMB polarization

Recent studies have shown that in realistic models of inflation cosmic string formation seems quite natural in a post-inflationary preheating phase [Tkachev et al., 1998, Kasuya & Kawasaki, 1998]. So, even if the gross features on CMB maps are produced by a standard (e.g., inflationary) mechanism, the presence of defects, most particularly cosmic strings, could eventually leave a distinctive signature. One such feature could be found resorting to CMB polarization: the lens effect of a string on the small scale E-type polarization of the CMB induces a significant amount of B-type polarization along the line-of-sight [Zaldarriaga & Seljak, 1998; Benabed & Bernardeau 2000]. This is an effect analogous to the Kaiser-Stebbins effect for temperature maps.

In the inflationary scenario, scalar density perturbations generate a scalar polarization pattern, given by E-type polarization, while tensor modes have the ability to induce both E and B types of polarization. However, tensor modes contribute little on very small angular scales in these models. So, if one considers, say, a standard ΛCDM model, only scalar primary perturbations will be present without defects. But if a few strings are left from a very early epoch, by studying the patch of the sky where they are localized, a distinctive signature could come to light.

In the small angular scale limit, in real space and in terms of the Stokes parameters Q and U one can express the E and B fields as follows

$$E \equiv \Delta^{-1}[(\partial x^2 - \partial y^2)Q + 2\partial x \partial y U], \qquad (69)$$
$$B \equiv \Delta^{-1}[(\partial x^2 - \partial y^2)U - 2\partial x \partial y Q]. \qquad (70)$$

---

[19] See http://www.stanford.edu/~schurch and http://astro.caltech.edu/~lgg/bicep_front.htm

The polarization vector is parallel transported along the geodesics. The lens affects the polarization by displacing the apparent position of the polarized light source. Hence, the observed Stokes parameters $\hat{Q}$ and $\hat{U}$ are given in terms of the *primary* (unlensed) ones by: $\hat{Q}(\vec{\alpha}) = Q(\vec{\alpha} + \vec{\xi})$ and $\hat{U}(\vec{\alpha}) = U(\vec{\alpha} + \vec{\xi})$. The displacement $\vec{\xi}$ is given by the integration of the gravitational potential along the line–of–sights. Of course, here the 'potential' acting as lens is the cosmic string whose effect on the polarization field we want to study.

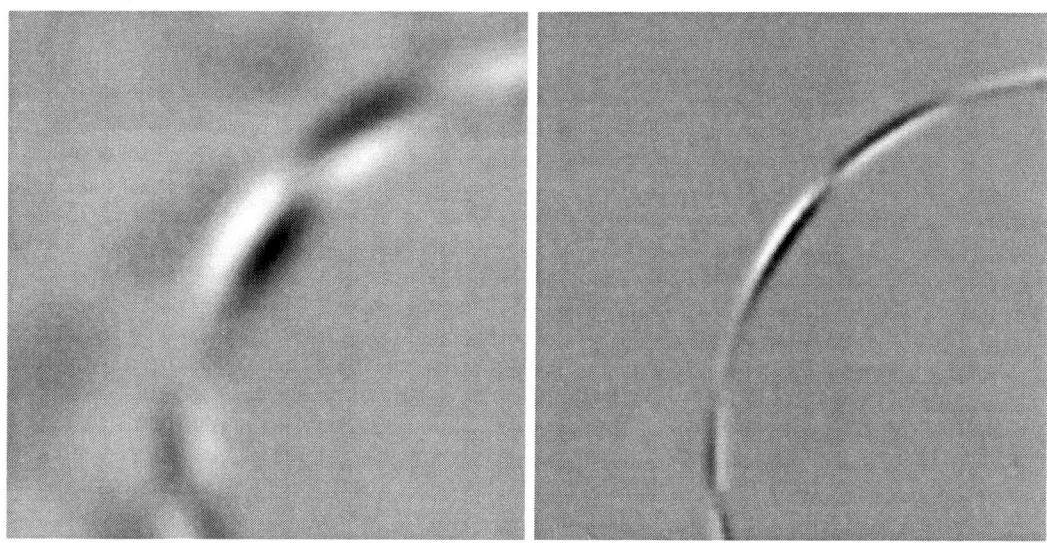

FIG. 17: *Simulations for the B field in the case of a circular loop. The angular size of the figure is $50' \times 50'$. The resolution is 5' (left) and 1.2' (right). The discontinuity in the B field is sharper the better the resolution. Weak lensing of CMB photons passing relatively apart from the position of the string core are apparent as faint patches outside of the string loop on the left panel. [Benabed & Bernardeau 2000].*

In the case of a straight string which is aligned along the y axis, the deflection angle (or half of the deficit angle) is $4\pi G\mu$ [Vilenkin & Shellard, 2000] and this yields a displacement $\xi_x = \pm \xi_0$ with

$$\xi_0 = 4\pi G\mu \mathcal{D}_{\text{lss,s}}/\mathcal{D}_{\text{s,us}} \qquad (71)$$

with no displacement along the y axis. $\mathcal{D}_{\text{lss,s}}$ and $\mathcal{D}_{\text{s,us}}$ are the cosmological angular distances between the last scattering surface and the string, and between the string and us, respectively. They can be computed, in an Einstein-de Sitter universe (critical density, just dust and no Λ), from

$$\mathcal{D}(z_1, z_2) = \frac{2c}{H_0} \frac{1}{1+z_2} [(1+z_1)^{-1/2} - (1+z_2)^{-1/2}] \qquad (72)$$

by taking $z_1 = 0$ for us and $z_2 \simeq 1000$ for the last scattering surface; see [Bartelmann & Schneider, 2001]. For the usual case in which the redshift of the string z_s is well below the z_{lss} one has $\mathcal{D}_{\text{lss,s}}/\mathcal{D}_{\text{lss,us}} \simeq 1/\sqrt{1+z_s}$. Taking this ratio of order $1/2$ (*i.e.*, distance from us to the last scattering surface equal to twice that from the string to the last scattering surface) yields $z_s \simeq 3$. Plugging in some numbers, for typical GUT strings on has $G\mu \simeq 10^{-6}$ and so the typical expected displacement is about less than 10 arc seconds. Benabed & Bernardeau [2000] compute the resulting B component of the polarization and find that the effect is entirely due to the discontinuity induced by the string, being nonzero just along the string itself. This clearly limits the observability of the effect to extremely high resolution detectors, possibly post-Planck ones.

The situation for circular strings is different. As shown by de Laix & Vachaspati [1996] the lens effect of such a string, when facing the observer, is equivalent to the one of a static linear mass distribution. Considering then a loop centered at the origin of the coordinate system, the displacement field can be expressed very simply: observing in a direction through the loop, $\vec{\xi}$ has to vanish, while outside of the loop the displacement decreases as α_l/α, *i.e.*, inversely proportional to the angle. One then has [Benabed & Bernardeau, 2000]

$$\vec{\xi}(\vec{\alpha}) = -2\xi_0 \frac{\alpha_l}{\alpha^2} \vec{\alpha} \quad \text{with} \quad \alpha > \alpha_l, \qquad (73)$$

where α_l is the loop radius.

This ansatz for the displacement, once plugged into the above equations, yields the B field shown in both panels of Figure 17. A weak lensing effect is barely distinguishable outside the string loop, while the strong lensing of those photons traveling close enough to the string is the most clear signature, specially for the high resolution simulation. One can check that the hot and cold spots along the string profile have roughly the same size as for the polarization field in the absence of the string loop. The simulations performed show a clear feature in the maps, although limited to low resolutions this can well be confused with other secondary polarization sources. It is well known that point radio sources and synchrotron emission from our galaxy may contribute to the foreground [de Zotti et al. 1999] and are polarized at a 10 % level. Also lensing from large scale structure and dust could add to the problem.

VI. COSMIC DEFECTS IN PERSPECTIVE

Cosmic defects have proved very interesting and fruitful in high–energy physics and astrophysics. Their generic production in grand unified theories has made defects an active field of research for over two decades. Many of the interesting subjects now associated with defects were only briefly mentioned in these notes, like the internal structure of defects –leading to persistent currents in their cores– and, as a consequence, the possible generation of primordial magnetic fields. Also, primordial gravitational waves, extremely high–energy phenomena associated to cosmic rays and uhecrons, electroweak baryogenesis and, finally, the very active condensed-matter-cosmology interface, dubbed cosmos in the lab, equally –and unjustly– received no attention [compensate for this with references like Vilenkin & Shellard, 2000; Hindmarsh & Kibble, 1995; Kibble, 2002; Gangui, 2001b, for example – yes, in that order.]. With regards to the most transparent test of current cosmology, namely the CMB and matter power spectra, (not so) recent investigations have pointed out severe problems in virtually all models where cosmic defects are the main source of the seeds of structure in the universe[20]. In the case of cosmic strings, however, these bad[21] news were reached by the use of non–negligible, albeit well-founded, approximations in order to cope with the limited range of realistic defect simulations. Although the whole method of unequal time correlators employed by most of the groups can be regarded as a good approximation to reality during both the matter and radiation eras, the important transition in between must be looked at more carefully, as the above–mentioned correlators do not scale as expected. Recent, full Boltzmann analyses aiming to solve this handicap are in progress [e.g., Landriau & Shellard, 2002] and already producing interesting results.

Acknowledgments

I'd like to thank my collaborators in some of the topics covered in these lectures for their insights and remarks. Thanks also to the other speakers and students for the many discussions during this very instructive time we spent together, and to the members of the L.O.C., particularly Prof. Mário Novello and Santiago Perez Bergliaffa from CBPF, for their superb job in organizing this charming school. Finally, I'd like to acknowledge CONICET, UBA and FUNDACIÓN ANTORCHAS for their financial support.

---

[1] Abrikosov, A. A. [1957], *Sov. Phys. JETP* **5**, 1174 [*Zh. Eksp. Teor. Fiz.* **32**, 1442 (1957)]
[2] Achúcarro, A. & Vachaspati, T. [2000], *Phys. Rept.* **327**, 347-426; *Phys. Rept.* **327**, 427. [hep-ph/9904229]
[3] Albrecht, A., Battye, R. & Robinson, J. [1997], *Phys. Rev. Lett.* **79**, 4736.; *Phys. Rev.* **D59**, 023508 (1998).
[4] Albrecht, A., Coulson, D., Ferreira, P. & Magueijo, J. [1995], Imperial Preprint /TP/94–95/30.
[5] Alexander, S., Brandenberger, R., Easther, R. & Sornborger, A. [1999], hep-ph/9903254.
[6] Avelino, P.P. & Martins, C.J.A.P. [2000], *Phys. Rev. Lett.* **85**, 1370.
[7] Avelino, P.P., Shellard, E.P.S., Wu, J.H.P. & Allen, B. [1998], *Astrophys. J.* **507**, L101.
[8] Bartelmann, M. & Schneider, P. [2001], *Phys. Rep.* **340**, 291.
[9] Battye, R. A., Bucher, M. & Spergel, D. [1999], astro-ph/9908047 .
[10] Battye, R. & Weller, J. [2000], *Phys. Rev.* **D61**, 043501.
[11] Benabed, K. & Bernardeau, F. [2000], *Phys. Rev.* **D61**, 123510.
[12] Bennett, D.P. & Rhie, S.H. [1990], *Phys. Rev. Lett.* **65**, 1709.

---

[20] *"There are no strings on me"* –Pinocchio
[21] bad or good, depending on which side you are.

[13] Bennett, D.P. & Rhie, S.H. [1993], *Ap. J. Lett.* **406**, L7.
[14] Bennett, D.P., Rhie, S.H. & Weinberg, D.H. [1993], preprint.
[15] Borrill, J., et al. [1994], *Phys. Rev.* **D50**, 2469.
[16] Bouchet, F.R., Peter, P., Riazuelo, A. & Sakellariadou, M. [2001], preprint astro-ph/0005022.
[17] Brandenberger, R. [1993], *Topological Defects and Structure Formation*, EPFL lectures, Lausanne, Switzerland.
[18] Bunkov, Y. & Godfrin, H. [2000], (editors) Proceedings of the NATO-ASI on topological defects and non-equilibrium dynamics of symmetry breaking phase transitions (Kluwer, Dordrecht).
[19] Callan, C. & Coleman, S. [1977], *Phys. Rev.* **D16**, 1762.
[20] Carter, B, [1990], *Phys. Rev.* **D41**, 3869.
[21] Carter, B. [1997], Tlaxcala lecture notes, hep-th/9705172.
[22] Carter, B., Peter, P. & Gangui, A. [1997], *Phys. Rev.* **D55**, 4647. [hep-ph/9609401];
[23] Contaldi, C., Hindmarsh, M. & Magueijo, J. [1999], *Phys. Rev. Lett.* **82** 2034.
[24] Copeland, E. [1993], in *The physical universe: The interface between cosmology, astrophysics and particle physics*, eds. Barrow, J.D., et al. (Sringer–Verlag).
[25] Cowie, L. & Hu, E. [1987], *Ap. J.* **318** L33.
[26] Davis, R. L. [1987], *Phys. Rev. D* **35**, 3705.
[27] Davis, R.L. & Shellard, E.P.S. [1988], *Phys. Lett.* **B207**, 404.
[28] de Bernardis, P. et al. [2000], Nature 404, 995 [astro-ph/0004404]
[29] de Laix, A.A. & Vachaspati, T. [1996], *Phys. Rev.* **D54**, 4780.
[30] de Zotti, G. et al.[1999], astro–ph/9908058.
[31] Dolan, L. & Jackiw, R. [1974], *Phys. Rev.* **D9**, 3320.
[32] Durrer, R. [2000], in Moriond meeting on Energy Densities in the Universe, astro-ph/0003363 .
[33] Durrer, R., Gangui, A. & Sakellariadou, M. [1996], Phys. Rev. Lett. **76**, 579. [astro-ph/9507035].
[34] Durrer, R., Kunz, M. & Melchiorri, A. [2002], Phys.Rept. 364 (2002) 1-81. [astro-ph/0110348].
[35] Durrer, R. & Zhou, Z.H. [1995], Zürich University Preprint, ZH–TH19/95, astro–ph/9508016.
[36] Dvali, G., Liu, H. & Vachaspati, T. [1998], *Phys. Rev. Lett.* **80**, 2281.
[37] Gangui, A. [2001], *Science* **291**, 837.
[38] Gangui, A. [2001b], Topological Defects in Cosmology, Lecture Notes for the First Bolivian School on Cosmology, in press [astro-ph/0110285].
[39] Gangui, A., Lucchin, F., Matarrese, S. & Mollerach, S. [1994], *Ap. J.* **430**, 447.
[40] Gangui, A. & Martin, J. [2000a], *Mon. Not. R. Astron. Soc.* **313**, 323. [astro-ph/9908009]
[41] Gangui, A. & Martin, J. [2000b], *Phys. Rev.* **D62**, 103004. [astro-ph/0001361]
[42] Gangui, A. & Mollerach, S. [1996], *Phys. Rev.* **D54**, 4750-4756. [astro-ph/9601069]
[43] Gangui, A. & Peter, P. [1998], Cosmological and Astrophysical Implications of Superconducting Cosmic Strings and Vortons, in Proceedings of the XXXIIIrd Rencontres de Moriond on 'Fundamental Parameters in Cosmology', pages 19-24, Editors: J. Tran Thanh Van et al., Editions Frontières, 1998.
[44] Gangui, A., Peter, P. & Boehm, C. [1998], *Phys. Rev.* **D57**, 2580. [hep-ph/9705204]
[45] Gangui, A., Pogosian, L. & Winitzki, S. [2001a], *New Astronomy Reviews* 46, 681-691 (2002). [astro-ph/0112145]
[46] Gangui, A., Pogosian, L. & Winitzki, S. [2001b], *Phys. Rev.* **D64**, 043001.
[47] Goto, T. [1971], *Prog. Theor. Phys.* **46**, 1560.
[48] Gott, R. [1985], *Ap. J.*, **288**, 422.
[49] Guth, A.H. & Weinberg, E. [1983], *Nucl. Phys.* **B212**, 321.
[50] Hanany, S., et al, Astrophys. J. Lett. accepted (2000), astro-ph/0005123; Jaffe, A.H., et al, astro-ph/0007333.
[51] Halverson, N.W. et al, preprint astro-ph/0104489.
[52] Hedman, M., Barkats, D., Gundersen, J. Staggs, S. & Winstein, B. [2001], *Ap. J.*, **548**, L111-L114 [astro-ph/0010592]; Hedman, M. et al. [2002], Astrophys. J. 573, L73 (2002).
[53] Higgs, P. [1964], *Phys. Lett.* **12**, 132.
[54] Hilton, P.J. [1953], *Introduction to homotopy theory* (Cambridge: Cambridge University Press).
[55] Hindmarsh, M.B. & Kibble, T.W.B. [1995], *Rept. Prog. Phys.* **58**, 477-562 [hep-ph/9411342].
[56] Hivon, E. & Kamionkowski, M. [2002], *Science* 298, 1349.
[57] Hu, W., Spergel, D. & White, M. [1997], *Phys. Rev.* **D55**, 3288-3302.
[58] Kaiser, N. & Stebbins, A. [1984], *Nature* **310**, 391.
[59] Kasuya, S. & Kawasaki, M, [1998], *Phys. Rev.* **D58**, 083516.
[60] Keating, B. et al. [2001], Astrophys. J. 560, L1 (2001).
[61] Kibble, T.W.B. [1976], *J. Phys.* **A9**, 1387.
[62] Kibble, T.W.B. [1980], *Phys. Rep.* **67**, 183.
[63] Kibble, T.W.B. [1985], *Nucl. Phys.* **B252**, 227; **B261**, 750 (1986).
[64] Kibble, T.W.B. [2002], Lectures at NATO ASI "Patterns of symmetry breaking", Cracow [cond-mat/0211110].
[65] Kibble, T.W.B., Lazarides, G. & Shafi, Q. [1982], *Phys. Lett.* **B113**, 237.
[66] Kirzhnits, D.A. & Linde, A.D. [1974], *Sov. Phys. JETP* **40**, 628.
[67] Kirzhnits, D.A. & Linde, A.D. [1976], *Ann. Phys.* **101**, 195.
[68] Kolb, E.W. & Turner, M.S. [1990] *The Early Universe* (New York: Addison–Wesley).
[69] Komatsu, E. & Spergel, D. [2000], preprint astro-ph/0005036;
[70] Kovac, J. et al. [2002], Nature 420, 772-787. [astro-ph/0209478]

[71] Lemperière, Y. & Shellard, E.P.S. [2002], preprint hep-ph/0207199.
[72] Landriau, M. & Shellard, E.P.S. [2002], preprint astro-ph/0208540.
[73] Langacker, P. & Pi, S.-Y. [1980], *Phys. Rev. Lett.* **45**, 1.
[74] Langer, S. [1992], in *Solids far from equilibrium*, Godrèche, C., ed. (Cambridge: Cambridge University Press).
[75] Leitch, E.M. et al. [2002], Nature 420, 763-771. astro-ph/0209476
[76] Liddle, A. [1995], *Phys. Rev.* **D51**, 5347-5351.
[77] Linde, A.D. [1983b], *Nucl. Phys.* **B216**, 421.
[78] Linet, B. [1986], *Phys. Rev.* **D33** 1833.
[79] Magueijo, J., Albrecht, A., Ferreira, P. & Coulson, D. [1996], *Phys. Rev.* **D54**, 3727-3744.
[80] Magueijo, J. & Brandenberger, R. [2000], Iran lectures on cosmic defects, astro-ph/0002030.
[81] Magueijo, J. & Ferreira, P. [1997], *Phys. Rev.* **D55**, 3358.
[82] Martins, C.J.A.P. & Shellard, E.P.S. [1996], *Phys. Rev.* **D54** 2535.
[83] Martins, C.J.A.P. & Shellard, E.P.S. [2000], hep-ph/0003298
[84] Mermin, M. [1979], *Rev. Mod. Phys.* **51**, 591.
[85] Nambu, Y. [1970], in Proc. Int. Conf. on Symmetries and Quark Models, ed. Chand, R. (New York: Gordon and Breach).
[86] Notzold, D. [1991], *Phys. Rev.* **D43**, R961.
[87] Peebles, P.J.E. & Ratra, B. [2002], Rev. Mod. Phys. (in press) astro-ph/0207347.
[88] Pen, U.-L., Seljak, U. & Turok, N. [1997], *Phys. Rev. Lett.* **79**, 1611-1614.
[89] Pen, U.-L., Spergel, D.N. & Turok, N. [1994], *Phys. Rev.* **D49**, 692.
[90] Perlmutter, S., et al. [1999], Astrophys. J. **517**, 565.
[91] Pogosian, L. [2001], Int. J. Mod. Phys. A16S1C. astro-ph/0009307.
[92] Pogosian, L. & Vachaspati, T. [1999], *Phys. Rev.* **D60**, 083504.
[93] Pogosian, L. & Vachaspati, T. [2000], *Phys. Rev.* **D62**, 105005.
[94] Preskill, J. [1979], *Phys. Rev. Lett.* **43**, 1365.
[95] Rajaraman, R. [1982], *Solitons and instantons* (Amsterdam: North–Holland).
[96] Riess, A.G, et al. [1998], Astron.J. 116, 1009-1038. astro-ph/9805201
[97] Sachs, R. & Wolfe, A. [1967], *Ap. J.* **147**, 73.
[98] Scaramella, R. & Vittorio, N. [1991], *Ap. J.* **375**, 439.
[99] Scaramella, R. & Vittorio, N. [1993], *M.N.R.A.S.* **263**, L17.
[100] Seljak, U., Pen, U.-L. & Turok, N. [1997], *Phys.Rev.Lett.* **79** 1615-1618.
[101] Seljak, U. & Zaldarriaga, M. [1996], *Astrophys. J.* **469**, 437.
[102] Seljak, U. & Zaldarriaga, M. [1997], *Phys. Rev. Lett.* **78**, 2054.
[103] Silk, J. & Vilenkin, A. [1984], *Phys. Rev. Lett.* **53**, 1700.
[104] Steenrod, N. [1951], *Topology of Fibre Bundles* (Princeton: Princeton University Press).
[105] Tkachev, I., Khlebnikov, S., Kofman, L. & Linde, A. [1998], *Phys.Lett.* **B440**, 262-268.
[106] Turok, N. [1989], *Phys. Rev. Lett.* **63**, 2625.
[107] Turok, N. [1996], *Phys. Rev. Lett.* **77**, 4138-4141; *Phys. Rev.* **D54**, 3686-3689
[108] Turok, N., Pen, U.-L. & Seljak, U. [1998], *Phys. Rev.* **D58**, 023506.
[109] Turok, N. & Spergel, D.N. [1990], *Phys. Rev. Lett.* **64**, 2736.
[110] Turok, N. & Zadrozny, J. [1990], *Phys. Rev. Lett.* **65**, 2331.
[111] Vachaspati, T. [2001], ICTP summer school lectures, hep-ph/0101270.
[112] Vachaspati, T. & Vilenkin, A. [1991], *Phys. Rev. Lett.* **67**, 1057.
[113] Veeraraghavan, S. & Stebbins, A. [1990], *Ap. J.* **365**, 37.
[114] Vilenkin, A. [1981], *Phys. Rev.* **D23**, 852.
[115] Vilenkin, A. [1983], *Phys. Rev.* **D27**, 2848.
[116] Vilenkin, A. [1984], *Phys. Rev. Lett.* **53**, 1016.
[117] Vilenkin, A. [1985], *Phys. Rep.* **121**, 263.
[118] Vilenkin, A [1990], *Phys. Rev.* **D41**, 3038.
[119] Vilenkin, A. & Shellard, E.P.S. [2000], *Cosmic Strings and other Topological Defects*, 2nd edition, (Cambridge: CUP).
[120] Vollick, D.N. [1992], *Phys. Rev.* **D45**, 1884.
[121] Walker, P.N., et al. [1991], *Ap. J.* **376**, 51.
Wang, L. & Kamionkowski, M. [2000], *Phys. Rev.* **D61**, 063504.
[122] Weinberg, S. [1974], *Phys. Rev.* **D9**, 3357.
[123] Wu, J.-H.P. et al, preprint astro-ph/9812156.
[124] Zaldarriaga, M. & Harari, D. [1995], Phys. Rev. D52, 3276.
[125] Zaldarriaga, M. & Seljak, U. [1998], *Phys. Rev.* **D58**, 023003.
[126] Zel'dovich, Ya. B., I. Yu. Kobzarev & L. B. Okun [1974], Zh. Eksp. Teor. Fiz. **67**, 3 [Sov. Phys. JETP **40**, 1 (1975)].
[127] A complete and updated version of these lecture notes with colour figures can be found at www.iafe.uba.ar/relatividad/gangui/xescola/

On the possible role of massive neutrinos in cosmological structure formation

Massimiliano Lattanzi,[1,2] Remo Ruffini,[1,2] and Gregory Vereshchagin[1,3]

[1]*ICRA — International Center for Relativistic Astrophysics.*
[2]*Dipartimento di Fisica, Università di Roma "La Sapienza", Piazzale Aldo Moro 5, I-00185 Roma, Italy.*
[3]*Belorussian State University, Theoretical Physics Department, Skorina ave. 4, 220050, Minsk, Republic of Belarus.*

In addition to the problem of galaxy formation, one of the greatest open questions of cosmology is represented by the existence of an asymmetry between matter and antimatter in the baryonic component of the Universe. We believe that a net lepton number for the three neutrino species can be used to understand this asymmetry. This also implies an asymmetry in the matter-antimatter component of the leptons. The existence of a nonnull lepton number for the neutrinos can easily explain a cosmological abundance of neutrinos consistent with the one needed to explain both the rotation curves of galaxies and the flatness of the Universe. Some propedeutic results are presented in order to attack this problem.

Contents

I. Evidence for dark matter and the possible role of neutrinos	264
II. Large scale structure	266
A. The cosmological principle	266
B. Two-point correlation function	267
C. Observed galaxy distribution	268
D. Power law clustering and fractals	268
III. Gravitational instability	269
A. Horizon scale and mass evolution	269
B. Self-gravitating ideal fluid: linear theory	271
1. Fluid equations and background solutions	271
2. Perturbed quantities	272
3. Linearized perturbations equations	272
4. The Jeans criterion	273
5. Multi-component system	273
C. Applications	274
1. Einstein-de Sitter Universe	274
2. Mixture of radiation and dark matter	274
D. Initial spectrum of perturbations	274
E. Damping of perturbations	275
1. Silk damping	275
2. Free streaming	275
F. Structure formation at late times	276
1. Nonlinear clustering	276
2. Structure formation scenarios	276
3. HDM models	276
4. CDM models	277
IV. Neutrinos and structure formation	277
A. Neutrino decoupling	277
1. The redshifted statistics	277
2. Energy density of neutrinos	279
3. Recent constraints on the neutrino mass m_ν and degeneracy parameter ξ_ν	279
4. The Jeans mass of neutrinos	280
B. Subsequent fragmentation model	281
1. Nonlinear model of spherical collapse	282
2. Successive fragmentation	283

| 3. The fractal model | 284 |
| References | 285 |

I. EVIDENCE FOR DARK MATTER AND THE POSSIBLE ROLE OF NEUTRINOS

The most popular model of the Universe being currently discussed in the literature is one usually indicated by the "ΛCDM model" (see e.g. E.W. Kolb, in these same proceedings [1]), which implies that almost 75% of the energy density of the Universe is due to a cosmological term, while only 1% is due to neutrinos [2]–[14]. This result seems to be in conflict with our current knowledge of the rest of physics. We will give in the following some propedeutical considerations to reconsider this problem.

Many theoretical considerations and observational facts make it clear (see [15] and references therein) that luminous matter alone cannot account for the whole matter content of the Universe. Among them there are the considerations on cosmological nucleosynthesis [16]–[18] as well as the measurements of the the cosmic background radiation (CBR) anisotropy spectrum [19]–[20]. In both cases the fit is consistent with a cosmological model in which just a fraction smaller than 10% of the total density is due to baryons.

Strong evidence for the presence of dark matter is directly given by the rotation curves of galaxies [21]. If we assume for simplicity a spherical or ellipsoidal mass distribution for a galaxy, the orbital velocity at a radius r is given by Newton's equation of motion:

$$v^2 = \frac{GM(r)}{r}, \qquad (1)$$

where G is the gravitational constant and $M(r)$ is the mass contained in a sphere of radius r. The peculiar velocity of stars beyond the visible edge of the galaxy should then decrease as $1/r$. What is observed instead is that the velocity stays nearly constant with increasing r. This requires a halo of invisible dark matter to be present outside the radius of the visible matter. From observations it follows that the halo radius can be 10 times larger than the radius of visible part of the galaxy. Then from Eq. (1) it follows that M_{halo} is at least 10 times larger than the galactic mass M_{gal}. We now assume that galactic halos are composed of neutral fermions of mass m_x and apply a gravitational Thomas-Fermi model to this system. The equation for the dimensionless gravitational potential χ as a function of a radial coordinate x is

$$\frac{d^2\chi}{dx^2} = -\frac{\chi^{3/2}}{\sqrt{x}}, \qquad (2)$$

with the initial condition $\chi(0) = 0$. For any number of particles N this equation can be used to compute the radius R of the system. Since the total mass is $M = Nm_x$, this defines a relation between M, R and m_x which allows one to estimate m_x, since both the value of the total mass and radius are known. Using for the radius of the halo $R_{halo} = 10R_{gal}$ and for the mass of the halo $M_{halo} = 10M_{gal}$ we obtain the particle mass $m_x \approx 4$eV. This value, although obtained by a simple argument, is in good agreement with the recent constraints on the electron neutrino mass obtained from the spectrum of tritium beta decay ($m_\nu < 2.5$ eV [22]).

This agreement can even be improved if one takes into account different families of neutrinos. In this case, the effect of the additional degrees of freedom can be expressed in terms of the effective mass $m_{eff} = \left(\sum_{i=1}^{n} m_i^4\right)^{1/4}$. If we consider three families with nearly the same mass, they provide a factor $3^{1/4} \simeq 1.32$ to improve the bound to $m_\nu \simeq 3$ eV.

Neutrinos were considered as the best candidate for dark matter about thirty years ago. Indeed, it was shown by Gerstein and Zel'dovich in 1966 [23] that if these particles have a small mass $m_\nu \sim 30$ eV, they provide a large energy density contribution up to the value $\Omega_\nu \sim 1$. It is in effect easy to show that the density parameter of a single family of mass $m_\nu \ll 1$ Mev would be, assuming a null chemical potential:

$$\Omega_\nu h^2 = \frac{m_\nu}{93 \, \text{eV}}, \qquad (3)$$

where h is the Hubble constant in units of 100 km sec^{-1} Mpc^{-1}. The generalization to the case of a nonnull chemical potential is (see e.g. Ruffini & Song [24]):

$$\Omega_\nu h^2 = \frac{m_\nu}{93 \, \text{eV}} A(\xi), \qquad (4)$$

where ξ is the dimensionless chemical potential (so called degeneracy parameter) at decoupling and

$$A(\xi) \equiv \frac{1}{4\eta(3)} \left[\frac{1}{3} |\xi|^3 + 4\eta(2)|\xi| + 4 \sum_{k=1}^{\infty} (-1)^{k+1} \frac{e^{-k|\xi|}}{k^3} \right]. \tag{5}$$

Here $\eta(n)$ denotes the Riemann η function of the index n. This allows to obtain even higher values of Ω_ν. In fact we have carried out a more detailed analysis, taking into account three different neutrino flavours with the same mass but different chemical potentials. Using the limits that primordial nucleosynthesis imposes on the degeneracy parameters of the e, μ and τ neutrinos [16], we found that if $m_\nu \simeq 2$ eV and $\xi_e \simeq 0.4$ then $\Omega_\nu \simeq 1$ [25].

However, in 1979 Tremaine and Gunn [26] claimed that massive neutrinos cannot be considered as a dark matter candidate. Their paper was very influential and turned most cosmologists away from neutrinos as cosmologically important particles. In their paper, Tremaine and Gunn establish an upper limit to the neutrino mass of $m_\nu \lesssim 1.2$ eV. They obtain this limit from arguments based on the velocity dispersion within galaxies and clusters, under the hypothesis that $m_\nu \lesssim 1$ MeV, so that neutrinos are ultrarelativistic at decoupling. They also obtain a lower bound $m_\nu \gtrsim 20$ eV from an argument based on phase space density considerations and on the rotation curves of galaxies. In this very strange situation with wildly contradictory constraints, they possibly see a way out by avoiding the fact that neutrinos are ultrarelativistic at decoupling and they conclude that massive galactic halos cannot be composed of stable neutral leptons of mass $\lesssim 1$ MeV.

While the Gunn and Tremaine result deeply influenced astrophysicists against the possible role neutrinos in cosmology, especially in the U.S.A., in 1977 Lee and Weinberg [27] turned their attention to massive neutrinos with $m_\nu > 2$ GeV, showing that such particles could provide a large contribution to the energy density of the Universe, in spite of a much smaller value of their number density. This paper was among the first to consider very massive particles as candidates for dark matter. This very interesting work, together with the Gunn and Tremaine purported difficulties for the neutrino scenario, induced some cosmologists to turn their attention to very massive particles, thus marking the birth of cold dark matter models.

A clear counterexample to the Gunn and Tremaine bound, which was indeed derived from a nontransparent mixture of quantum limits on classical Maxwell-Boltzmann statistics, was given by Gao and Ruffini [28]. They established a different upper bound on the neutrino mass from the assumption that galactic halos are composed of degenerate neutrinos: $m_\nu \lesssim 15$ eV. This result was further developed and confirmed by Arbolino and Ruffini [29]. They explicitly showed that rotation curves for galaxies in perfect agreement with the observations can be obtained for neutrino masses of the order of 9 eV. This limit could be lowered further if semidegenerate configurations for the neutrino halo were to be considered (see e.g. Merafina and Ruffini [30]-[31]); Ingrosso, Merafina, Ruffini and Strafella [32]).

Today, quite apart from the rotation curves of galaxies, the recent determination of the neutrino masses would appear to be in contradiction with an assumption of $\Omega_\nu \sim 1$. However, this is only an apparent difficulty, since for semidegenerate distributions (see Eq. (4)) this equality can indeed be fulfilled and important consequences on the matter-antimatter asymmetry in the leptonic component of the Universe can be inferred.

One of the most interesting features of neutrino cosmology is that they establish a natural cutoff for the largest possible structure in the Universe, related to the maximum value of the Jeans mass when the neutrinos become nonrelativistic:

$$M_J(z_{nr}) = 1.475 \cdot 10^{17} M_\odot g_\nu^{-\frac{1}{2}} N_\nu^{-\frac{1}{2}} \left(\frac{m_\nu}{10 eV} \right)^{-2} A(\xi)^{\frac{5}{4}} B(\xi)^{\frac{3}{4}}, \tag{6}$$

where z_{nr} is the redshift at which neutrinos enter the nonrelativistic regime, g_ν is the number of quantum degrees of freedom, N_ν is the number of neutrino families, $A(\xi)$ is as defined in Eq. (5), and

$$B(\xi) \equiv \frac{1}{48\eta(5)} \left[\frac{1}{5} \xi^5 + 8\eta(2)\xi^3 + 48\eta(4)\xi + 48 \sum_{n=1}^{\infty} (-1)^{n+1} \frac{e^{-n\xi}}{n^5} \right]. \tag{7}$$

This mass appears to be essential in determining the upper cutoff for a possible fractal structure of the Universe. Real difficulties still exist today in understanding the details of the fragmentation of these masses of $10^{17} M_\odot$ and the development of smaller structures all the way down to galaxies. In this lecture we illustrate some of this basic problems which still need additional work before a detailed correspondence with observations can be obtained.

II. LARGE SCALE STRUCTURE

A. The cosmological principle

There have been three distinct moments in the development of the so called *cosmological principle* which is at the very basis of our approach to the analysis of the Universe. The first formulation of the cosmological principle can be simply stated:

All the events in the Universe are equivalent. (8)

Such a cosmological principle was enunciated a few years after the introduction of the field equations of general relativity by Einstein himself [33] in the quest for visualizing a Universe the most democratic with respect to any special point and any possible moment of time: a Universe everlasting in time and totally homogenous in the spatial directions. No solution fulfilling such a cosmological principle could be found, and Einstein was so strongly confident of the validity of this principle that he modified his field equations of general relativity by introducing a cosmological constant Λ. George Gamow said that Einstein later on considered that the biggest mistake in his life.

It was through the work of Alexander Alexandrovich Friedmann [34]-[35] that a new cosmological principle was advanced:

All the points in the Universe are equivalent. (9)

As long as we look at our 'neighbour' Universe, this statement is certainly false, because the distribution of matter is far from homogeneous: there are planets, stars, and going to larger scales, galaxies and clusters of galaxies separated by almost empty regions. However, the Friedmann principle should apply when we average this distribution over a volume containing a large enough number of galaxies. For such a spatially homogeneous Universe Friedmann [34] found in 1922 explicit analytic solutions of the Einstein equations of general relativity. A remarkable property of these solutions is that they describe a non-static Universe. At that time, there was no observational evidence for the temporal evolution of the whole Universe. The first evidence came in 1929 from the observation by Hubble [36] of the recession of the nebulae. Hubble was the first trying to study the spatial distribution of objects as large as the galaxies, at that time thought to be the largest self-gravitating systems to exist. The Hubble law, interpreted within the framework of Friedmann cosmology, implied that the galaxy distribution is close to homogenous on the large-scale average [37]-[40].

It was through the above mentioned work of Hubble together with the later remarkable work of George Gamow and his collaborators (1946–1949) [41]-[50] who postulated an initially hot Universe, and the detailed work of Fermi and Turkievich in the same years [51] introducing the first computation of cosmological nucleosynthesis, that the Friedmann Universe has grown to become the standard paradigm in cosmology following the discovery of the CBR by Penzias & Wilson in 1965 [52].

In effect, one of the strongest predictions of Big Bang model is the presence of a background microwave radiation, a relic of the early Universe. This radiation is highly isotropic, reflecting through the coupling with matter the high isotropy and homogeneity of the primeval plasma. This tells us that the cosmological principle, and then the Friedmann picture, safely apply to the early Universe. Homogeneity on very large scales is confirmed by present day observations of in particular:

- X-ray background [53],

- radio sources [54],

- gamma ray bursts distribution [55],

- galaxies and clusters of galaxies [56].

So much so for the very large scales, but what about structures like galaxies and clusters of galaxies? More and more they appear to be distributed without any apparent homogeneity, but on the contrary showing regularities in an apparent hierarchical distribution of galaxies, clusters of galaxies and superclusters of galaxies separated by large voids [57]-[61] (see Fig. 1). Slowly but more and more clearly the presence of a fractal distribution in the Universe has started to surface and with it a new cosmological principle which can be simply expressed:

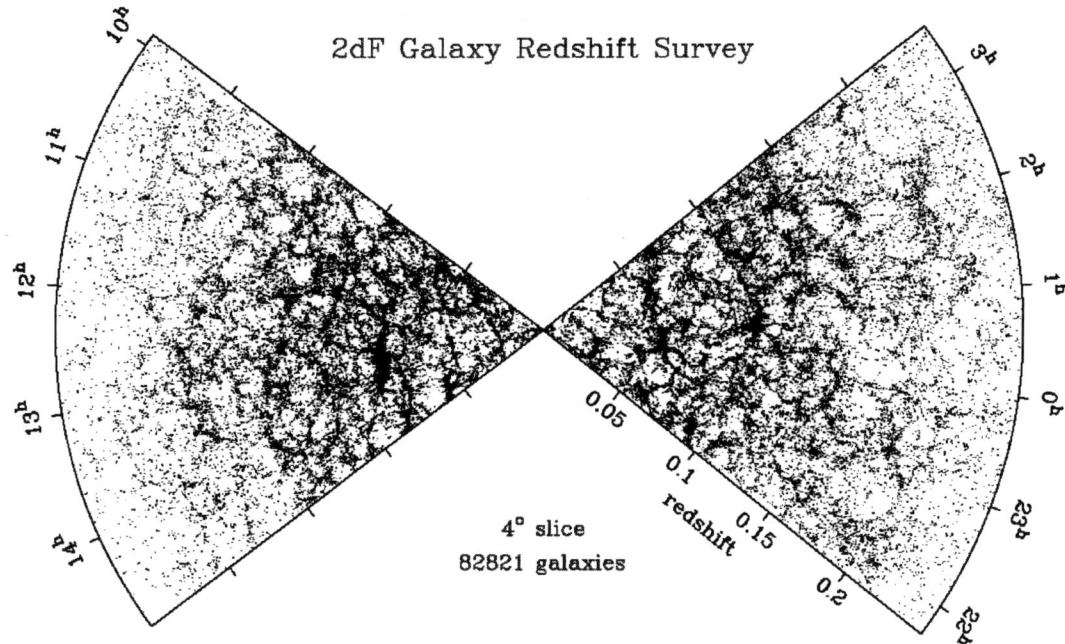

Figure 1: The distribution of galaxies in the 2dFGRS (from [56]). Courtesy of J.A. Peacock and the 2dFGRS Team.

$$\textit{All the observers in the Universe are equivalent.} \qquad (10)$$

We shall recall in the following a few basic points which are essential in arriving at this new principle and make possible the verification of its possible validity.

B. Two-point correlation function

The statistical description of clustering is based on the concept of correlation, namely, more precisely, on the probability of finding an object in the vicinity of another one. The standard way to quantify this probability is to define the two-point correlation function $\xi(\vec{x})$ [63].

Consider a distribution of objects in space described by the number density function $n(\vec{x})$. The probability that an object is found in an infinitesimal volume δV centered around the point $\vec{x}$ is proportional to the volume itself:

$$\delta P \propto \delta V. \qquad (11)$$

In the absence of structure, the joint probability of finding two objects in two different infinitesimal volumes δV_1 and δV_2, centered respectively around $\vec{x}_1$ and $\vec{x}_2$, is given by the product of the two probabilities:

$$\delta P = \delta P_1 \delta P_2 \propto \delta V_1 \delta V_2. \qquad (12)$$

On the other hand, if objects have a tendence to cluster, we will find an excess probability:

$$\delta P \propto \delta V_1 \delta V_2 \cdot (1 + \xi(\vec{x}_1, \vec{x}_2)). \qquad (13)$$

According to the cosmological principle, we don't expect the correlation function to depend either on the position or on the direction, but only on separation beetween volumes: $\xi(\vec{x}_1, \vec{x}_2) = \xi(r_{12})$, where $r_{12} \equiv |\vec{x}_1 - \vec{x}_2|$.

An equivalent definition of the two-point correlation function is the following:

$$\xi(r_{12}) = <\delta(\vec{x}_1)\delta(\vec{x}_2)>, \qquad (14)$$

where $<...>$ denotes averaging over all pairs of points in space separated by a distance r_{12}, and $\delta(\vec{x}) \equiv (n(\vec{x}) - \bar{n})/\bar{n}$.

C. Observed galaxy distribution

Observational data coming from galactic surveys are usually expressed in the form of a correlation function $\xi(\pi, \sigma)$ in redshift space, where π is a separation along the line of sight and σ is a angular separation on the plane of the sky between two galaxies. It is then possible to obtain the real-space correlation function $\xi(r)$; this step is never a trivial one, but we will not go into details since it is beyond the scope of this review.

Peebles [63] has shown that the distribution of galaxies can be described by a two point correlation function with a simple power law form:

$$\xi_g(r) = \left(\frac{r}{r_g}\right)^{-1.77}, \quad r < 10 h^{-1} \text{Mpc}, \tag{15}$$

where h is the present day Hubble parameter measured in units of $100 \frac{km}{s\,Mpc}$. The correlation length r_g determines the typical distance between objects. For galaxies, it has been estimated to be $\simeq 5 h^{-1}$Mpc.

For clusters of galaxies the same power law was found first by Bahcall and Soneira [64] and then Klypin and Kopylov [65]

$$\xi_c(r) = \left(\frac{r}{r_c}\right)^{-1.8}, \quad 5h^{-1} < r < 150 h^{-1} \text{Mpc} \tag{16}$$

with different correlation lengths, namely $r_c \simeq 25 h^{-1}$Mpc. Furthermore, Bahcall and Burgett [66] found a correlation function for superclusters of galaxies with the same power law.

Recent observations support these conclusions. The first results from the Sloan Digital Sky Survey (SDSS) on galaxy clustering [67] for about 30,000 galaxies give a real-space correlation function of

$$\xi_g(r) = \left(\frac{r}{r_0}\right)^{-1.75 \pm 0.03}, \quad 0.1 h^{-1} < r < 16 h^{-1} \text{Mpc}, \tag{17}$$

where $r_0 \simeq 6.1 \pm 0.2 h^{-1}$Mpc. The geometry of samples in SDSS is quite close to the Las Campanas Redshift Survey [68] and the results are very similar, but with much better resolution.

The largest data set today is a 2dF Galaxy Redshift Survey [69] (see fig.1) that consists of approximately 250,000 galaxy redshifts. Their results are:

$$\xi_g = \left(\frac{r}{r_0}\right)^{-1.87} \qquad r_0 \sim (6 \div 10) h^{-1} \text{Mpc},$$

$$\xi_g = \left(\frac{r}{r_0}\right)^{-1.76} \qquad r_0 \sim (3 \div 6) h^{-1} \text{Mpc}.$$

Their measurements are in agreement with previous surveys. However, having much smaller statistical errors they were able to find a slight difference in the power law exponent as well as in the correlation length for distances or redshifts, colors and types of galaxies. The result can be found at [70].

D. Power law clustering and fractals

It is clear that once a correlation function is given, the density of objects around any randomly chosen member of the system is:

$$n(r) \propto 1 + \xi(r). \tag{18}$$

If the correlation function has a power law behaviour with exponent γ then:

$$\xi(r) \propto r^{-\gamma}. \tag{19}$$

As is the case for galaxies and clusters of galaxies, where $\gamma \simeq 1.8$, the number of objects in a given volume scales in a similar way:

$$N(r) \propto r^{3-\gamma}. \tag{20}$$

So for noninteger γ, the number of objects scales with a fractional power of the radius of the volume under consideration. This behaviour is typical of fractal sets.

A fractal is a set in which 'mass' and 'radius' are linked by a fractional power law [71]:

$$M(r) \propto r^{D_F}, \qquad (21)$$

where D_F is the fractional or Hausdorff dimension of the set. So galaxies seem to show, at least up to scales of about 100 Mpc, a fractal distribution with $D_F \simeq 1.2$.

A crucial characteristic of a fractal distribution is the presence of fluctuations at all length scales and consequently the impossibility of defining an average value for the density. In a simple fractal set each observer at a matter point belonging to the set observes the same matter distribution as any other observer belonging to the set. In this sense, the fractal naturally leads to the formulation of the cosmological principle expressed by the statement (10). Three very serious issues, however, arise for the compatibility of a fractal structure with that of a cosmological Friedmann model. (1) There must necessarily be a cutoff in the fractal distribution (see Fig. 2)in order to recover the overrall homogeneity observed at the large scales up to $z = \simeq 10^3$ in the CMB. (2) Such a cutoff must occur homogeneously all over the Universe. (3) The dimension of that cutoff must automatically originate as the characteristic length determined by the microphysical properties of the dark matter composing the Universe. These were the three basic thoughts which motivated the cellular model of the Universe introduced by Ruffini in the eighties [72]-[89]. In this model fractals arise from successive fragmentation of primordial structures, the so called 'elementary cells', formed via gravitational instability in the neutrino component of the matter of the Universe. We shall further expand on this idea in the following paragraphs.

III. GRAVITATIONAL INSTABILITY

Gravitational instability is usually considered to be the basic mechanism of structure formation in the Universe (see for example [15]). It is believed that small inhomogeneities are already present at some initial time in the early Universe. Such small perturbations will grow due to gravitational attraction, because overdense regions will accrete matter from the neighbouring regions, increasing the density contrast.

One of the simplest examples showing the process of gravitational instability is a perfect fluid model. If density distribution in a self-gravitating fluid is slightly nonuniform, i.e. small density perturbations exist, they will tend to grow. When the density contrast is small, the linear approximation can be used. The main advantage of linear theory is that perturbations on different scales evolve independently.

It is the main result of this theory that the growth of perturbations are damped by the Hubble expansion, which leads to a power law behavior for the time dependence of density perturbations. For example in the Einstein-de Sitter model which is believed to describe our Universe after recombination, perturbation amplitudes grow like $(1+z)^{-1}$. Only during the nonlinear stage with large density contrast does the evolution become faster. At the nonlinear stage, however, perturbations grow much faster, leading to the formation of gravitationally bound objects.

The theory of linear density perturbations in a homogeneous medium was first developed by Jeans [90]-[91]. His study was motivated by the intention to explain the mechanism of star formation. We describe this theory below. First, however, its range of validity, i.e. the evolution of cosmological horizon, is discussed.

The linear perturbations in the expanding homogeneous and isotropic Friedman Universe were studied by Lifshitz [92] using a relativistic treatment. Relativistic theory, however, is necessary when the scale of perturbations is greater than the horizon, or when relativistic regimes are attained in matter condensation. In the most interesting cases such as perturbations in dark matter well inside the horizon after the equivalence epoch (when the energy densities of radiation and other components are equal), it is sufficient to consider nonrelativistic theory based on Newtonian gravity. Bonnor [93] (see also [94]) was the first to study evolution of spherically symmetric perturbations in Newtonian cosmology.

The theory of linear density perturbations in the Newtonian treatment is developed in detail in various textbooks, see e.g. [95]-[98].

A. Horizon scale and mass evolution

The Newtonian treatment is only applicable on scales smaller than the horizon scale $\lambda_H = cH^{-1}$. The associated mass scale, defined as the mass contained within a sphere of radius $\lambda_H/2$, where H is the Hubble parameter, is given

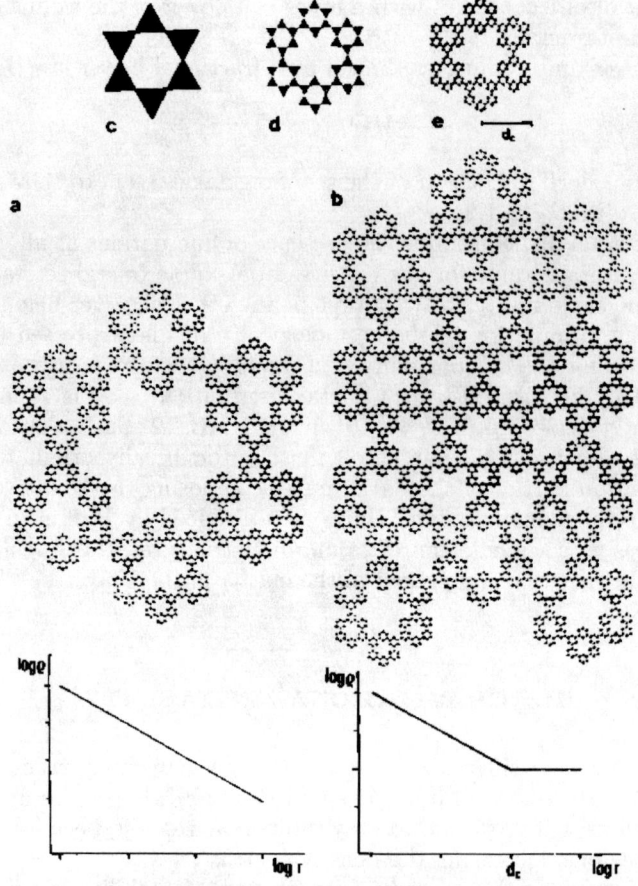

Figure 2: Two fractal structures **a** and **b** of dimension $d \simeq 1.6$ generated by the same algorithm as **c,d** and **e** and endowed with the same lower cutoff, are compared and contrasted. **a** has no upper cutoff, while **b** has an upper cutoff at a distance d_c, shown in **e**. In **a** the density decreases with distance as $\rho \propto r^{-1.6}$, while in **b**, for $r < d_c$ the distribution is self-similar and equivalent to the one in **a** but for $r > d_c$ the distribution becomes homogeneous and ρ stays constant with distance (from [75]).

by

$$M_H = \frac{4}{3}\pi\rho \left(\frac{\lambda_H}{2}\right)^3. \tag{22}$$

Beyond this scale events are causally disconnected and thus any correlation breaks down outside the horizon. Thus structures cannot form on scales larger than λ_H. M_H monotonically increases with time because the distance that light travels increases with time. There are several regimes, separated by the moment of equivalence in energy densities of radiation and nonrelativistic matter:

$$M_H \propto \begin{cases} a^3 & z > z_{eq} \\ a^{3/2} & z < z_{eq}. \end{cases} \tag{23}$$

Today the horizon scale is approximately 3000Mpc, which corresponds to a mass scale $M \sim 10^{22} M_\odot$ for an $\Omega = 1$ Universe. At recombination the total mass inside the horizon was then approximately $(1/z_{rec})^{-3/2} \simeq 10^{17} M_\odot$, where $M_\odot = 2\,10^{30}$ kg is a solar mass.

B. Self-gravitating ideal fluid: linear theory

1. Fluid equations and background solutions

Consider a perfect fluid with density ρ and pressure p in Euclidean space with a Cartesian ("physical") coordinate system r_i[1]. The fluid has a velocity field v_i; the gravitational potential Φ is induced by the mass density ρ distribution. All these quantities are related through the continuity, Euler and Poisson equations respectively. For a nonrelativistic fluid, i.e. for a fluid with $p << \rho c^2$, they read [63] [95] [97] [99]:

$$\frac{\partial \rho}{\partial t} + \partial_i(\rho v_i) = 0, \tag{24}$$

$$\frac{\partial v_i}{\partial t} + v_j \partial_j v_i + \frac{1}{\rho}\partial_i p + \partial_i \Phi = 0, \tag{25}$$

$$\partial^2 \Phi - 4\pi G \rho = 0, \tag{26}$$

where $\partial^2 = \partial_i \partial_i$. A cosmologically important solution of equations (24-26) is the one describing a spatially uniform fluid with zero pressure ('dust') on an expanding background [95] [99]:

$$v_i^0 = H(t) r_i, \tag{27}$$

$$\frac{d\rho_0}{dt} + 3H\rho_0 = 0, \tag{28}$$

$$p_0 = p_0(t), \tag{29}$$

$$\Phi_0 = \frac{2}{3}\pi G \rho_0 r^2, \tag{30}$$

$$\frac{dH}{dt} + H^2 = -\frac{4}{3}\pi G \rho_0, \tag{31}$$

where $r^2 = r_i r_i$ and all quantities depend only on time.

In a comoving coordinate system, with

$$r_\alpha = a(t) x_\alpha, \tag{32}$$

where $a(t)$ is the scale factor, the relation between coordinate differences Δr_i and Δx_α is

$$\frac{d\Delta r_\alpha}{dt} = a\frac{d\Delta x_\alpha}{dt} + \frac{da}{dt}\Delta x_\alpha = a\frac{d\Delta x_\alpha}{dt} + H(t)\Delta r_\alpha, \tag{33}$$

where

$$H(t) = \frac{1}{a}\frac{da}{dt}. \tag{34}$$

Correspondingly, for velocity fields we have:

$$v_\alpha(r_\beta, t) = u_\alpha(x_\beta, t) + H r_\alpha = u_\alpha(x_\beta, t) + v_\alpha^0. \tag{35}$$

Thus the solution (27-31) represents a uniform distribution of the fluid with zero peculiar velocity $u_\alpha^0 = 0$ and zero pressure $p_0 = 0$. Pressure and density are linked through the equation of state $p = p(\rho)$. The three equations (24-26) together with the equation of state are a complete set, allowing one to study the temporal evolution of the density and velocity distributions as well as of the pressure and gravitational potential.

---

[1] Greek indices denote comoving coordinates, Latin indices denote physical coordinates, both take the values 1,2,3; the Einstein summation rule is adopted.

2. Perturbed quantities

As well known, solutions (27-31) represent an isotropic and homogeneous distribution of matter. In order to study density perturbations in the linear approximation assume that

$$\rho(r_i, t) = \rho_0(t)\left[1 + \delta(r_i, t)\right], \tag{36}$$

$$v_i(r_j, t) = v_i^0(r_j, t) + \delta v_i(r_j, t), \tag{37}$$

$$\Phi(r_i, t) = \Phi_0(r_i, t) + \delta\Phi(r_i, t), \tag{38}$$

$$p(r_i, t) = \delta p(r_i, t), \tag{39}$$

where $\delta \equiv \frac{\rho - \rho_0}{\rho_0}$. Here all perturbed quantities δ, δv_i, δp and $\delta\Phi$ are assumed to be much smaller than the background quantities. All zero order values are given by (27-31). Assume also, that spatial together with temporal derivatives of perturbed quantities have the same order of magnitude as the quantities themselves.

Note that it is not necessary for the smallness condition on the perturbed quantities to hold over all of space. In particular, there could be a region in space where $|\delta v_i| > |v_i^0|$ [100]. In this case the standard linearization procedure leads to different perturbation equations and consequently to different solutions representing a time dependent density contrast $\delta(r_i, t)$.

3. Linearized perturbations equations

Rewriting (24-26) in comoving coordinates:

$$\frac{\partial \rho}{\partial t} + 3H\rho + \frac{1}{a}\rho\partial_\alpha u_\alpha + \frac{1}{a}u_\alpha\partial_\alpha\rho = 0, \tag{40}$$

$$\frac{d^2 a}{dt^2}x_\alpha + \frac{\partial u_\alpha}{\partial t} + Hu_\alpha + \frac{1}{a}u_\beta\partial_\beta u_\alpha + \frac{1}{a\rho}\partial_\alpha p + \frac{1}{a}\partial_\alpha\Phi = 0, \tag{41}$$

$$\partial^2\Phi - 4\pi Ga^2\rho = 0. \tag{42}$$

Here all quantities, except for H, depend on the comoving coordinates[2] x_α and the time t. Eqs. (40-42) written in physical coordinates can be found in [100] for example. One arrives at the above results from (24-26) by using the transformation laws $(\partial/\partial t)_{phys} = (\partial/\partial t)_{com} - Hx_\alpha\partial_\alpha$ and $(\partial_\alpha)_{phys} = (1/a)(\partial_\alpha)_{com}$.

In order to obtain equations for the density contrast δ in the linear approximation substitute (36-39) into equations (40-42). Taking into account that spatial as well as temporal derivatives of perturbed quantities have the same order of smallness as the perturbed quantities themselves and using (27-31), the perturbation equations read

$$\frac{\partial \delta}{\partial t} + \frac{1}{a}\partial_\alpha\delta u_\alpha = 0, \tag{43}$$

$$\frac{\partial \delta u_\alpha}{\partial t} + H\delta u_\alpha + \frac{1}{a}\partial_\alpha\delta p + \frac{1}{a}\partial_\alpha\delta\Phi = 0, \tag{44}$$

$$\partial^2\delta\Phi - 4\pi Ga^2\rho_0\delta = 0, \tag{45}$$

where $\delta u_\alpha(x_\beta, t)$ is first order quantity, because the unperturbed value is $u_\alpha^0(x_\beta, t) = 0$.

---

[2] If one also assumes that Hubble parameter can be disturbed (have spatial dependence) then the system of equations becomes overdefined. There is another approach [101], however, where $\partial_\alpha\delta$ and $\partial_\alpha H$ are taken as independent variables in order to study density perturbations.

The simplest way to find the equation governing density perturbations is to take the time derivative of Eq. (43) and use the divergence of Eq. (44) together with Eq. (45). After rather long calculations one finds the final expression:

$$\frac{\partial^2 \delta}{\partial t^2} + 2H\frac{\partial \delta}{\partial t} - \frac{v_s^2}{a^2}\partial^2 \delta - 4\pi G \rho_0 \delta = 0, \qquad (46)$$

where the relation

$$v_s^2 = \frac{dp}{d\rho} \qquad (47)$$

is assumed.

4. The Jeans criterion

Eq. (46) governs the dynamics of density perturbations. It is a wave-like second order partial differential equation. Thus it is natural to introduce the Fourier transform

$$\delta = \sum_k h(t) e^{i k_\alpha x_\alpha} \qquad (48)$$

in order to split perturbations on different scales.

Eq. (46) can be rewritten in k-space, taking into account that $\partial_\alpha \delta \to i k_\alpha h$:

$$\frac{d^2 h}{dt^2} = -2H\frac{dh}{dt} + (4\pi G \rho_0 - \frac{v_s^2 k^2}{a^2})h, \qquad (49)$$

where k_α is the comoving wavevector and $k = \sqrt{k_\alpha k_\alpha}$ is the corresponding wavenumber. The comoving wavelength of the perturbative mode is given by $l = 2\pi/k$, while the proper (physical) wavelength is simply $\lambda = al$.

The Jeans criterion (49) is governed by the wavelength

$$\lambda_J = v_s \sqrt{\frac{\pi}{G\rho_0}}, \qquad (50)$$

where λ_J separates gravitationally stable scales from unstable ones. Fluctuations on scales well above λ_J grow via gravitational instability, while on scales smaller than λ_J the pressure overwhelms gravity and perturbations do not grow.

The first term on the right-hand side of (49) comes from the general expansion. In the static world initially considered by Jeans, such a term is absent, leading to the exponential growth of perturbations. In expanding space perturbations grow with time according to a power law.

A very important quantity usually associated with the Jeans length is the Jeans mass (50)

$$M_J = \frac{4}{3}\pi \rho \left(\frac{\lambda_J}{2}\right)^3, \qquad (51)$$

defined as the mass contained within a sphere of radius $\lambda_J/2$, where ρ is density of the perturbed component.

5. Multi-component system

Perturbations for a given mode in a single component evolve according to (49). When several components such as Cold Dark Matter (CDM), Hot Dark Matter (HDM), baryons and radiation are present simultaneously, it is possible to generalize (46). Assuming gravitational interaction between components only, we arrive at

$$\frac{d^2 h_i}{dt^2} = -2H\frac{dh_i}{dt} + (4\pi G\rho_0 \sum_j \epsilon_j h_j - \frac{(v_s^2)_i k^2}{a^2}h_i), \qquad (52)$$

where the index i refers to the component under consideration, the sum is over all components and $\epsilon_i = \rho_i / \sum_j \rho_j$. Notice that any smoothly distributed component (like the cosmological constant) does not contibute to the right-hand side of (52).

C. Applications

Some important cases of matter content for the Universe will be considered below. First we discuss perturbation dynamics in the dominant nonrelativistic component (baryonic or not). Second example is dark matter perturbations in the presence of a dominant radiation component.

1. Einstein-de Sitter Universe

First consider the dust dominated $\Omega = 1$ Universe. This condition (Ω is the density parameter) corresponds to a flat-type cosmological model, namely the Friedman solution of Einstein's General Relativity equations with curvature parameter $k = 0$. This model is thought to provide a good description of our Universe after recombination. To zero order $a \sim t^{2/3}$, $H = 2/3t$ and $\rho_0 = 1/6\pi G t^2$. For perturbations well inside the horizon we have

$$t^2 \frac{d^2 h}{dt^2} + \frac{4}{3} t \frac{dh}{dt} - \frac{2}{3}\left[1 - \left(\frac{\lambda_J}{\lambda}\right)^2\right] h = 0. \tag{53}$$

For modes well inside the horizon and still larger than the Jeans length the solution is

$$h(k,t) = h_1(k) \left(\frac{t}{t_0}\right)^{2/3} + h_2(k) \left(\frac{t}{t_0}\right)^{-1}. \tag{54}$$

As expected there are two solutions, one growing and one decaying. At late time, however, only the growing mode is important. Perturbations evolve proportionally to the scale factor or as $(1+z)^{-1}$, where z is the redshift defined by:

$$1 + z = \frac{a_0}{a(t)}, \tag{55}$$

and a_0 denotes the value of a scale factor today.

Perturbations on scales smaller than the Jeans length cease to grow and oscillate with time.

2. Mixture of radiation and dark matter

Consider the radiation dominated Universe where $a \sim t^{1/2}$ and $H = 1/2t$. The second component to be considered is a collisionless dark matter with $v_{DM} = 0$. We still can use the Newtonian treatment on scales much smaller than the horizon size.

Since the small scale photon distribution is smooth and the energy density is dominated by the radiation, the equation governing dark matter instability reduces to

$$t \frac{d^2 h_{DM}}{dt^2} + \frac{dh_{DM}}{dt} = 0. \tag{56}$$

This has the solution

$$h_{DM}(k,t) = h_1(k) \log\left(\frac{t}{t_0}\right) + h_2(k). \tag{57}$$

Perturbations in dark matter component inside the horizon experience a slow logarithmic growth. This is the well known Meszaros effect [102].

D. Initial spectrum of perturbations

In the first example we have shown that perturbations on scales between horizon size and Jeans length $\lambda_J \ll \lambda \ll \lambda_H$ grow as $\delta \propto (1+z)^{-1}$. In order to study perturbation dynamics one also needs to know the initial values of the perturbations at some moment in early Universe.

One great possibility is provided by the CBR anisotropy measurements, because density inhomogeneities when photons were coupled to baryons can be extracted from temperature fluctuations observed in the CBR. Since $\frac{\delta T}{T} \simeq 10^{-5}$ it is usually assumed that at the moment of recombination $\delta \simeq 10^{-4}$ [15].

Perturbation amplitudes on different scales are usually represented by a power spectrum $P(k)$, which is the Fourier transform of a previously introduced correlation function [96]

$$\xi(r) = \frac{V}{(2\pi)^3} \int P(k) \frac{\sin kr}{kr} 4\pi k^2 dk. \qquad (58)$$

There is no evidence that the initial spectrum contained any preferred scale, so it should be a featureless power law

$$P(k) \propto k^n, \qquad (59)$$

where the index n governs the balance between perturbation amplitudes on large and small scales. The value $n = 0$ corresponds to white noise that has the same amplitude for every mass scale. The value $n = 1$ corresponds to a so-called Harrison-Zel'dovich scale invariant spectrum. Term 'scale invariance' means that perturbations had the same amplitude at the moment of horizon crossing.

E. Damping of perturbations

In addition to the Jeans scale some other cosmologically important scales appear in the theory of structure formation. They are related to physical processes that cannot be described within the perfect fluid approximation. However, fortunately such processes take place on limited scale intervals and outside such intervals the fluid description is still possible. We will discuss some dissipative effects, such as collisional damping of baryonic perturbations and free streaming of collisionless light particles.

1. Silk damping

Close to recombination the coupling between photons and baryons makes it possible for the former to erase perturbations of the latter. This is because at that time the free mean path of photons becomes larger, so they can travel from overdense into underdense regions dragging baryons with them, thus smoothing inhomogeneities in the primeval plasma. This effect was discovered by Silk [103]. The physical scale associated with it is [15]

$$l_S \simeq 3.5 (\Omega h^2)^{-3/4} \, \text{Mpc}, \qquad (60)$$

which gives a mass scale

$$M_S \simeq 6.2 \, 10^{12} (\Omega h^2)^{-5/4} \, M_\odot. \qquad (61)$$

This scale is close to the mass of a typical galaxy $10^{11} M_\odot$. However, Silk damping only affects baryonic perturbations. Moreover, it is important only around recombination when the coupling is still sufficiently strong to make photons drag baryons with them.

2. Free streaming

Another dissipative process is Landau damping or free streaming that originates from the free motion of collisionless particles on small scales. They can travel far if the velocity dispersion is large. This is important after particles decouple from the plasma and until they become nonrelativistic.

The discovery of free streaming was a dramatic moment for structure formation scenarios based upon Hot Dark Matter (HDM) models [104]. The name Hot Dark Matter means that particles composing such matter were ultrarelativistic at equivalence. Thus their velocity dispersion was near the speed of light c. The maximum distance scale travelled by collisionless particles from decoupling can be estimated to be [97]

$$l_{FS} \simeq 0.5 \left(\frac{m_{DM}}{1\,\text{keV}}\right)^{-4/3} (\Omega_{DM} h^2)^{1/3} \, \text{Mpc}, \qquad (62)$$

and a corresponding mass scale is of the order of superclusters of galaxies or even larger if the particle mass is $m_{DM} < 30\,\text{eV}$.

Cold thermal relics which compose the Cold Dark Matter (CDM) are slow enough so that free streaming can be neglected on cosmologically important scales. Therefore Landau damping affects only light particles like neutrinos with $m_\nu \sim 10\,\text{eV}$, and in the Universe dominated by HDM, all perturbations on scales smaller than superclusters of galaxies are erased. At the same time, after particles become nonrelativistic at z_{nr}, their velocity dispersion becomes small enough to make free streaming negligible.

F. Structure formation at late times

1. Nonlinear clustering

In previous sections we dealt with the linear evolution of cosmological perturbations. From long after recombination and even earlier until almost recent times, such a treatment is justified to describe the growth of inhomogeneity in the Universe because the condition $\delta \ll 1$ is sutisfied. In recent times, say at $z \sim 10$, the nonlinear behaviour of perturbations becomes important. During the nonlinear stage gravitationally bound objects such as galaxies form. Nonlinear evolution is a rapid process, where not only gravitational effects are important. In particular during the formation of galaxies, various dissipation and relaxation processes take place [96].

We will not go into the details of galaxy formation here. The LSS formation is the subject of this section. Here the main interaction remains gravity. However, the theoretical description based on linear equations for an ideal fluid becomes inadequate.

Usually N-body simulations are employed to study nonlinear clustering [96]. However, some useful simplified models are still possible even in the nonlinear stage, because numerical simulations provide limited physical insight into the physics of gravitational clustering. Among nonlinear approximations, the most famous are the Zel'dovich approximation [105] and the spherical collapse model (see for example [63]). The key point in the Zel'dovich model is that during collapse in an almost spherically symmetric overdense region, gravitational interaction amplifies asymmetry. Therefore, the final structure acquires a preferred direction and the final collapsed body will look like a 'pancake'. In the spherical collapse model, on the other hand, it is assumed that spherical symmetry remains valid during the entire period of collapse. It allows the splitting of the spherical overdense region into concentric shells and the evolution of each shell can be studied separately, which sufficiently simplifies the problem. This model will be described in detail in the next chapter.

2. Structure formation scenarios

Historically two different pictures of structure formation were considered, namely the HDM (see [106]) and CDM models (see [107]). We will discuss them briefly below.

3. HDM models

The neutrino dominated Universe with $m_\nu \sim 10\,\text{eV}$ is a typical HDM model. The HDM model is associated with so-called "top-down" scenario, where structures form on large scales first. This is so because the Jeans mass for HDM is of the order of the supercluster mass or even higher. At the same time, free streaming erases perturbations on smaller scales. Thus only when perturbations reach the nonlinear regime on large scales can they induce fragmentation on smaller scales.

Usually it is assumed that large scale perturbations become nonspherical according to the Zel'dovich model and thus the LSS look like a "net" of density condensations separated by huge voids. Simulations agree with such a picture.

The HDM model is in good agreement with observational data on scales larger than 10 Mpc. On smaller scales, however, HDM simulations can agree with observed the correlation function of galaxies only if the epoch of pancaking takes place at $z \simeq 1$ or less, which is too late, because we can see galaxies and quasars with much greater z.

The crucial cosmological property of HDM with neutrinos, as was mentioned above, is the damping of perturbations on small scales due to free streaming. Neutrino dominated models ($\Omega_\nu \sim 1$) alone could not describe the real Universe because on scales smaller than ~ 100 Mpc no structure appears at all.

One important prediction of HDM models with neutrinos is the existence of large smooth halos around galaxies. At the end of collapse, during formation of galaxies, the baryonic component can dissipate its energy via collisions, but the neutrino component cannot. Thus neutrinos remain less condensed than baryons, forming large galactic halos.

4. CDM models

CDM models do not have trouble with free streaming, because their particles have negligible velocities at decoupling. Moreover, the Jeans mass for typical CDM model lays well below $10^6 M_\odot$. Thus, perturbations start to develop on small scales simultaneously with perturbations on large scales.

In the CDM models the important feature is the weak growth experienced by perturbations between horizon crossing and equivalence (see §III C 2). This means that the density contrast increases when we move to smaller scales, or that the perturbations spectrum has more small-scale power.

After collapse the first structures in CDM models virialize through violent relaxation [108] into gravitationally bound objects that form galactic halos. Structures form in a self-similar manner from small to large scales, in other words according to a 'bottom-up' scenario.

Pure CDM models, however, fail to predict the observed correlation function for galaxies on large scales. If one wants to retain the CDM hypothesis, the simplest way is to reduce the matter density. This shifts matter-radiation equivalence to a later epoch, resulting in redistibution of power in the spectrum of perturbations in favour of larger scales.

Today the ΛCDM model with $\Omega_{tot} = 1$ and $\Omega_\Lambda = 0.7$ is considered to be the best fit to the full set of observational data (see E.W. Kolb, in these proceedings [1]).

IV. NEUTRINOS AND STRUCTURE FORMATION

In previous chapter we described the evolution of perturbations and we saw that the nature of dark matter particles is crucial to determining the way structure formation develops. In spite of the fact that a lot of candidates for CDM particles exist, there is presently no experimental evidence of such particles. On the other hand, neutrinos are the only candidates for DM known to exist.

'Light' neutrinos ($m_\nu \ll 1\text{MeV}$) [109], namely neutrinos that decouple while still in their ultrarelativistic regime (see below), may provide a significant contribution to the energy density of the Universe ($\Omega_\nu \sim 1$). Models with light neutrinos were extensively studied in the eighties; a large literature exists on this subject [110].

The key prediction of the cosmological model with neutrinos is a cellular structure on large scales (see Fig. 1). The qualitative drawing of cellular structure of the Universe is represented in Fig. 3. Ruffini and collaborators have studied such models with particular attention to the problem of clustering on large scales and its relation to the fractal distribution of matter. In the following, we will outline some of these ideas.

A. Neutrino decoupling

The cosmological evolution of a gas of particles can be split into two very different regimes. At early times, the particles are in thermal equilibrium with the cosmological plasma; this corresponds to the situation in which the rate $\Gamma = <\sigma v n>$ of the reactions supposed to maintain the equilibrium (such as $\nu_e + \bar{\nu}_e \leftrightarrow e^+ + e^- \leftrightarrow 2\gamma$ in the case of electronic neutrinos) is much greater than the expansion rate, given by the Hubble parameter. The gas then evolves through a sequence of thermodynamic equilibrium states, described by the usual Fermi-Dirac statistics:

$$f(p) = \frac{1}{\exp\left[(E(p) - \mu)/k_B T\right] + 1}, \tag{63}$$

where p, μ and T are the momentum, chemical potential and temperature of neutrinos respectively, and k_B is the Boltzmann constant.

However, as the Universe expands and cools, the collision rate Γ becomes smaller than the expansion rate; this means that the mean free path is greater than the Hubble radius, so we can consider the gas to be expanding without collisions. It is customary to describe the transition beetween the two regimes by saying that the gas has decoupled from the cosmological plasma.

1. The redshifted statistics

Since in a spatially homogeneous and isotropic Universe described by the Robertson-Walker metric, the product of the three-momentum $p(t)$ of a free particle times the scale factor $a(t)$ is a constant of the motion:

$$p(t) \cdot a(t) = \text{const}, \tag{64}$$

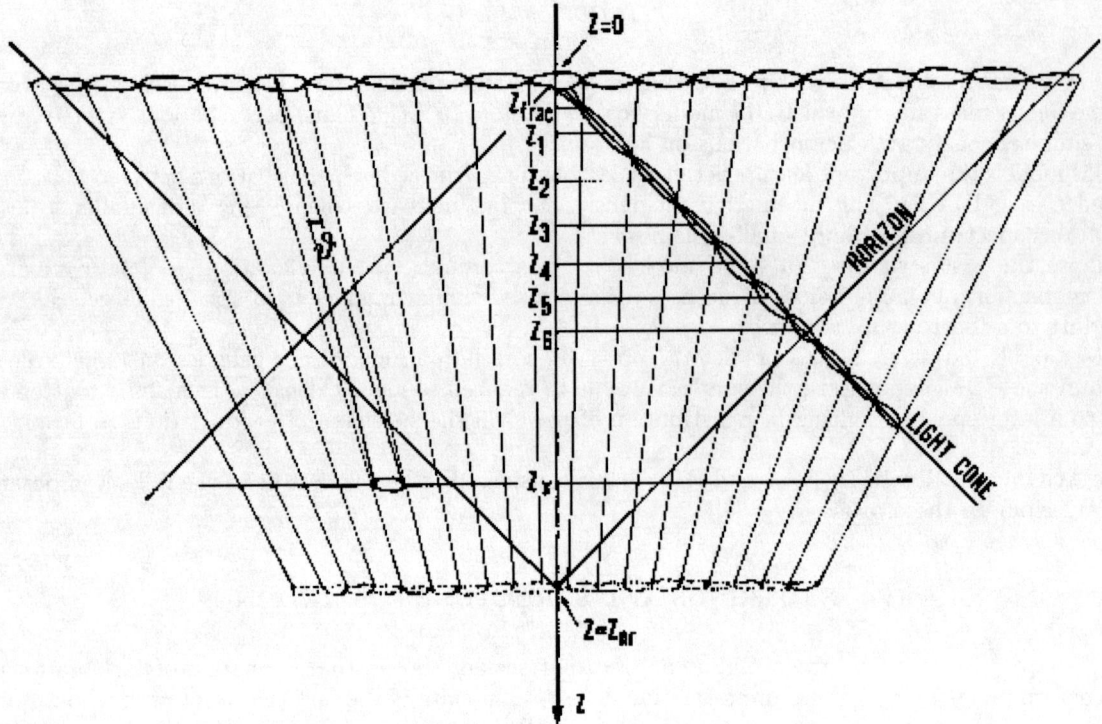

Figure 3: Qualitative drawing of the evolution of the cellular structure of the Universe. The vertical axis represents decreasing values of the redshift. $z = 0$ represents the cosmological observer today, embedded in his own 'elementary cell'; $z_1 z_2, ..., z_n$ are the redshifts of the centers of the successive 'elementary cells' seen by this observer. $z = z_\gamma$ represents the surface of last scattering (decoupling) of the CBR. θ is the angular size subtended today by an 'elementary cell' at the time of decoupling. Finally, $z = z_{nr}$ is the time at which the cells formed via gravitational instability.

each particle in the gas changes its momentum according to this relation. This fact, together with Liouville's theorem, implies that the distribuition function after the decoupling time t_d (defined as the time at which $\Gamma = H$) is given by [72]:

$$f(p, t > t_d) = f\left(\frac{a(t)}{a_d}p, t_d\right) = \frac{1}{\exp\left[(E\left(\frac{a(t)}{a_d}p\right) - \mu_d)/k_B T_d\right] + 1}, \qquad (65)$$

where the subscript d denotes quantities evaluated at the decoupling time.

Now we turn our attention to the special case of neutrinos with $m_\nu \lesssim 10\,\mathrm{eV}$. The ratio Γ/H, as a function of the cosmological temperature, can be evaluated using quantum field theory [15]

$$\frac{\Gamma}{H} \simeq \left(\frac{T}{1\,\mathrm{MeV}}\right)^3 \qquad (66)$$

as long as $T \gg m$. Therefore, neutrinos decouple from the cosmological plasma when $T = T_d \simeq 1\,\mathrm{MeV}$. Since $kT_d \gg mc^2$, many of the particles satisfy $pc \gg mc^2$ and then, when performing the integration over the distribution function (65), we can safely approximate:

$$f(p, t > t_d) = f\left(\frac{a(t)}{a_d}p, t_d\right) \simeq \frac{1}{\exp\left[\left(\frac{a(t)}{a_d}pc - \mu_d\right)/k_B T_d\right] + 1}, \qquad (67)$$

since the tail of the distribution function for which $mc^2 \gg pc$ gives little contribution.

In the following, we compute the mean value of physical quantities over this distribution. It will be useful to consider two limiting regimes, namely the nonrelativistic one and the ultrarelativistic one. They correspond to two approximations for the single particle energy [72]:

$$\begin{aligned} E \simeq mc^2 & \qquad kT \ll mc^2, \qquad \text{NR,} \\ E \simeq pc & \qquad kT \gg mc^2, \qquad \text{UR.} \end{aligned} \qquad (68)$$

We emphasize the fact that this substitution must be performed only in the function to be integrated, and not in the distribution function. The approximation (67) depends only on the fact that the particles are ultrarelativistic at the time of decoupling, and then it is valid even when $kT \ll mc^2$.

Then, with a suitable substitution of variables, all the relevant integrals can be recast into a very simple dimensionless form:

$$I_n(\xi) \equiv \int_0^\infty \frac{y^n dy}{\exp[(y-\xi)]+1}, \qquad (69)$$

where $\xi \equiv \mu_d/kT_d$ is the dimensionless chemical potential or degeneracy parameter. These integrals can be expressed using Riemann zeta and related functions [111].

2. Energy density of neutrinos

The present density parameter of neutrinos can be easily evaluated using the method outlined in the previous section. The energy density is given by:

$$\rho_{\nu+\bar{\nu}}(t_0) = \frac{g}{h_P^3} \int_0^\infty E(p) f(p,t_0) d^3p, \qquad (70)$$

where g is the number of helicity states and h_P is Planc's constant. By normalization with respect to the critical density $\rho_c = 1.054 \, h^2 \cdot 10^4 \, \frac{\text{eV}}{\text{cm}^3}$, we obtain [24] [75]:

$$\Omega_{\nu+\bar{\nu}} h^2 \simeq 1.10 \cdot 10^{-1} \, g \, \frac{m}{10\,\text{eV}} \, A(\xi), \qquad (71)$$

where $A(\xi)$ is defined as follows

$$A(\xi) \equiv \frac{I_2(\xi)+I_2(-\xi)}{2I_2(0)} = \frac{1}{4\eta(3)} \left[\frac{1}{3}|\xi|^3 + 4\eta(2)|\xi| + 4\sum_{k=1}^\infty (-1)^{k+1} \frac{e^{-k|\xi|}}{k^3} \right], \qquad (72)$$

and $\eta(n)$ is the Riemann eta function of index n.

The term $I_2(-\xi)$ appears because we have to take into consideration the presence of antiparticles, for which the relation $\xi_{\bar{\nu}} = -\xi_\nu$ holds. This result follows from the fact that if we consider a reaction such as

$$\nu + \bar{\nu} \longleftrightarrow \ldots \longleftrightarrow \gamma + \gamma, \qquad (73)$$

then since the chemical potentials of the initial and final states have to be equal and the chemical potential of the latter is equal to zero, it follows that $\xi_{\bar{\nu}} = -\xi_\nu$.

3. Recent constraints on the neutrino mass m_ν and degeneracy parameter ξ_ν

We discuss below very briefly recent bounds on the chemical potential and mass of neutrinos [112] that can be used to compute the recent bound on $\Omega_{\nu+\bar{\nu}}$.

a. Neutrino mass A recent laboratory limit on the electron neutrino mass comes from tritium β decay [22]. These data give limits

$$m_{\nu_e} < 2.5 \, \text{eV}. \qquad (74)$$

At the same time, no direct measurements or constraints on muonic and tauonic neutrino masses exist. Moreover, it is still unknown whether neutrinos are Majorana or Dirac particles. Very recent data from neutrinoless double β decay [113] give also lower bound on Majorana mass:

$$(0.05 \leq m_{\nu_{ee}} \leq 0.86) \, \text{eV}. \qquad (75)$$

b. Chemical potential The first constraints on the neutrino degeneracy parameter from cosmological nucleosynthesis were obtained in [114]. It was shown later [16] that a small value of ξ_e coupled with large values of $|\xi_{\mu,\tau}|$ can lead to cosmological nucleosynthesis abundances which are consistent with observations. It is found in particular that

$$0 \leq \xi_e \lesssim 1.5, \qquad (76)$$

with the additional constraint $F(\xi_\mu) + F(\xi_\tau) \approx F(10\xi_e)$, where $F(\xi) \equiv \xi^2 + \xi^4/2\pi^2$. In particular this implies $|\xi_{\mu,\tau}| \lesssim 10\xi_e$.

Recent data both from cosmological nucleosynthesis and CMBR [118] strongly constrain neutrino degeneracy parameters. Orito et al. [115] give surprisingly wide constraints, $\xi_e < 1.4$ and $|\xi_{\mu,\tau}| < 40$. Other papers give essentially stronger constraints using additional assumptions [116]–[118]

$$\begin{aligned}\xi_e &< 0.3 \\ |\xi_{\mu,\tau}| &< 2.6.\end{aligned} \qquad (77)$$

c. Neutrino oscillations When one consider different chemical potentials for all neutrino flavors at the epoch prior to cosmological nucleosynthesis, neutrino oscillations equalize chemical potentials [119] if there is enough time for the relaxation process [120]. On the basis of large mixing angle solution of the solar neutrino problem which is favored by recent data [121], cosmological nucleosynthesis considerations constrain the degeneracy parameters of all neutrino flavors [122]:

$$|\xi| \leq 0.07. \qquad (78)$$

However, the situation when flavor equilibrium is not achieved before cosmological nucleosynthesis is also possible. Thus in the following we consider quite high values of the degeneracy parameter and assume its value positive without loss of generality.

The main result that comes from considering oscillations is that masses of different neutrino species are nearly equal: $m_{\nu_e} \simeq m_{\nu_\mu} \simeq m_{\nu_\tau}$.

One can see that quite high values of the neutrino energy density are still possible if we assume the recent constraints discussed above. In particular, if one consider two Dirac neutrino flavors with equal masses and chemical potentials (ν_e gives very small contribution to Ω_ν), $\xi_\mu = \xi_\tau \lesssim 2.6$ [117] and $m_{\nu_\mu} = m_{\nu_\tau} \leq 2.5 \,\text{eV}$ [22], one gets the upper bound

$$\Omega_{\nu+\bar{\nu}} \leq 0.45. \qquad (79)$$

4. The Jeans mass of neutrinos

In neutrino dominated Universe the first possible structure occurs when these particles become nonrelativistic, since at earlier times free streaming erases all perturbations. At this epoch the cosmological redshift has the value [75]

$$1 + z_{nr} = 1.698 \, 10^4 \left(\frac{m_\nu}{10 eV}\right) A(\xi)^{\frac{1}{2}} B(\xi)^{-\frac{1}{2}}, \qquad (80)$$

where

$$B(\xi) \equiv \frac{I_3(\xi) + I_3(-\xi)}{I_3(0)} = \frac{1}{48\eta(5)} \left[\frac{1}{5}\xi^5 + 8\eta(2)\xi^3 + 48\eta(4)\xi + 48 \sum_{n=1}^{\infty} (-1)^{n+1} \frac{e^{-n\xi}}{n^5}\right]. \qquad (81)$$

The basic mechanism of fragmentation of the initial inhomogeneities in an expanding Universe is the Jeans instability described in the previous section. However, in the calculation of the Jeans length of nonrelativistic collisionless neutrinos, we cannot use the velocity of sound obtained from the classical formula (47). In fact, since the particles are collisionless, their effective pressure is zero and this would lead to a vanishing Jeans length, meaning that even the smallest perturbation would be unstable. This is not the case since in the absence of pressure, another mechanism works against gravitational collapse, namely the free streaming of particles (see §III E 2). The characteristic velocity associated with this process is simply the dispersion velocity $\sqrt{<v^2>/3}$, where the factor 3 comes from averaging over spatial directions. Thus, we have to make the substitution $v_s^2 \to <v^2>/3$ [24]. The correct expression for $<v^2>$ can be obtained using the method described above:

$$<v^2> = \begin{cases} c^2 & z > z_{nr}, \\ 12\frac{\eta_R(5)}{\eta_R(3)} \left(\frac{kT_{\nu 0}}{m_\nu}\right)^2 \frac{B(\xi)}{A(\xi)} & z < z_{nr}, \end{cases} \qquad (82)$$

where $T_{\nu 0} = 1.97\,\text{K}$ is the present temperature of neutrinos.

As a result, the Jeans mass grows in the UR regime and decreases in the NR regime [123]. The evolution of the Jeans mass of neutrinos for $m_\nu = 2.5\,\text{eV}$ and $\xi = 2.5$ with redshift z is described by Fig. 4. It is clear that for such

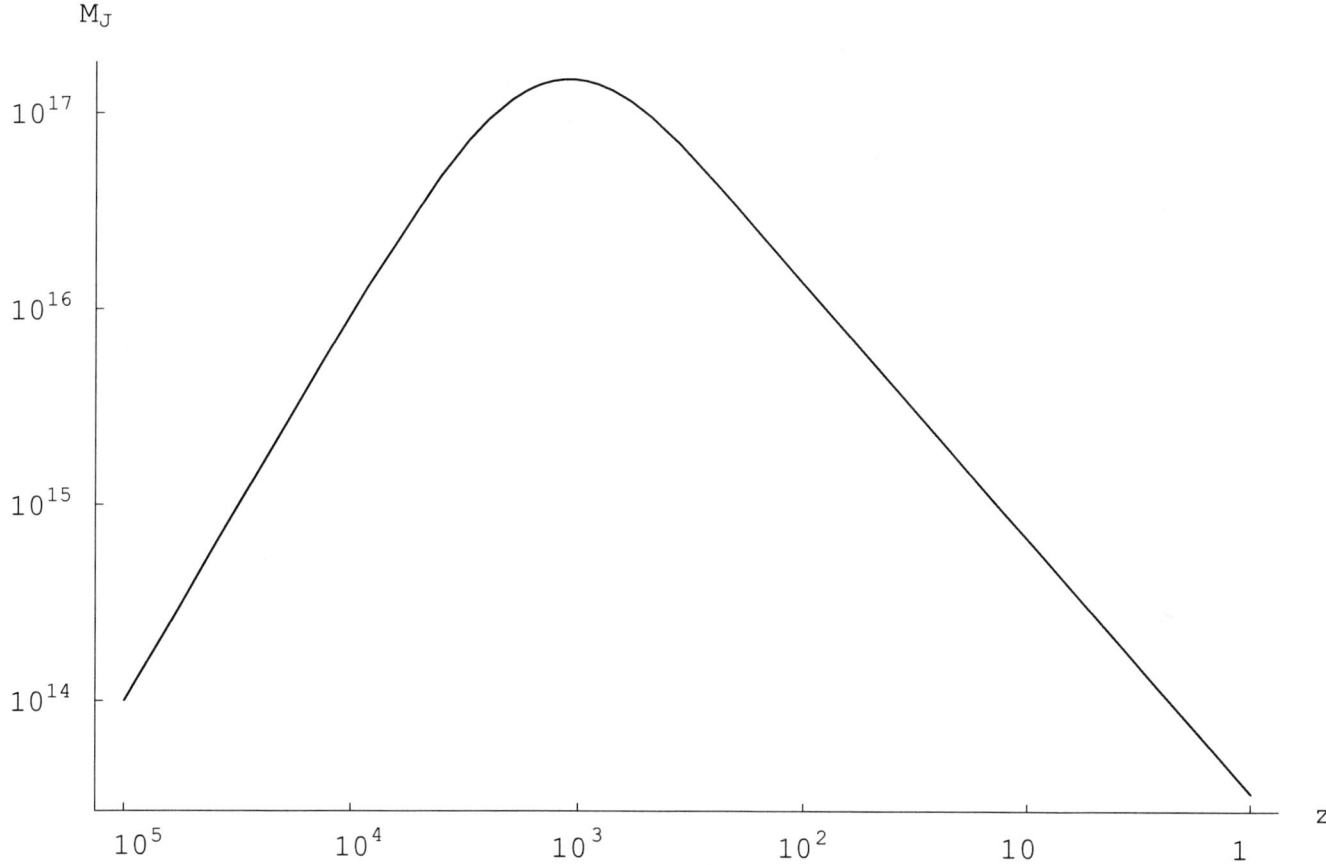

Figure 4: The Jeans mass dependence on redshift for neutrinos with mass $m_\nu = 2.5\text{eV}$ and degeneracy parameter $\xi = 2.5$.

values of the neutrino mass the peak of the Jeans mass lies above $10^{17}\,M_\odot$ and the corresponding comoving Jeans length is $\lambda_0 > 100\,\text{Mpc}$. On the other hand, the value of the Jeans mass today is still larger than the mass of massive galaxy $10^{12}\,M_\odot$.

Finally, the maximum value of Jeans mass at the moment (80) is [24]

$$M_J(z_{nr}) = 1.475\,10^{17}\,M_\odot g_\nu^{-\frac{1}{2}} N_\nu^{-\frac{1}{2}} \left(\frac{m_\nu}{10eV}\right)^{-2} A(\xi)^{-\frac{5}{4}} B(\xi)^{\frac{3}{4}}. \tag{83}$$

The peak of the Jeans mass as a function of the degeneracy parameter for different fixed values of the energy density as well as with constant mass $m_\nu = 2.5\,\text{eV}$ is shown at Fig.5.

By comparing different curves with a fixed value of ξ one can find the well known result that the Jeans mass increases with decreasing of neutrino mass. With the growth of degeneracy parameter, however, the neutrino mass decreases in the beginning, and its different values correspond to different points at the same curve.

The space above the dashed line at Fig. 5 represents the region in which the neutrino mass is less than $2.5\,\text{eV}$. It iss interesting to note that this value of m_ν is still sufficient to get $\Omega_\nu = 1$ with $\xi \approx 4$.

B. Subsequent fragmentation model

In this section we will describe a model of structure formation which explains the observed fractal distribution matter. The key point of this model is the existence of upper and lower cutoffs in the fractal. The upper cutoff appears due to causality of the elementary cell, and the lower cutoff corresponds to the time when the dark matter ceases to dominate in the formation of the structures [62].

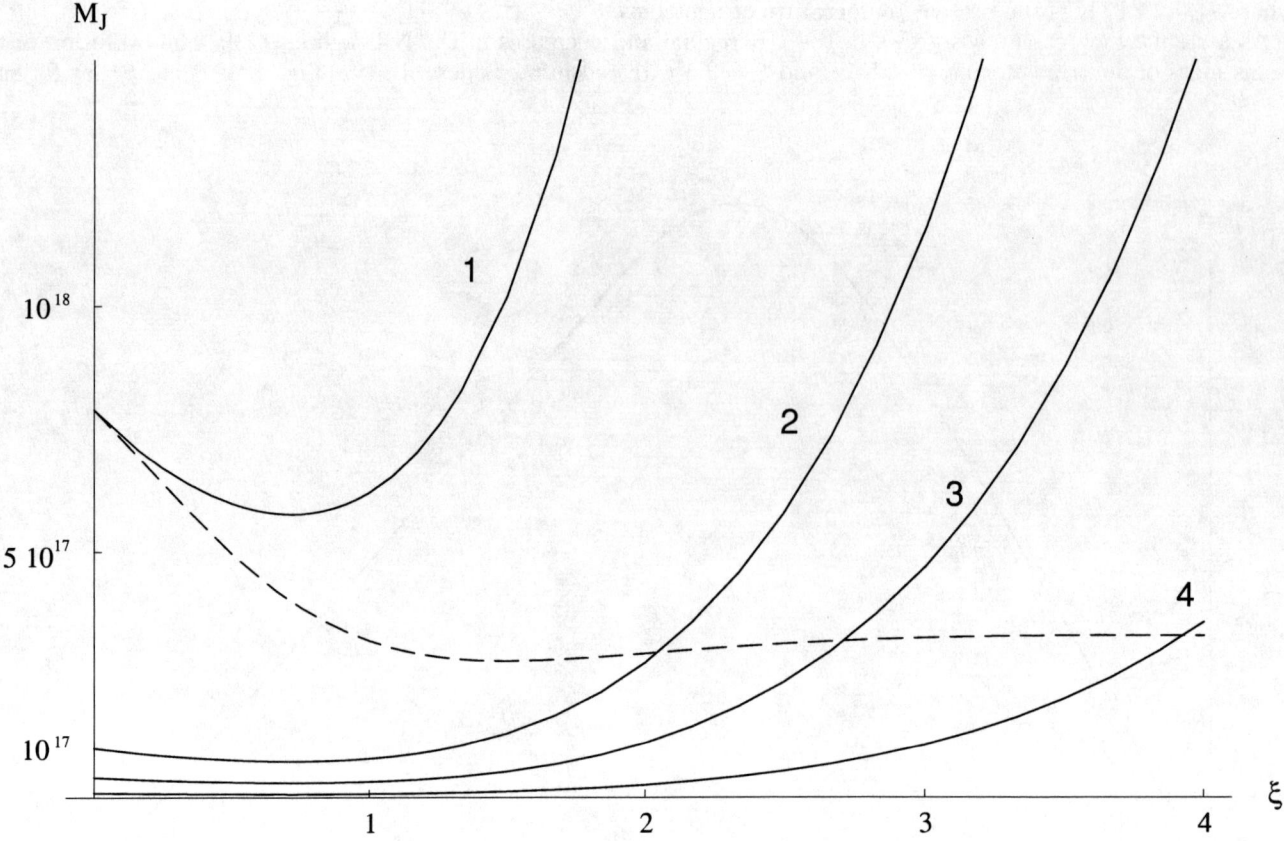

Figure 5: The Jeans mass dependence on the degeneracy parameter with a fixed value of energy density, curves (1–4). Curve (1) corresponds to energy density $\Omega_\nu = 0.11$. Curve (2) corresponds to $\Omega_\nu = 0.3$. Curve (3) represents the neutrino energy density $\Omega_\nu = 0.5$ and finally curve (4) gives Jeans mass for $\Omega_\nu = 1$. The dashed line represents Jeans mass dependence on the degeneracy parameter with fixed neutrino mass $m_\nu = 2.5\,\text{eV}$.

1. Nonlinear model of spherical collapse

Following Ruffini et al. [81] we consider the first spherical perturbation in the Friedmann Universe. For simplicity we neglect the interaction between neighboring perturbations, and treat a given elementary cell formed at the epoch z_{nr} as a spherically symmetric region. If the mass of an elementary cell is sufficiently larger than the Jeans mass, the mass density will dominate the pressure $p \ll \rho$ and then the sphere will expand freely from the expansion of the Friedmann background Universe. Then the equations governing the dynamical evolution of this shell can be obtained by solving the Einstein field equations in the Friedmann Universe in the metric

$$ds^2 = -dt^2 + e^{2\Lambda(\chi,t)}d\chi^2 + r^2(\chi,t)(d\theta^2 + sin^2\theta d\phi^2), \tag{84}$$

where χ is a comoving radial coordinate and $r(\chi, t)$ is the radius of a 2-sphere. This metric is very similar to the Robertson-Walker one, because the spherical overdense (underdense) region behaves like a closed (open) sub-universe.

The solution of the Einstein equations with the above metric gives an equation for the total energy of a test particle:

$$\dot{r}_1^2 - \frac{2Gm_1\chi}{r_1} + \epsilon_1(\chi) = 0, \tag{85}$$

where $m_1(\chi)$ and $\epsilon_1(\chi)$ are respectively the gravitational mass and the negative total energy of the test particle at radius r_1 of a 2-sphere representing the first perturbed region.

We assume that $\epsilon_1(\chi)$ is constant in time since there is no dissipation and that at the moment of separation from the expansion of the background the expansion rates are the same $(\dot{a}/a)_1 = (\dot{r}/r)_1$. If the mean mass density of the sphere is defined by

$$\bar{\rho}_1(t) \equiv \frac{m_1(\chi)}{\frac{4\pi}{3} r_1^3(\chi, t)}, \tag{86}$$

then we can write the energy $\epsilon_1(\chi)$ as

$$\epsilon_1(\chi) = \frac{8\pi}{3} G \rho_U(t_1) R_1^2 \bar{\delta}_1 = (H_1 R_1)^2 \bar{\delta}_1. \tag{87}$$

Here we have assumed a Friedmann Universe with $k = 0$ as a background. $R_1 = r(\chi, t_1)$, $\bar{\rho}_1(t_1)$, and $\rho_U(t_1)$ are respectively the radius of the sphere, its mean density, and the mean density of the Universe at the epoch of separation $t = t_1$. Furthermore, $\bar{\delta}_1 \equiv \bar{\rho}_1(t_1)/\rho_U(t_1) - 1$ and H_1 is the Hubble parameter at $z = z_1$. Eq. (85) can be solved in parametric form: $r_1 = r_1(\Theta_1)$, $t = t(\Theta_1)$, where the parameter Θ_1 is a conformal time which has an initial value θ_1 related to the initial value of the density contrast $\bar{\delta}_1$ by

$$\theta_1 = \frac{1}{\sqrt{\epsilon}} \arccos\left(\frac{1 + \bar{\delta}_1 - 2\epsilon\bar{\delta}_1}{1 + \bar{\delta}_1}\right), \tag{88}$$

where $\epsilon = -1, +1$ corresponds to underdense and overdense regions respectively, being the analog of the curvature parameter k in the usual Robertson-Walker metric. Using the relation between cosmic time t and redshift z in a flat ($k = 0$) Friedmann Universe, we arrive at the following expression for z as a function of Θ:

$$1 + z = (1 + z_1) \left[1 + \frac{3}{4}\left(\frac{1 + \bar{\delta}_1}{\bar{\delta}_1^{3/2}}\right)(\Theta_1 - \sin\Theta_1 - \theta_1 + \sin\theta_1)\right]^{-2/3}. \tag{89}$$

At the same time, an expression for $\bar{\delta}_1$ can be obtained:

$$\bar{\delta}_1(t) = 8 \left(\frac{1 + z_1}{1 + z}\right)^3 \frac{\bar{\delta}_1^3}{(1 + \bar{\delta}_1)^2} (1 - \cos\Theta_1)^{-3} - 1. \tag{90}$$

This set of equations determines the evolution of the elementary cell.

2. Successive fragmentation

After the separation of the elementary cell from the Hubble flow, the Jeans mass will continue to fall and perturbations of smaller masses will be able to detach themselves from the expansion either of the cosmological background or of the larger parent elementary cell. The perturbed regions in the background will follow the evolution given above, but the perturbed regions in the parent elementary cell clearly follow a different evolution.

Here we discuss the simplified case of a spherically symmetric pertubation separating from the expansion flow of its parent cell. We can proceed exactly as we did for the initial elementary cell perturbations, remembering, however, that now the perturbation detaches itself from the still expanding parent cell and not from the Hubble flow. Thus the density and evolutionary state of the parent cell, and not the one of the Friedmann background, will enter into our calculation of the evolution of successive perturbations.

For the perturbation of n we thus have

$$\bar{\delta}_n(t) = 8 \left(\frac{1 + z_n}{1 + z}\right)^3 \frac{(\bar{\delta}_n + \Delta_n)^3}{(1 + \bar{\delta}_n)^2} (1 - \cos\Theta_n)^{-3} - 1, \tag{91}$$

where Δ_n is a parameter in which the whole history of the previous fragmentation is summarized. It has the expression

$$\Delta_n = 1 - 4 \left(\frac{1 + z_{n-1}}{1 + z_n}\right)^3 \frac{(\bar{\delta}_{n-1} + \Delta_{n-1})^3}{(1 + \bar{\delta}_{n-1})^2} \frac{\sin^2 \Theta_{n-1}^n}{(1 - \cos\Theta_{n-1}^n)^4}, \tag{92}$$

where Θ_{n-1}^n is the value of conformal time when the n-th fragmentation appears inside the $(n-1)$-th one, and $\Delta_1 = 0$ for the first fragmentation.

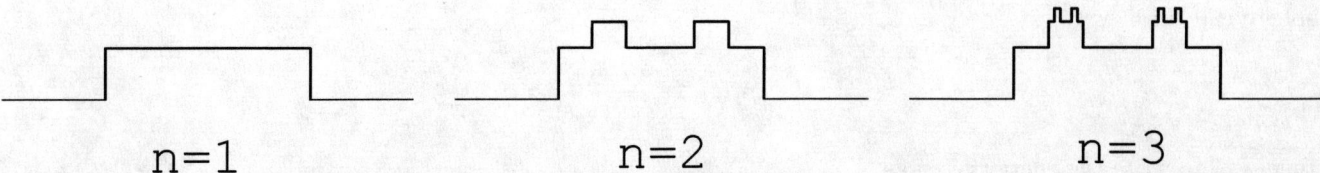

Figure 6: Qualitative illustration of subsequent fragmentation process in the spherical model. The final picture obtained from such a mechanism looks like a typical fractal.

3. The fractal model

To explain in a simple way the mechanism of this model, we start with a simple twofold scenario in which each condensation gives birth to two daughter condensations as soon as the value of the Jeans mass of the parent cell drops to half of the initial value. In short, we take a condensation of mass M_J, then reach the redshift at which the Jeans mass has become $M_J/2$. At that epoch inside the initial condensation two new daughter condensations originate, each of mass $M_J/2$. As we continue this process, we will have four granddaughter condensations, then eight, sixteen and so on. This process can easily be generalized to the occurence of N fragments at each step.

Since the mass of the initial condensation as well as the behaviour of the Jeans mass and length are given, the only free parameter at each successive step is the amplitude of the perturbation $\bar{\delta}_n$. Our goal is to reproduce at the end of the fragmentation process the expected fractal distribution and, therefore, to select at each step the suitable perturbation $\bar{\delta}_n$ for this purpose. It has been shown [71] that such a process leads to a system with fractal dimension D_F given by

$$D_F = \frac{\log N}{\log \lambda}, \tag{93}$$

where $\lambda = r_{i-1}/r_i$ is a constant. In our case we assume $D_F = 1.2$ [77].

In the spherical model decribed above the density need not be uniformly distributed inside the radius r_n: any spherically symmetric perturbation will clearly evolve at a given radius r_n in the same way as a uniform sphere containing the same mass. We assume then without loss of generality that

$$\delta_n(r) = \begin{cases} \bar{\delta}_n, & r < r_n, \\ 0, & r > r_n. \end{cases} \tag{94}$$

The process of fragmentation is shown qualitatively in Fig. 6. This picture is reminiscent of a fractal. Note that the position of each new fragment inside previous one is not important. Moreover, predictions of the model are quite insensitive to the number N. The key point is that even for a random number of fragments appearing at each step the resulting density distribution is still a fractal. One difficulty with this model is that it predicts too large a density contrast $\bar{\delta}_0$ today at the galactic scale.

In order to avoid such high values of the density contrast the authors introduced a suitable lagging time factor τ. This factor is a function of N that introduces a time delay in the formation of each daughter condensation. It was shown that a valid phenomenological relation leading to realistic values of density contrast today is

$$\tau(N) = \frac{N^2}{4} + \frac{3}{4}N. \tag{95}$$

Clearly, the existence of the lagging factor τ is not in contradiction with the Jeans instability picture and only means that the fragmentation occurs somewhat later than the time at which the necessary condition is fulfilled.

In the model under consideration the only free parameters are the initial density contrasts $\bar{\delta}_n$ at every step. These are chosen in such a way that Eq. (93) is satisfied at each step. Thus the natural question arises, what is the form of the initial spectrum? It can be obtained by following backward in time the evolution of perturbations. The result is quite surprising and simple. The spectral index of initial spectrum at z_{nr} (see (59)) is $n = 0$, which corresponds to white noise. Note also that agreement with the observed CBR anisotropy can be obtained within the framework of this model [85].

---

[1] E.W. Kolb, *Proceedings of the X Brazilian School of Cosmology and Gravitation*, Rio de Janeiro (2003).
[2] C.L. Bennett et al., astro–ph/0302207.
[3] C.L. Bennett et al., astro–ph/0302208.
[4] D.N. Spergel et al., astro–ph/0302209.
[5] A. Kogut et al., astro–ph/0302213.
[6] L. Page et al., astro–ph/0302214.
[7] C. Barnes et al., astro–ph/0302215.
[8] G. Hinshaw et al., astro–ph/0302217.
[9] L. Verde et al., astro–ph/0302218.
[10] L. Page et al., astro–ph/0302220.
[11] G. Hinshaw et al., astro–ph/0302222.
[12] E. Komatsu et al., astro–ph/0302223.
[13] N. Jarosik et al., astro–ph/0302224.
[14] H.V. Peiris et al., astro–ph/0302225.
[15] E.W. Kolb and M.S. Turner, *The Early Universe*, Addison-Wesley, (1990)
[16] A. Bianconi, H.W. Lee and R. Ruffini, *Astron. Astrophys.* **241** (1991) 343.
[17] V.F. Shvartsmann, *JETP Lett.*, **9**. (1969), 184.
[18] A.M. Boesgaard e G. Steigmann, *Ann. Rev. Astron. Astrophys.*, **23**, (1985) 319.
[19] P. de Bernardis et al., *Astrophys.J.* **564** (2002), 559.
[20] C.B. Netterfield et al., *Astrophys.J.* **571** (2002), 604.
[21] R. Sancisi and T.S. van Albada, in J. Kormendy and G.R. Knapp, *Dark Matter in The Universe*, (Reidel, Dordrecht, 1987).
[22] C.V. Weinheimer et al. *Phys.Lett.* B **460** (1999), 219;
Lobashev V.M. et al., ibid, 227.
[23] S.S. Gerstein and Ya.B. Zel'dovich, *Pis'ma ZhETF* **4** (1966) 174; *JETP Letters* **4** (1966) 120.
[24] R. Ruffini, D.J. Song, *Gamow Cosmology: Enrico Fermi Course* 86 (1986) 370.
[25] M. Lattanzi and R. Ruffini, to appear in *Proceedings of XI ICRA Network Workshop: Fermi and Astrophysics*.
[26] S. Tremaine and J.E. Gunn, *Phys. Rev. Lett.* **42** 407 (1979)
[27] B.W. Lee and S. Weinberg, *Phys. Rev. Lett.* **39** 165 (1977)
[28] J.G. Gao and R. Ruffini, *Phys. Lett.* **97B** 388 (1980)
[29] M.V. Arbolino and R. Ruffini, *Astron. Astrophys.* **192** (1988) 107.
[30] M. Merafina and R. Ruffini, *Astron. Astrophys.* **221** (1989) 4.
[31] M. Merafina and R. Ruffini, *Astron. Astrophys.* **227** (1990) 415.
[32] G. Ingrosso, M. Merafina, R. Ruffini and F. Strafella *Astron. Astrophys.* **258** (1992) 223.
[33] A. Einstein, in *Kosmologishe Betrachtungen zur allgemeinen Relativitätstheorie*, Sitzungsberichte der Preussischen Akad. d. Wissenschaften (1917)
[34] A. Friedmann, *Z. Phys.* **10**, (1922) 377.
[35] A. Friedmann, *Z. Phys.* **21**, (1924) 326.
[36] E. Hubble, *Proc. Natl. Acad. Sci* **15**, (1929) 169.
[37] H. Weyl, *Space Time Matter*, Dover Publications, New York (1952).
[38] E. Lemaître, *Ann. Soc. Sci. Bruxelles* I A, **47**, (1927) 49
[39] E. Lemaître, *Nature*, **127**, (1931) 706.
[40] E. Lemaître, *Nature*, **128**, (1931) 704.
[41] G. Gamow, *Phys. Rev.* **70**, (1946) 572.
[42] R.A. Alpher, H. Bethe and G. Gamow, *Phys. Rev.* **73**, (1948) 803.
[43] G. Gamow, *Phys. Rev.* **74**, 505 (1948).
[44] R.A. Alpher and R.C. Herman, *Nature* **162** (1948) 774.
[45] R.A. Alpher, R.C. Herman and G. Gamow, *Phys. Rev.* **74**, (1948) 1198.
[46] R.A. Alpher, R.C. Herman and G. Gamow, *Phys. Rev.* **75**, (1949) 332A.
[47] R.A. Alpher, R.C. Herman and G. Gamow, *Phys. Rev.* **74**, (1949) 701.
[48] G. Gamow, *Rev. Mod. Phys.* **21**, (1949) 367.
[49] R.A. Alpher, *Phys. Rev.* **74** (1948) 1577.
[50] R.A. Alpher and R.C. Herman, *Phys. Rev.* **75** (1949) 1089.
[51] R.A. Alpher and R.C. Herman, *Rev. Mod. Phys* **22** (1950) 153.
[52] A.A. Penzias e R.W. Wilson, *Astrophys. J.*, **142**, (1965) 419.
[53] A.C. Fabian and X. Barcons, *Ann.Rev.Astron.Astrophys.* **30**, (1992) 429.

[54] A.S. Webster, in *Radio Astronomy and Cosmology, IAU Symp. No. 74*, ed. Jauncey, D.L., Reidel, Dordrecht.
[55] W.S. Paciesas et al., *Astrophys. J. Suppl. Series*, **122**, (1999) 465.
[56] J.A. Peacock et al., to appear in T. Shanks and N. Metcalfe, *A New Era in Cosmology*, (ASP Conference Proceedings, 2002).
Avalaible on astro-ph/0204239.
[57] P.H. Coleman and L. Pietronero, *Phys. Reports*, **213**, (1992) 311.
[58] X. Luo and D.N. Schramm, *Science*, **256**, (1992) 513.
[59] E.M. de Gouveia dal Pino, et al., *Astrophys.J.*, **442**, (1995) L45.
[60] R. Durrer and S. Labini, *Astron.Astrophys.*, **339**, (1998) L85.
[61] J. Gaite, et al., *Astrophys.J.*, **552**, (1999) L5.
[62] R. Ruffini, in V.G. Gurzadyan and R. Ruffini, *The Chaotic Universe - Proceedings of the Second ICRA Network Workshop*, 656 (2000).
[63] P.J.E. Peebles, *The Large Scale Structure of the Universe*, Princeton Univ. Press, Princeton, (1980)
[64] N.A. Bahcall and R.M. Soniera, *Astrophys.J.*, **270**, (1983) 20.
[65] A. Klypin and A.I. Kopylov, *Sov.Astron.Letters*, **9**, (1983) 41.
[66] N.A. Bahcall and W.S. Burgett, *Astrophys.J.*, **300**, (1986) L35
[67] I. Zehavi et al, *Astrophys.J.*, **571**, (2002) 172.
[68] S.A. Shektman et al, *Astrophys.J.*, **470**, (1996) 172.
[69] J.A. Peacock et al, *Nature*, **410**, (2001) 169.
[70] P. Norberg et al, *MNRAS*, **328**, (2001) 64.
[71] B.B. Mandelbrot, *The fractal geometry of nature*, Freeman and co., New York, (1983)
[72] R. Ruffini, D.J. Song and L. Stella, *Astron. Astrophys.* **125** (1983) 265.
[73] R. Ruffini and D.J. Song, *Astron. Astrophys.*, **179**, (1987) 3.
[74] R. Ruffini and D.J. Song, in *Proceedings of the LXXXVI Course*, International School of Varenna (1987).
[75] R. Ruffini, D.J. Song and S. Taraglio, *Astron. Astrophys.*, **190**, (1988) 1.
[76] D. Calzetti, J. Einasto, M. Giavalisco, R. Ruffini and E. Saar, *Astrophys. Space Science* **137**, (1987) 101.
[77] D. Calzetti, M. Giavalisco and R. Ruffini, *Astron. Astrophys.* **198**, (1988) 1.
[78] R. Ruffini, D.J. Song and W.R. Stoeger, *Il Nuovo Cimento B*, **102**, (1988) 159.
[79] D. Calzetti, M. Giavalisco and R. Ruffini, *Astron. Astrophys.* **226**, (1989) 1.
[80] R. Fabbri and R. Ruffini, *Astron. Astrophys.* **228**, (1990) 1.
[81] R. Ruffini, D.J. Song and S. Taraglio, *Astron. Astrophys.*, **232**, (1990) 7.
[82] Long Long Fengm Hou Jun Mo and R. Ruffini, *Astron. Astrophys.* **243**, (1991) 283.
[83] D. Calzetti, M. Giavalisco, R. Ruffini, S. Taraglio and N.A. Bahcall, *Astron. Astrophys.* **245**, (1991) 1.
[84] D. Calzetti, M. Giavalisco, R. Ruffini and G. Wiedenmann, *Astron. Astrophys.* **251**, (1991) 385.
[85] R. Fabbri and R. Ruffini, *Astron. Astrophys.* **254**, (1992) 7.
[86] D. Calzetti, Yi-Peng Jing and R. Ruffini, *Astron. Astrophys.* **247**, (1991) 1.
[87] S. Torres, R. Fabbri and R. Ruffini, *Astron. Astrophys.* **287**, (1994) 15.
[88] M. Capalbi, S. Filippi, J.G. Gao, R. Ruffini and L.A. Sanchez, in *Proc. Seventh Marcel Grossman Meeting*, Stanford, USA (1994)
[89] C. Sigismondi, S. Filippi, R. Ruffini and L.A. Sanchez, *International Journal of Modern Physics D*, **10**, (2001) 663.
[90] J.H. Jeans, *Philosophical Transactions* **199A** (1902) 49
[91] J.H. Jeans, *Asntronomy and Cosmology*, Cambridge, (1929)
[92] E.M. Lifshitz, *Journal of Physics, (USSR)* **10** (1946) 116
[93] W.B. Bonnor, *Mon. Not. Roy. Astron. Soc.* **117** (1957) 104
[94] D.J. Heath, *Astrophys. and Sp. Sci.* **175** (1991) 35
[95] S. Weinberg, *Gravitation and Cosmology*, Wiley, New York, (1972)
[96] J.A. Peacock, *Cosmological Physics*, Cambridge Univ. Press, Cambridge, (1999)
[97] T. Padmanabhan, *Structure Formation in the Universe*, Cambridge Univ. Press, Cambridge, (1993)
[98] Ya.B.Zeldovich and I.D. Novikov, *The Structure and Evolution of the Universe*, Nauka, Moscow, 1975, (Univ. of Chicago Press, (1983))
[99] A. Raychaudhuri, *Zeitschrift fur Astrophysik*, **43**, (1957) 161.
[100] A. Meszaros, *Astron. Astrophys.* **278** (1993) 1.
[101] G.F.R. Ellis and M. Bruni, *Phys. Rev.* **40D** (1989) 1804; G.F.R. Ellis, *MNRAS* **243** (1990) 509
[102] P. Meszaros, *Astron. Astrophys.* **37** (1974) 225.
[103] J. Silk, *Nature* **215**, (1967) 1155.
[104] J.R. Bond and A.S. Szalay, *Ap. J.* **274** (1983) 443.
[105] Ya.B. Zel'dovich, *Astron.Astrophys.* **5**, (1970) 84.
[106] J.R. Peacock, astro-ph/0007165
[107] J.R. Peacock, astro-ph/0205391
[108] D. Lynden-Bell, *MNRAS* **136**, (1967) 101; F.S. Shu *Astrophys.J.* **225**, (1978) 83.
[109] A.D. Dolgov, *Phys. Reports* **370**, (2002) 333.
[110] Ya.B. Zel'dovich, R.A. Syunayaev *Pis'ma Astron. Zh.* **6** (1980) 451; A.G. Doroshkevich and M.Yu. Khlopov, *Astron. Zh.* **58** (1981) 913;

P.J.E. Peebles, *Ap. J.* **258** (1982) 415.

[111] L. D. Landau and E. M. Lifshitz, *The Classical Theory of Fields*, Oxford, Pergamon Press (1975).

[112] R. Ruffini, M. Lattanzi, C. Sigismondi, G. Vereshchagin, accepted by *Spacetime and Substance*.

[113] S. Pakvasa and P. Roy, hep-ph/0203188;
H. Klapdor-Kleingrothaus and U. Sarkar, *Mod.Phys.Lett.* A **16** (2001), 2449.

[114] A.G. Doroshkevish, I.D. Novikov, R.A. Siunaiev, Y.B.Zel'dovich, in *Highlights of Astronomy,* de Jader ed., (1971) 318;
W.A.Fowler, *The astrophysical aspects of the weak interactions*, Academia Nazionale dei Lincei, Roma, **157** (1971) 115;
G. Beaudet and P. Goret, *Astron. Astrophys.* **49** (1976) 415.

[115] M. Orito, T. Kajino, G.J. Matthews and Y. Wang, astro-ph/0005446.

[116] J.P. Kneller, R.J. Scherrer, G. Steigman and T.P. Walker, *Phys.Rev.* D **64** (2002) 123506.

[117] M. Orito, T. Kajino, G.J. Matthews and Y. Wang, *Phys.Rev.* D **65** (2002) 123504.

[118] S.H. Hansen, G. Mangano, A. Melchiorri, G. Miele and O. Pisanti, *Phys.Rev.* D **65** (2002) 023511.

[119] M. J. Savage, R.A. Malaney and G.M. Fuller, *Ap. J.* **368** (1991) 1.

[120] K.N. Abazajian et al. astro-ph/0203442.

[121] Q.R. Ahmad et al. *Phys. Rev. Lett.* **87** (2001) 071301;
Recent results from SNO collaboration are available at
http://www.sno.phy.queensu.ca/sno/results_04_02.

[122] A.D. Dolgov et al. hep-ph/0201287.

[123] J.R. Bond, G. Efstathiou, and J. Silk, *Phys. Rev. Lett.* **45** (1980).

Effective Geometry

M. Novello and S. E. Perez Bergliaffa

Centro Brasileiro de Pesquisas Físicas,

Rua Dr. Xavier Sigaud 150,

Urca 22290-180 Rio de Janeiro, RJ – Brazil

Abstract

We introduce the concept of effective geometry by studying several systems in which it arises naturally. As an example of the power and conciseness of the method, it is shown that a flowing dielectric medium with a linear response to an external electric field can be used to generate an analog geometry that has many of the formal properties of a Schwarzschild black hole for light rays, in spite of birefringence. The surface gravity of this analog black hole has a contribution that depends only on the dielectric properties of the fluid (in addition to the usual term dependent on the acceleration). This term may be give a hint to a new mechanism to increase the temperature of Hawking radiation.

I. INTRODUCTION

In recent years, there has been a lot of interest in models that mimic in the laboratory some features of gravitation [1]. These models are built using systems that sometimes look (deceivingly) simple, and are very different in nature: ordinary nonviscous fluids, superfluids, flowing and non-flowing dielectrics, non-linear electromagnetism in vacuum, and Bose-Einstein condensates (see [2] for a complete list of references). The underlying physics in all these cases is the same: the behaviour of the fluctuations around a background solution is governed by an "effective metric". More precisely, the particles associated to the perturbations do not follow geodesics of the background spacetime but of a Lorentzian geometry described by the effective metric, which depends on the background solution. This allows a rather complete analogy of some kinematical aspects of general relativity [3], but not of its dynamical features (see however [2, 4]).

Using this analogy, the geometrical tools of General Relativity can be used to study some condensed matter systems [5]. More important perhaps is the fact that the analogy has permitted the simulation of several configurations of the gravitational field, such as wormholes and closed space-like curves for photons in nonlinear electrodynamics [7, 8], and warped spacetimes for phonons [9]. Particular attention has been paid to yet another configuration, namely analog black holes, because these would emit Hawking radiation exactly as gravitational black holes do, and they are obviously much easier to generate in the laboratory than their astrophysical counterpart. The fact that analog black holes emit thermal radiation was shown first by Unruh in the case of dumb black holes [10], and it is the prospect of observing this radiation (thus testing the hypothesis that the thermal emission is independent of the physics at arbitrarily short wavelengths [10]) that motivates the quest for a realization of analog black holes in the laboratory. Let us emphasize that the actual observation of the radiation is a difficult task from the point of view of the experiment, if only because of the extremely low temperatures involved. In the case of a quasi one-dimensional flow of a Bose-Einstein condensate for instance, the temperature of the radiation would be around 70 nK, which is comparable but lower than the temperature needed form the condensate [11].

We shall begin by presenting in Sect.II the basics of the idea of the effective geometry by giving a sketch of an example given many years ago by W. Gordon [12]. Then we shall move on to the more interesting case of nonlinear electromagnetism, where we introduce the mathematical tool of *surface discontinuity*. In Sect. III we shall analyze another example: photons in a flowing dielectric medium. We shall see that, in analogy to the most general nonlinear electromagnetic case, the photons experience bi-refringence *and* bi-metricity. Then we demonstrate in Sect. IV that it is possible to build a static and spherically symmetric analog black hole, generated by a *flowing* isotropic dielectric that depends on an applied electric field. We give a specific example in Sect.V, in which the radius of the horizon and the temperature depend on three parameters (the zeroth order permittivity, the charge that generates the external field, and the linear susceptibility) instead of depending only on the zeroth order permittivity. As we shall show in Sect.VI, another feature of this black hole is that there is a new term in the surface gravity (and hence in the temperature of Hawking radiation), in addition to the usual term proportional to the acceleration of the fluid. This new term depends exclusively on the dielectric properties of the fluid, and it might give an opportunity to get Hawking radiation with temperature higher than that reported up to date.

II. THE EFFECTIVE METRIC

Historically, the first example of the idea of effective metric was presented by W. Gordon in 1923 [12]. In modern language, the wave equation for the propagation of light in a moving nondispersive medium, with slowly varying refractive index n and 4-velocity u^μ is given by

$$\left[\partial_\alpha \partial^\alpha + (n^2 - 1)(u^\alpha \partial_\alpha)^2\right] F_{\mu\nu} = 0.$$

Note that in this equation the components of $F_{\mu\nu}$ are not coupled. Consequently, the propagation will be the same independently of the polarization. In other words, there is no bi-refringence in moving media *with constant refraction index* (we shall see later that this is not the case if n is a function of the coordinates). Taking the geometrical optics limit, with the eikonal *Ansatz*, given by $F_{\mu\nu} = \mathcal{F}_{\mu\nu} e^{i(\vec{k}\cdot\vec{x} - \omega t)}$, the Hamilton-Jacobi equation for light rays can be written as $g^{\mu\nu} k_\mu k_\nu = 0$ (see [13] for details), where

$$g^{\mu\nu} = \eta^{\mu\nu} + (n^2 - 1)u^\mu u^\nu \tag{1}$$

is the effective metric for this problem. It must be remarked that only photons in the geometric optics approximation move on geodesics of $g^{\mu\nu}$: the particles that compose the fluid couple instead to the background Minkowskian metric (in fact, the dynamics of the fluid is described by Euler's equation, and hence the background spacetime seen by the fluid particles is Newtonian).

Let us study now in detail the example of nonlinear electromagnetism. We start with the action

$$S = \int \sqrt{-\gamma}\, L(F)\, d^4x, \tag{2}$$

where the invariant F is given by $F \equiv F^{\mu\nu}F_{\mu\nu}$ [30], and L is an arbitrary function of F. Notice that γ is the determinant of the background metric, which we take in the following to be that of flat spacetime. However, the same techniques can be applied when the background is curved (see for instance [14]). Varying this action w.r.t. the potential A_μ, related to the field by the expression $F_{\mu\nu} = A_{\mu,\nu} - A_{\nu,\mu}$, we obtain the Euler-Lagrange equations of motion (EOM):

$$(\sqrt{-\gamma}\, L_F F^{\mu\nu})_{;\nu} = 0, \tag{3}$$

where L_F is the functional derivative $L_F \equiv \frac{\delta L}{\delta F}$. In the particular case of a linear dependence of the Lagrangian with the invariant F we recover Maxwell's EOM.

As mentioned in the Introduction, we would like to study the behaviour of perturbations of these EOM around a fixed background solution. In particular, we shall be interested in the causal structure inherent to the EOM (3). This structure is described by the characteristics of the EOM [15]. Instead of using the infinite-momentum limit of the eikonal approximation [17], we shall use a more elegant method set out by Hadamard [18]. In this method, the propagation of low-energy photons is studied by following the evolution of the wave front (*i.e.* the characteristic surface), through which the electromagnetic field is continuous but its first derivative is not. To be specific, let Σ be the surface of discontinuity defined by the equation

$$\Sigma(x^\mu) = \text{constant}.$$

The discontinuity of a function J through Σ will be represented by $[J]_\Sigma$, and its definition is

$$[J]_\Sigma \equiv \lim_{\delta \to 0^+} \left(J|_{\Sigma+\delta} - J|_{\Sigma-\delta} \right).$$

The discontinuities of the field and its first derivative are given by

$$[F_{\mu\nu}]_\Sigma = 0, \qquad [F_{\mu\nu,\lambda}]_\Sigma = f_{\mu\nu} k_\lambda, \tag{4}$$

where the vector k_λ is nothing but the normal to the surface Σ, that is, $k_\lambda = \Sigma_{,\lambda}$.

To set the stage for the nonlinear case, let us first discuss the causal properties of Maxwell's electrodynamics, for which $L_F =$const. The EOM then reduces to $F^{\mu\nu}_{,\nu} = 0$, and taking its discontinuity we get

$$f^{\mu\nu} k_\nu = 0. \tag{5}$$

The other Maxwell equation is given by $F^*_{\mu\nu}{}^{,\nu} = 0$ or equivalently,

$$F_{\mu\nu,\lambda} + F_{\nu\lambda,\mu} + F_{\lambda\mu,\nu} = 0. \tag{6}$$

The discontinuity of this equation yields

$$f_{\mu\nu} k_\lambda + f_{\nu\lambda} k_\mu + f_{\lambda\mu} k_\nu = 0. \tag{7}$$

Multiplying this equation by k^λ gives

$$f_{\mu\nu} k^2 + f_{\nu\lambda} k^\lambda k_\mu + f_{\lambda\mu} k^\lambda k_\nu = 0, \tag{8}$$

where $k^2 \equiv k_\mu k_\nu \gamma^{\mu\nu}$. Using the orthogonality condition from Eqn.(5) it follows that

$$f^{\mu\nu} k^2 = 0. \tag{9}$$

Since the tensor associated to the discontinuity cannot vanish (we are assuming that there is a true discontinuity!) we conclude that the surface of discontinuity is null w.r.t. the metric $\gamma^{\mu\nu}$. That is,

$$k_\mu k_\nu \gamma^{\mu\nu} = 0. \tag{10}$$

(compare with Eqn.(1)). It follows that $k_{\lambda,\mu} k^\lambda = 0$, and since the vector of discontinuity is a gradient,

$$k_{\mu,\lambda} k^\lambda = 0. \tag{11}$$

This shows that the propagation of discontinuities of the electromagnetic field, in the case of Maxwell's equations (which are linear), is along the null geodesics of the Minkowski background metric.

Let us apply the same technique to the case of a nonlinear Lagrangian for the electromagnetic field, given by $L(F)$. Taking the discontinuity of the EOM (3), we get

$$L_F f^{\mu\nu} k_\nu + 2a\, L_{FF}\, F^{\mu\nu} k_\nu = 0, \tag{12}$$

where we defined the quantity a by $F^{\alpha\beta} f_{\alpha\beta} \equiv a$. Note that contrary to the linear case in which the discontinuity tensor $f_{\mu\nu}$ is orthogonal to the propagation vector k^μ, here there is a complicated relation between the vector $f^{\mu\nu} k_\nu$ and quantities dependent on the background field. This is the origin of a more involved expression for the evolution of the discontinuity vector, as we shall see next. Multiplying equation (8) by $F^{\mu\nu}$ we obtain

$$a\, k^2 + F^{\mu\nu} f_{\nu\lambda} k^\lambda k_\mu + F^{\mu\nu} f_{\lambda\mu} k^\lambda k_\nu = 0. \tag{13}$$

Now we substitute in this equation the term $f^{\mu\nu} k_\nu$ from Eqn.(12), and we arrive at the expression

$$ak^2 - 2\frac{L_{FF}}{L_F} a(F^{\mu\lambda} k_\mu k_\lambda - F^{\lambda\mu} k_\mu k_\lambda), \tag{14}$$

which can be written as $g^{\mu\nu} k_\mu k_\nu = 0$, where

$$g^{\mu\nu} = \gamma^{\mu\nu} - 4\frac{L_{FF}}{L_F} F^{\mu\nu}. \tag{15}$$

We then conclude that

> The low-energy photons of a *nonlinear* theory of electrodynamics with $L = L(F)$ do not propagate on the null cones of the background metric but on the null cones of an *effective* metric, generated by the self-interaction of the electromagnetic field.

This statement is always true in case of Lagrangians depending only of the invariant F. For Lagrangians that depend also of F^*, there may be some special cases in which the propagation coincides with that in Minkowski [6]. Another feature of the more general case $L = L(F, F^*)$ is that bi-refringence is present. That is, each of the two polarization states of the photon has its own dispersion relation. In some special cases, there is also bi-metricity (one effective metric for each polarization state). Some more special cases (such as Born-Infeld electrodynamics) even exhibit only a single metric [16, 17]. Several of these features are present in our next example.

III. EFFECTIVE METRIC(S) IN THE PRESENCE OF A DIELECTRIC

Let us now move to another interesting case where the effective geometry is useful to study the causal properties of low-energy photons. We shall analyze the propagation of such photons in a nonlinear medium (see Ref.[23] for details and notation). Let us define first the antisymmetric tensors $F_{\mu\nu}$ and $P_{\mu\nu}$, which are convenient to represent the electromagnetic field when material media are present. These tensors can be expressed in terms of the strengths (E, H) and the excitations (D, B) of the electric and magnetic fields as

$$F_{\mu\nu} = v_\mu E_\nu - v_\nu E_\mu - \eta_{\mu\nu}{}^{\alpha\beta} v_\alpha B_\beta,$$
$$P_{\mu\nu} = v_\mu D_\nu - v_\nu D_\mu - \eta_{\mu\nu}{}^{\alpha\beta} v_\alpha H_\beta.$$

where v_μ represents the 4-velocity of an arbitrary observer (which we will take later as comoving with the fluid). The Levi-Civita tensor introduced above is defined in such way that $\eta^{0123} = +1$ in Cartesian coordinates. Since the electric and magnetic fields are spacelike vectors, we shall use the notation $E^\alpha E_\alpha \equiv -E^2$, $H^\alpha H_\alpha \equiv -H^2$. We will consider here media with properties determined only by the tensors $\epsilon_{\alpha\beta}$ and $\mu_{\alpha\beta}$ (i.e. media with null magneto-electric tensor), which relate the electromagnetic excitations to the field strengths by the constitutive laws,

$$D_\alpha = \epsilon_\alpha{}^\beta(E, H) E_\beta, \qquad B_\alpha = \mu_\alpha{}^\beta(E, H) H_\beta. \tag{16}$$

In order to get the effective metric, we shall use Hadamard's method [18] as in the previous section. By taking the discontinuity of the field equations $*F^{\mu\nu}{}_{,\nu} = 0$ and $P^{\mu\nu}{}_{,\nu} = 0$, and assuming that

$$\epsilon^{\mu\beta} = \epsilon(E)(\gamma^{\mu\beta} - v^\mu v^\beta), \tag{17}$$

and

$$\mu^{\mu\beta} = \mu_0(\gamma^{\mu\beta} - v^\mu v^\beta), \tag{18}$$

with $\mu_0 = $ const., we get the following equations:

$$\epsilon(k.e) - \frac{\epsilon'}{E}(E.e)(k.E) = 0, \tag{19}$$

$$\mu_0(k.h) = 0, \tag{20}$$

$$\epsilon(k.v)e^\mu - \frac{\epsilon'}{E} E^\alpha e_\alpha (k.v) E^\mu + \eta^{\mu\nu\alpha\beta} k_\nu v_\alpha h_\beta = 0, \tag{21}$$

$$\mu_0(k.v)h^\mu - \eta^{\mu\nu\alpha\beta} k_\nu v_\alpha e_\beta = 0, \tag{22}$$

where k^μ is the wave propagation vector, ϵ' is the derivative of ϵ w.r.t. E, and

$$[E_{\mu,\lambda}]_\Sigma = e_\mu k_\lambda, \qquad [H_{\mu,\lambda}]_\Sigma = h_\mu k_\lambda.$$

Note in particular that Eqn.(19) shows that the vectors k^μ and e^μ are not always orthogonal, as would be the case if ϵ' was zero. Substituting Eqn.(22) in (21), we get

$$Z^{\mu\beta} e_\beta = 0, \tag{23}$$

where the matrix Z is given by

$$Z^{\mu\beta} = \left[k^2 + (k.v)^2(\mu_0 \epsilon - 1)\right]\gamma^{\mu\beta} - \mu_0 \frac{\epsilon'}{E}(k.v)^2 E^\mu E^\beta + (v.k)(v^\mu k^\beta + k^\mu v^\beta) - \left[\epsilon\mu_0(k.v) + k^2\right]v^\mu v^\beta - k^\mu k^\beta. \tag{24}$$

Non-trivial solutions of Eqn.(23) can be found only for cases in which $\det |Z^{\mu\beta}| = 0$ (this condition is a generalization of the well-known Fresnel equation [19]).

Eqn.(23) can be solved by expanding e_ν as a linear combination of the four linearly independent vectors v_ν, E_ν, k_ν and $\eta_{\alpha\beta\mu\nu}v^\alpha E^\beta k^\mu$ [31]. That is,

$$e_\nu = \alpha E_\nu + \beta \eta_{\alpha\lambda\mu\nu} v^\alpha E^\lambda k^\mu + \gamma k_\nu + \delta v_\nu. \tag{25}$$

Notice that taking the discontinuity of $E^\mu{}_{,\lambda}$ we can show that $(e.v) = 0$. This restriction imposes a relation between the coefficients of Eqn.(25):

$$\delta = -\gamma(k.v)$$

With the expression given in Eqn.(25), Eqn.(23) reads

$$\alpha \left[k^2 - (1 - \mu_0\, (\epsilon\, E)')\,(k.v)^2 \right] - \gamma \left[\mu_0 (k.v)^2 \frac{1}{E}\, \epsilon' E^\alpha k_\alpha \right] = 0,$$
$$\alpha E^\mu k_\mu + \gamma(1 - \mu_0 \epsilon)(k.v)^2 + \delta(k.v) = 0,$$
$$\alpha(k.v) E^\mu k_\mu + \gamma(k.v)k^2 + \delta\left[k^2 + \mu_0 \epsilon\, (k.v)^2 \right] = 0,$$
$$\beta\left[k^2 - (1 - \mu_0 \epsilon)(k.v)^2 \right] = 0.$$

The solution of this system results in the following dispersion relations:

$$k_-^2 = (k.v)^2 \left[1 - \mu_0(\epsilon\, E)' \right] + \frac{1}{\epsilon E}\, \epsilon'\, E^\alpha E^\beta k_\alpha k_\beta, \tag{26}$$
$$k_+^2 = [1 - \mu_0 \epsilon(E)](k.v)^2. \tag{27}$$

They correspond to the propagation modes

$$e_\nu^- = \rho^- \left\{ \mu_0\, \epsilon(k.v)^2 E_\nu + E^\alpha k_\alpha [k_\nu - (k.v)v_\nu] \right\}, \tag{28}$$
$$e_\nu^+ = \rho^+\, \eta_{\alpha\lambda\mu\nu} v^\alpha E^\lambda k^\mu, \tag{29}$$

where ρ^- and ρ^+ are arbitrary constants. The labels "+" and "−" refer to the ordinary and extraordinary rays, respectively. Eqns.(26) and (27) govern the propagation of photons in the medium characterized by $\mu = \mu_0 =$const., and $\epsilon = \epsilon(E)$. They can be rewritten as $g_\pm^{\mu\nu} k_\mu k_\nu = 0$, where we have defined the effective geometries

$$g_{(-)}^{\mu\nu} = \gamma^{\mu\nu} - \left[1 - \mu_0\, (\epsilon\, E)'\right] v^\mu v^\nu - \frac{1}{\epsilon E}\, \epsilon'\, E^\mu E^\nu, \tag{30}$$
$$g_{(+)}^{\mu\nu} = \gamma^{\mu\nu} - [1 - \mu_0\, \epsilon] v^\mu v^\nu. \tag{31}$$

The metric given by Eqn.(30) was derived first in [22], while the second metric very much resembles the metric obtained by Gordon [12] (see Eqn.(1). The difference is that in the case under consideration, ϵ is a function of the modulus of the external electric field, while Gordon worked with a constant permeability.

We see then that in this example each polarization state has its own dispersion relation (Eqns.(26) and (27)), so there is bi-refringence. There is also bi-metricity, because each type of photon moves according a different metric (see Eqns.(30) and (31)).

IV. THE ANALOG BLACK HOLE

We shall show in this section that the system described by the effective metrics given by Eqns.(30)-(31) can be used to produce an analog black hole. It will be convenient to rewrite at this point the inverse of the effective metric given by Eqn.(30) using a different notation:

$$g_{\mu\nu}^{(-)} = \gamma_{\mu\nu} - \frac{v_\mu v_\nu}{c^2}(1 - f) + \frac{\xi}{1+\xi}\, l_\mu l_\nu, \tag{32}$$

where we have defined the quantities

$$f \equiv \frac{1}{c^2 \mu_0 \epsilon (1+\xi)}, \quad \xi \equiv \frac{\epsilon' E}{\epsilon}, \quad l_\mu \equiv \frac{E_\mu}{E}.$$

Note that $\epsilon = \epsilon(E)$. We have introduced here the velocity of light c, which was set to 1 before. Taking a Minkowskian background in spherical coordinates, and

$$v_\mu = (v_0, v_1, 0, 0), \qquad E_\mu = (E_0, E_1, 0, 0), \tag{33}$$

we get for the effective metric described by Eqn.(32),

$$g_{00}^{(-)} = 1 - \frac{v_0^2}{c^2}(1-f) + \frac{\xi}{1+\xi} l_0^2, \tag{34}$$

$$g_{11}^{(-)} = -1 - \frac{v_1^2}{c^2}(1-f) + \frac{\xi}{1+\xi} l_1^2, \tag{35}$$

$$g_{01}^{(-)} = -\frac{v_0 v_1}{c^2}(1-f) + \frac{\xi}{1+\xi} l_0 l_1, \tag{36}$$

and $g_{22}^{(-)}$ and $g_{33}^{(-)}$ as in Minkowski spacetime. The vectors v_μ and l_μ satisfy the constraints

$$v_0^2 - v_1^2 = c^2, \tag{37}$$

$$l_0^2 - l_1^2 = -1, \tag{38}$$

$$v_0 l_0 - v_1 l_1 = 0. \tag{39}$$

This system of equations can be solved in terms of v_1, and the result is

$$v_0^2 = c^2 + v_1^2, \tag{40}$$

$$l_0^2 = \frac{v_1^2}{c^2}, \qquad l_1^2 = \frac{c^2 + v_1^2}{c^2}. \tag{41}$$

Now we can rewrite the metric in terms of $\beta \equiv v_1/c$, a definition which coincides with the usual one for small values of v_1. The explicit expression of the metric coefficients is:

$$g_{00}^{(-)} = \frac{1 - \beta^2(c^2\mu_0\epsilon - 1)}{c^2\mu_0(\epsilon + \epsilon' E)}, \tag{42}$$

$$g_{01}^{(-)} = \beta\sqrt{1+\beta^2}\,\frac{1 - c^2\mu_0\epsilon}{c^2\mu_0(\epsilon + \epsilon' E)}, \tag{43}$$

$$g_{11}^{(-)} = \frac{\beta^2 - c^2\mu_0\epsilon(1+\beta^2)}{c^2\mu_0(\epsilon + \epsilon' E)}. \tag{44}$$

From Eqn.(42) it is easily seen that, depending on the function $\epsilon(E)$, this metric has a horizon at $r = r_h$, given by the condition $g_{00}(r_h) = 0$ or equivalently,

$$\left. \left(c^2\mu_0\epsilon - \frac{1}{\beta^2} \right) \right|_{r_h} = 1. \tag{45}$$

The metric given above resembles the form of Schwarzschild's solution in Painlevé-Gullstrand coordinates [24, 25]:

$$ds^2 = \left(1 - \frac{2GM}{r}\right) dt^2 \pm 2\sqrt{\frac{2GM}{r}}\, dr\, dt - dr^2 - r^2 d\Omega^2. \tag{46}$$

With the coordinate transformation

$$dt_P = dt_S \mp \frac{\sqrt{2GM/r}}{1 - \frac{2GM}{r}} \, dr, \qquad (47)$$

the line element given in Eqn.(46) can be written in Schwarzschild's coordinates. The "+" sign covers the future horizon and the black hole singularity.

The effective metric given by Eqns.(42)-(44) looks like the metric in Eqn.(46) [32]. In fact, it can be written in Schwarzschild's coordinates, with the coordinate change

$$dt_{PG} = dt_S - \frac{g_{01}(r)}{g_{00}(r)} dr. \qquad (48)$$

Using this transformation with the metric coefficients given in Eqns.(42) and (43), we get the expression of $g_{11}^{(-)}$ in Schwarzschild coordinates:

$$g_{11}^{(-)} = -\frac{\epsilon(E)}{(1 - \beta^2[c^2\mu_0\epsilon(E) - 1])(\epsilon(E) + \epsilon(E)'E)}. \qquad (49)$$

Note that $g_{01}^{(-)}$ is zero in the new coordinate system, while $g_{00}^{(-)}$ is still given by Eqn.(42). Consequently, the position of the horizon does not change, and is still given by Eqn.(45).

Working in Painlevé-Gullstrand coordinates, we have shown that the metric for the "−" polarization describes a Schwarzschild black hole if Eqn.(45) has a solution. Afterwards we have rewritten the "−" metric in more familiar coordinates. By means of similar calculations, it can be shown that photons with the other polarization "see" the metric (in Schwarzschild coordinates) given by

$$g_{00}^{(+)} = \frac{1 + \beta^2(1 - c^2\mu_0\epsilon(E))}{c^2\mu_0\epsilon(E)}, \qquad (50)$$

$$g_{11}^{(+)} = -\frac{1}{1 + \beta^2(1 - c^2\mu_0\epsilon(E))}. \qquad (51)$$

It is important to stress then that *the horizon is located at r_h given by Eqn.(45) for photons with any polarization.* Moreover, the motion of the photons in both geometries will be qualitatively the same, as we shall show below.

V. AN EXAMPLE

We have not specified up to now the functions $\epsilon(E)$ and $E(r)$ that determine the dependence of the coefficients of the effective metrics with the coordinate r. From now on we assume a linear $\epsilon(E)$, a type of behaviour which is exhibited for instance by electrorheological fluids [29]. Specifically, we take

$$\epsilon(E) = \epsilon_0(\overline{\chi} + \chi^{(2)}E(r)), \qquad (52)$$

with $\overline{\chi} = 1 + \chi^{(1)}$. The nontrivial Maxwell's equation then reads

$$\left(\sqrt{-\gamma}\,\epsilon(r)F^{01}\right)_{,1} = 0. \qquad (53)$$

Taking into account that $(F^{01})^2 = \frac{E^2}{c^2}$, we get

$$F^{01} = \frac{-\overline{\chi} \pm \sqrt{\overline{\chi}^2 + 4\chi^{(2)}Q/\epsilon_0 r^2}}{2c\chi^{(2)}}. \qquad (54)$$

as a solution of Eqn.(53) for a point source in a flat background in spherical coordinates. Let us consider a particular combination of parameters: $\chi^{(2)} > 0$, $Q > 0$ and the "+" sign in front of the square root in F^{01}, in such a way that $E > 0$ for all r. To get more manageable expressions for the metric, it is convenient to define the function $\sigma(r)$:

$$E(r) \equiv \frac{\overline{\chi}}{2\chi^{(2)}}\,\sigma(r) \qquad (55)$$

where
$$\sigma(r) = -1 + \frac{1}{r}\sqrt{r^2 + q} \tag{56}$$

and
$$q = \frac{4\chi^{(2)}Q}{\epsilon_0 \overline{\chi}^2}. \tag{57}$$

In terms of σ, the metrics take the form

$$ds^2_{(-)} = \frac{2 - \beta^2 \left[\overline{\chi}\left(\sigma(r) + 2\right) - 2\right]}{2\,\overline{\chi}\,(1 + \sigma(r))} d\tau^2 - \frac{2 + \sigma(r)}{\left[2 - \beta^2\left(\overline{\chi}\left(\sigma(r) + 2\right) - 2\right)\right](1 + \sigma(r))} dr^2 - r^2 d\Omega^2, \tag{58}$$

$$ds^2_{(+)} = \frac{2 - \beta^2 \left[\overline{\chi}\left(\sigma(r) + 2\right) - 2\right]}{\overline{\chi}\,(2 + \sigma(r))} d\tau^2 - \frac{2}{2 + \beta^2\left[2 - \overline{\chi}\left(\sigma(r) + 2\right)\right]} dr^2 - r^2 d\Omega^2. \tag{59}$$

Notice that the (t, r) sectors of these metrics are related by the following expression:

$$ds^2_{(+)} = \Phi(r)\,ds^2_{(-)} \tag{60}$$

where the conformal factor Φ is given by:

$$\Phi = 2\,\frac{1 + \sigma(r)}{2 + \sigma(r)}$$

We shall study next some features of the effective black hole metrics. It is important to remark that up to this point, the velocity of the fluid v_1 is completely arbitrary; it can even be a function of the coordinate r. We shall assume in the following that v_1 is a constant. This assumption, which will be lifted in Sect.VI, may seem rather restrictive but it helps to display the main features of the effective metrics in an easy way.

To study the motion of the photons in these geometries, we can use the technique of the effective potential. Standard manipulations (see for instance [26]) show that in the case of a static and spherically symmetric metric, the effective potential is given by

$$V(r) = \varepsilon^2 \left(1 + \frac{1}{g_{00}(r)\,g_{11}(r)}\right) - \frac{L^2}{r^2 g_{11}(r)} \tag{61}$$

where ε is the energy and L the angular momentum of the photon.

In terms of $\sigma(r)$, and of the impact parameter $b^2 = L^2/\varepsilon^2$, the "small" effective potential $v(r) \equiv V(r)/\varepsilon^2$ for the metric Eqn.(58) in Schwarzschild coordinates can be written as follows:

$$v^{(-)}(r) = 1 - \frac{2(1 + \sigma(r))^2}{2 + \sigma(r)} - \frac{b^2}{r^2}\frac{(2 - \beta^2 \sigma(r))(1 + \sigma(r))}{2 + \sigma(r)} \tag{62}$$

A short calculation shows that $v^{(-)}$ is a monotonically decreasing function of β. Consequently, we shall choose a convenient value of it, for the sake of illustrating the features of the effective potential. Figures (1) and (2) show the plots of the potential for the $(-)$ metric for several values of the relevant parameters.

The effective potential for the Gordon-like metric can be obtained in the same way. From Eqns.(61) and (59) we get

$$v^{(+)}(r) = 1 - \frac{2 + \sigma(r)}{2} + \frac{b^2}{2r^2}\left[2 - \beta^2 \sigma(r)\right]. \tag{63}$$

The plots in Figures (3) and (4) show the dependence of $v^{(+)}(r)$ on the different parameters.

We see from these plots that, in the case of a constant flux velocity, the shape of the effective potential for both metrics qualitatively agrees with that for photons moving on the geometry of a Schwarzschild black hole (see for instance Ref.[26], pag. 143).

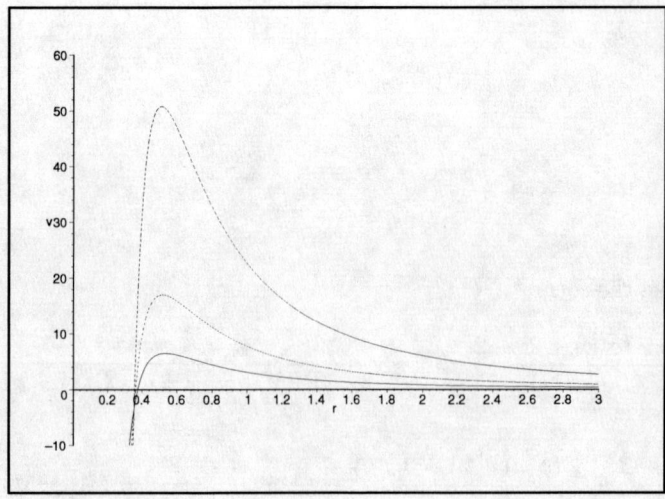

FIG. 1: Plot of the effective potential $v^{(-)}(r)$ for $q = 1$, $b = 1, 3, 5$ (starting from the lowest curve), and $\beta = 0.5$.

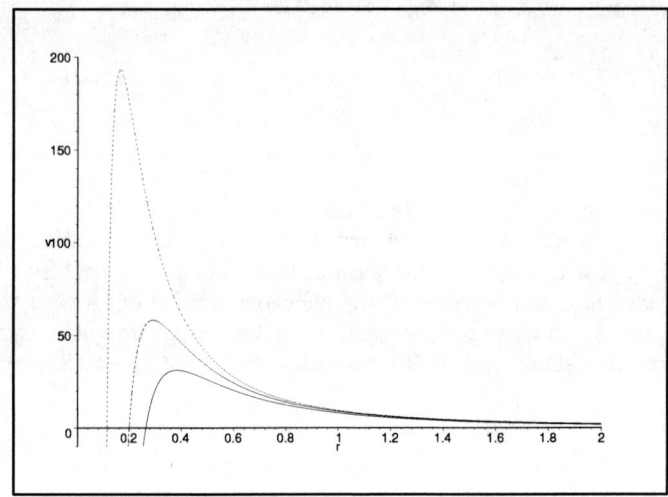

FIG. 2: Plot of the effective potential $v^{(-)}(r)$ for $b = 3$, and $q = 1, 3, 5$ (starting from the lowest curve), and $\beta = 0.5$.

VI. SURFACE GRAVITY AND TEMPERATURE

Let us now go back to the more general case of $\beta = \beta(r)$, and calculate the "surface gravity" of our analog black hole. We present first the results for the constant permittivity case. By setting $\epsilon'(E) \equiv 0$ in the metrics Eqns.(30) and (31), we regain the example of constant index of refraction studied for instance in [21]. It is easy to show from Eqn.(45) that the horizon of the black hole in this case is given by

$$\beta^2(r_h) = \frac{1}{\bar{\chi} - 1}. \tag{64}$$

The "surface gravity" of a spherically symmetric analog black hole in Schwarszchild coordinates is given by [5]

$$\kappa = \frac{c^2}{2} \lim_{r \to r_h} \frac{g_{00,r}}{\sqrt{|g_{11}| \, g_{00}}}. \tag{65}$$

For the metrics Eqns.(30) and (31) with $\epsilon = \epsilon_0 \bar{\chi}$ and r_h given by Eqn.(64), the analog surface gravity is

$$\kappa = -\frac{c^2}{2} \frac{1 - \bar{\chi}}{\sqrt{\bar{\chi}}} \left. (\beta^2)' \right|_{r_h}. \tag{66}$$

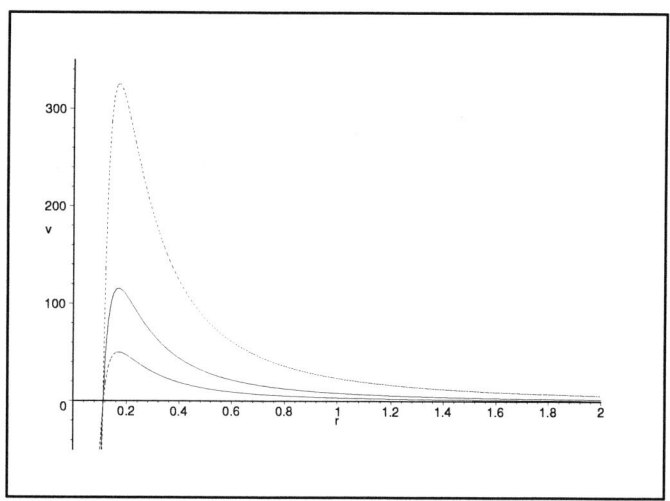

FIG. 3: Plot of the effective potential for the Gordon-like metric, for $q = 1$, $b = 1, 3, 5$ (starting from the lowest curve), and $\beta = 0.5$.

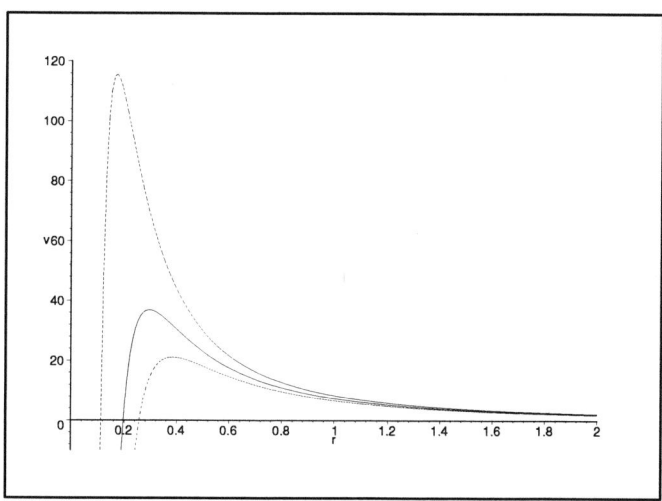

FIG. 4: Plot of the effective potential for the Gordon-like metric, for $b = 3$, $q = 1, 3, 5$.

This equation can be rewritten in terms of the velocity of light in the medium and the refraction index, respectively given by

$$c_m^2 = \frac{1}{\mu_0 \epsilon}, \qquad n = \frac{c}{c_m}. \tag{67}$$

The result is

$$\kappa = \frac{c^2}{2} \frac{1-n^2}{n} (\beta^2)_{,r} \tag{68}$$

In this expression we can see the influence of the dielectric properties of the fluid (through the index of refraction of the medium) and also of its dynamics through the physical acceleration in the radial direction, given by

$$a_r|_{r_h} = \frac{c^2}{2} (\beta^2)'|_{r_h},$$

for $\beta^2(r_h) \ll 1$. This acceleration is a quantity that must be determined solving the equations of motion of the fluid. (Notice that if β is set equal to 0 in Eqns.(42)-(44), we cease to have a black hole (this situation was analyzed in [16])).

Going back the the more general case of a linear permittivity, described by the metrics given by Eqns.(58) and (59), and considering that $\beta(r_h) \ll 1$, the radius of the horizon is [33]:

$$r_h^2 = \frac{q\bar{\chi}^2}{4}\beta^4(r_h). \tag{69}$$

Using the expressions given above, the result for the surface gravity of the "−" black hole for $\beta(r_h) \ll 1$ is

$$\kappa^{(-)} = \frac{c^2}{\beta}\left(\frac{1}{\bar{\chi}\sqrt{q}} - \frac{1}{2}(\beta^2)'\right)\bigg|_{r_h}. \tag{70}$$

This equation differs from the surface gravity of the case of constant permittivity (Eqn.(66)) by the presence of a new term that does not depend on the acceleration of the fluid. To see where this new term comes from, we can go back to the definition of the surface gravity given in Eqn.(65), and use the fact that in the high frequency limit the velocity of light and the index of refraction in a medium of variable ϵ are still given by Eqn.(67), replacing the constant permittivity by $\epsilon = \epsilon(E)$. The result is

$$\kappa = \left(\frac{c^2}{2}\frac{1-n^2(E)}{n(E)}(\beta^2)_{,r} + \frac{n(E)\epsilon(E)}{\epsilon(E) + \epsilon(E)'E}(c_m^2)_{,r}\right)\bigg|_{r_h} \tag{71}$$

In this expression, the first term is the generalization of the case $\epsilon = $ const. (compare with Eqn.(68)), which mixes the acceleration of the fluid with its dielectric properties. On the other hand, the second term, which is the new term displayed in Eqn.(70), is related to the radial variation of the velocity of light in the medium. It is important to point out that the result exhibited in Eqn.(71) is parallel to that of dumb holes: Unruh found in that case [10] that the surface gravity for constant speed of sound is proportional to the acceleration of the fluid (as in the first term of Eqn.(71)). This was generalized by Visser [5], who showed that for a position-dependent velocity of sound a second term appears, coming from the gradients of the speed of sound, in analogy with the second term of Eqn.(71).

It is easy to show that the these results also apply to the black hole described by the Gordon-like metric. This is not surprising though, because of the conformal relation between the two metrics, given by Eqn.(60) [28].

Let us remark once more that the concept of temperature, and indeed that of effective geometry is valid in this context only for low-energy photons, *i.e.* photons with wavelengths long compared to the intermolecular spacing in the fluid. For shorter wavelengths, there would be corrections to the propagation dictated by the effective metric. However, results for other systems (such as dumb black holes [10] and Bose-Einstein condensates) suggest that the phenomenon of Hawking radiation is robust (*i.e.* independent of this "high-energy" physics). Consequently, it makes sense to talk about the temperature of the radiation in these systems.

At first sight it may seem that by choosing an appropriate material and a convenient value of the charge we could obtain a high value of the temperature of the radiation, given by

$$T \equiv \frac{\hbar}{2\pi k_B c}\kappa \approx 4 \times 10^{-21} \kappa \ \text{Ks}^2/\text{m}. \tag{72}$$

However, the equation for the surface gravity can be rewritten as [34]

$$\kappa = c^2 \left(\frac{\beta}{2r} - \beta_{,r}\right)\bigg|_{r_h}.$$

We see then that, because $\beta(r_h) \ll 1$, the new term appearing in κ is bound to be very small. In spite of this result, the emergence in the surface gravity of the term due to the variable velocity of light suggests that it may be worth to study if some media with nonlinear dependence on an external electromagnetic field can be used to generate analog black holes whose Hawking radiation could be measured in laboratory.

Acknowledgements

The authors would like to thank CNPq and FAPERj for financial support.

---

[1] "Artificial Black Holes", Proceedings of the Workshop "Analog Models of General Relativity" (held at the Centro Brasileiro de Pesquisas Fisicas, Brazil, Oct. 2000), M. Novello, M. Visser, and G. Volovik (Eds), World Scientific (2002).

[2] A complete list of references can be found in C. Barcelo, S. Liberati, and M. Visser, Class. Quantum Grav. **18**, 3595 (2001).

[3] M. Visser, Phys. Rev. Lett. **80**, 3439 (1998).

[4] C. Barcelo, S. Liberati, M. Visser, Class. Quantum Grav. **19**, 2961 (2002).

[5] See M. Visser, Class. Quantum Grav. **15**, 1767 (1998) and references therein.

[6] V. A. De Lorenci, Renato Klippert, M. Novello, J.M. Salim, Phys. Lett. **B482**, 134 (2000).

[7] F. Baldovin, M. Novello, S.E. Perez Bergliaffa, and J.M. Salim, Class. Quant. Grav. **17**, 3265 (2000).

[8] M. Novello, J. M. Salim, V. A. De Lorenci, and E. Elbaz, Phys. Rev. D **63**, 103516 (2001).

[9] "Warped spacetime for phonons moving in a perfect nonrelativistic fluid", U. Fischer and Matt Visser, `gr-qc/0211029`.

[10] W. Unruh, Phys. Rev. Lett. **46**, 1351 (1981), and Phys. Rev. D **51**, 2827 (1995).

[11] *Towards the observation of Hawking radiation in Bose-Einstein condensates*, C. Barcelo, S. Liberati, and M. Visser, `gr-qc/0110036`.

[12] W. Gordon, Ann. Phys. (Leipzig), **72**, 421 (1923).

[13] U. Leonhardt and P. Piwnicki, Phys. Rev. A **60**, 4301 (1999).

[14] M. Novello, S.E. Perez Bergliaffa, J.M. Salim, Class. Quant. Grav. **17**, 3821 (2000).

[15] *Methods of Mathematical Physics*, R. Courant and D. Hilbert, Wiley, John and Sons (1990).

[16] M. Novello, V. De Lorenci, J. Salim, and R. Klippert, Phys. Rev D**61**, 45001 (2000).

[17] *Bi-refringence versus bi-metricity*, M. Visser, C. Barcelo, and S. Liberati. Contribution to the Festschrift in honor of Mario Novello, to be published. `gr-qc/0204017`.

[18] *Leçons sur la propagation des ondes et les equations de l'hydrodynamique*, J. Hadamard, Ed. Dunod, Paris, 1958.

[19] See for instance *Electrodynamique des milieux continus*, L. Landau and E. Lifshitz, Ed. Mir, Moscow (1969).

[20] C. Barcelo, S. Liberati, and M. Visser, Class. Quant. Grav. **18**, 3595 (2001).

[21] U. Leonhardt and P. Piwnicki, Phys. Rev. Lett. **84**, 822 (2000) and Phys. Rev. A **60**, 4301 (1999). See the subsequent discussion in M. Visser, Phys. Rev. Lett. **85**, 5252 (2000) and U. Leonhardt and P. Piwnicki, Phys. Rev. Lett. **85**, 5253 (2000), R. Schutzhold, G. Plunien, and G. Soff, Phys. Rev. Lett. **88**, 061101 (2002).

[22] M. Novello, J. Salim, Phys. Rev. D **63**, 083511 (2001).

[23] *Analog black holes in flowing dielectrics*, M. Novello, S. Perez Bergliaffa, J. Salim, V. De Lorenci, and R. Klippert, `gr-qc/0201061`. To be published in Class. Quantum Grav.

[24] P. Painlevé, *C. R. Acad. Sci. (Paris)*, **173**, 677 (1921).

[25] A. Gullstrand, *Arkiv. Mat. Astron. Fys.*, **16**, 1 (1922).

[26] *General Relativity*, R. M. Wald, The University of Chicago Press (1984).

[27] *Essential and inessential features of Hawking radiation*, M. Visser, `hep-th/0106111`.

[28] T. Jacobson and G. Kang, Class. Quantum Grav. **10**, L201 (1993).

[29] W. Wen, S. Men, and K. Lu, Phys. Rev. E **55**, 3015 (1997).

[30] We could have considered $L = L(F, F^*)$ instead, where $F^* \equiv F^*_{\mu\nu} F^{\mu\nu}$. This case is studied in [6].

[31] The particular instance in which the vectors used as a basis in Eqn.(25) are not linearly independent is discussed in [23].

[32] Note that a conformal factor to make $g_{11} = -1$ in Eq.(44) is needed. Consequently, the two metrics are actually conformally equivalent.

[33] Notice that we cannot take the limit $q \to 0$ in this expression or in any expression in which this one has been used.

[34] Note that this equation depends on $\chi^{(2)}$ through the expression for r_h, Eqn.(69).

LIST OF PARTICIPANTS IN THE Xth BRAZILIAN SCHOOL OF COSMOLOGY AND GRAVITATION

1. ARGENTINA

   - Alejandro Gangui – IAFE

2. BOLIVIA

   - Jose Nogales – UMSA

3. BRAZIL

   - Alexandre Yasuda Miguelote – UFF - RJ
   - Ana Helena de Campos – USP - SP
   - Carlos Molina Mendes – USP – SP
   - Carlos Romero – UFPB – PB
   - Davi Giugno – USP – SP
   - Edmundo Marinho do Monte – UFPB – PB
   - Eduardo Sergio Santini – CBPF – RJ
   - Fábio Dahia – UFPB – PB
   - Flávio Gimenes Alvarenga – UFES – ES
   - Galina Klimchitskaya – UFPB – PB
   - Glauber Tadaiesky – UFES – ES
   - Hector Leny Carrion Salazar – CBPF – RJ
   - Henrique Pereira de Oliveira – UERJ – RJ
   - Herman Julio Mosquera Cuesta – CBPF – RJ
   - Humberto Belich Junior – CBPF – RJ
   - Ilya Shapiro – UFJF – MG
   - José Ademir Sales Lima – UFRN – RN
   - José Martins Salim – CBPF – RJ
   - Julio César Fabris – UFES – ES
   - Léo Gouvêa Medeiros – IFT/UNESP – SP
   - Luís Carlos Bassalo Crispino – UFPA – PA
   - Luís Carlos Garcia de Andrade – UERJ – RJ
   - Luiz Alberto Rezende de Oliveira – CBPF – RJ
   - Mário Novello – CBPF – RJ

- Martín Makler – CBPF – RJ
- Nami Fux Svaiter – CBPF – RJ
- Nazira Abache Tomimura – UFF – RJ
- Nelson Pinto Neto – CBPF – RJ
- Regina Celia Arcuri – UFRJ – RJ
- Renato Klippert – EFEI – MG
- Reuven Opher – IAG/USP – SP
- Rodrigo Rocha Cuzinatto – IFT/UNESP – SP
- Rogério Rosenfeld – IFT/UNESP – SP
- Ronaldo Penna Neves – CBPF – RJ
- Santiago Estebán Perez Bergliaffa – CBPF – RJ
- Sergio Eduardo C. Jorás – UFRJ – RJ
- Urbano Lopes Franca Júnior – IFT/UNESP – SP
- Valdir B. Bezerra – UFPB – PB
- Vanessa Carvalho de Andrade – IFT/UNESP – SP
- Vitório Alberto de Lorenci – EFEI – MG
- Vladimir Mostepanenko – UFPB – PB
- William Santiago Hipolito Ricaldi – CBPF – RJ

4. CANADA

- Bojan Losic – UBC

5. CHILE

- Mauricio Cataldo – Univ. del Bio-Bio
- Sergio Del Campo – Univ. Catolica de Valparaiso
- Victor-Hugo Cardenas – Univ. Catolica de Valparaiso

6. DENMARK

- Jakob Hansen – TAC

7. ENGLAND

- Edmund John Copeland – Sussex Univ.
- Pier Stefano Corasaniti – Sussex Univ.

8. **FRANCE**

   - James Bartlett – College de France
   - Marc Lachièze-Rey – Saclay
   - Roland Triay – Univ. de Marseille

9. **GERMANY**

   - Matteo Carrera – Univ. of Freiburg

10. **IRELAND**

    - Thomas Waters – Dublin City Univ.

11. **ITALY**

    - Andrea Geralico – ICRA
    - Andrea Lunari – ICRA
    - Carlo Luciano Bianco – ICRA
    - Christian Cherubini – ICRA
    - Giovanni Imponente – ICRA
    - Massimiliano Lattanzi – ICRA
    - Remo Ruffini – ICRA
    - Serena Fagnocchi – Bologna Univ.
    - Vahe Gurzadyan – ICRA
    - Xue She Sang – ICRA

12. **POLAND**

    - Irina Dymnikova - Univ.of Warmia and Mazury in Olsztyn

13. **RUSSIA**

    - Kirill Bronnikov - Russian Institute of Metrological Service – RIMS
    - Vitaly N. Melnikov - Russian Institute of Metrological Service – RIMS

14. UNITED STATES

- Chang S. Chan – Princeton University
- Edward Kolb – Fermilab
- Jorge Pullin – Penn State Univ.
- Monica Maria Guica – Univ. of Chicago
- Patrick Richard Brady – Univ. of Wisconsin
- Stephen Harnish – Bluffton College

15. TURKEY

- Ozay Gurtug – Eastern Mediterraenan University – EMU

Author Index

B
Bartlett, J. G., 1
Bianco, C. L., 16

C
Chardonnet, P., 16
Copeland, E. J., 125

D
Dymnikova, I., 204

F
Fraschetti, F., 16

G
Gangui, A., 226
Gurzadyan, V. G., 108

L
Lachièze-Rey, M., 173
Lattanzi, M., 263

M
Melnikov, V. N., 187
Mostepanenko, V. M., 198

N
Novello, M., 288

P
Perez Bergliaffa, S. E., 288
Pullin, J., 141

R
Ruffini, R., 16, 263

S
Svaiter, N. F., 154

V
Vereshchagin, G., 263
Vitagliano, L., 16

X
Xue, S.-S., 16